B&T 9/23/92 54.95

CINCINM⁻ᵗ   RC

D0203271

697 E27 1992

Eastop, T. D.

Mechanical services for
 buildings

Mechanical Services for Buildings

697
E27
1992

CINCINNATI TECHNICAL COLLEGE LRC

Mechanical Services for Buildings

T D Eastop BSc PhD C Eng FIMechE FCIBSE

Formerly Head, School of Engineering, Wolverhampton Polytechnic

W E Watson BSc C Eng MIMechE MCIBSE

Engineering Consultant

Longman
Scientific &
Technical

Copublished in the United States with
John Wiley & Sons, Inc., New York

Longman Scientific & Technical,
Longman Group UK Limited,
Longman House, Burnt Mill, Harlow,
Essex CM20 2JE, England
and Associated Companies throughout the world.

Copublished in the United States with
John Wiley & Sons, Inc., 605 Third Avenue, New York, NY 10158

© Longman Group UK Limited 1992

All rights reserved; no part of this publication may be reproduced, stored in a retrieval
system, or transmitted in any form or by any means, electronic, mechanical, photocopying,
recording, or otherwise without either the prior written permission of the Publishers or a
licence permitting restricted copying in the United Kingdom issued by the Copyright Licensing
Agency Ltd, 90 Tottenham Court Road, London, W1P 9HE.

First published 1992

British Library Cataloguing in Publication Data
Eastop, T.D.
   Mechanical services for buildings.
   I. Title   II. Watson, W.E.
   696
   ISBN 0-582-05695-0

Library of Congress Cataloging-in-Publication Data
Eastop, T.D. (Thomas D.)
      Mechanical services for buildings / T.D. Eastop, W.E. Watson.
         p.   cm.
      Includes bibliographical references and index.
      ISBN 0-470-21790-1
      1. Buildings--Mechanical equipment.   2. Buildings--Environmental
engineering.   I. Watson, W.E. (William Edward), 1939–
II. Title.
TH6010.E27   1992
697--dc20   91-14026
                                                  CIP

Set in Lasercomp Times
Produced by Longman Group (FE) Ltd
Printed in Hong Kong

LIST OF CHAPTERS

CONTENTS

PREFACE

Building Services Engineering is a combination of art, science and engineering design applied to create internal environments, pleasing for the occupants and suitable for the objects and equipment used within. A well-designed building requires systems for heating, ventilating, air conditioning, lighting, electricity supply, lifts and escalators, telecommunications, water supply, drainage and sewage disposal; the whole system must be controlled, safety measures against fire installed, and noise kept to a minimum level. The occupants of the building must be comfortable, and in certain cases the environment must be biologically 'clean'. In order to achieve the required integration of services to provide the 'ideal environment' it is necessary for the Building Services designer to work closely throughout with the Architect, the Structural Engineer and the Builder.

It can be seen from the above that the Building Services Engineer is required to have a good understanding of a wide range of engineering subjects: in broad terms this would include energy conversion, psychrometry, refrigeration, pumps and fans, electric motors and generators, electricity distribution, lighting, heat and mass transfer, acoustics, control systems and building technology.

Any textbook for Building Services Engineers must be selective; this book is mainly concerned with advanced topics in Heating, Ventilating and Air Conditioning. It is assumed that the reader already has a basic knowledge of Thermodynamics, Heat Transfer, and Fluid Mechanics.

The emphasis throughout is on CIBSE design procedures as laid down in the *Guide to Current Practice*. The aim is to teach through Worked Examples using, wherever possible, examination questions from the Engineering Council, CIBSE, and various Polytechnic and University degree papers.

The first chapter provides an overall basic coverage of heating, psychrometry, air conditioning, and ventilation; some of these topics are developed further in later chapters. Chapter 2 extends the work of Chapter 1 on heating systems to high-pressure hot water systems and boilers. Chapters 3 and 4 introduce the environmental network method of allowing for internal radiation between surfaces and show how this method can be used to solve problems in both steady and transient cases for heating and air conditioning. Thermal comfort is discussed and the Fanger method introduced. Chapter 5 considers the transfer of moisture within structures and between spaces, and Chapter 6 gives a basic coverage of mass transfer applied to the design of cooling towers, humidification and de-humidification equipment. Refrigeration is covered in some detail in Chapter 7, and Chapter 8 brings some of the previous work together in a coverage of air-conditioned buildings. Finally, in Chapter 9 the important topic of energy efficiency is considered.

Much of the work of the Architect and Building Services Engineer in practice is now computerized, but although the former manual design methods have been replaced in design offices by commercially available computer software it is important that students continue to be taught first principles and the theoretical and practical basis of the design methods. Honours degree courses and Masters' courses, as well as introducing the students to commercial software, would normally cover the advanced numerical analysis and computing skills required to enable students to write their own programs to solve real problems. Because of the breadth of coverage of the present book, the authors have made no attempt to include computer analysis; in many parts of the text it is clear that a computer may be used to facilitate repetitive calculations and/or to input data (e.g. weather information), and/or for iterative solutions, and/or for more complex analysis (e.g. transient heat transfer in buildings); it is assumed that this will be done through course work and projects.

The book is aimed at students on B Eng courses in Building Services Engineering and Environmental Engineering, certain M Sc/M Eng courses, and those students on Civil Engineering, Mechanical Engineering, Combined Engineering or Building degree courses who are studying Building Services as an option. Some of the material will also be found suitable for students of Architecture, and students on Higher Certificate and Higher Diploma courses in Building Services.

The symbols, particularly in the case of those sections involving heat transfer, have been chosen to agree with decisions taken by the 8th International Heat Transfer Conference in San Francisco in 1986. For example, α is now used for heat transfer coefficient instead of h; λ for thermal conductivity instead of k; N_{tu} for number of transfer units instead of NTU; $\Delta \bar{t}_{ln}$ for logarithmic mean temperature difference instead of LMTD. The paper by Dr Y R Mayhew, Use of physical quantities, units, mathematics, and nomenclature in heat transfer publications *Heat Transfer Engineering* 1988 **9**(4) is recommended for further reading, and the authors are grateful to Dr Mayhew for many helpful discussions on symbols and units, and the presentation of material in graphs and tables.

The authors are also indebted to Mr R A Briggs of Yates, Edge & Partners for his help and encouragement.

TDE
WEW
1991

ACKNOWLEDGEMENTS

We are grateful to the following for permission to reproduce copyright material:

Building Services Publications Ltd for Table 8.5; The Chartered Institution of Building Services Engineers for Figs 1.8, 1.20, 2.11, 2.12, 2.13 and Tables 1.2, 2.1, 4.1, 4.2, 4.3, 4.4, 4.5, 4.6, 8.1, 8.2 and data in Tables 1.1, 1.2, 1.3, 3.1, 3.2, 3.4, 4.7, 9.2, 9.4 and examination questions; College of Technology (Dublin) for examination questions; Engineering Council for examination questions; Council of the Institution of Mechanical Engineers for Fig. 6.15 (Carey & Williamson, 1950); McGraw-Hill Publishing Company for Figs 3.12, 3.13, 3.14, 3.16, 3.17 and Tables 3.3, 3.5 (Fanger, 1972) copyright (1972) McGraw-Hill Publishing Company; Newcastle upon Tyne Polytechnic for examination questions; South Bank Polytechnic for Table 8.3 and examination questions; University of Glasgow for examination questions; The University of Liverpool for examination questions; The University of Manchester Institute of Science & Technology (UMIST) for examination questions; Wolverhampton Polytechnic for examination questions.

Whilst every effort has been made to trace the owners of copyright material, in a few cases this has proved impossible and we take this opportunity to offer our apologies to any copyright holders whose rights we may have unwittingly infringed.

NOMENCLATURE

A	surface area
A_c	cross-sectional area
A_o	surface area of fabric conducting to outside air
a	area per unit volume; index for wind speed; constant
B	overall mass transfer coefficient
b	slope of saturation enthalpy–temperature line
C	constant; concentration
CHP	combined heat and power
COP	coefficient of performance
CV	calorific value
C_d	discharge coefficient; drag coefficient
C_p	non-dimensional pressure coefficient
C_v	ventilation conductance
c	specific heat; concentration
clo	unit of clothing
c_p	specific heat at constant pressure
c_{pma}	specific heat of air per unit mass of dry air
c_v	specific heat at constant volume
D	diffusion coefficient
d	diameter
E	energy; effectiveness of a heat exchanger
$\dot{E}$	energy rate
F	geometric factor
F_{au}, F_{av}	factors defined by Eqns [3.9], [3.10]
F_u, F_v	factors defined by Eqn [3.16]
F_1, F_2	factors defined by Eqn [3.18]
F_{ay}	factor defined by Eqn [4.8]
F_y	factor defined by Eqn [4.9]
f	decrement factor; friction factor; diversity factor; fractional fuel saving; frequency of occurrence; percentage of hours
f_r	response factor defined by Eqn [4.17]
Gr	Grashof number
g	gravitational acceleration
H	enthalpy; heat loss coefficient
h	specific enthalpy
Δh	specific enthalpy difference

$\Delta \bar{h}_{\text{ln}}$	logarithmic mean specific enthalpy difference
I	electric current; solar irradiance
$\bar{I}_t$	mean solar irradiance
I'_t	peak solar irradiance
$\tilde{I}_t$	swing of solar irradiance about mean
j_{H}	Colburn factor for heat transfer
j_{M}	Colburn factor for mass transfer
K	constant
K_s	parameter for wind speed
k	pressure loss factor; absolute roughness
L	length
$\dot{L}$	latent heat gain
LPPD	Lowest Possible Percentage Dissatisfied
Le	Lewis number
MCR	maximum continuous rating
m	mass
$\dot{m}$	mass flow rate
$\tilde{m}$	molar mass
N	rotational speed; number of hours of operation
Nu	Nusselt number
N_{P}	number of people
N_{tu}	number of transfer units
N_{w}	number of working days
n	air changes per hour; number of kilomoles per unit volume; number of cylinders
$\dot{n}$	number of kilomoles per unit area per unit time
P	permeance
P_t	fan total pressure increase
Pr	Prandtl number
PMV	Predicted Mean Vote
PPD	Predicted Percentage Dissatisfied
p	absolute pressure
$\bar{p}_{\text{ln}}$	logarithmic mean pressure
Δp	pressure loss
Q	heat
$\dot{Q}$	rate of heat transfer
$\tilde{Q}$	swing about the mean heat transfer rate
$\dot{q}$	rate of heat transfer per unit area
R	specific gas constant; thermal resistance
$\tilde{R}$	molar gas constant
RRL	room ratio line, $\dot{S}/(\dot{S} + \dot{L})$
RDF	refuse-derived fuel
Re	Reynolds number
r	radius; vapour resistivity
r_{P}	pressure ratio
$\dot{S}$	sensible heat gain
$\bar{S}$	mean solar gain factor
$\tilde{S}$	fluctuating solar gain factor
Sc	Schmidt number
Sh	Sherwood number

St	Stanton number
S_h	scale of enthalpy axis
S_ω	scale of specific humidity axis
s	specific entropy; number of sections; surface factor
T	absolute temperature
TDS	total dissolved solids
t	temperature
$\bar{t}$	mean temperature
$\tilde{t}$	swing about the mean temperature
Δt	temperature difference
$\Delta \bar{t}$	mean temperature difference
$\Delta \bar{t}_{ln}$	logarithmic mean temperature difference
U	internal energy; overall heat transfer coefficient; thermal transmittance
u	specific internal energy; velocity
V	volume; voltage
$\dot{V}$	volume flow rate
v	specific volume
W	work
$\dot{W}$	rate of work transfer; power
x	dryness fraction; general fraction; thickness
Y	thermal admittance
y	difference function; ratio $\dot{m}_{a1}/(\dot{m}_{a1} + \dot{m}_{a2})$
z	height

Greek symbols

α	heat transfer coefficient; absorptivity for radiation; angle
β	mass transfer coefficient defined by Eqn [6.8]; angle
β_G	mass transfer coefficient defined by Eqn [6.9]
β_P	mass transfer coefficient defined by Eqn [6.10]
β_ω	mass transfer coefficient defined by Eqn [6.11]
γ	ratio of specific heats, c_p/c_v; proportion of heat input at environmental point
δ	proportion of heat at air point
ε	emissivity
η	efficiency; dynamic viscosity
θ	angle
κ	thermal diffusivity
λ	thermal conductivity
μ	percentage saturation; permeability
ν	kinematic viscosity
ρ	density
σ	Stefan–Boltzmann constant; heat: power ratio; slip factor
τ	time; shear stress in a fluid
τ_a	time lead for admittance
τ_c	time constant
τ_f	time lag for decrement factor
τ_s	time lag for surface factor
Φ	phase angle between current and voltage
ϕ	relative humidity
ω	specific humidity

Subscripts

A	absorber
ADP	apparatus dew point
a	dry air; air point
ac	air to dry resultant
ai	inside air
ao	outside air; air in the free stream
B	boiler; building
b	base surface; body; blade
C	cold fluid; casual gain; condenser
c	condensate; convective; dry resultant
d	dew point; design
db	dry bulb
D	diffuser
E	evaporator
e	environmental point; exit; evaporation; equilibrium condition
ec	dry resultant to environmental
ei	inside environmental
eo	outside environmental
F	flow; fuel; fin; fabric; fluid; feed
f	saturated liquid; friction; flow
fg	change of phase, liquid–gas, at constant pressure
G	gas; glazing; grid; generator
Ga	gain through glass at air point
Gc	gain conducted through glass
Ge	gain through glass at environmental point
g	saturated vapour
H	hot fluid; heat
HP	heat pump
HWS	hot water supply
I	induced
i	inlet; a constituent in a mixture; inside surface; initial; input
KE	kinetic energy
L	long-wave; lighting
ln	logarithmic
M	matrix
MAN	manufacturer
m	mean
mr	mean radiant
max	maximum
min	minimum
N	nozzle; natural ventilation
O	orthogonal
o	overall; free stream; outside surface; zero time; operating
P	plant; process steam; planar; primary
R	return; refrigerant
REJ	rejected
r	radiation; relative
ref	refrigeration
S	supply

s	vapour; isentropic; surface; stored; swept
T	total; thermal
TT	total thermal
TV	total vapour
t	total pressure
V	ventilation
v	velocity pressure; volumetric
W	work; wind
wb	wet bulb
w	water; wall

vapour; isentropic surface; stored; wept

T	total enthalpy
TT	total thermal
TV	total vapour
t	total pressure
V	ventilation
v	velocity; pressure; volume
w	work; wind
wb	wet bulb
w, wall	water; wall

1 BASIC HEATING, AIR CONDITIONING AND VENTILATION

This chapter covers some of the elementary theory for heating and air conditioning such as heat transfer from buildings and psychrometry. Some simple applications on heating and air conditioning systems, including pipe and duct sizing, are also given. More advanced work is also included on fan–duct systems and natural and mechanical ventilation.

It is assumed that the reader has a basic understanding of Thermodynamics and Fluid Mechanics; the texts by Eastop and McConkey[1.1] and Douglas et al.[1.2] give an adequate grounding in these subjects.

1.1 BUILDING HEAT LOSSES AND GAINS

To calculate the steady-state heat transfer through the fabric of a building the simple one-dimensional theory can be used. Since winter conditions tend to govern building design, at least in the UK, it is usual to assume that the rate of heat transfer from a building to the outside is positive, i.e.

$$\text{Fabric heat loss, } Q_F = UA\Delta t \qquad [1.1]$$

where U is the *thermal transmittance*; A is the surface area of the fabric; Δt is the temperature difference between the inside and outside air.

The thermal transmittance (sometimes referred to as the U-value) is given by

$$\frac{1}{U} = \frac{1}{\alpha_i} + \Sigma(R) + \frac{1}{\alpha_o} \qquad [1.2]$$

where α_i and α_o are the heat transfer coefficients for the inside and outside surfaces; the summation term is the sum of the resistances of the individual layers of the fabric (note that the reciprocal of this summation is known as the *thermal conductance*).

The Chartered Institution of Building Service Engineers (CIBSE) gives thermal transmittances for a wide range of building constructions.[1.3]

In any building there is an infiltration and exfiltration of air due to unavoidable gaps in the construction; the rate of air movement into and out of any space within a building depends on the pressure differences, which in turn are affected by wind direction and velocity outside the building. A certain number of air changes are required to maintain fresh, comfortable conditions inside any occupied space; in buildings which are mechanically ventilated or fully air-conditioned, air changes are more easily controlled.[1.3]

In the steady state the mass flow rate of air out of any space, $\dot{m}$, is equal to the mass flow rate into the space.

Also,

$$\dot{m} = \rho_i \dot{V}_i = \rho_o \dot{V}_o \qquad [1.3]$$

where ρ_i and ρ_o are the densities of the air inside and outside the space or building; u_i and u_o are the volumetric flow rates of air out of the space and into the space.

The volume flow rate out of the space, $\dot{V}_i$, is frequently expressed as an *air change rate*, n, given by

$$\text{Air change rate, } n = \dot{V}_i/V \qquad [1.4]$$

where V is the volume of the space.

The net heat transfer rate due to the air movement is

$$\text{Ventilation heat transfer rate} = \dot{m}(c_{pi}t_i - c_{po}t_o)$$

A mean value of specific heat at constant pressure, c_p, can be taken over the temperature range from outside to inside, i.e.

$$\text{Ventilation heat transfer rate} = \dot{m}c_p(t_i - t_o) \qquad [1.5]$$

In winter the air from outside is heated as it enters the space and then leaves the space at the room temperature, so the net energy transfer is known as the *ventilation loss*, $\dot{Q}_V$. Then substituting from Eqn [1.3] in Eqn [1.5], we have:

$$\text{Ventilation loss, } \dot{Q}_V = \rho_i \dot{V}_i c_p(t_i - t_o) \qquad [1.6]$$

or, substituting from Eqn [1.4],

$$\dot{Q}_V = \rho_i c_p n V (t_i - t_o) \qquad [1.7]$$

For atmospheric air over the normal inside temperature and humidity range it can be shown that the mean value of the product $\rho_i c_p$ is approximately equal to $1200 \, \text{J/m}^3 \, \text{K}$. Hence as a good approximation for normal ventilation conditions we can write:

$$\text{Ventilation loss, } \frac{\dot{Q}_V}{[\text{W}]} = \frac{1}{3} \frac{n}{[\text{h}^{-1}]} \frac{V}{[\text{m}^3]} \frac{\Delta t}{[\text{K}]} \qquad [1.8]$$

or

$$\frac{\dot{Q}_V}{[\text{W}]} = 1200 \frac{\dot{V}_i}{[\text{m}^3/\text{s}]} \frac{\Delta t}{[\text{K}]} \qquad [1.9]$$

Example 1.1

Figure 1.1 shows a simplified vertical section through a building 30 m long consisting of offices and stores separated by a corridor. The offices are maintained at 21 °C and the stores at 5 °C when the ambient air temperature is −5 °C. The corridor is unheated. The ventilation is from ambient to the offices, from the offices to the corridor, and from the corridor to ambient. There is a negligible flow of air between the corridor and the store. Using the data below, calculate:

(i)   the temperature in the corridor;
(ii) the heat input required for the offices;
(iii) the heat input required for the stores.

Figure 1.1 Vertical section through building

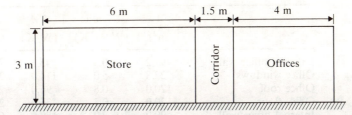

Data

Thermal transmittance of store external wall, 0.7 W/m² K; thermal transmittance of corridor external wall, 0.5 W/m² K; thermal transmittance of both internal walls, 2.0 W/m² K; thermal transmittance of the floor, 0.3 W/m² K; thermal transmittance of the external office walls (excluding glazing), 0.5 W/m² K; thermal transmittance of the office windows, 5.0 W/m² K; area of external office walls glazed, 20%; thermal conductance of the roof, 1.0 W/m² K; external heat transfer coefficient for the outside surface of the roof, 20 W/m² K; internal surface resistance of the roof, 0.2 m² K/W; air changes per hour in the office, 1.0.

Take the mean value of (ρc_p) as 1200 J/m³ K throughout.

Solution

(i) For the roof the thermal transmittance can be found using Eqn 1.2, i.e.

$$1/U = 1/(20) + 1/(1.0) + 0.2 = 1.25 \text{ m}^2 \text{ K/W}$$
$$\therefore \quad U = 1/1.25 = 0.8 \text{ W/m}^2 \text{ K}$$

Let the temperature of the corridor be t °C.

For the offices the ventilation heat loss rate can be found using Eqn [1.8], i.e.

$$\dot{Q}_v = 1 \times 30 \times 3 \times 4(21 + 5)/3 = 3120 \text{ W}$$

From Eqn [1.4] the volume flow rate from the offices to the corridor is given by $(1 \times 30 \times 3 \times 4)/3600 = 0.1 \text{ m}^3/\text{s}$.

Therefore, assuming the same value of (ρc_p) for the corridor, the net ventilation heat transfer rate from the offices to the corridor is

$$\dot{Q}_v = 1200 \times 0.1(t - 21) = -120(21 - t) \text{ W}$$

(Note that in this case the space has a ventilation gain since the corridor temperature, t, is less than the office temperature.) The fabric heat losses can now be found; a tabular method can be used as shown overleaf.

Since the corridor is unheated then the net heat transfer for the corridor is zero. Therefore, equating gains against losses, referring to the table

$$180(21 - t) + 120(21 - t) = 36(t + 5) + 15(t + 5)$$
$$+ 180(t - 5) + 4.5(t + 5)$$

$$\therefore \qquad \text{Corridor temperature, } t = 13.37 \text{ °C}$$

(ii) The heat input required for the offices is given by the net heat loss from the offices, i.e. referring to the table overleaf,

$$\text{Heat input} = 1185.6 + 2964 + 2496 + 936 + 180(21 - 13.37)$$
$$+ 3120$$
$$= 12.08 \text{ kW}$$

	A	U-value	UA	Δt	Q_F
	(m^2)	$(W/m^2\,K)$	(W/K)	(K)	(W)
External office wall	91.2	0.5	45.6	26	1185.6
Office windows	22.8	5.0	114.0	26	2964.0
Office roof	120.0	0.8	96.0	26	2496.0
Office floor	120.0	0.3	36.0	26	936.0
Internal office wall	90.0	2.0	180.0	$(21-t)$	$180(21-t)$
External store wall	126.0	0.7	88.2	10	882.0
Store roof	180.0	0.8	144.0	10	1440.0
Store floor	180.0	0.3	54.0	10	540.00
Corridor roof	45.0	0.8	36.0	$(t+5)$	$36(t+5)$
Corridor floor	45.0	0.3	15.0	$(t+5)$	$15(t+5)$
Corridor to store	90.0	2.0	180.0	$(t-5)$	$180(t-5)$
Corridor to outside	9.0	0.5	4.5	$(t+5)$	$4.5(t+5)$
Office ventilation loss					3120.0
Corridor ventilation loss					$-120(21-t)$

(iii)   Similarly, the heat input required for the store is the net heat loss from the store, i.e.

$$\text{Heat input} = 882 + 1440 + 540 - 180(13.37 - 5) = 1.36\,\text{kW}$$

Example 1.2

A detached house has a floor area of 15 m by 10 m and a pitched roof with an angle of 35 °, the ridge running parallel to the long side; the height to the eaves is 5.5 m. Double-glazed windows occupy 12% of the wall area of the long sides.

The loft space is separated from the rooms below by a 10 mm thick plasterboard ceiling and a 100 mm layer of glass fibre quilt. Natural ventilation causes an air change rate for the house, excluding the loft space, of 1 per hour; the air change rate for the loft space is 4 per hour. It may be assumed that there is no air movement between the house and the loft space.

The average temperature of the air inside the house is to be kept at 20 °C. Using the data below, calculate:

(i)   the lowest outside temperature at which the loft space air temperature remains above freezing point;

(ii)   the required heat input to the house for an external design temperature of $-5.5\,°C$.

Data

Thermal transmittance of tiled roof, 3.6 W/m² K; thermal transmittance of all exterior walls, 0.56 W/m² K; thermal transmittance of floor, 0.45 W/m² K; thermal transmittance of windows, 3.3 W/m² K; thermal conductivity of glass fibre quilt, 0.04 W/m K; thermal conductivity of plasterboard, 0.16 W/m K; heat transfer coefficient for inside loft space, 10 W/m² K; heat transfer coefficient for ceiling surface, 8 W/m² K; for the inside of the house, $\rho c_p = 1200\,\text{J/m}^3\,\text{K}$; for the loft space, $\rho c_p = 1300\,\text{J/m}^3\,\text{K}$

(Wolverhampton Polytechnic)

Solution

(i)   When the loft space is at $0\,°C$ let the outside air temperature be $t_o\,°C$. In the steady state the net rate of heat transfer from the loft is zero.

For a roof pitch angle of $35\,°$ the height from the eaves to the ridge is given by $10 \tan 35\,°/2 = 3.5$ m, and the length of the sloping sides as $5/\cos 35\,° = 6.1$ m.

Then,

$$\text{Area of sloping roof} = 2 \times 6.1 \times 15 = 183\ \text{m}^2$$

Area of external walls excluding loft space

$$= (1 - 0.12)(2 \times 15 \times 5.5) + (2 \times 10 \times 5.5) = 255.2\ \text{m}^2$$

$$\text{Area of external walls for loft space} = 2 \times 0.5 \times 10 \times 3.5 = 35\ \text{m}^2$$

$$\text{Area of windows} = 0.12 \times 2 \times 15 \times 5.5 = 19.8\ \text{m}^2$$

$$\text{Area of floor} = \text{Area of ceiling between house and loft}$$
$$= 15 \times 10 = 150\ \text{m}^2$$

The thermal transmittance for heat transfer between the house and the loft space is given by Eqn [1.2], i.e.

$$1/U = 1/8 + (0.01/0.16) + (0.1/0.04) + 1/10$$
$$= 2.7875\ \text{m}^2\ \text{K/W}$$
$$\therefore \qquad U = 0.36\ \text{W/m}^2\ \text{K}$$

The ventilation loss for the loft space is given by Eqn [1.7], i.e.

$$\dot{Q}_V = 1300 \times 4(0.5 \times 10 \times 3.5 \times 15)(0 - t_o)/3600$$
$$= 379.2(0 - t_o)\ \text{W}$$

Then, applying an energy balance to the loft space, we have

$$379.2(0 - t_o) + 183 \times 3.6(0 - t_o) + 35 \times 0.56(0 - t_o)$$
$$= 150 \times 0.36(20 - 0)$$

i.e.   Outside temperature, $t_o = -1.02\,°C$

(ii)   The outside temperature is now $-5.5\,°C$. The ventilation loss for the house is given by Eqn [1.8],

$$\dot{Q}_V = 1 \times (10 \times 5.5 \times 15)(20 + 5.5)/3 = 7013\ \text{W}$$

For the loft space there will be an energy balance as before between the heat transferred by conduction from the house, and the heat loss through the end walls and sloping roof and by ventilation. The air temperature in the loft space will reach a new equilibrium value, t, for the outside temperature of $-5.5\,°C$.

Applying an energy balance to the loft space as before, we have
$$379.2(t + 5.5) + 183 \times 3.6(t + 5.5) + 35 \times 0.56(t + 5.5)$$
$$= 150 \times 0.36(20 - t)$$

i.e.   Temperature of loft space, $t = -4.26\,°C$

To find the total heat input required to maintain the house at $20\,°C$ for the outside temperature of $-5.5\,°C$, the loft space can now be treated as exterior to the house.

Then, using a tabular method as in the previous example:

	Area	U	UA	Δt	Q_F
	(m²)	(W/m² K)	(W/K)	(K)	(W)
External walls below eaves	255.2	0.56	142.9	25.5	3644
Windows	19.8	3.30	65.3	25.5	1665
Floor	150.0	0.45	67.5	25.5	1721
Ceiling to loft	150.0	0.36	54.0	(20 + 4.26)	1310
Total for fabric					8340
Ventilation loss for house					7013
Total heat loss					15 353

From the table,

$$\text{Total heat loss} = 15\,353 \text{ W} = 15.353 \text{ kW}$$

Note: For pitched roofs with loft spaces an overall thermal transmittance is frequently quoted to be used with the area of the ceiling between the building and the loft spaces, and the overall temperature difference from inside the building to the outside. The ventilation loss in the loft space is then incorporated into an equivalent air space resistance. For example, in Example 1.2 above, the equivalent thermal transmittance of the pitched roof, U, is given by:

$$U = 1310/150(20 + 5.5) = 0.34 \text{ W/m}^2 \text{ K}$$

Values of thermal transmittances for typical roof constructions are given by CIBSE.[1.3]

1.2 HEATING SYSTEMS

There are a large number of ways in which a building may be heated and it is not the intention in this book to give detailed descriptions of heating systems. The references at the end of the chapter should be consulted for full information on the various methods of providing heating; manufacturers' catalogues can then be used to obtain practical details and prices.

Heat Emission

Any surface at a temperature above the surrounding air loses heat by convection, and if the surface temperature is above the mean temperature of the surrounding surfaces then heat is also transferred by radiation. A proper calculation of heat emission from a radiator, say, would involve applying natural convection heat transfer equations of the form: $Nu = f\{(Gr), (Pr)\}$. For radiation the heat transfer depends on the fourth power of the surface temperatures but also is a function of the geometry of the surroundings and of the emissivities of the various surfaces. For a detailed analysis of heat transfer from surfaces see a text such as that by Welty;[1.4] CIBSE[1.5] summarizes the generally accepted empirical expressions for heat transfer applicable to the Building Services field.

For practical design calculations it is a good approximation to use expressions for heat emission based on practical measurements of the heating device.

For example, for a convection-type radiator using either hot water or steam, British Standard tests (BS 3528)[1.6] specify the following:

$$\text{Heat emission, } \dot{Q} = Cs(t_m - t_{ai})^{1.3} \qquad [1.10]$$

where C is a constant; s is the number of sections or units of length of the radiator; t_m is the mean temperature of the surface of the radiator; t_{ai} is the mean temperature of the air in the room.

Manufacturers in the UK normally quote heat emissions for a temperature difference, $t_m - t_{ai}$, of 60 K; the actual heat emission is then given by

$$\dot{Q} = \dot{Q}_{MAN}\left\{\frac{\Delta t}{60}\right\}^{1.3} \qquad [1.11]$$

(where $\dot{Q}_{MAN}$ is the heat emission quoted by the manufacturer for a temperature difference of 60 K; Δt is the actual temperature difference.)

Equations [1.10] and [1.11] apply to exposed radiators, the most common example being those used in low-pressure hot water heating circuits. When pipes or tubes carrying hot water are placed within cabinets and the heat is transferred from the water to the air in the space by the movement of air through inlet and outlet grills, the units are known as natural convection heaters. In this case tests indicate[1.6] that the heat emission can be taken as

$$\dot{Q} \propto (\Delta t)^{1.4} \qquad [1.12]$$

For fan–coil units in which the air movement across the tubes is caused by a fan integral with the cabinet, heat transfer is by forced convection and hence the heat emission is given by:

$$\text{Heat emission, } \dot{Q} = UA\Delta\bar{t}_{ln}$$

where U is the overall heat transfer coefficient between the water and the air; A is the effective heat transfer area; $\Delta\bar{t}_{ln}$ is the logarithmic mean temperature difference given by

$$\Delta\bar{t}_{ln} = \frac{(t_{w1} - t_{a2}) - (t_{w2} - t_{a1})}{\ln\{(t_{w1} - t_{a2})/(t_{w2} - t_{a1})\}} \qquad [1.13]$$

where t_{w1} and t_{w2} are the temperatures of the water at inlet and outlet to the heater; t_{a1} and t_{a2} are the temperatures of the air at inlet and outlet from the heater.

British Standard 4856[1.7] indicates that, for given air and water flows,

$$\text{Heat emission, } \dot{Q} \propto (t_m - t_{ai}) \qquad [1.14]$$

where t_m is the mean water temperature; t_{ai} is the mean air temperature in the room.

For fan–coil heaters using wet steam at a temperature t_g, the value of $\Delta\bar{t}_{ln}$ reduces to,

$$\Delta\bar{t}_{ln} = \frac{(t_{a2} - t_{a1})}{\ln\{(t_g - t_{a1})/(t_g - t_{a2})\}}$$

For a fixed air flow, Eqn 1.14 then applies with t_g substituted for t_m.

For a fan–coil heater using hot water, the value of the overall heat transfer coefficient depends on both the air flow rate and the water flow rate. For a unit heater using wet steam, the heat transfer coefficient on the steam side is very large compared with that on the air side. For air flowing across the tube, the heat transfer coefficient for a given temperature range is proportional to the air velocity to the power 0.5, and since for a fan the volume flow rate is proportional to the fan speed, N, we can write:

$$\text{Heat emission, } \dot{Q} \propto (t_g - t_{ai})\sqrt{N} \qquad [1.15]$$

For any given heater of the above types, tests can be done to establish constants for the equations [1.12], [1.14] and [1.15] or, knowing the manufacturer's data for one particular temperature difference and fan speed, heat emissions can be calculated for other values of these parameters.

Example 1.3

A space is heated by three radiators placed in series on a single-pipe, low-pressure hot water heating circuit; each radiator is required to give an output of 3 kW at a design outside temperature of $-2\,°C$ when the space is maintained at $20\,°C$. Each radiator is made up of sections which emit 100 W for a temperature difference of 60 K between the mean surface temperature of the radiator and the mean room air temperature. The temperature of the water from the boiler is fixed at $80\,°C$ and the return temperature to the boiler is $62\,°C$ under design conditions.

Two alternative control system may be used as shown in Figs 1.2(a) and (b); in Fig. 1.2(a) a flow mixing valve at point A mixes water from the boiler with returned water from the heating circuit; in Fig. 1.2(b) a diverter valve at point B receives some of the boiler water which mixes with the water returning from the heating circuit before passing back to the boiler inlet. In the first case the mass flow rate is kept constant through the radiators and control is achieved by varying the temperature, t_F; in the second case the temperature of the supply to the radiators is always the boiler temperature (neglecting heat losses), and the mass flow rate is varied.

Figure 1.2 Low-pressure hot water heating

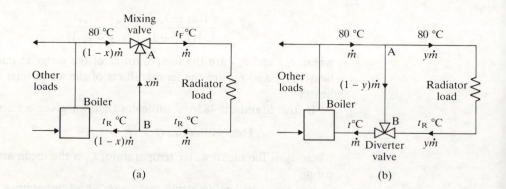

(a)   (b)

Neglecting all losses, calculate:
(i)   the number of sections required for each radiator;
(ii) the temperature and mass flow rate of the water supplied to the radiators when the outside temperature is $8\,°C$ and a mixing valve is used;
(iii) the mass flow rate of water to the radiators when the outside temperature is $8\,°C$ and a diverter valve is used.

For water assume as an approximation that the specific heat is 4.2 kJ/kg K.

Solution

(i) The size of the radiators is fixed at the outside design temperature of $-2\,°C$.

Each radiator will have a water temperature drop of $(80 - 62)/3 = 6\,K$, and hence each will be at a different mean surface temperature.

For the first radiator: $t_m = \{80 + (80 - 6)\}/2 = 77\,°C$

Then using Eqn [1.11],

$$\text{Heat emission per section} = 100 \left\{ \frac{77}{60} \right\}^{1.3} = 138.31\,W$$

i.e. Required number of sections $= 3000/138.31 = 22$, say

For the second radiator: $t_m = \{74 + (74 - 6)\}/2 = 71\,°C$

$$\text{Heat emission per section} = 100 \left\{ \frac{71}{60} \right\}^{1.3} = 124.46\,W$$

i.e. Required number of sections $= 3000/124.46 = 24$, say

For the third radiator: $t_m = (68 + 62)/2 = 65\,°C$

$$\text{Heat emission per section} = 100 \left\{ \frac{65}{60} \right\}^{1.3} = 110.97$$

i.e. Required number of sections $= 3000/110.97 = 27$, say

(ii) The heat loss from the space is $9\,kW$ when the outside temperature is $-2\,°C$ and therefore the heat loss when the outside temperature is $8\,°C$ is given by

$$Q = 9(20 - 8)/(20 + 2) = 4.909\,kW$$

Then using Eqn [1.10],

$$4.909 = 9 \left\{ \frac{t_m - 20}{71 - 20} \right\}^{1.3}$$

i.e. $$(t_m - 20)^{1.3} = \frac{4.909}{9}(71 - 20)^{1.3}$$

$$\therefore \quad t_m = 52\,°C$$

For a design outside temperature there is no flow required between A and B in Fig. 1.2; in Fig. 1.2(a) $x = 0$, and in Fig. 1.2(b) $y = 1$.
Hence,

$$\text{Heat load} = \dot{m}c(80 - 62) = (3 \times 3)\,kW$$

i.e. Design mass flow rate to radiators, $\dot{m} = (3 \times 3)/\{4.2(80 - 62)\}$
$$= 0.119\,kg/s$$

This flow is maintained through the radiators when a mixing valve is used, as in Fig. 1.2(a).

Referring to Fig. 1.2(a), we have

$$0.119 \times 4.2(t_F - t_R) = 4.909$$
$$\therefore \quad (t_F - t_R) = 9.82\,K$$

Also,

$$(t_F + t_R)/2 = t_M = 52 \text{ K}$$
$$\therefore \quad t_F = (9.82 + 104)/2 = 56.9 \,°C$$
and
$$t_R = 56.9 - 9.82 = 47.1 \,°C$$

The flow of water from the boiler is given by

$$(1 - x)\dot{m} = 9.82\dot{m}/(80 - 47.1) = 0.3\dot{m} = 0.036 \text{ kg/s}$$

i.e.   for an outside temperature of 8 °C, $x = 0.7$

(iii)   In this case the flow temperature is the same as the boiler temperature of 80 °C. As before, the mean temperature of the radiator surfaces is 52 °C; therefore,

$$t_R = 52 - (80 - 52) = 24 \,°C$$

Then the mass flow rate through the radiators is given by

$$y\dot{m} = 4.909/4.2(80 - 24) = 0.021 \text{ kg/s}$$

i.e.   for an outside temperature of 8 °C, $y = 0.177$

Note that the return temperature to the boiler in this case is given by $80 - \{4.909/4.2 \times 0.119\} = 70.2 \,°C$.

Example 1.4
A factory building is heated by a number of fan–coil unit heaters using hot water. As shown in Fig. 1.3, each heater draws fresh air from outside the building where it mixes with air re-circulated from the space; the fan then blows the mixed air across the heater bank into the space. The relative amounts of re-circulated air and fresh air are controlled by dampers, and a motorized valve in the hot water circuit controls the heat input to the heater bank. The fan gives a constant flow rate of 0.25 m³/s at all damper settings.

Figure 1.3   Room heated by air heaters

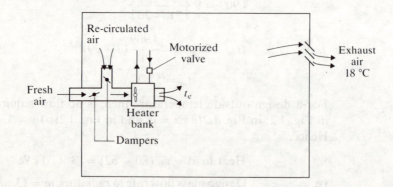

The system is designed for an inside air temperature of 18 °C and an outside temperature of −2 °C. At the design conditions the volume ratio of fresh air to re-circulated air is 1 to 3, and the fabric heat loss for the space served by each heater is 3 kW. As the outside temperature rises it is proposed to maintain the inside temperature at 18 °C by reducing the volume of re-circulated air while keeping the flow of hot water unaltered. Assuming that the re-circulated air is at the room air temperature of 18 °C, and neglecting the temperature rise across the fan, calculate:

(i)   the heat output from the heater at the design conditions;
(ii) the temperature of the air leaving each unit at the design conditions;
(iii) the highest outside temperature for which it is possible to maintain the inside temperature at 18 °C without reducing the flow of hot water to the unit;
(iv) the temperature of the air leaving each unit for the conditions in (iii).

Take the product (ρc_p) as 1.2 kJ/m³ K throughout.

Solution

(i)   At the design conditions, $\dot{Q}_F = 3$ kW (given), and

$$\text{Ventilation loss, } \dot{Q}_V = \frac{1.2 \times 0.25(18 + 2)}{3} = 2 \text{ kW}$$

i.e.   Total heat input of heater $= 3 + 2 = 5$ kW

(ii)   The temperature of the air in the unit after mixing, t, is given by:

$$t = \left\{ \frac{1}{3} \times -2 \right\} + \left\{ \frac{2}{3} \times 18 \right\} = 11.33 \text{ °C}$$

Then, for the heater bank,

$$1.2 \times 0.25(t_e - 11.33) = 5 \text{ kW}$$

where t_e is the air temperature at exit from the heater bank, i.e.

$$t_e = 28 \text{ °C}$$

Note: An alternative method is to equate the fabric loss to the heat required to cool the air from the heater to the room temperature of 18 °C:

$$3 = 1.2 \times 0.25(t_e - 18)$$
$$\therefore \qquad t_e = 28 \text{ °C, as before.}$$

(iii)   For the outside temperatures the control is to be achieved by keeping the heater output constant at 5 kW and adjusting the mixed air temperature such that the air is heated by a sufficient amount to balance the heat losses. At a certain outside temperature the re-circulation dampers will be closed and any further adjustment must then be made by reducing the hot water flow to the heater using the motorized valve.

Let the outside temperature above which the motorized valve will need to be used be t_o.

At this temperature the fabric loss is given by

$$\dot{Q}_F = 3 \left(\frac{18 - t_o}{18 + 2} \right) = 0.15(18 - t_o) \text{ kW}$$

The air to the fan is all fresh air and the volume flow from the fan remains constant throughout at 0.25 m³/s, therefore

$$\text{Ventilation loss, } \dot{Q}_V = 1.2 \times 0.25(18 - t_o) = 0.3(18 - t_o) \text{ kW}$$

Hence,

$$(0.15 + 0.3)(18 - t_o) = 5 \text{ kW}$$
$$\therefore \qquad t_o = 6.9 \text{ °C}$$

Above this temperature the motorized valve must reduce the water flow to reduce the heat input below 5 kW.

(iv) Applying an energy balance to the heater bank:

$$5 = 1.2 \times 0.25(t_e - 6.9)$$

i.e. $t_e = 23.6\ ^\circ C$

or, since $\dot{Q}_F = 0.15(18 - 6.9) = 1.67$ kW, then

$$t_e = 18 + (1.67/1.2 \times 0.25) = 23.6\ ^\circ C$$

as before.

Pipe Sizing

For water flowing in a piped system, the main pressure loss is due to friction in the pipes.

For the steady flow of a fluid flowing at a velocity u, in a pipe of diameter d, the shear stress at the pipe wall, τ, sets up a resisting force which must be balanced by a pressure difference in the fluid.

Referring to Fig. 1.4,

$$\tau(\pi d L) = (p_1 - p_2)\pi d^2/4$$

i.e. $$p_1 - p_2 = \Delta p_f = \frac{4L\tau}{d}$$

Figure 1.4   Fluid flow in a pipe

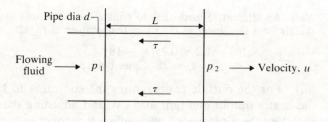

A friction factor, f, can be defined as

$$f = \tau/(\tfrac{1}{2}\rho u^2) \qquad [1.16]$$

Then substituting in the equation for pressure loss, we obtain the D'Arcy equation

$$\Delta p_f = \frac{4fL}{d}(\tfrac{1}{2}\rho u^2) \qquad [1.17]$$

For laminar flow ($Re < 2000$), the friction factor is given by $16/Re$. For fully turbulent flow, and for the transitional regime between fully laminar and fully turbulent flow, the friction factor is found to be a function of the Reynolds number, Re, and the relative roughness of the pipe wall; the *Moody chart* gives a plot of f against Re for various values of relative roughness.[1.2,1.5] Values of absolute roughness, k, for pipes and ducts normally used for building services fluids are given in Ref. 1.5; the relative roughness is defined as k/d.

For the transitional flow regime, for $Re > 3000$, Colebrook and White[1.8] give an equation as follows:

$$1/\sqrt{f} = -4\log_{10}\{(k/3.71d) + (1.26/Re\sqrt{f})\}$$

This equation converges to an expression for turbulent flow in smooth pipes when k tends to zero, and when Re tends to infinity it gives an expression for f which depends only on k/d. It is used in the Moody chart and is also used by CIBSE[1.5] in drawing up tables of pressure drop for various pipe sizes, fluid velocities and mass flow rates, for various grades of pipe materials. The tables are for a mean water temperature of 75 °C.

In any fluid system there are also pressure losses due to valves, pipe bends, junctions, etc. In the case of a fluid such as water the friction loss is found to be relatively greater than the losses due to the fittings, bends, etc. The concept of *equivalent length* is therefore introduced. This is defined as the length of straight pipe which would give a friction pressure loss equivalent to one velocity head, $\frac{1}{2}\rho u^2$. Referring to Eqn [1.17], the equivalent length to give a friction loss of one velocity head, $\frac{1}{2}\rho u^2$, is given by

$$l_e = d/4f \qquad [1.18]$$

Values of l_e are given for various pressure drops and pipe diameters in the CIBSE tables.[1.5]

For air the friction pressure loss in a duct is relatively lower and is of the same order as the pressure loss due to the fittings. It is therefore more convenient to express the fittings loss in terms of a pressure loss factor, k, given by

$$\Delta p = k(\tfrac{1}{2}\rho u^2) \qquad [1.19]$$

Pressure loss factors are given by CIBSE for both pipe and duct fittings. In the case of a water system, the equivalent length for the particular fitting is then kl_e.

For any given low-pressure hot water heating system the longest run to a radiator or unit is called the *index circuit* and is the one used to size the pump; the pipe sizes of the various branches are then found from the pressures available at the relevant off-takes. Pipe sizing is usually done by assuming initially a nominal pressure drop per unit length or a particular value of fluid velocity; the higher the pressure drop or fluid velocity chosen, then the higher is the pumping power but the lower is the cost of fuel used in providing the heat loss from the pipe surface; the discounted capital costs must be offset by the running costs in a suitably short time interval. It can be seen that the complex calculations involved are best tackled using computer programs.

The example below shows a simple manual method for preliminary pipe sizing.

Example 1.5

A low-pressure hot water heating system with flow and return temperatures of 80 °C and 70 °C is shown diagrammatically in plan view in Fig. 1.5. The lengths of the various pipe sections, which are of medium-grade steel, are given in the table below.

Pipe section	1	2	3	4	5	6	7	8	9
Length (flow and return)/(m)	40	25	25	25	30	15	15	15	15

Figure 1.5 Low-pressure hot water heating system

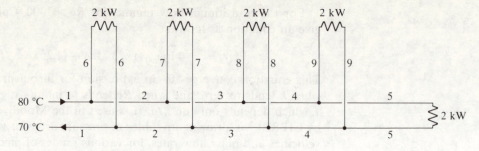

The heat emission from each radiator is 2 kW.

Assuming that fittings, bends, etc., account for 20% of the total pressure drop in any section, make a preliminary pipe sizing for the system taking an average pressure drop of 300 Pa/m, and calculate the approximate pressure rise required from the pump; use Table 1.1 for medium grade steel extracted from Ref. 1.5 (by permission of CIBSE). Neglect pipe heat losses, and take the specific heat of water as 4.2 kJ/kg K.

Solution
The index circuit consists of pipe sections 1, 2, 3, 4, and 5.

$$\text{Total heat emission} = 5 \times 2 = 10 \text{ kW}$$

Water flow rate in section $1 = 10/\{4.2 \times (80 - 70)\} = 0.238$ kg/s
Water flow rate in section $2 = 8/(4.2 \times 10) = 0.190$ kg/s
Water flow rate in section $3 = 6/(4.2 \times 10) = 0.143$ kg/s
Water flow rate in section $4 = 4/(4.2 \times 10) = 0.095$ kg/s
Water flow rate in section $5 = 2/(4.2 \times 10) = 0.048$ kg/s

From tables, for an average pressure drop of 300 Pa/m, the nearest diameter is chosen to give the required flow rate. We then have:

Pipe section	1	2	3	4	5
Mass flow rate/(kg/s)	0.238	0.190	0.143	0.095	0.048
Diameter/(mm)	20	20	20	15	10

The actual pressure drops in each section can be found by reading the pressure drop per metre length at the known mass flow rate and pipe diameter, and multiplying it by the equivalent length of the particular pipe section. In this case the equivalent length is to be taken as 20% more than the actual length to allow for fittings, etc.

Therefore, for the index circuit, we have:

$$\Delta p_1 = 280 \times 1.2 \times 40 = 13\,440 \text{ Pa} = 0.134 \text{ bar}$$
$$\Delta p_2 = 180 \times 1.2 \times 25 = 5400 \text{ Pa} = 0.054 \text{ bar}$$
$$\Delta p_3 = 107 \times 1.2 \times 25 = 3210 \text{ Pa} = 0.032 \text{ bar}$$
$$\Delta p_4 = 220 \times 1.2 \times 25 = 6600 \text{ Pa} = 0.066 \text{ bar}$$
$$\Delta p_5 = 227 \times 1.2 \times 30 = 8172 \text{ Pa} = 0.082 \text{ bar}$$

i.e.   Total pressure rise across pump = 0.368 bar

For the various branches the pressure drops available can now be found, i.e.

$$\Delta p_6 = 0.368 - 0.134 = 0.234 \text{ bar} = 23\,400 \text{ Pa}$$

Table 1.1 Pressure drop due to friction in medium grade steel pipes for water at 75 °C

Δp_l		10 mm		15 mm		20 mm		25 mm		32 mm		40 mm		50 mm			Δp_l
(Pa/m)	v	M	l_e	M	l_e	M	l_e	M	l_e	M	l_e	M	l_e	M	l_e	v	(Pa/m)
80.0		0.027	0.3	0.055	0.5	0.122	0.7	0.228	1.0	0.480	1.4	0.720	1.8	1.35	2.4		80.0
82.5		0.028	0.3	0.056	0.5	0.124	0.7	0.232	1.0	0.488	1.4	0.732	1.8	1.37	2.4		82.5
85.0		0.028	0.3	0.057	0.5	0.126	0.7	0.236	1.0	0.496	1.4	0.743	1.8	1.39	2.4		85.0
87.5		0.029	0.3	0.058	0.5	0.128	0.7	0.240	1.0	0.503	1.4	0.755	1.8	1.41	2.4		87.5
90.0		0.029	0.3	0.059	0.5	0.130	0.7	0.243	1.0	0.511	1.4	0.766	1.8	1.43	2.4		90.0
92.5		0.029	0.3	0.060	0.5	0.132	0.7	0.247	1.0	0.518	1.5	0.778	1.8	1.45	2.4		92.5
95.0		0.030	0.3	0.061	0.5	0.134	0.7	0.251	1.0	0.526	1.5	0.789	1.8	1.48	2.4		95.0
97.5		0.030	0.3	0.062	0.5	0.136	0.7	0.254	1.0	0.533	1.5	0.800	1.8	1.50	2.4		97.5
100.0		0.031	0.3	0.062	0.5	0.138	0.7	0.258	1.0	0.540	1.5	0.810	1.8	1.52	2.4		100.0
120.0	0.30	0.034	0.3	0.069	0.5	0.152	0.7	0.284	1.0	0.595	1.5	0.893	1.8	1.67	2.4		120.0
140.0		0.037	0.3	0.075	0.5	0.165	0.8	0.308	1.0	0.646	1.5	0.968	1.8	1.81	2.5		140.0
160.0		0.040	0.4	0.081	0.5	0.178	0.8	0.331	1.0	0.693	1.5	1.04	1.8	1.94	2.5		160.0
180.0		0.042	0.4	0.086	0.5	0.189	0.8	0.353	1.0	0.738	1.5	1.11	1.8	2.06	2.5	1.0	180.0
200.0		0.045	0.4	0.091	0.5	0.200	0.8	0.373	1.1	0.780	1.5	1.17	1.9	2.18	2.5		200.0
220.0		0.047	0.4	0.096	0.5	0.211	0.8	0.392	1.1	0.820	1.5	1.28	1.9	2.29	2.5		220.0
240.0		0.050	0.4	0.100	0.5	0.221	0.8	0.411	1.1	0.858	1.5	1.29	1.9	2.40	2.5		240.0
260.0		0.052	0.4	0.105	0.5	0.230	0.8	0.428	1.1	0.895	1.5	1.34	1.9	2.50	2.5		260.0
280.0		0.054	0.4	0.109	0.5	0.239	0.8	0.445	1.1	0.931	1.5	1.39	1.9	2.60	2.6		280.0
300.0		0.056	0.4	0.113	0.5	0.248	0.8	0.462	1.1	0.965	1.5	1.44	1.9	2.69	2.6		300.0
320.0	0.50	0.058	0.4	0.117	0.5	0.257	0.8	0.478	1.1	0.998	1.6	1.49	1.9	2.78	2.6		320.0
340.0		0.060	0.4	0.121	0.5	0.265	0.8	0.493	1.1	1.03	1.6	1.54	1.9	2.87	2.6		340.0
360.0		0.062	0.4	0.125	0.5	0.273	0.8	0.508	1.1	1.06	1.6	1.59	1.9	2.96	2.6		360.0
380.0		0.064	0.4	0.128	0.5	0.281	0.8	0.523	1.1	1.09	1.6	1.63	1.9	3.04	2.6		380.0
400.0		0.065	0.4	0.132	0.5	0.289	0.8	0.537	1.1	1.12	1.6	1.68	1.9	3.12	2.6		400.0
420.0		0.067	0.4	0.135	0.5	0.297	0.8	0.551	1.1	1.15	1.6	1.72	1.9	3.20	2.6	1.5	420.0
440.0		0.069	0.4	0.139	0.5	0.304	0.8	0.564	1.1	1.18	1.6	1.76	1.9	3.28	2.6		440.0
460.0		0.070	0.4	0.142	0.5	0.311	0.8	0.578	1.1	1.21	1.6	1.80	1.9	3.36	2.6		460.0
480.0		0.072	0.4	0.145	0.5	0.318	0.8	0.591	1.1	1.23	1.6	1.84	1.9	3.43	2.6		480.0
500.0		0.074	0.4	0.148	0.5	0.325	0.8	0.603	1.1	1.25	1.6	1.88	1.9	3.51	2.6		500.0
520.0		0.075	0.4	0.151	0.5	0.332	0.8	0.616	1.1	1.29	1.6	1.92	1.9	3.58	2.6		520.0
540.0		0.077	0.4	0.154	0.6	0.338	0.8	0.628	1.1	1.31	1.6	1.96	1.9	3.65	2.6		540.0
560.0		0.078	0.4	0.157	0.6	0.345	0.8	0.640	1.1	1.34	1.6	2.00	1.9	3.72	2.6		560.0
580.0		0.080	0.4	0.160	0.6	0.351	0.8	0.652	1.1	1.36	1.6	2.03	1.9	3.78	2.6		580.0
600.0		0.081	0.4	0.163	0.6	0.355	0.8	0.664	1.1	1.38	1.6	2.07	1.9	3.85	2.6		600.0
620.0		0.082	0.4	0.166	0.6	0.364	0.8	0.675	1.1	1.41	1.6	2.10	1.9	3.92	2.6		620.0
640.0		0.084	0.4	0.169	0.6	0.370	0.8	0.686	1.1	1.43	1.6	2.14	1.9	3.98	2.6		640.0
660.0		0.085	0.4	0.172	0.6	0.376	0.8	0.697	1.1	1.45	1.6	2.17	1.9	4.04	2.6		660.0
680.0		0.087	0.4	0.174	0.6	0.382	0.8	0.708	1.1	1.48	1.6	2.21	1.9	4.11	2.6		680.0
700.0		0.088	0.4	0.177	0.6	0.388	0.8	0.719	1.1	1.50	1.6	2.24	1.9	4.17	2.6		700.0
720.0		0.089	0.4	0.180	0.6	0.393	0.8	0.730	1.1	1.52	1.6	2.27	1.9	4.23	2.6		720.0
740.0		0.091	0.4	0.182	0.6	0.399	0.8	0.740	1.1	1.54	1.6	2.31	2.0	4.29	2.6	2.0	740.0
760.0		0.092	0.4	0.185	0.6	0.405	0.8	0.750	1.1	1.56	1.6	2.34	2.0	4.35	2.6		760.0
780.0		0.093	0.4	0.187	0.6	0.410	0.8	0.761	1.1	1.59	1.6	2.37	2.0	4.41	2.6		780.0
800.0		0.094	0.4	0.190	0.6	0.416	0.8	0.771	1.1	1.61	1.6	2.40	2.0	4.46	2.6		800.0
820.0		0.096	0.4	0.192	0.6	0.421	0.8	0.780	1.1	1.63	1.6	2.43	2.0	4.52	2.6		820.0
840.0		0.097	0.4	0.195	0.6	0.426	0.8	0.790	1.1	1.65	1.6	2.46	2.0	4.58	2.6		840.0
860.0		0.098	0.4	0.197	0.6	0.431	0.8	0.800	1.1	1.67	1.6	2.49	2.0	4.63	2.6		860.0
880.0		0.099	0.4	0.200	0.6	0.437	0.8	0.810	1.1	1.69	1.6	2.52	2.0	4.69	2.6		880.0
900.0		0.100	0.4	0.202	0.6	0.442	0.8	0.819	1.1	1.71	1.6	2.55	2.0	4.74	2.6		900.0
920.0		0.102	0.4	0.204	0.6	0.447	0.8	0.828	1.1	1.73	1.6	2.58	2.0	4.80	2.6		920.0
940.0		0.103	0.4	0.207	0.6	0.452	0.8	0.838	1.1	1.75	1.6	2.61	2.0	4.85	2.6		940.0
960.0		0.104	0.4	0.209	0.6	0.457	0.8	0.847	1.1	1.76	1.6	2.64	2.0	4.90	2.6		960.0
980.0		0.105	0.4	0.211	0.6	0.462	0.8	0.856	1.1	1.78	1.6	2.66	2.0	4.95	2.6		980.0
1 000.0		0.106	0.4	0.213	0.6	0.467	0.8	0.865	1.1	1.80	1.6	2.69	2.0	5.00	2.6		1 000.0
1 100.0		0.112	0.4	0.224	0.6	0.490	0.8	0.909	1.1	1.89	1.6	2.83	2.0	5.26	2.7		1 100.0
1 200.0	1.0	0.117	0.4	0.235	0.6	0.513	0.8	0.950	1.1	1.98	1.6	2.96	2.0	5.49	2.7		1 200.0
1 300.0		0.122	0.4	0.245	0.6	0.535	0.8	0.990	1.1	2.06	1.6	3.08	2.0	5.72	2.7		1 300.0
1 400.0		0.127	0.4	0.254	0.6	0.555	0.8	1.03	1.1	2.14	1.6	3.20	2.0	5.94	2.7		1 400.0
1 500.0		0.131	0.4	0.263	0.6	0.576	0.8	1.07	1.1	2.22	1.6	3.31	2.0	6.16	2.7		1 500.0
1 600.0		0.136	0.4	0.272	0.6	0.595	0.9	1.10	1.1	2.29	1.6	3.42	2.0	6.36	2.7	3.0	1 600.0
1 700.0		0.140	0.4	0.281	0.6	0.614	0.9	1.14	1.2	2.37	1.6	3.53	2.0	6.56	2.7		1 700.0
1 800.0		0.144	0.4	0.290	0.6	0.632	0.9	1.17	1.2	2.44	1.6	3.64	2.0	6.76	2.7		1 800.0
1 900.0		0.148	0.4	0.298	0.6	0.650	0.9	1.20	1.2	2.50	1.6	3.74	2.0	6.94	2.7		1 900.0

Therefore for this section the pressure drop per unit length is approximately $23\,400/(1.2 \times 15) = 1300$ Pa/m. The mass flow rate required for this section is $2/(4.2 \times 10) = 0.048$ kg/s. Hence from Table 1.1 the diameter required is 10 mm. The actual pressure drop available will then be approximately $220 \times 1.2 \times 15 = 3960$ Pa; the excess pressure drop available (i.e. $23\,400 - 3960 = 19\,440$ Pa $= 0.194$ bar) can be taken up using a valve in section 6. Sections 7, 8 and 9 are sized in a similar way; the results are summarized in the table below.

Pipe section	6	7	8	9
Pressure drop available/(Pa/m)	1300	1000	822	456
Diameter/(mm)	10	10	10	10
Excess pressure/(bar)	0.194	0.140	0.108	0.043

Example 1.6
For the system of Example 1.5, shown in Fig. 1.5, re-calculate the required pipe diameters assuming estimated pipe emissions as shown in the following table.

Pipe section	1	2	3	4	5	6	7	8	9
Heat emission/(kW)	0.20	0.10	0.10	0.10	0.05	0.03	0.03	0.03	0.03

Solution
The most convenient method of pipe sizing allowing for pipe emission is that using a direct proportioning of the mains emission. In this method an equivalent mass flow rate is found which compensates for the increased flow required due to the heat loss from each pipe section. Starting from the boiler, an equivalent mass flow rate is found for section 1; the equivalent mass flow rates for sections 2 and 6 are then found as a proportion of the equivalent flow rate in section 1 as follows:

Total mass flow rate in section 1

$$= \frac{\{(5 \times 2) + (0.2 + 0.1 + 0.1 + 0.1 + 0.05 + 0.03 + 0.03 + 0.03 + 0.03)\}}{4.2 \times 10}$$

$$= 10.67/42 = 0.254 \text{ kg/s}$$

Equivalent mass flow rate in section 1 due to pipe emission, $\dot{m}_{e1}$, is given by

$$\dot{m}_{e1} = 0.2/(4.2 \times 10) = 0.2/42 \text{ kg/s}$$

Mass flow rates in sections 6 and 2, neglecting the emission in section 1, are given by

$$\dot{m}_6 = (2 + 0.03)/42 \text{ kg/s} = 2.03/42 \text{ kg/s}$$

and $$\dot{m}_2 = (10.67 - 2.03 - 0.2)/42 = 8.44/42 \text{ kg/s}$$

Therefore, by proportioning, we have:

$$\dot{m}_{e2} = \frac{8.44}{(2.03 + 8.44)} \times \frac{0.2}{42} = \frac{0.161}{42} \text{ kg/s}$$

and $$\dot{m}_{e6} = \frac{0.2 - 0.161}{42} = \frac{0.039}{42} \text{ kg/s}$$

Hence,

$$\dot{m}'_6 = \dot{m}_6 + \dot{m}_{e6} = (2.03 + 0.039)/42 = 0.049 \text{ kg/s}$$

and $\qquad \dot{m}'_2 = 0.254 - 0.049 = 0.205 \text{ kg/s}$

(or $\dot{m}'_2 = (8.44 + 0.161)/42 = 0.205 \text{ kg/s}$).

Similarly, for the remaining sections:

$$\dot{m}_7 = 2.03/42 \text{ kg/s}$$

and $\qquad \dot{m}_3 = (8.44 - 2.03 - 0.10)/42 = 6.31/42 \text{ kg/s}$

$$\dot{m}_{e3} = \frac{6.31}{2.03 + 6.31} \times \frac{(0.161 + 0.1)}{42} = \frac{0.198}{42} \text{ kg/s}$$

$$\dot{m}_{e7} = \frac{0.261 - 0.198}{42} = \frac{0.063}{42} \text{ kg/s}$$

Therefore, $\qquad \dot{m}'_7 = (2.03 + 0.198)/42 = 0.053 \text{ kg/s}$

and $\qquad \dot{m}'_3 = 0.205 - 0.053 = 0.152 \text{ kg/s}$

Then, $\qquad \dot{m}_8 = 2.03/42 \text{ kg/s}$

and $\qquad \dot{m}_4 = (6.31 - 2.03 - 0.10)/42 = 4.18/42 \text{ kg/s}$

$$\dot{m}_{e4} = \frac{4.18}{2.03 \times 4.18} \times \frac{(0.198 + 0.1)}{42} = \frac{0.201}{42} \text{ kg/s}$$

$$\dot{m}_{e8} = (0.298 - 0.201)/42 = 0.097/42 \text{ kg/s}$$

$\therefore \qquad \dot{m}'_8 = (2.03 + 0.097)/42 = 0.051 \text{ kg/s}$

$$\dot{m}'_4 = 0.152 - 0.051 = 0.101 \text{ kg/s}$$

Finally, $\qquad \dot{m}_9 = 2.03/42 \text{ kg/s}$

and $\qquad \dot{m}_5 = 2.05/42 \text{ kg/s}$

Also, $\qquad \dot{m}_{e5} = \frac{2.05}{2.03 + 2.05} \times \frac{0.251}{42} = \frac{0.126}{42} \text{ kg/s}$

$$\dot{m}_{e9} = (0.251 - 0.126)/42 = 0.125/42 \text{ kg/s}$$

Therefore,

$$\dot{m}'_9 = (2.03 + 0.125)/42 = 0.051 \text{ kg/s}$$
$$\dot{m}'_5 = 0.101 - 0.051 = 0.050 \text{ kg/s}$$

Then, using Table 1.1, we can read the diameters for the index circuit shown in the following table.

Pipe section	1	2	3	4	5
Diameter/(mm)	25	20	20	15	10

As shown in Example 1.5, we now calculate the actual pressure drops available to each branch, i.e.

$$\Delta p_1 = 97.5 \times 1.2 \times 40 = 4680 \text{ Pa}$$
$$\Delta p_2 = 210 \times 1.2 \times 25 = 6300 \text{ Pa}$$

$$\Delta p_3 = 120 \times 1.2 \times 25 = 3600 \text{ Pa}$$
$$\Delta p_4 = 244 \times 1.2 \times 25 = 7320 \text{ Pa}$$
$$\Delta p_5 = 240 \times 1.2 \times 30 = 8640 \text{ Pa}$$

Therefore,

$$\text{Pump pressure rise required} = 30\,540 \text{ Pa} = 0.305 \text{ bar}$$

Also, for the branches:

$$\text{Pressure drop for section 6} = 30\,540 - 4680 = 25\,860 \text{ Pa}$$

i.e.

$$\text{Pressure drop per unit length} = 25\,860/(1.2 \times 15) = 1437 \text{ Pa/m}$$

Therefore, from Table 1.1,

$$\text{Diameter for pipe 6} = 10 \text{ mm}$$

The excess pressure to be absorbed is then given by the actual pressure drop available minus the pressure drop for the flow and the diameter chosen, i.e.

$$\text{Excess pressure} = (1437 - 233) \times 15 = 18\,060 \text{ Pa} = 0.181 \text{ bar}$$

For branch section 7,

$$\text{Pressure drop available} = (25\,860 - 6300)/(1.2 \times 15)$$
$$= 1087 \text{ Pa/m}$$

i.e.

$$\text{Diameter for section 7} = 10 \text{ mm}; \text{ Excess pressure} = 0.122 \text{ bar}$$

$$\text{Pressure drop available for section 8} = (19\,560 - 3600)/(1.2 \times 15)$$
$$= 887 \text{ Pa/m}$$

i.e.

$$\text{Diameter for section 8} = 10 \text{ mm}; \text{ Excess pressure} = 0.093 \text{ bar}$$

$$\text{Pressure drop for section 9} = 8640/(1.2 \times 15) = 480 \text{ Pa/m}$$

i.e.

$$\text{Diameter for section 9} = 10 \text{ mm}; \text{ Excess pressure} = 0.03 \text{ bar}$$

The preliminary design can be checked using pressure loss values for the actual fittings, junctions, etc. More accurate values of heat emission from the various pipe lengths can be found once the diameters are chosen, and hence more accurate mass flow rates can be calculated based on the actual temperature differences at each junction. The process is iterative and therefore, as stated earlier, is more easily done using a computer; commercial software is widely available.

1.3 PSYCHROMETRY

The air we breathe contains a certain percentage of superheated water vapour; if this were not so we would find it difficult to survive. In order to study air conditioning we must therefore understand the scientific principles of air–water vapour mixtures.

Humidity

For normal atmospheric conditions the water vapour in air is at a low partial pressure and may be taken to act as a perfect gas. We can therefore use the laws of mixtures of perfect gases, such as the Gibbs–Dalton law.[1.1]

Taking a volume, V, of an air/water vapour mixture at a pressure p, and an absolute temperature T, and letting the partial pressures of the dry air and water vapour be p_a and p_s, we have

$$p_a + p_s = p \qquad [1.20]$$

and

$$m_a = p_a V / R_a T \qquad m_s = p_s V / R_s T \qquad [1.21]$$

where m_a and m_s are the masses of dry air and water vapour in the mixture; R_a is the specific gas constant for dry air; R_s is the specific gas constant for superheated water vapour, defined as the molar gas constant divided by the molecular mass of water vapour.

Specific humidity

The *specific humidity*, or *moisture content*, ω, is defined as the ratio of the mass of water vapour in the mixture to the mass of dry air in the mixture (it is also sometimes known as *humidity ratio*). That is,

$$\text{Specific humidity, } \omega = m_s / m_a \qquad [1.22]$$

Substituting from Eqn [1.21] we then have

$$\omega = R_a p_s / R_s p_a$$

and then using Eqn [1.20] we have

$$\omega = \frac{287.1 p_s}{461.9 (p - p_s)}$$

where the value of the gas constant for air is taken as 287.1 J/kg K, and the gas constant for superheated water vapour as $\tilde{R}/M_s = 8314.5/18 = 461.9$ J/kg K; i.e.

$$\omega = \frac{0.622 p_s}{(p - p_s)} \qquad [1.23]$$

Dew point

When a mixture is saturated with water vapour, then any reduction in temperature at constant pressure will cause the water vapour to condense. The temperature of the mixture at this condition is called the *dew point temperature*.

Relative humidity

Specific humidity is a measure of the amount of water vapour present in a mixture but it gives no indication of how near the mixture is to saturation. The *relative humidity*, ϕ, is defined as the ratio of the mass of water vapour in a given volume of the mixture to the mass of water vapour in the same volume of a saturated mixture at the same temperature. Using the perfect gas equation (Eqn [1.21]) we have

$$\text{Relative humidity, } \phi = m_s / m_g = p_s / p_g \qquad [1.24]$$

where p_g is the saturation pressure of water vapour at the temperature of the mixture.

The relative humidity varies from zero for dry air to unity for air which is completely saturated; it is usually expressed as a percentage.

Percentage saturation

An alternative way of defining how close a mixture is to saturation, is to use the ratio of the specific humidity of the mixture to the specific humidity of the saturated mixture at the same temperature, expressed as a percentage, i.e.

$$\text{Percentage saturation, } \mu = 100\omega/\omega_g \qquad [1.25]$$

Note: 'percentage saturation' is the name used by CIBSE in its *Guide to Current Practice*,[1.5] but it would be more logical to call the ratio ω/ω_g 'relative saturation', or 'degree of saturation' (the name used by Threlkeld[1.9]); to conform with common practice in the UK the term 'percentage saturation' will be used throughout the book.

Substituting from Eqn [1.23] in Eqn [1.25] we have:

$$\mu = 100\omega/\omega_g = \frac{100 \times 0.622p_s(p - p_g)}{(p - p_s)0.622p_g}$$

Then substituting from Eqn [1.24],

$$\mu = 100\phi \frac{(p - p_g)}{(p - p_s)} \qquad [1.26]$$

When the partial pressure of water vapour is small compared with the pressure of the mixture, the ratio of bracketed terms in Eqn [1.26] is approximately unity. The percentage saturation is therefore approximately equal to the percentage relative humidity. The CIBSE tables[1.5] give both percentage saturation and percentage relative humidity over a range of temperatures from $-10\,°C$ to $60\,°C$; the difference in the values is greatest at high temperatures and low humidities but for the conditions encountered in air conditioning the difference lies in the range 0.1% to about 2%, (e.g. at $20\,°C$ and 30% percentage saturation, the percentage relative humidity is 30.5%, a difference of 1.7%).

Measurement of humidity

Instruments for measuring the amount of water vapour in air are called *hygrometers* or *psychrometers*; Ruskin[1.10] gives details of the various methods. In one method evaporation from a wetted sleeve around a thermometer bulb to the air surrounding the bulb causes a reduction in the temperature reading compared with that from a dry bulb placed in the same air. The amount of evaporation is found to depend on the humidity of the surrounding air and hence the difference in temperatures between the wet bulb and the dry bulb can be used to calculate the percentage saturation. A wet bulb will give a different reading if it is in still air or in an air stream above about 2 m/s; these are normally referred to as '*screen*' and '*sling*' wet bulb temperatures. (The word sling comes from the *sling psychrometer*, in which the thermometers are mounted on a rotating arm which is whirled round to provide the necessary air flow across the wet bulb.)

Adiabatic Saturation

If air flows across a surface which is saturated with water at temperature t, and a reservoir of water also at temperature t continuously replaces the water evaporated into the air, then for a very long surface the air will eventually become saturated with water vapour at the temperature t, provided there is no heat transfer to or from the surroundings. This process is known as *adiabatic saturation*, and the temperature t is called the *thermodynamic wet bulb temperature*; this is the ideal equivalent of the measured wet bulb temperature using a wetted sleeve. During the above process the air dry bulb temperature decreases, the humidity increases, but the wet bulb temperature remains constant at the value t.

Specific Enthalpy

In air conditioning systems it is usually possible to neglect velocity changes, and hence the processes are flow processes in which either heat or work is transferred by an enthalpy difference.

From the Gibbs–Dalton law, the enthalpy, H, of a mixture of air and water vapour is given by

$$H = m_a h_a + m_s h_s$$

where h_a and h_s are the specific enthalpies of the dry air and water vapour in the mixture.

It is usual to express the enthalpy of the mixture per unit mass of dry air. Thus, specific enthalpy of mixture per unit mass of dry air, h, is given by

$$h = \frac{H}{m_a} = h_a + \omega h_s \qquad [1.27]$$

Taking the datum as 0 °C for both air and water vapour and assuming mean specific heats, for a mixture at temperature t we can write:

$$h = c_{pa} t + \omega \{ (h_g \text{ at } p_s) + c_{ps}(t - t_g) \} \qquad [1.28]$$

where c_{pa} and c_{ps} are the mean specific heats of dry air and superheated water vapour.

Over the range of conditions for air conditioning it can be shown that the term $\{ (h_g \text{ at } p_s) - c_{ps} t_g \}$ in Eqn [1.28] is approximately 2500 kJ/kg. An approximate expression for the enthalpy of the mixture is therefore

$$\frac{h}{(\text{kJ/kg})} = 1.005 \frac{t}{(°\text{C})} + \omega \left(2500 + 1.88 \frac{t}{(°\text{C})} \right) \qquad [1.29]$$

where the mean specific heats of dry air and superheated water vapour are taken as 1.005 kJ/kg K and 1.880 kJ/kg K.

This is a useful approximation when the temperature, t, of the mixture is unknown. A more direct approximation for the enthalpy of the mixture is given by taking the enthalpy of superheated water vapour as approximately equal to the enthalpy of saturated vapour at the temperature of the mixture. Referring to Eqn [1.27],

$$h = c_{pa}t + \omega(h_g \text{ at } t) \tag{1.30}$$

Accurate values of the enthalpy of the mixture per unit mass of dry air are given in the CIBSE tables.[1.5]

Specific Heat

The specific heat of the mixture is sometimes required; this is usually expressed per unit mass of dry air, i.e.

$$\text{Specific heat of mixture, } c_{pma} = c_{pa} + \frac{m_s c_{ps}}{m_a}$$

$$\frac{c_{pma}}{\text{(kJ/kg K)}} = 1.005 + 1.88\omega \tag{1.31}$$

For heating or cooling of a mixture of air and water vapour the heat supplied or rejected is given by the change of enthalpy. In cases where there is no vapour added or condensed out during the process the specific humidity remains constant and, referring to Eqn [1.29], the enthalpy change is given by:

$$\frac{h_1 - h_2}{\text{(kJ/kg)}} = 1.005 \frac{(t_1 - t_2)}{(°C)} + 1.88\omega \frac{(t_1 - t_2)}{(°C)}$$

where $\omega = \omega_1 = \omega_2$.

Substituting from Eqn [1.31] it can then be seen that, for this case,

$$h_1 - h_2 = c_{pma}(t_1 - t_2)$$

Note: This equation is true only when the specific humidity is constant during the process; for Building Services applications c_{pma} is normally within the range 1.01 to 1.03 kJ/kg K and many problems are simplified by taking a mean value of 1.02 kJ/kg K. Such processes are known as *sensible heating* or *sensible cooling*: the term 'sensible' was formerly used in Thermodynamics for energy added or rejected in the absence of a phase change, but is no longer recommended. Since the term is still widely used in air conditioning practice it will be employed in the present text.

Specific Volume

As with the specific heat and enthalpy, the specific volume of the mixture is normally expressed as the volume per unit mass of dry air, i.e.

$$\text{Specific volume of mixture, } v = V/m_a = R_a T/p_a \tag{1.32}$$

Then, for a flow process,

$$\text{Mass flow rate of dry air, } \dot{m}_a = \dot{V}/v \tag{1.33}$$

where $\dot{V}$ is the volume flow rate of the mixture.

Example 1.7

Air in a space is at 20 °C dry bulb with a percentage saturation of 50% when the pressure is 1.02 bar.

Calculate, using Rogers and Mayhew tables,[1.11]
(i)   the specific humidity;
(ii)   the relative humidity;
(iii)   the dew point temperature;
(iv)   the specific volume per unit mass of dry air;
(v)   the percentage saturation and the relative humidity for the same temperature and specific humidity when the barometric pressure falls to 1 bar.

Solution
(i)   At 20 °C the saturation pressure of water vapour from tables is 0.023 37 bar. Then, using Eqn [1.23], we have:

$$\text{Specific humidity of saturated air, } \omega_g = \frac{0.622 \times 0.023\,37}{(1.02 - 0.023\,37)}$$
$$= 0.014\,59$$

Then, using Eqn [1.25],

$$\text{Percentage saturation} = 50 = 100\omega/0.014\,59$$

i.e.
$$\omega = 0.014\,59 \times 0.5 = 0.007\,29$$

(ii)   The partial pressure of the water vapour can then be found from Eqn [1.23]:

$$0.007\,29 = 0.622p_s/(1.02 - p_s)$$
$$\therefore \quad p_s = 0.011\,82 \text{ bar}$$

Then, using Eqn [1.24],

$$\text{Relative humidity} = 0.011\,82/0.023\,37 = 0.5058 = 50.58\%$$

(iii)   The partial pressure of the water vapour is 0.011 82 bar and the dew point temperature is the temperature to which the mixture must be cooled for condensation to begin at this pressure, i.e.

$$\text{Dew point temperature} = t_g \text{ at } 0.011\,82 \text{ bar}$$

Therefore, interpolating from tables,

$$\text{Dew point temperature} = 9.4 \,°C$$

(iv)   From Eqn [1.32],

$$\text{Specific volume per unit mass of dry air} = R_a T/p_a.$$

Also,
$$p_a = 1.02 - 0.011\,82 = 1.008\,18 \text{ bar}$$
$$\therefore \quad v = 0.2871 \times 10^3 \times (20 + 273)/1.008\,18 \times 10^5$$
$$= 0.8344 \text{ m}^3/\text{kg}$$

(v)   The specific humidity at saturation is now given by Eqn [1.23] as

$$\omega_g = 0.622 \times 0.023\,37/(1 - 0.023\,37) = 0.014\,88$$

Then,

$$\text{Percentage saturation} = 100 \times 0.007\,293/0.014\,88$$
$$= 49.00\%$$

Also, using Eqn [1.23], we have

$$0.007\,293 = 0.622 p_s/(1 - p_s)$$

i.e. $p_s = 0.011\,589$ bar

Then, from Eqn [1.24],

Relative humidity $= 0.011\,589/0.02\,337 = 0.4959 = 49.59\%$

It will be noted from the above example that the difference between percentage saturation and relative humidity is only about 1.2% for both values of total pressure. The change of total pressure from 1.02 bar to 1 bar, on the other hand, causes a change of about 2% in both percentage saturation and relative humidity for the same values of specific humidity and dry bulb temperature. It is normal practice to take the total pressure as the standard atmospheric pressure of 1.013 25 bar, and this is the value used in the CIBSE Psychrometric Tables and Chart.[1.5] The Psychrometric Chart is considered in detail in the next section. The errors incurred in taking the pressure as 1.013 25 bar are within the range of errors due to assuming that the perfect gas laws apply to the vapour, and are certainly within the errors involved in the reading of values from the chart.

1.4 PSYCHROMETRIC CHART

A psychrometric chart is a plot of specific humidity against specific enthalpy per unit mass of dry air, for air containing water vapour.

It has been found that greater accuracy can be obtained with less crushing of the lines if the specific enthalpy axis is skewed at an angle of β to the vertical as shown in Fig. 1.6. A line of constant enthalpy on the chart is therefore at an angle $(180° - \beta)$ to the horizontal.

Figure 1.6 Skewed plot of specific humidity vs specific enthalpy

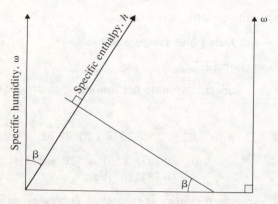

For the CIBSE chart,[1.5] the scales chosen for the two axes are: for specific enthalpy, $S_h = 2.4$ mm per kJ/kg; for specific humidity, $S_\omega = 7250$ mm per kg/kg.

Note: CIBSE uses 'moisture content' instead of specific humidity, and gives it the symbol g.

Dry Bulb Temperatures

The angle β is chosen such that lines of constant dry bulb temperature are almost vertical. One line of dry bulb temperature (30 °C in the CIBSE chart) is chosen to be exactly vertical and this fixes the value of β, as shown below.

For any process from state 1 to state 2 on the chart, let the angle of the process line to the horizontal be θ, as shown in Fig. 1.7. The angle, α, of the process line to the constant enthalpy line through state 2 is given by:

$$\alpha + \theta = 180° - \beta \qquad \therefore \qquad \alpha = 180° - (\beta + \theta)$$

Figure 1.7 Process 1–2 on the psychrometric chart

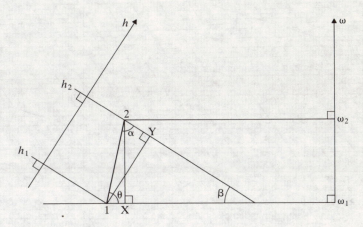

Now,

$$\text{line } 1Y = (h_2 - h_1)S_h$$

and

$$\text{line } 2\text{–X} = (\omega_2 - \omega_1)S_\omega$$

Then,

$$\text{line } 1\text{–2} = \text{line } 1Y/\sin \alpha = \text{line } 1Y/\sin (\beta + \theta)$$

and

$$\text{line } 1\text{–2} = \text{line } 2\text{–X}/\sin \theta$$

i.e.

$$\text{line } 1\text{–Y}/\sin(\beta + \theta) = \text{line } 2\text{–X}/\sin \theta$$

$\therefore$

$$\frac{(h_2 - h_1)S_h}{(\omega_2 - \omega_1)S_\omega} = \frac{\sin(\beta + \theta)}{\sin \theta} = \frac{\sin \beta \cos \theta + \cos \beta \sin \theta}{\sin \theta}$$

i.e.

$$\cot \theta + \cot \beta = \frac{(h_2 - h_1)S_h}{(\omega_2 - \omega_1)S_\omega \sin \beta} \qquad [1.34]$$

For the line of constant temperature, $t = 30$ °C, to be vertical, then $\theta = 90°$ for a constant temperature process at 30 °C. Using Eqn [1.30] for the specific enthalpy, then at a constant temperature of 30 °C we have

$$h_2 - h_1 = (\omega_2 - \omega_1)(h_g \text{ at } 30 \,°\text{C}) = (\omega_2 - \omega_1) \times 2555.7$$

Figure 1.8 CIBSE
psychrometric chart
(reproduced by courtesy
of CIBSE)

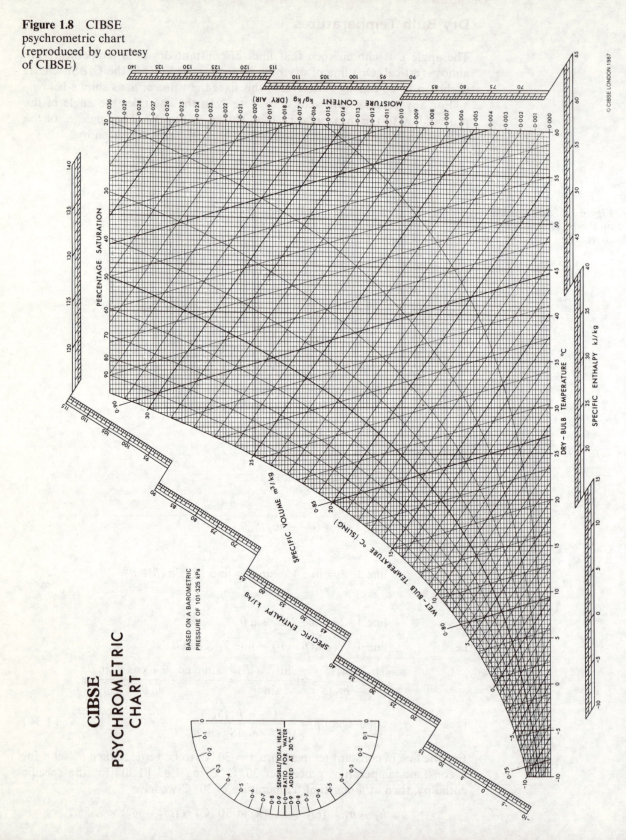

Substituting in Eqn [1.34],

$$\cot 90° + \cot \beta = \frac{2555.7 \times 2.4}{7250 \times \sin \beta}$$

where $S_h = 2.4$ mm per kJ/kg, and $S_\omega = 7250$ mm per kg/kg, are the scales chosen for the CIBSE chart.

Then, since $\cot 90° = 0$, the angle β is calculated to be 32.22°, which gives an equation for any process line as:

$$\cot \theta = 0.000\,621 \frac{(h_2 - h_1)}{(\omega_2 - \omega_1)} - 1.587 \qquad [1.35]$$

Figure 1.8 is a reduced-size version of the CIBSE chart of Ref. 1.5, reproduced here by permission of CIBSE.

Room Ratio Line (or Room Condition Line)

In summer, conditioned air enters a space at state 1 and leaves at state 2 as shown in Figs 1.9(a) and (b). The rise in temperature $(t_2 - t_1)$ is due to energy gains from people, equipment, etc., as well as from fabric gains from the outside air. The increase in specific humidity $(\omega_2 - \omega_1)$ is mainly due to moisture evaporating from people in the room, and the moisture may be taken to be added as a saturated vapour.

Figure 1.9 Room air condition line (summer)

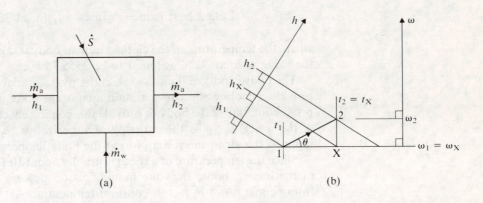

Let the mass flow rate of dry air entering and leaving the room be $\dot{m}_a$, let the mass flow rate of evaporated moisture entering the room be $\dot{m}_w$, at a temperature t_w, and let the rate of heat transfer to the room be $\dot{S}$.

We then have

$$\dot{S} + \dot{m}_w(h_g \text{ at } t_w) = \dot{m}_a(h_2 - h_1)$$

and

$$\dot{m}_w = \dot{m}_a(\omega_2 - \omega_1)$$

Therefore,

$$\frac{\dot{S}}{\dot{m}_w} + (h_g \text{ at } t_w) = \frac{(h_2 - h_1)}{(\omega_2 - \omega_1)} \qquad [1.36]$$

For a given state 2 for the air in the room, the condition line can be drawn through point 2 with the slope given by Eqn [1.35]. The air entering at state 1 is then fixed by the mass flow rate, $\dot{m}_a$, and the temperature increase $(t_2 - t_1)$.

Some psychrometric charts provide a protractor which gives lines with values of slope given by Eqn [1.34]. In the UK the practice adopted by the CIBSE is slightly different. The term *sensible heat gain* is used for enthalpy change in the absence of a change in specific humidity, and the term *latent heat gain* is used for enthalpy change due to the evaporation of water vapour at constant dry bulb temperature. Latent heat gain plus sensible heat gain is known as *total heat gain*, the change of enthalpy of the air in the room.

The use of 'heat gain' is contrary to the accepted Thermodynamic definition of 'heat' as a transitory form of energy; also, the term 'latent heat' has now been replaced by *enthalpy of vaporization*. The terms 'sensible heat gain' and 'latent heat gain' are used extensively in the Building Services industry in the UK, and by the professional body, the CIBSE, and hence will be employed in this book.

Referring to Fig. 1.9(b),

$$\text{Sensible heat gain, } \dot{S} = \dot{m}_a(h_X - h_1)$$
$$= \dot{m}_a c_{pma}(t_2 - t_1) \qquad [1.37]$$

Usually the latent heat gain is mainly, or entirely, from people; hence using the approximation for the specific enthalpy, Eqn [1.30], it can be written as

$$\text{Latent heat gain} = \dot{m}_a(\omega_2 - \omega_1)(h_g \text{ at } 30\,°C)$$

where the temperature of the clothed human body is taken as 30 °C; from tables the value of h_g at 30 °C is found to be 2555.7 kJ/kg.

The body produces liquid sweat, enthalpy h_w, and evaporates this in order to provide a cooling effect to maintain a constant body temperature. The energy production rate of the body to provide this cooling effect is equal to the enthalpy of the water, h_w, plus the enthalpy of vaporization, h_{fg}, giving a total energy input to the air in the room of h_g at the body temperature as stated above.

Since the temperature of a room is usually about 10 K below the temperature of the human body, then the humidification process due to moisture added from people is not in fact at constant temperature. At a constant temperature t_2, we have

$$\text{Latent heat gain, } \dot{L} = \dot{m}_a(\omega_2 - \omega_1)(h_g \text{ at } t_2) \qquad [1.38]$$

At a normal room temperature of 20 °C the value of h_g is 2537.6 kJ/kg. Using this value, rather than h_g at 30 °C, gives a difference in the latent heat gain of less than 1 %.

Assuming that the latent heat gain occurs at constant temperature, we have

$$\text{Latent heat gain, } \dot{L} = \dot{m}_a(h_2 - h_X)$$

$$\therefore \quad \text{Total heat gain} = \dot{S} + \dot{L} = \dot{m}_a(h_X - h_1 + h_2 - h_X) = \dot{m}_a(h_2 - h_1)$$

Let the ratio $\dot{S}/(\dot{S} + \dot{L})$ be RRL; then $\dot{L}/(\dot{S} + \dot{L}) = 1 - RRL$. That is,

$$1 - RRL = \frac{(\omega_2 - \omega_1)}{(h_2 - h_1)} \times 2555.7$$

Substituting in Eqn [1.35],

$$\cot \theta = \frac{0.000\,621 \times 2555.7}{(1 - RRL)} - 1.587 = \frac{1.587 RRL}{(1 - RRL)}$$

or $\qquad \tan \theta = 0.63(1 - RRL)/RRL$ [1.39]

A protractor is constructed on the CIBSE chart, giving values of RRL based on Eqn [1.39]: note that when $\dot{S}/(\dot{S} + \dot{L}) = RRL = 1$, there is no latent heat gain and $\theta = 0$; when $RRL = 0$, there is no sensible heat gain and $\theta = 90°$. Summer conditions use the bottom quadrant of the protractor; the known sensible heat gains are divided by the sum of the sensible and latent heat gains to give RRL and hence the slope of the condition line can be transferred from the protractor to the chart.

In winter there are net heat losses instead of gains although there is still a moisture addition from the people present. A typical winter room condition line is shown in Fig. 1.10.

Figure 1.10 Room air condition line (winter)

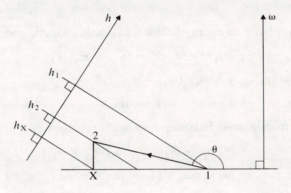

In this case,

$$\dot{S} = -\dot{m}_a(h_1 - h_x)$$

As before, $\qquad \dot{L} = \dot{m}_a(h_2 - h_x) = \dot{m}_a(\omega_2 - \omega_1) \times 2555.7$

Also (as before),

$$\dot{S} + \dot{L} = \dot{m}_a(h_2 - h_1)$$

In winter $\dot{S}$ will usually have a negative value but $\dot{L}$ will be positive. The slope of the condition line is now in the top quadrant of the protractor on the chart, and the implication is that the values marked on this quadrant are also the ratios of sensible heat to total heat gains. This is not so. In order to obtain the slope from the quadrant it is necessary to take the ratio of the sensible heat loss to the sum of the sensible heat loss and the latent heat gain, i.e.

$$RRL = \frac{(h_1 - h_x)}{(h_1 - h_x) + (h_2 - h_x)}$$

Using the protractor is not accurate: the value of *RRL* cannot be fixed accurately on the protractor circle unless it is exactly equal to one of the values given on the circle; also, the line must be transferred to the chart using setsquares, which can introduce further inaccuracy.

Adiabatic Mixing

An adiabatic mixing process is shown diagrammatically in Fig. 1.11(a).

Figure 1.11 Adiabatic mixing

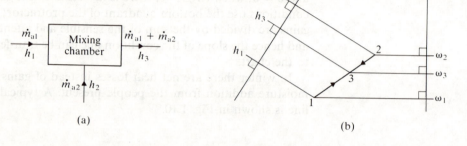

(a)

(b)

Applying an energy balance, neglecting kinetic energy changes, we have

$$\dot{m}_{a1}h_1 + \dot{m}_{a2}h_2 = (\dot{m}_{a1} + \dot{m}_{a2})h_3$$

Defining $y = \dot{m}_{a1}/(\dot{m}_{a1} + \dot{m}_{a2})$

Then $h_3 = yh_1 + (1 - y)h_2$

and from a mass balance,

$$\omega_1 \dot{m}_{a1} + \omega_2 \dot{m}_{a2} = \omega_3(\dot{m}_{a1} + \dot{m}_{a2})$$

i.e. $\omega_3 = y\omega_1 + (1 - y)\omega_2$

It follows from the two equations that

$$\frac{h_2 - h_3}{h_2 - h_1} = \frac{\omega_2 - \omega_3}{\omega_2 - \omega_1} = y \qquad\qquad [1.40]$$

Therefore the state of the mixed air at 3 lies on a straight line joining states 1 and 2 as shown on Fig. 1.11(b); the larger the value of y the nearer point 3 is to point 1.

Note: Since the lines of dry bulb temperature on the chart are very close to the vertical it is a good approximation to write $y = (t_2 - t_3)/(t_2 - t_1)$; the error involved in using this expression is usually less than the error in reading the enthalpy values from the chart, particularly since the differences in enthalpy are small.

De-humidification

In Section 1.3 the dew point temperature is defined as the temperature at which condensation of vapour will just begin when air is cooled at constant pressure.

For any state 1 on the chart the dew point is therefore the point, d, where a horizontal line through 1 cuts the saturation line (see Fig. 1.12). The specific humidity of the air can therefore be reduced by cooling it below its dew point temperature, t_d.

Figure 1.12 Dew point

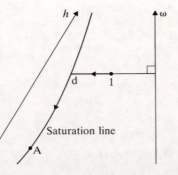

De-humidification can be carried out in two main ways: by physical adsorption using solid or liquid sorbents such as silica gel, lithium chloride/water solution, etc.; or by cooling the air below its dew point temperature. In Building Services engineering practice we are concerned mainly with the latter method; for further information on adsorption consult the book by Threlkeld,[1.9] or the relevant manufacturers' literature. In theory, cooling the air below its dew point gives a process such as that shown by 1–d–A on Fig. 1.12. In practice the cooling is effected by a refrigerant in a coil, or by a chilled water spray, and the air is cooled locally below its dew point at the surface of each tube in the coil, or on each water droplet in the spray. The temperature and specific humidity therefore drop continuously as the air flows through the de-humidifier and the process therefore follows a path similar to the broken line shown in Fig. 1.13. The actual heat and mass transfer processes involved are considered in more detail in Chapter 6.

Figure 1.13 De-humidification

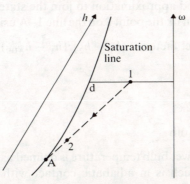

An overall energy and mass balance can be made on the system; referring to Figs 1.14(a) and (b), we have

$$\dot{Q}_{REJ} = \dot{m}_a(h_1 - h_2) - \dot{m}_w h_w$$

Figure 1.14 De-humidification by coil and spray

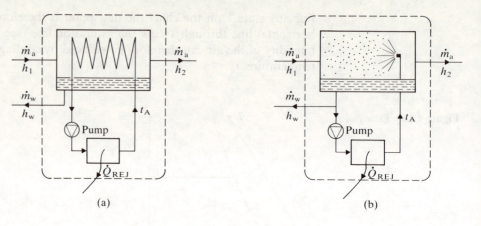

(a)   (b)

and

$$\dot{m}_w = \dot{m}_a(\omega_1 - \omega_2)$$

i.e.

$$\dot{Q}_{REJ} = \dot{m}_a\{(h_1 - h_2) - h_w(\omega_1 - \omega_2)\} \qquad [1.41]$$

For an infinite number of coil rows (see Fig. 1.14a) the air would reach the state A, but in practice the air leaving the coil is at a state point 2. The point A is known as the *apparatus dew point* (ADP) and in practice the apparatus dew point measurement is taken as the mean temperature of the outer surface of the coil. A de-humidification process is normally defined by either:

$$\text{Coil contact factor} = (h_1 - h_2)/(h_1 - h_{ADP}) \qquad [1.42]$$

or

$$\text{Coil by-pass factor} = (h_2 - h_{ADP})/(h_1 - h_{ADP}) \qquad [1.43]$$

In the case of a chilled water spray de-humidifier (see Fig. 1.14b), condensation of the water vapour in the air occurs at the surface of the individual water droplets and the apparatus dew point temperature is taken as the average temperature of the spray water. The process follows a line similar to that shown in Fig. 1.13; the state A is only attainable by the air if the spray chamber is infinitely long so in practice the air at exit is at the state 2. In using the chart it is a sufficiently good approximation to join the state point 1 to the apparatus dew point, A, and then fix the point 2 on the line 1–A using the contact factor, i.e.

$$\text{Contact factor} = (h_1 - h_2)/(h_1 - h_{ADP}) \simeq \text{line } 1\text{–}2/\text{line } 1\text{–}A$$

Humidification

(a) Constant wet bulb

The thermodynamic wet bulb temperature is defined as the temperature attained by an air stream which is in adiabatic contact with an infinitely long water surface. In practice, when air passes through a well-insulated chamber which has a water surface continuously replenished by a pump, as shown diagrammatically in Fig. 1.15(a), then the process approximates to constant wet bulb; the CIBSE chart, (page 20), uses 'sling' values.

The process is shown on a psychrometric chart in Fig. 1.15(b), from which

Figure 1.15
Humidification at
constant wet bulb

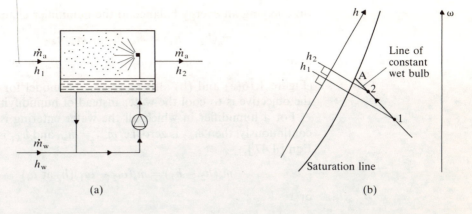

(a)   (b)

it can be seen that the slope of a constant wet bulb line is slightly steeper than
that of a constant enthalpy line. Applying an energy balance, we have

$$\dot{m}_a(h_2 - h_1) = \dot{m}_a(\omega_2 - \omega_1)h_w$$

i.e. $$h_2 = h_1 + (\omega_2 - \omega_1)h_w \qquad [1.44]$$

where h_w is the enthalpy of water at the wet bulb temperature at state 1.

In an ideal process the air at exit would reach state point A at a dry bulb
temperature equal to the wet bulb temperature (see Fig. 1.15b). A humidification
efficiency (sometimes called washer efficiency) can then be defined as

$$\text{Humidification efficiency} = (h_2 - h_1)/(h_A - h_1) \qquad [1.45]$$

(b) Pumped re-circulation with heating

When a pumped re-circulation system with a heater to heat the water to the
desired temperature is used (see Fig. 1.16a), then the process is as shown on
the psychrometric chart in Fig. 1.16(b); the exact path of the process depends
on the heat and mass transfer processes taking place—a more detailed analysis
is given in Chapter 5.

Applying an overall energy balance to Fig. 1.16(a), we have

$$\dot{Q} = \dot{m}_a\{(h_2 - h_1) - h_w(\omega_2 - \omega_1)\} \qquad [1.46]$$

Figure 1.16
Humidification with
heated water spray

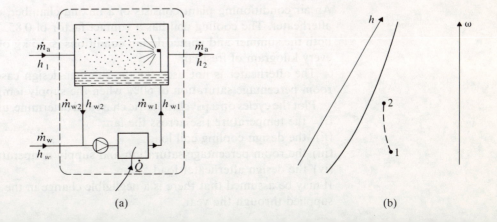

(a)   (b)

or, applying an energy balance to the humidifier chamber only,

$$\dot{m}_a(h_2 - h_1) = \dot{m}_{w1} h_{w1} - \dot{m}_{w2} h_{w2} \qquad [1.47]$$

and

$$\dot{m}_{w1} = \dot{m}_{w2} + (\omega_2 - \omega_1)\dot{m}_a = \dot{m}_{w2} + \dot{m}_w \qquad [1.48]$$

(Figure 1.16(a) and (b) also represent the model for a cooling tower, where the objective is to cool the water instead of humidifying the air.)

For a humidifier in which all the water entering is completely evaporated continuously, then $\dot{m}_{w2}$ is zero, i.e. $\dot{m}_{w1} = \dot{m}_w$, and h_{w1} is equal to h_g at t_w. From Eqn [1.47],

$$\dot{m}_a(h_2 - h_1) = \dot{m}_a(\omega_2 - \omega_1)(h_g \text{ at } t_w)$$

or

$$\frac{(h_2 - h_1)}{(\omega_2 - \omega_1)} = h_g \text{ at } t_w \qquad [1.49]$$

The slope of the line on the chart therefore depends on the temperature of the water spray at entry; the slope can be found from Eqn [1.35].

(c) Steam humidification

The analysis for this case is the same as for the case above of a water spray continuously and completely evaporated, and Eqn [1.49] can be re-written as

$$\frac{(h_2 - h_1)}{(\omega_2 - \omega_1)} = h_g \qquad [1.50]$$

where h_g is the specific enthalpy of the steam injected.

It was shown earlier that for a vertical line on the chart the ratio of enthalpy difference to specific humidity difference is 2555.7 kJ/kg. The specific enthalpy of saturated steam at atmospheric pressure is 2675.8 kJ/kg and hence, for steam humidification with wet steam at atmospheric pressure, the process line is almost vertical.

Air conditioning systems are considered in detail in Chapter 8. The following example of a simple system is given to illustrate the use of the psychrometric chart.

Example 1.8

An air conditioning plant consists of a mixing chamber, cooling coil, fan and afterheater. The cooling coil has a contact factor of 0.85. The mixing ratio for both the summer and winter design conditions is 2.5 kg of re-circulated air for every kilogram of fresh air.

The afterheater is not used for the summer design case, which results in a room percentage saturation of 60% when the supply temperature is 15 °C.

Plot the cycles on a psychrometric chart and determine, using the data below:

(i)   the temperature rise across the fan;

(ii) the design cooling coil load;

(iii) the room percentage saturation and supply temperature in winter;

(iv) the design afterheater load.

It may be assumed that there is a negligible change in the mass flow rate of air supplied through the year.

Data

Room temperature, 21 °C (winter and summer); summer design condition, 28 °C dry bulb, 23 °C wet bulb; winter design condition, −2 °C saturated; summer loads, 40 kW sensible gain, 10 kW latent gain; winter loads, 30 kW sensible loss, 10 kW latent gain.

Take the specific heat of the air per unit mass of dry air as approximately 1.02 kJ/kg K; the enthalpy of dry saturated water vapour at 30 °C is 2555.7 kJ/kg.

(Wolverhampton Polytechnic)

Solution

The system is shown diagrammatically in Fig. 1.17; in this case the afterheater is used only in winter and the cooling coil is used only in summer.

Using the state point numbers of Fig. 1.17, the processes for summer are then as shown in the sketch of the psychrometric chart given in Fig. 1.18; in summer points 6 and 1 coincide.

Figure 1.17 Simple air conditioning system

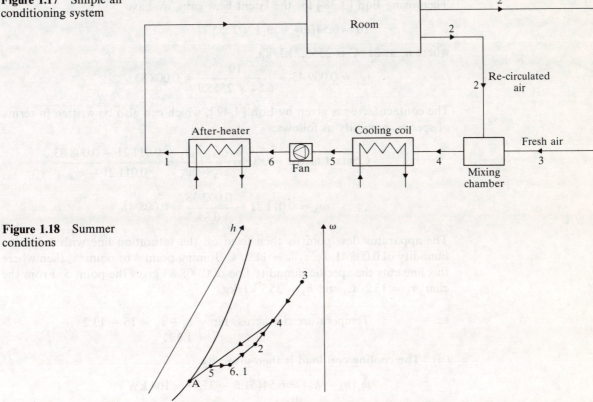

Figure 1.18 Summer conditions

(i) Referring to the summer conditions, point 2 is fixed at 21 °C and a percentage saturation of 60%; therefore, from the chart,

$$h_2 = 45.1 \text{ kJ/kg} \quad \text{and} \quad \omega_2 = 0.009\,43$$

Point 3 is also fixed at 28 °C dry bulb and 23 °C wet bulb; therefore, from the chart,

$$h_3 = 67.8 \text{ kJ/kg} \quad \text{and} \quad \omega_3 = 0.01560$$

Using Eqn [1.40] for adiabatic mixing between states 2 and 3, where $y = 2.5/(2.5 + 1) = 0.7143$ (given), we have:

$$h_4 = h_3 - 0.7143(h_3 - h_2) = 67.8 - 0.7143(67.8 - 45.1)$$
$$= 51.6 \text{ kJ/kg}$$

Alternatively, a quicker method is to measure the length of line 3–2 and then multiply this length by (2.5/3.5) to give the length of the line 3–4, thus fixing point 4.

From the chart the specific humidity at point 4 is found to be

$$\omega_4 = 0.011\,21$$

The mass flow rate, m_a, can be found using Eqn [1.37],

i.e.   Sensible heat gain $= 40 \text{ kW} = \dot{m}_a c_{pa}(t_2 - t_1)$

and an approximate value of c_{pa} is given as 1.02 kJ/kg K.

$$\therefore \quad \dot{m}_a = 40/1.02(21 - 15) = 6.54 \text{ kg/s}$$

Then using Eqn [1.38] for the latent heat gain, we have

$$10 = 6.54(\omega_2 - \omega_1) \times 2555.7$$

where h_g at 21 °C is 2555.7 kJ/kg,

i.e. $$\omega_1 = 0.009\,43 - \frac{10}{6.54 \times 2555.7} = 0.008\,83$$

The contact factor is given by Eqn [1.42], which can also be written in terms of specific humidity as follows:

$$\text{Contact factor} = 0.85 = \frac{\omega_4 - \omega_5}{\omega_4 - \omega_A} = \frac{0.011\,21 - 0.008\,83}{0.011\,21 - \omega_A}$$

$$\therefore \quad \omega_A = 0.011\,21 - \frac{0.002\,38}{0.85} = 0.008\,41$$

The apparatus dew point is then fixed on the saturation line with a specific humidity of 0.008 41, i.e. $t_{ADP} = 11.5$ °C. Joining point 4 to point A, then where this line cuts the specific humidity line of 0.008 83 gives the point 5. From the chart $t_5 = 13.2$ °C, and $h_5 = 35.7$ kJ/kg,

i.e.   Temperature rise across fan $= t_6 - t_5 = 15 - 13.2$
$$= 1.8 \text{ K}$$

(ii)   The cooling coil load is then given by:

$$\dot{m}_a(h_4 - h_5) = 6.54(51.6 - 35.7) = 104 \text{ kW}$$

(iii)   In winter the processes are as shown on a sketch of the psychrometric chart in Fig. 1.19 using the same state point numbers as Fig. 1.17; points 4 and 5 coincide since the cooling coil is not used in winter.

It is stated that the mass flow rate of air to the room is the same throughout the year, i.e. $\dot{m}_a = 6.54$ kg/s. Therefore, for the given latent heat gain,

$$10 = 6.54(\omega_2 - \omega_1) \times 2555.7$$

$$\therefore \quad \omega_2 - \omega_1 = 0.000\,60 = \omega_2 - \omega_4$$

Figure 1.19 Winter
conditions

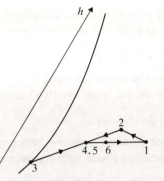

Point 3 is given as $-2\,°C$ saturated; hence, from the chart,

$$\omega_3 = 0.003\,21$$

Then using Eqn [1.40] for adiabatic mixing, we have

$$\omega_2 - \omega_3 = 0.000\,60 \times (2.5 + 1) = 0.002\,10$$

i.e.

$$\omega_2 = 0.003\,21 + 0.002\,10 = 0.005\,31$$

and

$$\omega_4 = \omega_1 = 0.005\,31 - 0.000\,60 = 0.004\,71$$

Point 2 can now be fixed on the chart where the specific humidity line of 0.005 31 cuts the dry bulb temperature line of 21 °C. Then joining point 2 to point 3 fixes point 4 at the specific humidity of 0.004 71, giving $t_4 = 14.3\,°C$.

Using the approximate equation for the sensible heat loss, we have

$$6.54 \times 1.02(t_1 - 21) = 30$$

i.e.

$$t_1 = 21 + 4.5 = 25.5\,°C$$

Point 1 is fixed where the dry bulb temperature line of 25.5 °C cuts the specific humidity line of 0.004 71, i.e.

$$\text{Percentage saturation} = 34\%$$

(iv)   Assuming that the temperature rise across the fan of 1.8 K is the same as before, then

$$t_6 = t_4 + 1.8 = 14.3 + 1.8 = 16.1\,°C$$

Then the afterheater load is given by the approximate expression:

$$\text{Afterheater load} = 6.54 \times 1.02(25.5 - 16.1) = 62.7\,\text{kW}$$

1.5 FAN–DUCT SYSTEMS

For any air conditioning system it is necessary to arrive at suitable sizes for all ducts and to choose a fan which will give the necessary volume flow rate at the pressure drop required. A full theoretical treatment of fluid flow and fan construction and characteristics is given by Douglas et al.;[1.2] a concise coverage of fan–system matching is given in this section.

For a flowing fluid the total pressure is defined as the pressure which the fluid would attain if brought to rest without losses, i.e.

$$\text{Total pressure, } p_t = p + p_v$$

where $p_v = \frac{1}{2}\rho u^2$ and p is the static pressure.

For any fan–duct system between any two sections 1 and 2 we can write:

$$p_{t1} + P_t = p_{t2} + \Sigma \Delta p$$

where P_t is the difference in total pressure between the fan outlet and inlet; $\Sigma \Delta p$ is the total of the pressure losses between sections 1 and 2 due to friction,

Figure 1.20 Pressure drop due to friction for air flow in circular ducts

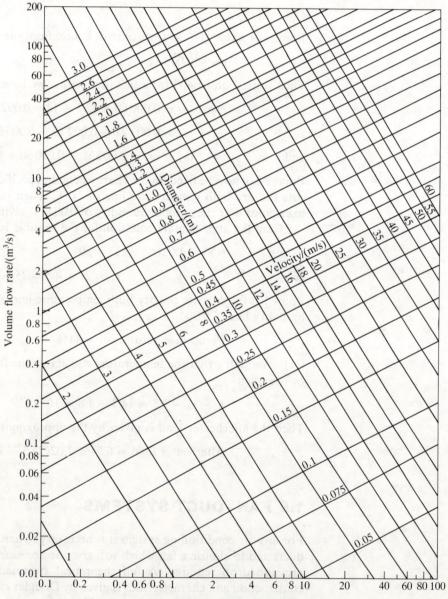

bends, and fittings. For bends, off-takes, expanders, dampers, etc., the loss of pressure can always be expressed as a factor, k, multiplied by the velocity pressure, p_v (see Section 1.2, Eqn [1.19]), i.e.

$$\Delta p = k p_v = \tfrac{1}{2} k \rho u^2 = K \dot{V}^2 \qquad [1.51]$$

where $\dot{V}$ is the volume flow rate, and K is a constant.

The friction loss is found most easily using a friction chart such as the one given by CIBSE[1.5] which is shown on a reduced scale in Fig. 1.20, by permission of CIBSE. The chart is for commercially available mild steel, galvanized round ducts of a given absolute roughness; it applies to air only at a standard density of 1.2 kg/m^3. Correction factors are given[1.5] for ducts of other materials and for air at different conditions; for rectangular ducts an equivalent diameter can be found.

For a fan, assuming a compressibility coefficient of unity, the air power is given by

Fan air power input, $\dot{W} = \dot{V} P_t$

The fan efficiency is then

$$\text{Fan efficiency, } \eta = \frac{\text{air power input}}{\text{shaft power input}}$$

Similar expressions for power and efficiency can be found using the static pressure difference across the fan instead of the total pressure difference.

Fan characteristics of pressure difference, power and efficiency against flow rate can be plotted as shown in Figs 1.21(a), (b) and (c); it can be seen that the characteristics vary depending on whether the blades at fan exit are in the radial direction or are inclined in the direction of rotation (forward-inclined),

Figure 1.21 Centrifugal fan characteristics: (a) forward inclined blades; (b) radial blades; (c) backward-inclined blades

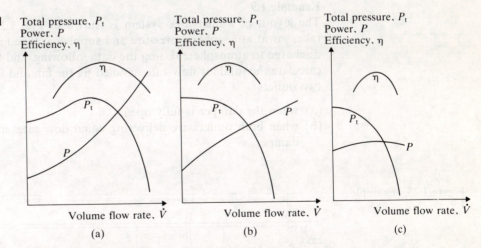

or in the opposite direction to rotation (backward-inclined). Note that for the case of backward-inclined blades, the power reaches a turning point and therefore the electric motor can be safely rated at maximum power without any danger of overloading. For a fan and a system of ducts, the pressure drop characteristic of the duct system intersects the fan characteristic at the operation point, as shown in Fig. 1.22.

Figure 1.22 Fan
system operating point

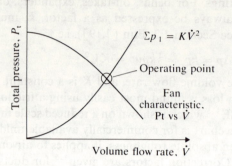

For two fans running in series the combined characteristic can be found by
adding together the pressure differences at a given flow; for two fans running
in parallel the combined characteristic is found by adding together the flows
at given values of the pressure difference.

For geometrically similar fans *laws of similarity* apply as follows:[1,2]

$$\left.\begin{array}{l} \dot{V}/Nd^3 = \text{constant}; \\ P_t/N^2d^2\rho = \text{constant}; \\ \dot{W}/N^3d^5\rho = \text{constant} \end{array}\right\} \qquad [1.52]$$

It follows that for a fan of a given diameter, d, delivering air at standard density,
ρ, the volume flow rate, $\dot{V}$, is directly proportional to the fan speed, N; the
pressure difference across the fan, P_t, is proportional to the square of the fan
speed; and the power input, $\dot{W}$, is proportional to the cube of the fan speed.

Example 1.9
The layout of a ductwork system is shown in Fig. 1.23. The centrifugal fan
takes air at atmospheric pressure and supplies it through two branches which
discharge to atmosphere. Using the data following, and the duct friction chart,
calculate the total air flow rate handled by the fan and the flow rates from the
two outlets:

(a) when the damper is fully open;
(b) when both outlets are delivering equal flow rates after adjustment of the
    damper.

Figure 1.23 Example
1.9

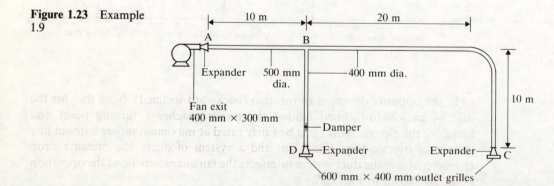

Data

Velocity pressure factors:

Bend, 0.3

Branch: flow to main, 0.2 (applied to downstream velocity pressure); flow to branch, 0.5 (applied to velocity pressure in off-take)

Discharge grille, 0.4

Expander, 0.25 (applied to maximum velocity pressure)

Damper (fully open), 0.2

Fan characteristic: $P_t = 200 - 12\dot{V}^2$

(Polytechnic of the South Bank)

Solution

(a)   For the necessary velocity pressures we have:

$$p_v = \tfrac{1}{2} \times 1.2 \times (\dot{V}/A)^2$$

where A is the relevant cross-sectional area. That is,

For duct AB,

$$A = (\pi \times 0.5^2)/4 = 0.196 \text{ m}^2$$

$$\therefore \quad p_{vAB} = \frac{0.6\dot{V}^2}{(0.196)^2} = 15.56\dot{V}^2$$

For duct BC,

$$A = (\pi \times 0.4^2)/4 = 0.126 \text{ m}^2$$

$$\therefore \quad p_{vBC} = \frac{0.6\dot{V}^2}{(0.126)^2} = 38.00\dot{V}^2$$

Similarly for duct BD,

$$p_{vBD} = 38.00\dot{V}^2$$

For fan outlet,

$$A = 0.4 \times 0.3 = 0.12 \text{ m}^2$$

$$\therefore \quad p_{ve} = \frac{0.6\dot{V}^2}{(0.12)^2} = 41.67\dot{V}^2$$

For grille outlets,

$$A = 0.6 \times 0.4 = 0.24 \text{ m}^2$$

$$\therefore \quad p_{vg} = \frac{0.6\dot{V}^2}{(0.24)^2} = 10.42\dot{V}^2$$

For each duct section the friction loss can be expressed as a constant multiplied by the square of the volume flow rate. By taking a value of pressure drop for an *arbitrarily* chosen flow rate from the CIBSE chart, the value of the constant for any particular duct size can be found.

For example, for AB, from the friction chart (Fig. 1.20), at a diameter of 0.5 m, taking a flow rate of 4 m³/s, say, we have $\Delta p_f = 8$ Pa/m. Then since friction loss, $\Delta p_f = K\dot{V}^2$,

$$\therefore \quad K = 8 \times 10/(4)^2 = 5 \quad \text{and} \quad \Delta p_{fAB} = 5(\dot{V}_{AB})^2$$

For BC, from the chart, at a diameter of 0.4 m and $\dot{V} = 4\ \text{m}^3/\text{s}$, say, $\Delta p_f = 26\ \text{Pa/m}$, i.e.

$$K = 26 \times 30/(4)^2 = 48.8$$
$$\therefore \qquad \Delta p_{fBC} = 48.8(\dot{V}_{BC})^2$$

For BD, from the chart, at a diameter of 0.4 m and $\dot{V} = 4\ \text{m}^3/\text{s}$, say, $\Delta p_f = 26\ \text{Pa/m}$, i.e.

$$K = 26 \times 10/(4)^2 = 16.3$$
$$\therefore \qquad \Delta p_{fBD} = 16.3(\dot{V}_{BD})^2$$

Hence, for duct BC, we have:

$$\Delta p_{BC} = p_{vBC}\Sigma k_{BC} + (p_{vg}k_g)_{BC} + \Delta p_{fBC}$$

i.e.
$$\Delta p_{BC} = \{(0.2 + 0.3 + 0.25) \times 38(\dot{V}_{BC})^2\} + \{0.4 \times 10.42(\dot{V}_{BC})^2\}$$
$$+ 48.8(\dot{V}_{BC})^2$$
$$= 81.5(\dot{V}_{BC})^2$$

Also, for duct BD, we have:

$$\Delta p_{BD} = \{(0.5 + 0.2 + 0.25) \times 38(\dot{V}_{BD})^2\} + \{0.4 \times 10.42(\dot{V}_{BD})^2\}$$
$$+ 16.3(\dot{V}_{BD})^2$$
$$= 56.6(\dot{V}_{BD})^2$$

For part (a), the pressure drop in BC and BD must be equal with the damper open, i.e.

$$\frac{\dot{V}_{BC}}{\dot{V}_{BD}} = \left\{\frac{81.5}{56.6}\right\}^{1/2} = 1.2$$

For duct AB, we have

$$\Delta p_{AB} = \{0.25 \times 41.67(\dot{V}_{AB})^2\} + 5(\dot{V}_{AB})^2 = 15.4(\dot{V}_{AB})^2$$

We can now match the fan characteristic against either duct ABC or duct ABD, where $\dot{V}_{AB} = \dot{V}_{BD} + \dot{V}_{BC} = \dot{V}_{BD}(1 + 1.2) = 2.2\dot{V}_{BD}$, i.e.

$$\Delta p_{ABD} = 15.4(\dot{V}_{AB})^2 + 56.6(\dot{V}_{AB}/2.2)^2$$
$$= 27.1(\dot{V}_{AB})^2$$

Therefore, to match the system against the fan:

$$27.1(\dot{V}_{AB})^2 = 200 - 12(\dot{V}_{AB})^2$$

i.e.
$$\dot{V}_{AB} = (200/39.1)^{1/2} = 2.26\ \text{m}^3/\text{s}$$

and $\dot{V}_{BD} = 2.26/2.2 = 1.03\ \text{m}^3/\text{s}$, $\dot{V}_{BC} = 2.26 - 1.03 = 1.23\ \text{m}^3/\text{s}$

(b)   For this case, $\dot{V}_{BC} = \dot{V}_{BD} = \dot{V}_{AB}/2$

Matching the fan against duct ABC, we have:

$$200 - 12(\dot{V}_{AB})^2 = 15.4(\dot{V}_{AB})^2 + 81.5(\dot{V}_{BC})^2$$

i.e.
$$200 - 12(\dot{V}_{AB})^2 = \left(15.4 + \frac{81.5}{4}\right)(\dot{V}_{AB})^2 = 35.8(\dot{V}_{AB})^2$$

$$\therefore \qquad \dot{V}_{AB} = (200/47.8)^{1/2} = 2.05\ \text{m}^3/\text{s}$$

and $\dot{V}_{BC} = \dot{V}_{BD} = 1.025\ \text{m}^3/\text{s}$

Note that for ABD the new pressure loss characteristic is given by:

$$15.4(\dot{V}_{AB})^2 + 56.6(\dot{V}_{BD})^2 + (k - 0.2) \times 38(\dot{V}_{BD})^2$$

i.e. $\quad \Delta p_{ABD} = 15.4(\dot{V}_{AB})^2 + (49 + 38k)(\dot{V}_{BD})^2$

where k is the velocity pressure loss factor for the partially closed damper, i.e.

$$200 - 12(2.05)^2 = 15.4(2.05)^2 + (49 + 38k)(1.025)^2$$
$$\therefore \quad k = 0.84$$

Example 1.10

(a) A large process area is ventilated with fans in both a supply and return duct. The characteristics of both fans are given in the table below. The supply duct has a pressure loss of 180 Pa for an air mass flow rate of 3 m³/s, the extract duct loss is 169 Pa for a flow rate of 2.6 m³/s. Leakage of air from the process area is given by the formula, $\Delta p = 300\dot{V}^2$. Find the supply and extract volume flow rates and the pressure in the process area.

(b) Find the new supply and extract volume flow rates and the pressure in the process area when a structural fault develops resulting in additional leakage of air from the process area equivalent to an orifice of 0.1 m² area and with a coefficient of discharge of 0.6.

Take the density of air as 1.2 kg/m³ throughout the system.

Supply fan characteristics:

Volume flow/(m³/s)	0	1	2	3	4
Fan total pressure/(Pa)	470	475	450	370	180

Extract fan characteristics:

Volume flow/(m³/s)	0	1	2	3	3.5
Fan total pressure/(Pa)	380	385	350	225	8

(Polytechnic of the South Bank)

Solution

(a) The system is shown diagrammatically in Fig. 1.24.

Figure 1.24 Example 1.10

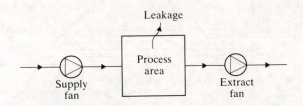
Leakage

Supply fan

Process area

Extract fan

For the supply ducting:

$$\Delta p_f = K\dot{V}^2 = K(3)^2 = 180 \qquad \therefore \quad K = 20$$

i.e. $\quad \Delta p_f = 20\dot{V}^2$ [1]

For the extract ducting:

$$\Delta p_f = \frac{169}{(2.6)^2} \dot{V}^2 = 25\dot{V}^2 \tag{2}$$

In this case the difference between the supply fan pressure and the system pressure loss for the supply side is equal to the pressure in the process room. Similarly the difference between the extract fan pressure and the system pressure loss for the extract ducting is equal to the pressure in the process room. It is therefore more convenient in cases of this kind to plot room pressure against volume flow rate.

Calculating the system resistances from Eqns [1] and [2] above and subtracting these from the fan pressures, we obtain the following table of values:

Supply:

Volume flow rate/(m³/s)	0	1	2	3	4
Fan total pressure/(Pa)	470	475	450	370	180
System pressure loss/(Pa)	0	20	80	180	320
Room pressure/(Pa)	470	455	370	190	−140

Extract:

Volume flow rate/(m³/s)	0	1	2	3	3.5
Fan total pressure/(Pa)	380	385	350	225	80.0
System pressure/(Pa)	0	25	100	225	306.3
Room pressure/(Pa)	−380	−360	−250	0	+226.3

The room pressures are then plotted against volume flow rate as shown in Fig. 1.25. When the room is well sealed, then the point where the two graphs intersect gives the room pressure, i.e. in this case the room pressure would be

Figure 1.25 Pressure vs volume flow rate for Example 1.9

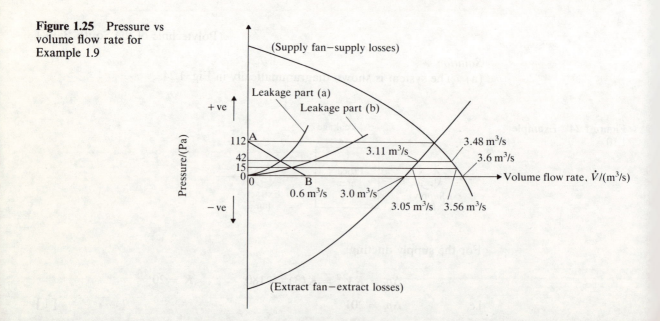

112 Pa. Since there is leakage from the room, the room pressure will be less than 112 Pa; a pressure line below 112 Pa cuts each of the curves at a different value of volume flow rate, the difference between the two values giving the leakage flow rate. The line AB can be drawn on the graph to represent the room pressures for various leakage flow rates. It can be seen that the room pressure curves are almost straight lines at the relevant part and hence the leakage flow rate varies approximately linearly with the room pressure. Line AB can therefore be drawn by fixing point A at 112 Pa at zero flow rate, and by fixing point B at a volume flow rate of $(3.6 - 3.0)$ m^3/s at zero pressure.

The leakage characteristic line, given as $\Delta p = 300\dot{V}^2$, can now be drawn on the graph as shown in Fig. 1.25. The point where this cuts line AB fixes the room pressure at 42 Pa. At this value of room pressure the supply and extract volume flow rates are then read from the curves as:

$$\text{Supply volume flow rate} = 3.48 \text{ m}^3/\text{s}$$

and $$\text{Extract volume flow rate} = 3.11 \text{ m}^3/\text{s}$$

(b)   For an orifice with a coefficient of discharge, C_d,

$$\text{Volume flow rate, } \dot{V} = C_d uA = C_d A(2\Delta p/\rho)^{1/2}$$

Therefore for this example,

$$\Delta p = \frac{1.2\dot{V}^2}{2(0.6 \times 0.1)^2} = 167\dot{V}^2$$

The two leakages are in parallel and hence the flows are added at the same pressure difference, i.e.

$$\text{Total leakage} = \sqrt{\Delta p}\left\{\frac{1}{\sqrt{300}} + \frac{1}{\sqrt{167}}\right\} = 0.135\sqrt{\Delta p}$$

$$\therefore \quad \Delta p = 54.8\dot{V}^2$$

This curve is then plotted on the graph as shown in Fig. 1.25 and the point where it cuts line AB gives the room pressure as 15 Pa. The supply and extract flows are then read off as 3.56 m^3/s and 3.05 m^3/s.

Duct Sizing

For a multiple zone air conditioning system, air is delivered through a number of different branch ducts from a main duct supplied by a suitable fan. To supply the required volume flow rate of air to each conditioned space it is necessary to choose the nearest commercial sizes for all ducts such that the various pressure losses exactly match the fan characteristic. Pressure losses for fittings are found by using the velocity pressure loss factors given by CIBSE,[1.5] and the friction losses from the chart (Fig. 1.20). Since these losses depend on the velocity pressure, which is not known until the area of the duct is known, a trial-and-error method is necessary. Computer programs are available which give a complete duct sizing for a given complex system, but it is still useful to be able to do a quick manual sizing to give preliminary design.

The main methods are as follows.

(a)   Equal velocity

A velocity is chosen to suit the system: this may be because of a requirement to keep noise down or because of pneumatic conveying requirements. The duct areas are then fixed from the known volume flow rates required.

This method is used for simple systems with few branches, or for pneumatic conveying.

(b)   Constant friction loss per unit length

A suitable velocity is chosen in the first section of the main duct and hence the duct area is fixed; the friction loss per unit length is then read from the chart for that section of the main. The remainder of the main duct is then sized using the same pressure loss per unit length. This results in a gradual reduction in velocity from plant room to terminal exit which can be acoustically beneficial in certain buildings, e.g. hospitals.

(c)   Static regain

This method is suitable for a system with a number of similar branches coming off a main at regular intervals. It uses the principle from Bernoulli's equation that the static pressure increases when the velocity pressure decreases. In the main across a branch off-take the area of the main duct downstream of the off-take is chosen such that the velocity pressure is reduced, thus giving an increase in static pressure; this increase then offsets the friction loss in the next section of the main. By this means the available pressure at each branch off-take can be made approximately equal.

In practice there is a loss of pressure across an off-take and hence the static regain is never 100%. Figure 1.26 shows part of a system with sections 1 and 2 just before and just after the first off-take and section 3 just before the second off-take. The pressure loss factor, k, in going across an off-take is usually given in terms of the velocity head downstream of the off-take, i.e.

$$p_1 + p_{v1} = p_2 + p_{v2} + kp_{v2}$$
$$\therefore \quad \text{Actual static regain, } p_2 - p_1 = p_{v1} - (1 + k)p_{v2} \qquad [1.53]$$

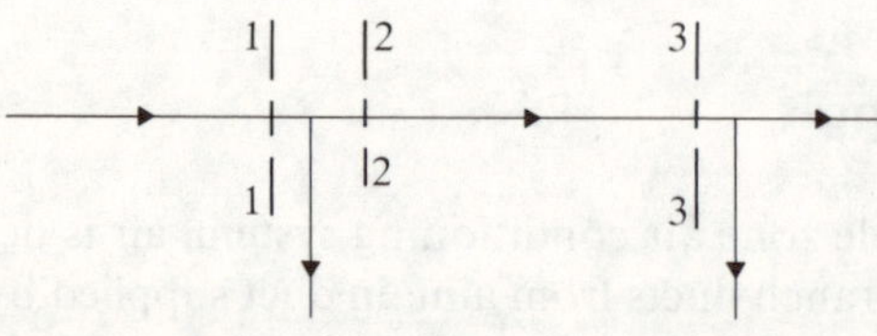

Figure 1.26 Air
distribution system

The actual static regain can be compared with the maximum possible static regain and expressed as a percentage, i.e.

$$\text{Percentage static regain} = 100\frac{\{p_{v1} - (1 + k)p_{v2}\}}{(p_{v1} - p_{v2})}$$

The friction loss between sections 2 and 3 should be made equal to the actual static regain, thus making the pressure at section 3 as close as possible to that at section 1.

In any system the flow must be adjusted by dampers in off-takes to give the exact flow rates required.

Example 1.11

For the air duct system shown in Fig. 1.27 determine, for (a) an equal pressure drop method, and (b) a static regain method:

(i) the diameters for duct lengths AB, BC and CD;
(ii) the required damper pressure drops to be allowed at the grilles B, C and D assuming that these discharge to atmosphere.

Figure 1.27
Example 1.11

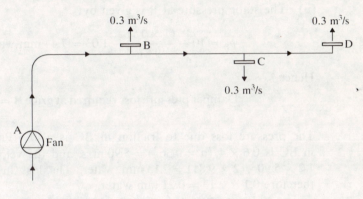

Data

Static pressure at fan outlet, 10 mm water; pressure loss factor for bend, 0.4; pressure loss factor for the main at each off-take, 0.1 (based on the velocity pressure downstream of the off-take); pressure drop for each off-take including grille, 4.5 mm water; length of duct AB, 15 m; length of duct BC, 9 m; length of duct CD, 6 m; flow from each off-take, 0.3 m³/s.

Solution

Note: In the solution that follows it is assumed that the velocity pressure in the main at any one off-take and the velocity pressure at exit from the grille at that off-take are approximately equal; this implies that the static pressure drop at each off-take from main to atmosphere is equal to the pressure loss at that off-take.

(a) (i) Flow rate in duct AB = 3 × 0.3 = 0.9 m³/s
The friction pressure drop in the main, ABCD, is approximately

$$(10 - 4.5 - \text{losses}) \times \frac{1000 \times 9.81}{1000}\ \text{Pa} = 9.81(5.5 - \text{losses})\ \text{Pa}$$

where the density of water is taken as 1000 kg/m³ and the acceleration due to gravity as 9.81 m/s².
Making an approximate allowance for the losses in the bend and across each off-take, we can take the initial pressure drop due to friction as 30 Pa, say. That is,

$$\text{Friction pressure drop per unit length} = 30/(15 + 9 + 6)$$
$$= 1\ \text{Pa/m}$$

Then from the CIBSE chart (Fig. 1.20), at 1 Pa/m and a flow rate of 0.9 m³/s, the nearest duct diameter is 0.42 m.

The velocity in AB is $0.9 \times 4/(\pi \times 0.42^2) = 6.50$ m/s, and hence

$$\text{Velocity pressure, } p_v = \frac{1.2 \times 6.50^2}{2} = 25.35 \text{ Pa}$$

$$= 25.35/9.81 = 2.58 \text{ mm water}$$

i.e.   Loss in bend $= 0.4 \times 2.58 = 1.03$ mm water

Then taking the same friction drop per unit length, of 1 Pa/m, for BC and CD we have from the chart:

Diameter for BC for a flow of 0.6 m^3/s $= 0.36$ m
Diameter for CD for a flow of 0.3 m^3/s $= 0.28$ m

(ii)   The static pressure at B is given by:

$$p_B = 10 - \frac{(1 \times 15)}{9.81} - 1.03 = 7.44 \text{ mm water}$$

Hence,

Damper pressure loss required at grille B $= 7.44 - 4.5$
$$= 2.94 \text{ mm water}$$

The pressure loss due to friction in BC is also 1 Pa/m. Also, the velocity in BC is $0.6 \times 4/(\pi \times 0.36^2) = 5.90$ m/s, and the velocity pressure for BC is $1.2 \times 5.90^2/(2 \times 9.81) = 2.13$ mm water. The loss in the main across B is therefore $0.1 \times 2.13 = 0.21$ mm water.

The static pressure at C is then given by:

$$p_C = 7.44 - 0.21 - \frac{(1 \times 9)}{9.81} = 6.31 \text{ mm water}$$

Hence,

Damper pressure loss required at grille C $= 6.31 - 4.5$
$$= 1.81 \text{ mm water}$$

The pressure loss due to friction in CD is 1 Pa/m. Also, the velocity in CD is $0.3 \times 4/(\pi \times 0.28^2) = 4.87$ m/s, and the velocity pressure for CD is $1.2 \times 4.87^2/(2 \times 9.81) = 1.45$ mm water. The pressure loss in the main across C is therefore $0.1 \times 1.45 = 0.15$ mm water.

The static pressure at D is therefore given by:

$$p_D = 6.31 - 0.15 - \frac{(1 \times 6)}{9.81} = 5.55 \text{ mm water}$$

Hence,

Damper pressure loss required at grille D $= 5.55 - 4.5$
$$= 1.05 \text{ mm water}$$

A check must then be done for the index circuit to see if the total pressure drop can be provided by the fan, i.e.

$$\text{Total pressure loss from A to D} = 1.03 + \frac{(1 \times 15)}{9.81} + 0.21$$

$$+ \frac{(1 \times 9)}{9.81} + 0.15 + \frac{(1 \times 6)}{9.81}$$

$$= 4.45 \text{ mm water}$$

The actual total pressure difference available is $\{(10 + 2.58) - (5.55 + 1.45)\} =$ 5.58 mm water. So the design is satisfactory with the use of an appropriate damper.

(b) The object of this method is to size the ducts so that the pressures at B, C and D are approximately the same. Duct AB can be sized initially to give the minimum damper pressure loss at grille B. Therefore we have

$$\text{Friction pressure loss for AB} = 9.81(10 - 4.5 - \text{loss in bend}) \text{ Pa}$$
$$= 30 \text{ Pa, say,}$$
$$= 30/15 = 2 \text{ Pa/m}$$

Then, from the chart, at 0.9 m³/s and 2 Pa/m the diameter is 0.36 m for duct AB.

The velocity in AB is therefore $0.9 \times 4/(\pi \times 0.36^2) = 8.84$ m/s, and the velocity pressure is 4.78 mm water, i.e.

$$\text{Loss due to bend} = 0.4 \times 4.78 = 1.91 \text{ mm water}$$

Hence the pressure at B is:

$$p_B = 10 - 1.91 - \frac{(15 \times 2)}{9.81} = 5.03 \text{ mm water}$$

Then,

$$\text{Damper pressure loss for grille B} = 5.03 - 4.5 = 0.53 \text{ mm water}$$

For the static regain method, the friction pressure loss in BC should be equal to the static regain across off-take B.

Try a diameter of 0.35 m for duct BC. That is,

$$\text{Friction pressure loss} = 1.2 \text{ Pa/m}$$
$$= 1.2 \times 9 = 10.8 \text{ Pa}$$
$$= 10.8/9.81 = 1.10 \text{ mm water}$$

The velocity pressure in AB is 4.78 mm water (see above). The velocity in BC is given by $0.6 \times 4/(\pi \times 0.35^2) = 6.24$ m/s, i.e.

$$\text{Velocity pressure for BC} = \frac{1.2 \times 6.24^2}{2 \times 9.81} = 2.38 \text{ mm water}$$

From Eqn [1.53]:

$$\text{Actual static regain} = p_{v1} - (1 + k)p_{v2}$$
$$= 4.78 - (1 + 0.1) \times 2.38$$
$$= 2.16 \text{ mm water}$$

i.e. $$\text{Static pressure in main after B} = 5.03 + 2.16$$
$$= 7.19 \text{ mm water}$$

Therefore,

$$p_C = 7.19 - 1.10 = 6.09 \text{ mm water}$$

i.e. $$\text{Damper pressure loss for grille C} = 6.09 - 4.5$$
$$= 1.59 \text{ mm water}$$

Try 0.3 m diameter for duct CD. That is,

$$\text{Friction pressure drop} = 0.72 \text{ Pa/m}$$

The velocity in CD is $0.3 \times 4/(\pi \times 0.3^2) = 4.24$ m/s. Hence,

$$\text{Velocity pressure for CD} = \frac{(1.2 \times 4.24^2)}{2 \times 9.81} = 1.10 \text{ mm water}$$

Using Eqn [1.53]:

$$\text{Actual static regain} = 2.38 - (1 + 0.1) \times 1.10$$
$$= 1.17 \text{ mm water}$$

Therefore,

$$\text{Static pressure in main after C} = 6.09 + 1.17 = 7.26 \text{ mm water}$$

Hence,

$$\text{Pressure in main at D} = 7.26 - \frac{(0.72 \times 6)}{9.81} = 6.82 \text{ mm water}$$

Therefore,

$$\text{Damper pressure loss at grille D} = 6.82 - 4.5 = 2.32 \text{ mm water}$$

Checking the pressure loss in the index circuit, we have:

Total pressure loss from A to D
$$= \frac{(2 \times 15)}{9.81} + 1.91 + (0.1 \times 2.38) + 1.10 + (0.1 \times 1.1)$$
$$+ \frac{(0.72 \times 6)}{9.81}$$
$$= 6.86 \text{ mm water}$$

The pressure difference available is $\{(10 + 4.78) - (6.82 + 1.10)\} = 6.86$ mm water.

Summary

The results given by the two methods are summarized in the table below.

Method	Duct diameters/(m)			Damper pressure drop/(mm H$_2$O)		
	AB	BC	CD	B	C	D
Equal pressure method	0.42	0.36	0.28	2.94	1.81	1.05
Static regain method	0.36	0.35	0.30	0.53	1.59	2.32

1.6 VENTILATION

Ventilation is necessary to remove any undesirable odours or contaminants in a living space and to ensure that the carbon dioxide produced is being replaced by an adequate supply of oxygen. In unoccupied zones ventilation may be necessary to ensure that there is no possibility of vapours forming an explosive mixture.

There are two main ways of ventilating a building: by *natural ventilation*, in which open windows, ventilators or shafts are used to allow air to enter and leave the building due to pressure differences between the outside and inside air; or by *mechanical ventilation*, using fans to control the amount of air entering and leaving. For further information on ventilation the reader is recommended to the CIBSE *Guide*,[1.3] and the books by McIntyre,[1.12] and Croome and Roberts.[1.13]

Natural Ventilation

There are two basic driving forces which cause air to enter a building: firstly, the pressure on the face of a building due to the rapid deceleration of the wind on the outside surface; secondly, the buoyancy effect due to the temperature differences between the outside and inside air, usually known as the stack effect.

Wind effect

Wind speed varies with height above the ground and can be shown to be governed by an equation of the form:

$$u = u_m K_s z^a \qquad [1.54]$$

where u is the mean wind speed at height z; u_m is the mean wind speed at a height of 10 m in open country; K_s and a are parameters dependent on the terrain; some values are given in Table 1.2, which is reproduced by permission of CIBSE.[1.3]

Table 1.2 Values of K_s and a in Eqn [1.54]

Terrain	K_s	a
Open, flat country	0.68	0.17
Country with scattered windbreaks	0.52	0.20
Urban	0.35	0.25
City	0.21	0.33

When wind flows across a building, the air is brought to rest on the front face of the building causing a pressure on the face of the building higher than the pressure of the undisturbed air stream. On the leeward side the pressure on the building surface is lower than the pressure of the undisturbed air stream. For a flat roof, the air pressure on the roof is lower than the undisturbed air stream; for a pitched roof the pressure on the roof is lower than the undisturbed air stream except when the pitch of the roof is greater than about 45°.

The air pressures on any particular building depend therefore on the wind direction and the wind speed and will vary from day to day. The flow pattern round a building is also influenced by other buildings: in most cases the pattern is complex, with vortices from one building affecting the flow around another building. Tests can be undertaken using models in wind tunnels with the atmospheric conditions carefully replicated. The Building Research Establishment has published results of such tests; some examples of flow around buildings are given by Croome and Roberts.[1.13]

A non-dimensional pressure coefficient, C_p, is defined as follows:

$$C_p = \frac{p - p_o}{\frac{1}{2}\rho u_r^2} \qquad [1.55]$$

CINCINNATI TECHNICAL COLLEGE LRC

where p is the mean pressure at any point on the building surface; p_o is the pressure in the undisturbed air stream; u_r is the mean wind speed at a height equal to the building height; ρ is the density of the air at the temperature of the outside air.

For the volumetric flow rate, $\dot{V}$, through a small ventilation opening we can write:

$$\dot{V} = C_d Au$$

where A is the cross-sectional area of the opening; u is the velocity of the air leaving the opening; C_d is the discharge coefficient.

The velocity, u, is given by $\rho u^2 / 2 = \Delta p$, where Δp is the pressure difference across the opening. The volume flow rate, $\dot{V}$, through the opening is therefore,

$$C_d Au = C_d A(2\Delta p/\rho)^{1/2}$$

For a number of openings in the same face of the building, since the pressure difference Δp is the same for each opening, and assuming the same value of C_d for each opening, we can write

$$\dot{V} = C_d(\Sigma A)(2\Delta p/\rho)^{1/2} \qquad [1.56]$$

where ΣA is the sum of all the areas on the same face of the building.

Assuming air enters a building through ventilation openings on one face of a building and leaves through openings on the opposite side, we have

$$\dot{V}_1 = C_d(\Sigma A_1)\{2(p_1 - p_i)/\rho_1\}^{1/2}$$

and

$$\dot{V}_2 = C_d(\Sigma A_2)\{2(p_i - p_2)/\rho_2\}^{1/2}$$

where suffices 1 and 2 refer to the inlets and outlets of the ventilating air.

In the steady state, $\dot{V}_1 = \dot{V}_2 = \dot{V}_w$; hence, re-arranging the equations and adding, (assuming also, $\rho_1 = \rho_2 = \rho$),

$$p_1 - p_2 = \frac{\rho \dot{V}_w^2}{2C_d^2}\left(\frac{1}{(\Sigma A_1)^2} + \frac{1}{(\Sigma A_2)^2}\right) \qquad [1]$$

From Eqn [1.55] we have

$$C_{p1} = (p_1 - p_o)/(\tfrac{1}{2}\rho u_r^2)$$

and

$$C_{p2} = (p_2 - p_o)/(\tfrac{1}{2}\rho u_r^2)$$

$$\therefore \qquad p_1 - p_2 = (C_{p1} - C_{p2}) \times \tfrac{1}{2}\rho u_r^2 = \Delta C_p \rho u_r^2/2$$

Substituting in Eqn [1] above we then have

$$\frac{\Delta C_p \rho u_r^2}{2} = \frac{\rho \dot{V}_w^2}{2C_d^2(\Sigma A_w)^2}$$

where

$$\frac{1}{(\Sigma A_w)^2} = \frac{1}{(\Sigma A_1)^2} + \frac{1}{(\Sigma A_2)^2} \qquad [1.57]$$

Then,

$$\dot{V}_w = C_d(\Sigma A_w)u_r(\Delta C_p)^{1/2} \qquad [1.58]$$

The value of C_d for a sharp-edged orifice is 0.61, and for most ventilation openings taking a value for C_d of 0.61 is sufficiently accurate. For greater accuracy, measurements of air flow rate against pressure difference for a particular type of ventilator can be made and values of an equivalent area, $C_d A$, determined.

Stack effect

Assume that a wall of a building has two ventilation openings of areas A_1 and A_2 separated by a height z, as shown in Fig. 1.28. It will be assumed initially that the effect of wind pressure is negligible. The mean air density outside the building is ρ_o and the mean density of the inside air is ρ_i. The outside pressure at section 1 is then given by

$$p_{o1} = p_{o2} + g\rho_o z \qquad [1]$$

where g is the acceleration due to gravity.

Figure 1.28 Natural ventilation

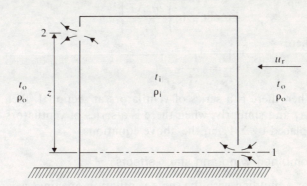

The inside pressure at section 1 is

$$p_{i1} = p_{i2} + g\rho_i z \qquad [2]$$

Then subtracting Eqn [2] from Eqn [1],

$$p_{o1} - p_{i1} = gz\Delta\rho - (p_{i2} - p_{o2}) \qquad [3]$$

where $\Delta\rho$ is the difference in density between the outside and inside air.

In the steady state the mass flow entering the building at section 1 is equal to the mass flow rate leaving the building at section 2, i.e.

$$\rho_o \dot{V}_1 = \rho_i \dot{V}_2$$
$$\rho_o C_d A_1 \{2(p_{o1} - p_{i1})/\rho_o\}^{1/2} = \rho_i C_d A_2 \{2(p_{i2} - p_{o2})/\rho_i\}^{1/2}$$
$$\therefore \quad p_{i2} - p_{o2} = (p_{o1} - p_{i1})\frac{\rho_o(A_1)^2}{\rho_i(A_2)^2}$$

Substituting in Eqn [3] above:

$$p_{o1} - p_{i1} = gz\Delta\rho - (p_{o1} - p_{i1})\frac{\rho_o(A_1)^2}{\rho_i(A_2)^2}$$

$$\therefore \quad p_{o1} - p_{i1} = \frac{gz\Delta\rho}{1 + (\rho_o A_1^2)/(\rho_i A_2^2)} \qquad [4]$$

Then the volume flow rate entering the building by natural ventilation is given by

$$\dot{V}_N = C_d A_1 \{2(p_{o1} - p_{i1})/\rho_o\}^{1/2}$$

Substituting for $(p_{o1} - p_{i1})$ from Eqn [4] above

$$\dot{V}_N = C_d A_1 \left\{ \frac{2gz\Delta\rho}{\rho_o\{1 + (\rho_o A_1^2)/(\rho_i A_2^2)\}} \right\}^{1/2}$$

$$= C_d \left\{ \frac{2gz\Delta\rho}{\rho_o\left(\dfrac{1}{A_1^2} + \dfrac{\rho_o}{\rho_i A_2^2}\right)} \right\}^{1/2}$$

Two assumptions can be made with good accuracy:

(i)   the density of air at the mean of the inside and outside temperatures, ρ, can be put equal to the densities, ρ_o and ρ_i;

(ii)   the ratio $\Delta\rho/\rho$ can be put equal to $\Delta T/T$, where T is the mean of the absolute temperatures of the inside and outside air, and ΔT is the difference between the mean temperatures inside and outside the building.

We therefore have

$$\dot{V}_N = C_d A_N \left\{ \frac{2gz\Delta T}{T} \right\}^{1/2} \tag{1.59}$$

where

$$\frac{1}{A_N^2} = \frac{1}{A_1^2} + \frac{1}{A_2^2}$$

When there is a series of ventilators at section 1, then A_1 can be replaced by ΣA_1 and similarly, when there is a series of ventilators at section 2, A_2 can be replaced by ΣA_2, in the above equations.

Combined wind and stack effects

For a building with the same ventilation openings as are shown in Fig. 1.28, it can be seen that when the wind is blowing against the right-hand wall the positive wind pressure will assist the stack effect at section 1, and the negative wind pressure on the left-hand side will also assist the stack effect at section 2. It is therefore a good approximation to assume that the two pressure drops are additive. Since the volumetric flow rates are proportional to the square root of the pressure difference, we can write:

$$\dot{V} = (\dot{V}_W^2 + \dot{V}_N^2)^{1/2} \tag{1.60}$$

For the case shown in Fig. 1.29, where there are ventilation openings at sections 1 and 2 on both faces of the buildings, the situation is more complex since the wind pressure opposes the stack effect at the stations A and B, but assists it at stations C and D.

Research at the Building Research Establishment shows that it is a good approximation under these conditions to assume that:

$$\dot{V} = \dot{V}_W \quad \text{when} \quad u_r A_W(\Delta C_p)^{1/2} > 0.26 A_N(z\Delta T)^{1/2}$$

and

$$\dot{V} = \dot{V}_N \quad \text{when} \quad u_r A_W(\Delta C_p)^{1/2} < 0.26 A_N(z\Delta T)^{1/2}$$

Note that the constant 0.26 in the above equations comes from Eqn [1.59], i.e.

$$\dot{V}_N = C_d A_N \left\{ 2gz\frac{\Delta T}{T} \right\}^{1/2} = C_d A_N \times 0.26 \times (z\Delta T)^{1/2}$$

Figure 1.29 Natural ventilation with combined wind and stack effect

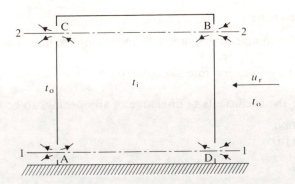

where $(2g/T)^{1/2} = (2 \times 9.81/290)^{1/2} = 0.26$; T is taken at a value of 290 K = 17 °C, and the acceleration due to gravity, g, is 9.81 m/s^2.

The equations above effectively mean that when $\dot{V}_W > \dot{V}_N$, then $\dot{V} = \dot{V}_W$, and that when $\dot{V}_W < \dot{V}_N$, then $\dot{V} = \dot{V}_N$.

In the case shown in Fig. 1.29 the combined area, A_W, from Eqn [1.57] is based on $\Sigma A_1 = A_D + A_B$, and $\Sigma A_2 = A_A + A_C$; the combined area, A_N, from Eqn [1.59] is based on $\Sigma A_1 = A_A + A_D$ and $\Sigma A_2 = A_B + A_C$.

Example 1.12

A large rectangular-plan building in an urban setting is orientated with one side perpendicular to the prevailing wind direction. At the top of the building, at a height of 50 m, the infiltration area A_2 on the windward side totals 1.2 m^2 and at ground level on the leeward side the infiltration area A_1 totals 0.6 m^2.

Assessment of the ventilation rate is required for three different cases:

Case (a) Exposure to wind effects only:
The wind velocity on a particular day may be assumed to follow the law:

$$u = 3.3z^{0.26} \text{ m/s}$$

where z is the vertical height in metres.

The coefficients of pressure on the upstream and downstream faces are $+0.75$ and -0.35, based on the velocity upstream and at the height of A_1.

Temperature effects may be ignored in this case.

Calculate:

(i) the overall static pressure difference along the flow path through the combination of A_1 and A_2;
(ii) the ventilation flow rate along this path.

Case (b) Stack effect only:
The same building has a mean internal temperature of 16 °C and a mean external temperature of 28 °C. Wind effects may be ignored in this case.

Calculate:

(i) the pressure difference providing the driving force for the ventilation;
(ii) the ventilation flow rate.

Case (c) Combined wind and stack effects:
The same building is now simultaneously subjected to the two effects described in Cases (a) and (b).

Calculate:

(i) the combined pressure difference along the flow path through the combination of A_1 and A_2;

(ii) the infiltration rate along the path.

Take the coefficients of discharge of all openings to be 0.61.

Solution
Case (a):

(i) $u_r = 3.3(50)^{0.26} = 9.125 \text{ m/s}$

The air density at the mean temperature of $(16 + 28)/2 = 22\,°C$ is approximately 1.2 kg/m^3.

Then using Eqn [1.55] and referring to Fig. 1.30,

$$p_2 - p_o = C_p \times \tfrac{1}{2}\rho u_r^2 = 0.75 \times 0.5 \times 1.2 \times (9.125)^2$$
$$= 37.47 \text{ N/m}^2$$

Figure 1.30 Diagram for Example 1.12

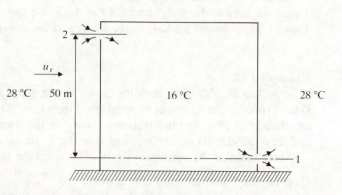

Similarly,

$$p_1 - p_o = -0.35 \times 0.5 \times 1.2 \times (9.125)^2$$
$$= -17.49 \text{ N/m}^2$$

i.e.   Overall pressure difference $= p_2 - p_1$
$$= 37.47 + 17.49$$
$$= 54.96 \text{ N/m}^2$$

(ii) The ventilation flow rate can then be found as

$$C_d(\Sigma A_w)\{2(p_2 - p_1)/\rho\}^{1/2}$$

or from Eqn [1.58],

$$C_d(\Sigma A_w)u_r(\Delta C_p)^{1/2}$$

where

$$\frac{1}{(\Sigma A_w)^2} = \frac{1}{(\Sigma A_1)^2} + \frac{1}{(\Sigma A_2)^2}$$

$$= \frac{1}{(0.6)^2} + \frac{1}{(1.2)^2} = 3.472$$

$$\therefore \qquad \Sigma A_w = 0.537 \text{ m}^2$$

i.e. $\qquad$ Ventilation rate $= 0.61 \times 0.537(2 \times 54.96/1.2)^{1/2}$
$$= 3.135 \, \text{m}^3/\text{s}$$

or $\qquad$ $0.61 \times 0.537 \times 9.125 \times (0.75 + 0.35)^{1/2} = 3.135 \, \text{m}^3/\text{s}$

Case (b):

(i) In this example the outside temperature is greater than the inside temperature and hence the ventilation flow enters the top of the building and flows down through it.

Therefore, referring to Fig. 1.30,

$$p_{o1} - p_{o2} = g\rho_o z \qquad \text{and} \qquad p_{i1} - p_{i2} = g\rho_i z$$

i.e. $\qquad$ Pressure difference providing ventilation flow

$$= (p_{o1} - p_{o2}) - (p_{i1} - p_{i2}) = gz(\rho_o - \rho_i)$$

$$= \frac{9.81 \times 50 \times 1.013\,25 \times 10^5}{287} \left(\frac{1}{289} - \frac{1}{301} \right)$$

$$= 23.89 \, \text{N}/\text{m}^2$$

where the atmospheric pressure is taken as $1.013\,25 \times 10^5 \, \text{N}/\text{m}^2$ and the gas constant for air is $287 \, \text{J}/\text{kg K}$.

(ii) In this case $\Sigma A_N = \Sigma A_W = 0.537 \, \text{m}^2$
Therefore,

$$\text{Ventilation rate} = C_d(\Sigma A_N)(2\Delta p/\rho)^{1/2}$$
$$= 0.61 \times 0.537 \times (2 \times 23.89/1.2)^{1/2}$$
$$= 2.067 \, \text{m}^3/\text{s}$$

or, from Eqn [1.59],

$$\text{Ventilation rate} = 0.61 \times 0.537 \times (2 \times 9.81 \times 12 \times 50/295)^{1/2}$$
$$= 2.067 \, \text{m}^3/\text{s}$$

Case (c):

(i) In this case, when the wind effect is combined with the stack effect the two pressure differences are additive since the wind pressure is positive at the top of the building and circulation due to the density differences is from the outside at the top into the building. That is,

$$\text{Combined pressure difference available for ventilation}$$
$$= 54.96 + 23.89 = 78.85 \, \text{N}/\text{m}^2$$

(ii) Then the ventilation rate is

$$0.61 \times 0.537 \times (2 \times 78.85/1.2)^{1/2} = 3.755 \, \text{m}^3/\text{s}$$

Note that since $\dot{V} \propto (\Delta p)^{1/2}$, the ventilation rate can always be found by proportion for the same areas; i.e. in this case we can write

$$\dot{V} = 2.067 \times (78.85/23.89)^{1/2} = 3.755 \, \text{m}^3/\text{s}$$

Example 1.13

Re-calculate the previous example assuming that there are additional ventilation openings of $1.2 \, \text{m}^2$ at the high point on the leeward side, and of $0.6 \, \text{m}^2$ at the low point on the windward side. All other data remain unchanged.

Figure 1.31 Diagram
for Example 1.13

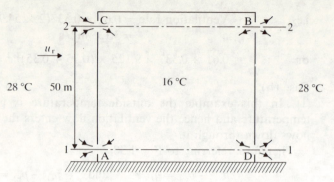

Solution

Case (a):

(i)   The ventilation system is now as shown in Fig. 1.31. In this case the total area on both the windward and leeward sides is $(1.2 + 0.6) = 1.8$ m^2, i.e.

$$\frac{1}{A_W^2} = \frac{1}{1.8^2} + \frac{1}{1.8^2} = 0.617$$

$$\therefore \quad A_W = 1.273 \text{ m}^2$$

The pressure difference is as before, now acting on two openings on each side, i.e.

Pressure difference $= 54.96$ N/m^2

(ii)   Ventilation rate $= 0.61 \times 1.273 \times (2 \times 54.96/1.2)^{1/2}$
$$= 7.432 \text{ m}^3/\text{s}$$

or,

Ventilation rate $= 3.135 \times 1.273/0.537 = 7.432$ m^3/s

Case (b):

(i)   In this case the combined area at the bottom is $(0.6 + 0.6) = 1.2$ m^2, and the combined area at the top is $(1.2 + 1.2) = 2.4$ m^2, i.e.

$$\frac{1}{A_N^2} = \frac{1}{1.2^2} + \frac{1}{2.4^2} = 0.868$$

$$\therefore \quad A_N = 1.073 \text{ m}^2$$

The pressure difference due to the density differences is the same as before, now acting on different areas, i.e.

Pressure difference for ventilation $= 23.89$ N/m^2

(ii)   Then the ventilation is given by

$$2.067 \times 1.073/0.537 = 4.130 \text{ m}^3/\text{s}$$

Case (c):

(i)   In this case the combined pressure difference of the previous example no longer has any meaning since at openings C and D the wind and stack effects combine but at openings A and B the effects are in opposition.

(ii)   The ventilation rate is recommended to be taken as the larger of the two effects. In this case, $\dot{V}_W = 7.432$ m^3/s and $\dot{V}_N = 4.13$ m^3/s; hence the ventilation rate is taken as 7.432 m^3/s.

Mechanical Ventilation

Due to imperfections in a building structure air tends to leak into a building through cracks round windows, doors and in the actual building fabric; this is known as *air infiltration*. Air infiltration is uncontrolled and undesirable; improved construction methods and better building design can reduce this to a minimum. A fully air-conditioned building should in theory be completely sealed so that all the air is introduced and conditioned in a controlled manner; just enough 'fresh' outside air can be introduced to satisfy the requirements for comfort and health.

In winter the ventilation air introduced from outside into a heated space must be heated and hence the heating costs are increased; a careful balance must be made between adequate ventilation and excessive heating bills. In a similar way, an air-conditioned building in summer will have an increased running cost if more outside air than is necessary is introduced. In later chapters more is said about ventilation energy loss and ways of allowing for a necessary flow of outside air into buildings.

In mechanically ventilated buildings one or more fans are used to give better control of the exact amounts of air entering and leaving the building (see Example 1.10, for instance). The air can be distributed into different parts of the building at different rates and at different temperatures as required. Mechanical ventilation covers the whole range from a simple extract system in a kitchen or bathroom, say, to a completely air-conditioned building. Texts such as that by Croome and Roberts[1.13] should be consulted for an extensive treatment.

One of the important functions of ventilation is to remove or reduce contaminants in air in spaces in buildings which are occupied by people, or in storage areas where an explosion risk may occur. In some areas with very stringent requirements for air purity (e.g. clean rooms, hospital operating theatres, etc.), a method of ventilation known as *displacement ventilation* or sometimes as piston air distribution, is used. In this method air is supplied at low velocity through a large number of openings in the ceiling or floor, say, and pushes across the cross-section of the space taking all the contaminants with it. More commonly, the ventilation system is that known as *dilution ventilation*, where air is introduced through jets into the room; the jets set up a mixing pattern in the room, thus diluting the contaminants in the space before extract. A more recent system has been introduced making use of heat sources within the premises to create convection currents; a brief discussion of this method is given by Martin.[1.15]

Contaminants in air

To ensure that people are not subjected to too high a level of contaminants, a Threshold Limiting Value (TLV) is defined, *either*

(i) as a time-weighted average concentration for a working day, or working week, to which workers may be subjected without adverse effects; *or*
(ii) as a maximum concentration to which workers can be subjected for a short time period, say up to 15 minutes.

Additionally, a ceiling value can be set which is the maximum TLV to which anyone should be subjected, even instantaneously.

With certain substances in air there is also a risk of an explosive mixture being formed if the concentration is allowed to rise too high; this can apply to occupied as well as to unoccupied areas. A *lower explosive limit* is defined as that concentration which must never be exceeded at any time.

Dilution of contaminants

For a space of volume, V, assume a volume flow rate production of contaminants of $\dot{V}_c$, and a constant dilution ventilation volume flow rate of $\dot{V}$. Let the initial concentration of contaminants in the air per unit volume be C_o at a time $\tau = 0$.

Assume also that the air entering the space has a concentration of contaminants of C_i per unit volume.

At any time τ the concentration of contaminants in the space is C per unit volume.

Therefore for a time interval $d\tau$, the contaminants entering the space less the contaminants leaving the space must equal the rate of increase of contaminants. Let the volume flow rate leaving the space be $(\dot{V} + \dot{V}_c)$ to allow for the production flow rate of contaminants. That is,

$$\{C_i\dot{V} + \dot{V}_c - C(\dot{V} + \dot{V}_c)\}\,d\tau = d(CV)$$

$$\int_{C_o}^{C} \frac{dC}{\{C_i\dot{V} + \dot{V}_c - C(\dot{V} + \dot{V}_c)\}} = \int_0^{\tau} \frac{d\tau}{V}$$

$$\therefore \quad \frac{-1}{(\dot{V} + \dot{V}_c)} \ln\left\{\frac{C_i\dot{V} + \dot{V}_c - C(\dot{V} + \dot{V}_c)}{C_i\dot{V} + \dot{V}_c - C_o(\dot{V} + \dot{V}_c)}\right\} = \frac{\tau}{V}$$

$$\frac{C_i\dot{V} + \dot{V}_c - C(\dot{V} + \dot{V}_c)}{C_i\dot{V} + \dot{V}_c - C_o(\dot{V} + \dot{V}_c)} = \exp\{-(\dot{V} + \dot{V}_c)\tau/V\}$$

Then the concentration, C, at any time τ is given by

$$C = \frac{(C_i\dot{V} + \dot{V}_c)}{(\dot{V} + \dot{V}_c)} + \left\{C_o - \frac{(C_i\dot{V} + \dot{V}_c)}{(\dot{V} + \dot{V}_c)}\right\}\exp[-(\dot{V} + \dot{V}_c)\tau/V]$$

$$[1.61]$$

In many cases in practice the production flow rate of contaminants is very small compared with the ventilation rate and therefore the term $(\dot{V} + \dot{V}_c)$ in Eqn [1.61] can be replaced by $\dot{V}$.

Various special cases can be easily derived from Eqn [1.61].

(a)   No contaminants in incoming air, $(C_i = 0)$, i.e.

$$C = \frac{\dot{V}_c}{(\dot{V} + \dot{V}_c)} + \left\{C_o - \frac{\dot{V}_c}{(\dot{V} + \dot{V}_c)}\right\}\exp[-(\dot{V} + \dot{V}_c)\tau/V]$$

$$[1.62]$$

(b)   Production of contaminants stopped at a time $\tau = 0$ when the contamination is C_o and where there are no contaminants in the incoming air $(\dot{V}_c = 0, C_i = 0)$, i.e.

$$\text{Concentration at time } \tau, \quad C = C_o\exp[-\dot{V}\tau/V] \qquad [1.63]$$

(c)   Space which is uncontaminated at time τ, $(C_o = 0)$, i.e.

$$C = \frac{(C_i\dot{V} + \dot{V}_c)}{(\dot{V} + \dot{V}_c)}\{1 - \exp[-(\dot{V} + \dot{V}_c)\tau/V]\} \qquad [1.64]$$

(d)   Sudden failure of the ventilation system at time $\tau = 0$ when concentration is C_o, $(\dot{V} = 0)$, i.e.

$$C = 1 + (C_o - 1)\exp[-\dot{V}_c\tau/V] \qquad [1.65]$$

(e)   The equilibrium condition is a special case. Assuming the equilibrium concentration is C_e, we have

$$C_i \dot{V} + \dot{V}_c - C_e(\dot{V} + \dot{V}_c) = 0$$

i.e.

$$C_e = \frac{C_i \dot{V} + \dot{V}_c}{\dot{V} + \dot{V}_c} \qquad [1.66]$$

Hence the ventilation flow rate required to restrict the concentration to C_e is given by

$$\dot{V} = \frac{\dot{V}_c(1 - C_e)}{(C_e - C_i)} \qquad [1.67]$$

Equations [1.61]–[1.65] are shown in Fig. 1.32 as concentration against time.

Figure 1.32 Equations for concentration against time

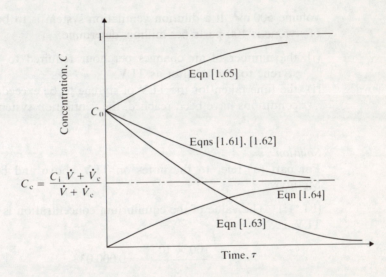

Note that Eqn [1.61] is the sum of two equations of the form of Eqns [1.63] and [1.64], i.e.

$$C = C_o \exp[-(\dot{V} + \dot{V}_c)\tau/V]$$
$$+ \frac{(C_i \dot{V} + \dot{V}_c)}{(\dot{V} + \dot{V}_c)} \{1 - \exp[-(\dot{V} + \dot{V}_c)\tau/V]\}$$

Denoting the first part by A and the second part by B, the curves of these equations are plotted as shown in Fig. 1.33.

Note that when $(C_i \dot{V} + \dot{V}_c)/(\dot{V} + \dot{V}_c) > C_o$ then the curve for Eqn [1.61] rises as in Fig. 1.33, but for the opposite case the curve falls as in Fig. 1.32.

The CIBSE Guide[1.16] gives further information on ventilation systems for specific purposes.

Example 1.14

(a)   With respect to toxic hazards, clearly explain the term Threshold Limit Value and its limitation as a safety standard.

(b)   Styrene vapour is released at a rate of 90 litre/h into a space of

Figure 1.33 Alternative format of Eqn [1.61]

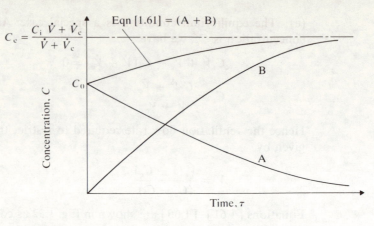

volume 500 m³. If a dilution ventilation system is to be used and if the TLV of styrene is 100 parts per million, determine:

(i) the number of air changes per hour required to limit concentration of styrene to one-third of its TLV;

(i) the time taken for the TLV of styrene to be exceeded if, after equilibrium conditions have been reached, the ventilation system fails.

(CIBSE)

Solution
For part (a) refer to the notes on TLV above and books such as that by McIntyre.[1.12]

(b) (i) The value of the equilibrium concentration is to be one-third of the TLV, i.e.

$$C_e = \frac{100 \times 10^{-6}}{3} = 0.000\,033$$

Then, using Eqn [1.67],

$$\text{Ventilation rate, } \dot{V} = \dot{V}_c \frac{(1 - C_e)}{C_e}$$

(where there are no contaminants entering with the air, i.e. $C_i = 0$). That is,

$$\dot{V} = \frac{90 \times 10^{-3} \times (1 - 0.000\,033)}{3600 \times 0.000\,033}$$
$$= 0.75 \text{ m}^3/\text{s}$$

Therefore,

$$\text{Number of air changes required per hour} = 0.75 \times 3600/500 = 5.4$$

(ii) The initial concentration of styrene is 0.000 033 when the ventilation system fails. It is required to find how long it will take for the value to rise to the TLV of 0.0001.

Using Eqn [1.65],

$$C = 1 + (C_o - 1) \exp[-\dot{V}_c \tau/V]$$

where C is the TLV of 0.0001 and C_o is 0.000 033, i.e.

$$\tau = \frac{V}{V_c} \ln\{(1 - C_o)/(1 - C)\}$$

$$= \frac{500 \times 3600}{90 \times 10^{-3}} \ln\left\{\frac{1 - 0.000\ 033}{1 - 0.0001}\right\} = 1333\ \text{s} = 0.37\ \text{h}$$

PROBLEMS

1.1 Figure 1.28 shows a cross-section through a store attached to a building. The store is 20 m long with external walls constructed of two layers of 105 mm brick with a cavity air gap. The building is normally occupied and is maintained at 21 °C at all times and the ventilation is from the building to the store and from the store to the outside. The store is unheated and normally unoccupied.

Figure 1.34 Problem 1.1

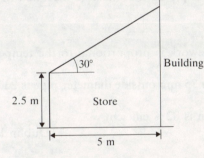

If the outside temperature is $-2\ °C$ calculate:
(i)   the temperature in the store;
(ii) the heat loss rate;
(iii) the temperature in the store if outside air had been used for ventilation (assume constant density);
and discuss the implications of using outside air to ventilate the store rather than air from the building.

Data
Thermal conductivity of brickwork, 1.1 W/m K; thermal resistance of air gap, 0.18 m² K/W; thermal transmittance from the building to the store, 1.0 W/m² K; thermal conductance of the roof, 0.8 W/m² K; thermal transmittance of the floor, 0.4 W/m² K; external surface heat transfer coefficient, 20 W/m² K; all internal surface heat transfer coefficients, 8 W/m² K; density of air, 1.2 kg/m³; specific heat of air, 1.02 kJ/kg K; ventilation rate of store, 0.5 air changes per hour.

(Wolverhampton Polytechnic)

(9.64 °C; 2.63 kW; 5.18 °C)

1.2 A building is maintained at 20 °C when it is $-1\ °C$ outside by a modulated low-pressure hot water heating system serving steel panel radiators. The flow and return temperatures to maintain these conditions are 78 °C and 68 °C respectively. The boiler temperature is constant at 82 °C.

(i) Calculate the proportions of boiler and by-pass water through the mixing valve when the outside temperature is (a) −2 °C, (b) 5 °C;

(ii) Sketch a typical piping arrangement for the system including the necessary controls for the mixing valve.

(Dublin Institute of Technology)

(5.62:1; 0.41:1)

1.3 A low-pressure hot water system heats a warehouse to 17 °C when it is −1 °C outside by using bare 75 mm outside diameter mild steel pipes installed at skirting level. The mean water temperature is 74 °C. A change in the product being stored requires that the space temperature be reduced to 14 °C, when −1 °C outside with the mean water temperature remaining constant. This is achieved by insulating the building to reduce the heat loss to 70% of its original value and by insulating some of the exposed pipework.

Calculate:

(i) the warehouse temperature after the building is insulated;

(ii) the percentage of piping that must be insulated to reduce the temperature to 14 °C.

Data

Heat loss from pipework is proportional to the temperature difference to the power 1.3.

Heat loss from bare 75 mm outside diameter pipe is 230 W/m at a temperature difference of 60 K.

Pipework insulation is 82% efficient.

(Dublin Institute of Technology)

(30.7 °C; 26.9%)

1.4 Explain the term 'equivalent length' as applied to a pipeline (not more than 50 words).

Water is pumped at the rate of 7 tonne per hour through a black steel pipe of bore 40 mm, absolute roughness 0.046 mm and equivalent length 45 m which rises by 10 m along its length. Determine:

(i) head loss due to friction;

(ii) pumping power required;

A Moody chart should be used.

(Dublin Institute of Technology)

(3.29 m; 0.254 kW)

1.5 (a) Calculate the pressure drop across AB which must be provided by the pump for the system shown in Fig. 1.35.

Data

Pipe diameter, 20 mm; total length of straight pipe, 14 m; vertical height, 4 m; supply temperature, 80 °C; return temperature, 70 °C; mass flow rate, 0.081 kg/s; material, heavy grade steel; density of water at 80 °C, 971.8 kg/m³; density of water at 70 °C, 977.5 kg/m³.

Pressure loss factors for fittings:

globe valve, 10; gate valve, 0.3; radiator, 2.5; bend, 0.4.

Figure 1.35 Problem 1.5(a)

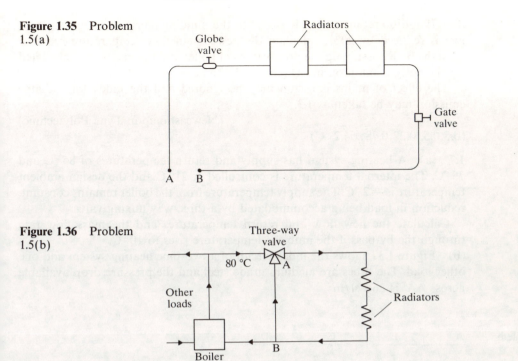

Figure 1.36 Problem 1.5(b)

From tables for heavy grade steel of 20 mm diameter:
pressure drop per unit length, 50 Pa/m; equivalent length, l_e, 0.6 m.
(b)   Figure 1.36 shows part of a low-pressure hot water heating system supplying two 3 kW radiators controlled by a flow-mixing valve at A. The boiler delivers water at a constant temperature of 80 °C and the design return temperature is 60 °C. The room design temperature is 21 °C and is kept constant. The outside design temperature is −2 °C. Each radiator is made up of sections which emit 100 W for a temperature difference of 55 K between the room and the radiator.
   Calculate:
(i)   the number of sections required for each radiator;
(ii) the supply and return temperatures to the radiators and the mass flow rate through AB when the outside temperature is 10 °C.
Take the specific heat of water as 4.2 kJ/kg K.

(Wolverhampton Polytechnic)

(971.3 Pa; 31, 40; 53.6 °C, 44 °C, 0.0524 kg/s)

1.6   A boiler plant operates in such a manner as to provide a constant-flow temperature of 82.2 °C throughout the heating season. A circulation is taken from the boiler headers, through a three-way mixing valve to feed radiators, sized to provide for a building internal temperature of 15.6 °C and −1.1 °C outside. The mixing valve is controlled by an external sensor to provide a circuit flow of 71.1 °C under design conditions, and to vary that temperature with rise in external temperature. The design temperature drop across the circuit is 11 K.
(a)   If the building heat requirement varies directly with changes in outside temperature, calculate the proportions of boiler and by-pass water which flow through the mixing valve for outside temperatures of −1.1 °C and 4.4 °C.

(b)   If additional insulation is added to the building fabric such that the heat loss is reduced by 20%, calculate the new design flow temperature required from the boiler assuming the proportions of boiler and by-pass water calculated for the design condition in (a) remain the same.

The effect of mains emission may be ignored and the index for radiator emission may be taken as 1.3.

(Newcastle upon Tyne Polytechnic)

(0.5:0.5, 0.22:0.78; 54.2 °C)

1.7 (a)   A heating system has supply and return temperatures of 85 °C and 75 °C. The internal temperature is controlled at 21 °C and the design ambient temperature is -2 °C. The supply temperature from the boiler remains constant, reduction in load being accommodated by a three-way mixing valve.

Calculate the new flow and return temperatures and the mass flow rate through the by-pass if the ambient temperature rises to 10 °C.

(b)   Figure 1.37 shows the index circuit of a two-pipe heating system and one other load. The tubes are medium grade steel and the pressure drop available across AA' is 26 kN/m².

Figure 1.37 Problem 1.7(b)

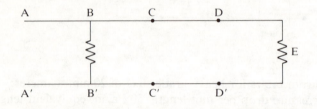

Select *preliminary* pipe sizes, using the table given as Table 1.1 (page 15).

Data

Pipe	AB	BC	CD	DED'	BB'
Mass flow rate/(kg/s)	1.00	0.80	0.69	0.40	0.20
Equivalent length/(m)	10	15	15	50	60

(Wolverhampton Polytechnic)

(40 mm, 40 mm, 32 mm, 32 mm, 20 mm)

1.8 (a)   Starting with the D'Arcy equation for the loss of fluid head in a pipe due to friction, show that the pressure drop per unit length is related to mass flow rate and diameter by the following equation:

$$\frac{\Delta p}{L} = (\text{a constant}) \times \frac{\dot{m}^2}{d^5}$$

(b)   Figure 1.38 shows part of a low-pressure hot water heating system supplying three 3 kW radiators in series. The boiler delivers water at a constant temperature of 80 °C and the design return temperature is 62 °C. Control is by a mixing valve. The room design temperature of 20 °C is maintained constant and the external design temperature is -2 °C. If each radiator is made up of

Figure 1.38 Problem 1.8(b)

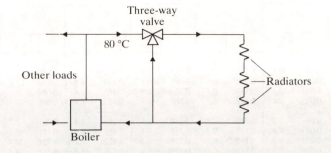

sections which emit 100 W for a temperature difference of 55 K between the room and the radiator, calculate:

(i) the number of sections required for each radiator;
(ii) the supply and return temperatures to the bank of radiators and the mass flow rate through the by-pass when the outside temperature is 8 °C.

(Wolverhampton Polytechnic)

(29, 33, 39; 56.9 °C, 47.1 °C, 0.084 kg/s)

1.9 Figure 1.39 shows a schematic lay-out of an LPHW heating circuit. Utilizing the data below, determine the following:

(i) the necessary flow rates required in each section of pipework making due allowance for pipework emission; the direct ratio method of proportioning mains loss may be used;
(ii) using these flow rates, size all pipework in the network on the basis of an approximate average pressure drop of 300 Pa/m; ignore critical sizing of branches;
(iii) determine the necessary pump duty required to provide for this circulation.

Use Table 1.3 overleaf for heavy grade steel.

Figure 1.39 Diagram for Problem 1.9

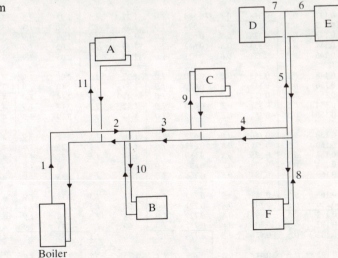

Data

Boiler flow temperature, 80 °C; boiler return temperature, 70 °C. Tube: heavy grade mild steel. Allow 30% of pipework length for fittings loss. Take specific heat of water as 4.2 kJ/kg K.

Table 1.3 Pressure drop due to friction in heavy grade steel pipes for water at 75 °C

Δp₁	v	10 mm M	10 mm lₑ	15 mm M	15 mm lₑ	20 mm M	20 mm lₑ	25 mm M	25 mm lₑ	32 mm M	32 mm lₑ	40 mm M	40 mm lₑ	50 mm M	50 mm lₑ	v	Δp₁
80.0		0.021	0.3	0.045	0.4	0.105	0.7	0.194	0.9	0.425	1.4	0.649	1.7	1.24	2.3		80.0
82.5		0.021	0.3	0.046	0.4	0.107	0.7	0.197	0.9	0.432	1.4	0.659	1.7	1.26	2.3		82.5
85.0		0.022	0.3	0.046	0.4	0.108	0.7	0.201	0.9	0.439	1.4	0.670	1.7	1.28	2.3		85.0
87.5		0.022	0.3	0.047	0.4	0.110	0.7	0.204	0.9	0.446	1.4	0.680	1.7	1.30	2.3		87.5
90.0		0.023	0.3	0.048	0.4	0.112	0.7	0.207	0.9	0.452	1.4	0.691	1.7	1.31	2.3		90.0
92.5		0.023	0.3	0.049	0.4	0.113	0.7	0.210	0.9	0.459	1.4	0.701	1.7	1.33	2.3		92.5
95.0		0.023	0.3	0.049	0.4	0.115	0.7	0.213	0.9	0.466	1.4	0.711	1.7	1.35	2.3		95.0
97.5		0.024	0.3	0.050	0.4	0.117	0.7	0.216	0.9	0.472	1.4	0.721	1.7	1.37	2.3		97.5
100		0.024	0.3	0.051	0.4	0.118	0.7	0.219	0.9	0.479	1.4	0.731	1.7	1.39	2.3		100
120		0.026	0.3	0.056	0.4	0.131	0.7	0.242	0.9	0.527	1.4	0.805	1.7	1.53	2.4		120
140	0.3	0.029	0.3	0.061	0.5	0.142	0.7	0.262	0.9	0.572	1.4	0.873	1.7	1.66	2.4		140
160		0.031	0.3	0.065	0.5	0.152	0.7	0.282	1.0	0.614	1.4	0.937	1.7	1.78	2.4		160
180		0.033	0.3	0.070	0.5	0.162	0.7	0.300	1.0	0.654	1.4	0.997	1.8	1.89	2.4		180
200		0.035	0.3	0.074	0.5	0.172	0.7	0.317	1.0	0.691	1.4	1.05	1.8	2.00	2.4	1.0	200
220		0.037	0.3	0.078	0.5	0.181	0.7	0.334	1.0	0.727	1.4	1.11	1.8	2.10	2.4		220
240		0.039	0.3	0.081	0.5	0.189	0.7	0.349	1.0	0.761	1.4	1.16	1.8	2.20	2.4		240
260		0.040	0.3	0.085	0.5	0.198	0.7	0.364	1.0	0.793	1.5	1.21	1.8	2.29	2.4		260
280		0.042	0.3	0.088	0.5	0.206	0.7	0.379	1.0	0.825	1.5	1.26	1.8	2.38	2.4		280
300		0.044	0.3	0.092	0.5	0.213	0.7	0.393	1.0	0.855	1.5	1.30	1.8	2.47	2.5		300
320		0.045	0.3	0.095	0.5	0.221	0.7	0.407	1.0	0.884	1.5	1.35	1.8	2.55	2.5		320
340		0.047	0.3	0.098	0.5	0.228	0.7	0.420	1.0	0.913	1.5	1.39	1.8	2.64	2.5		340
360		0.048	0.3	0.101	0.5	0.235	0.7	0.433	1.0	0.941	1.5	1.43	1.8	2.71	2.5		360
380	0.5	0.049	0.3	0.104	0.5	0.242	0.7	0.445	1.0	0.970	1.5	1.47	1.8	2.79	2.5		380
400		0.051	0.3	0.107	0.5	0.248	0.7	0.457	1.0	0.994	1.5	1.51	1.8	2.87	2.5		400
420		0.052	0.3	0.110	0.5	0.255	0.7	0.469	1.0	1.02	1.5	1.55	1.8	2.94	2.5	1.5	420
440		0.054	0.3	0.113	0.5	0.261	0.7	0.481	1.0	1.04	1.5	1.59	1.8	3.01	2.5	1.5	440
460		0.055	0.3	0.115	0.5	0.267	0.7	0.492	1.0	1.07	1.5	1.63	1.8	3.08	2.5		460
480		0.056	0.3	0.118	0.5	0.273	0.8	0.503	1.0	1.09	1.5	1.66	1.8	3.15	2.5		480
500		0.057	0.3	0.120	0.5	0.279	0.8	0.514	1.0	1.12	1.5	1.69	1.8	3.22	2.5		500
520		0.059	0.3	0.123	0.5	0.285	0.8	0.524	1.0	1.14	1.5	1.73	1.8	3.28	2.5		520
540		0.060	0.3	0.125	0.5	0.291	0.8	0.535	1.0	1.16	1.5	1.77	1.8	3.35	2.5		540
560		0.061	0.3	0.128	0.5	0.296	0.8	0.545	1.0	1.17	1.5	1.80	1.8	3.41	2.5		560
580		0.062	0.3	0.130	0.5	0.302	0.8	0.555	1.0	1.21	1.5	1.83	1.8	3.47	2.5		580
600		0.063	0.3	0.133	0.5	0.307	0.8	0.565	1.0	1.23	1.5	1.87	1.8	3.53	2.5		600
620		0.064	0.3	0.135	0.5	0.312	0.8	0.575	1.0	1.25	1.5	1.90	1.8	3.59	2.5		620
640		0.065	0.3	0.137	0.5	0.318	0.8	0.584	1.0	1.27	1.5	1.93	1.8	3.65	2.5		640
660		0.066	0.3	0.139	0.5	0.323	0.8	0.594	1.0	1.29	1.5	1.96	1.8	3.71	2.5		660
680		0.067	0.3	0.142	0.5	0.328	0.8	0.603	1.0	1.31	1.5	1.99	1.9	3.77	2.5		680
700		0.069	0.3	0.144	0.5	0.333	0.8	0.612	1.0	1.33	1.5	2.02	1.9	3.83	2.5		700
720		0.070	0.3	0.146	0.5	0.338	0.8	0.621	1.0	1.35	1.5	2.05	1.9	3.88	2.5		720
740		0.071	0.3	0.148	0.5	0.343	0.8	0.630	1.0	1.37	1.5	2.08	1.9	3.94	2.5		740
760		0.072	0.3	0.150	0.5	0.347	0.8	0.639	1.0	1.39	1.5	2.10	1.9	3.99	2.5	2.0	760
780		0.073	0.3	0.152	0.5	0.352	0.8	0.648	1.0	1.41	1.5	2.14	1.9	4.04	2.5		780
800		0.074	0.3	0.154	0.5	0.357	0.8	0.656	1.0	1.42	1.5	2.17	1.9	4.10	2.5		800
820		0.075	0.4	0.156	0.5	0.362	0.8	0.665	1.0	1.44	1.5	2.19	1.9	4.15	2.5		820
840		0.075	0.4	0.158	0.5	0.366	0.8	0.673	1.0	1.46	1.5	2.22	1.9	4.20	2.5		840
860		0.076	0.4	0.160	0.5	0.371	0.8	0.681	1.0	1.48	1.5	2.25	1.9	4.25	2.5		860
880		0.077	0.4	0.162	0.5	0.375	0.8	0.689	1.0	1.50	1.5	2.27	1.9	4.30	2.5		880
900		0.078	0.4	0.164	0.5	0.379	0.8	0.698	1.0	1.51	1.5	2.30	1.9	4.35	2.5		900
920		0.079	0.4	0.166	0.5	0.384	0.8	0.706	1.0	1.53	1.5	2.33	1.9	4.40	2.5		920
940		0.080	0.4	0.168	0.5	0.388	0.8	0.713	1.0	1.55	1.5	2.35	1.9	4.45	2.5		940
960		0.081	0.4	0.170	0.5	0.392	0.8	0.721	1.0	1.56	1.5	2.38	1.9	4.50	2.5		960
980		0.082	0.4	0.172	0.5	0.397	0.8	0.729	1.0	1.58	1.5	2.40	1.9	4.55	2.5		980
1 000		0.083	0.4	0.173	0.5	0.401	0.8	0.737	1.0	1.60	1.5	2.43	1.9	4.59	2.5		1 000
1 100		0.087	0.4	0.182	0.5	0.421	0.8	0.774	1.1	1.68	1.5	2.55	1.9	4.82	2.6		1 100
1 200		0.091	0.4	0.191	0.5	0.441	0.8	0.809	1.1	1.75	1.5	2.67	1.9	5.04	2.6		1 200
1 300	1.0	0.095	0.4	0.199	0.5	0.459	0.8	0.844	1.1	1.83	1.5	2.78	1.9	5.25	2.6		1 300
1 400		0.099	0.4	0.207	0.5	0.477	0.8	0.876	1.1	1.90	1.5	2.89	1.9	5.46	2.6		1 400
1 500		0.102	0.4	0.214	0.5	0.495	0.8	0.908	1.1	1.98	1.5	2.99	1.9	5.65	2.6		1 500
1 600		0.106	0.4	0.222	0.5	0.511	0.8	0.939	1.1	2.03	1.5	3.09	1.9	5.84	2.6		1 600
1 700		0.109	0.4	0.229	0.5	0.528	0.8	0.968	1.1	2.10	1.5	3.19	1.9	6.02	2.6	3.0	1 700
1 800		0.113	0.4	0.236	0.5	0.543	0.8	0.997	1.1	2.16	1.6	3.28	1.9				1 800
1 900		0.116	0.4	0.242	0.5	0.559	0.8	1.03	1.1	2.22	1.6	3.37	1.9				1 900
2 000		0.119	0.4	0.249	0.5	0.574	0.8	1.05	1.1	2.28	1.6	3.46	1.9				2 000

Pipe section	1	2	3	4	5	6	7	8	9	10	11
Length (flow and return)/(m)	10	20	20	30	25	10	5	20	6	12	20
Emission/(kW)	1.0	2.0	2.0	1.5	1.0	0.5	0.1	1.4	0.2	0.7	1.0

Load	A	B	C	D	E	F
Output/(kW)	15	6	8	6	17	15
Resistance/(kPa)	1.1	0.6	0.7	1.0	0.5	0.4

(Newcastle Polytechnic)

(1.84 kg/s, 1.46 kg/s, 1.28 kg/s, 1.08 kg/s, 0.57 kg/s, 0.42 kg/s, 0.15 kg/s, 0.40 kg/s, 0.20 kg/s, 0.18 kg/s, 0.39 kg/s; 50 mm, 50 mm, 40 mm, 40 mm, 32 mm, 32 mm, 20 mm, 32 mm, 20 mm, 20 mm, 25 mm; 0.249 bar)

1.10 The barometric pressure in a room is 1.022 bar and the average air temperature in the room is 20 °C. It is noted that when the temperature of the inside of the window surface is 5 °C, condensation just begins on the inside surface of the glass. Using property tables only (such as those of Ref. 1.11), calculate:

(i)   the relative humidity of the air in the room;
(ii) the specific humidity;
(iii) the percentage saturation;
(iv) the specific volume of the air per unit mass of dry air;
(v)   the specific enthalpy of the air per unit mass of dry air.

Take the specific heat of dry air as 1.005 kJ/kg K, and the specific heat of superheated steam at low pressure as 1.880 kJ/kg K.

(37.30%; 0.005 35; 36.75%; 0.830 m³/s; 33.68 kJ/kg)

1.11 A building is maintained at 21 °C throughout the year by an air conditioning plant consisting of a mixing chamber, a cooling coil, a fan and an afterheater.

For the summer design case minimum fresh air is used which results in 3.87 *volumes* of recirculated air being used for each *volume* of fresh air. The afterheater is switched off, the cooling coil contact factor is 0.8 and the room percentage saturation is 55%.

For the winter design case only the fan and afterheater are used, the percentage saturation being controlled by varying the recirculation ratio.

Using the psychrometric chart, calculate:

(i)   the temperature rise across the fan;
(ii) the cooling coil load;
(iii) the lowest external temperature (saturated) which will allow 45% saturation to be maintained in the room in winter, assuming the same total mass flow rate and minimum ventilation by mass as for the summer design;
(iv) the afterheater load for part (iii).

Data

Summer design: 28 °C dry bulb, 60% saturation.
Winter design: −2 °C saturated.
Summer room loads: 80 kW sensible gain, 20 kW latent gain.
Winter room loads: 50 kW sensible loss, 20 kW latent gain.
Summer supply temperature, 13.5 °C.
Take the specific heat of humid air per unit mass of dry air as 1.02 kJ/kg K,

and the specific enthalpy of saturated vapour added to the room as 2555.7 kJ/ kg K. Assume that the temperature rise across the fan is the same in winter as in summer.

(Wolverhampton Polytechnic)

(1.4 K; 161.5 kW; −1.5 °C; 81.6 kW)

1.12   Two air streams A and B mix at a pressure of 101.3 kPa in the mixing chamber of an air conditioning plant. The mixed air then passes through a heater battery. Calculate using only the data below:

(i)   the mass ratio in which the air streams from the ducts A and B are mixed;
(ii)   the precise dry bulb temperature of the mixed air;
(iii) the heat input required to the heater battery.

Also, assuming the fan is situated directly after the heater battery, describe qualitatively the effect on the mass flows of air through A and B, of the heater battery being switched 'on' and 'off'.

Data

Airstream A: 23 °C dry bulb, specific humidity 0.0110.
Airstream B: −2 °C dry bulb, specific humidity 0.0025.
Rate of flow of mixed air, 31 kg/s dry air; specific humidity of mixed air, 0.0072; specific heat of dry air, 1.001 kJ/kg K; specific heat of water vapour, 1.884 kJ/kg K; temperature rise through heater battery, 10 K.

(University of Liverpool)

(1.237:1; 11.92 °C; 314.5 kW)

1.13   An air-conditioned space is maintained at 22 °C dry bulb and 50% saturation by a plant supplying air at 15 °C when the outside air condition is 27 °C dry bulb and 21 °C wet bulb. The plant consists of mixing box, cooler battery, reheat battery and supply fan, and the air temperature leaving the cooler is 11.5 °C dry bulb.

Using the psychrometric chart, calculate the thermal loads at the batteries.

Describe two air (cooler) by-pass arrangements that would eliminate the reheat load. Calculate the load at the cooler battery in each case.

Data

Sensible heat gain, 50 kW; latent heat gain, 7.5 kW; 40% fresh air (by mass); cooling coil contact factor, 0.93 (constant); temperature rise through supply fan, 2 K; specific heat of humid air per unit mass of dry air, 1.02 kJ/kg K.

(Dublin Institute of Technology)

(129.5 kW, 10.7 kW; 119 kW)

1.14   A room is to be maintained at 22 °C dry bulb, 50% saturation. The sensible heat loss and latent heat gain are 60 kW and 12 kW respectively. The ventilation system, which handles 4.65 m³/s of outside air at an initial state of −1 °C, 100% saturation, consists of a preheater and air washer for humidification followed by a reheater. The air washer is 80% efficient.

Assuming adiabatic saturation for the air washer, determine:

(i)   the state of the air leaving the preheater;
(ii)   the state of the air leaving the air washer;
(iii) the preheater load;
(iv)   the reheater load;
(v)   the rate of make-up water to the air washer.

Take the specific heat at constant pressure of humid air per unit mass of dry air as 1.02 kJ/kg K.

(Dublin Institute of Technology)

(24 °C, 18% saturation; 14.3 °C, 74% saturation; 150.0 kW; 107.4 kW; 0.025 kg/s)

1.15 A ventilation system requires 1.750 m³/s of air at a fan static pressure of 250 N/m². The fan rotating at 10 rev/s has the following characteristics:

Volume flow/(m³/s)	1.0	1.5	2.0	2.5	3.0
Pressure/(N/m²)	400	415	368	275	87

Find:
(i)   the actual air delivered under these conditions;
(ii) the fan speed required to deliver 1.750 m³/s without using a damper;
(iii) the excess power to be absorbed if a damper is needed to maintain the flow of 1.750 m³/s.

(Newcastle upon Tyne Polytechnic)

(2.1 m³/s; 3.9 rev/s; 263 W)

1.16 (a)   Describe and discuss the possible unstable operation of fans running in parallel. What steps can a building services engineer take to avoid the problem?
(b)   The characteristics of a fan running at 20 rev/s are shown below. It is proposed to use one or more of these fans to deliver an air flow of 6.1 m³/s against a resistance of 500 Pa. The options available are (a) run one fan at increased speed, (b) run two fans in series. Determine which option would lead to the lowest running costs.

Volume flow rate/(m³/s)	3.0	4.0	5.0	6.0	7.0	8.0	9.0
Pressure/(N/m²)	492	480	445	385	300	180	45
Power consumption/(kW)	3.710	3.718	3.718	3.713	3.698	3.680	3.660

(Dublin Institute of Technology)

(4.88 kW; 5.23 kW)

1.17 A test on a fan running at 1000 rev/min gave the characteristics shown overleaf. The fan has a two-speed motor so that it can also run at 1500 rev/min.
   The fan is used for air flow in a duct which has a resistance of 38.1 mm of water at a flow rate of 1.42 m³/s. For some processes a filter is used which has a resistance of 12.7 mm of water at a flow rate of 1.42 m³/s and in this case the fan is run at the higher speed.
   Calculate:
(i)   the volume rate of air delivered when the filter is fitted;
(ii) the power required under these conditions;
(iii) the fan efficiency at the operating point;
(iv) the resistance required to be put in series with the system to reduce the flow rate by 0.47 m³/s.

Volume flow rate/(m³/s)	0.5	1.0	1.5	2.0	2.5	3.0
Pressure/(mm water)	46	52	53	48	37	19
Power consumption/(kW)	0.80	1.10	1.40	1.70	2.05	2.60

(2.18 m³/s; 4.65 kW; 55.1%; 45 mm water)

1.18 A fan running at 12 rev/s has the characteristics shown below. The fan has been selected for a system requiring 1.5 m³/s at a total pressure of 240 N/m². Determine:

(i)   the actual volume flow rate that would be obtained when connected to the system;

(ii) the power lost in a damper required to give the design volume flow rate;

(iii) the speed at which the fan should run without the damper to give the design volume flow rate;

(iv) the power used at the new speed compared with the power used when the damper is installed.

Volume flow rate/(m³/s)	0.5	1.0	1.5	2.0	2.5	3.0
Pressure/(N/m²)	397	373	330	270	182	68
Power/(kW)	0.57	0.61	0.65	0.71	0.77	0.84

(Dublin Institute of Technology)

(1.7 m³/s; 135 W; 10.6 rev/s; 0.46 kW, 0.65 kW)

1.19 An air conditioning system with interlinked fresh air, exhaust and re-circulation dampers is shown in Fig. 1.40. The pressure loss through the air handling unit and supply ductwork, AB, is 96 Pa. The extract ductwork from the room, BC, has a pressure loss of 64 Pa and the duct sections AC, CD, and DA each have a pressure loss of 32 Pa, including a fully open damper in each case. All pressure losses were calculated using the air flow rate of 0.8 m³/s.

Figure 1.40 Problem 1.19

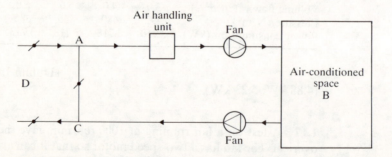

Assuming that both fans have the characteristics given below and that the room is air-tight, determine:

(i) the supply air volume when damper AC is closed and AD and DC are open;

(ii) the supply air volume with damper AC open and AD and DC closed.

Volume flow rate/(m³/s)	0	0.4	0.8	1.2	1.6	2.0	2.4
Pressure/(Pa)	140	140	135	125	108	78	65

(Dublin Institute of Technology)

(0.88 m³/s; 0.94 m³/s)

1.20 (a) Briefly explain the potential hazards presented by mechanical ventilation systems in a fire situation and how these may be overcome. Illustrate your answer with sketches.

(b) An escape staircase is to be provided with a pressurization system such that the pressure is at least 25 Pa above atmospheric pressure. The design leakage from the stairs is estimated to be 1500 litre/s at 25 Pa.

The centrifugal fan to be used in the supply system has the characteristics shown below when running at 1000 rev/min. The estimated pressure drop of the rest of the supply system is 200 Pa when passing 1600 litre/s. At what speed must the fan be run in order to provide the specified pressure?

Total pressure/(Pa)	250	234	212	183	145
Volume handled/(litre/s)	800	1000	1200	1400	1600

(University of Liverpool)

(1060 rev/min)

1.21 (a) Discuss the principles of the static regain method of duct sizing and give examples of where its use may be appropriate.

(b) An air-tight room has air supplied through a ductwork system having a pressure of 300 Pa at a flow of 3.8 m³/s, whilst air is extracted through a system having a pressure loss of 180 Pa at a flow of 3.8 m³/s. The supply and extract fans have the characteristics shown below.

What will be the pressure in the room and the supply and extract volume flows?

Volume flow rate/(m³/s)	2.83	3.30	3.78	4.25	4.72
Supply fan total pressure/(Pa)	430	410	380	340	290
Extract fan total pressure/(Pa)	390	350	300	245	170

(University of Liverpool)

(−27 Pa; 4.22 m³/s)

1.22 Figure 1.41 illustrates a centrifugal fan having the following characteristics when running at 15 rev/s.

Volume/(m³/s)	0.2	0.4	0.6	0.8	1.0
Fan total pressure/(Pa)	230	210	170	110	39
Fan total efficiency/(%)	69	81	83	70	39

Figure 1.41 Problem 1.22

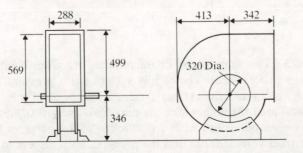

All dimensions in mm

(a) Plot the above characteristics. Determine the following characteristics and plot them on the same graph:

(i) fan power against volume flow rate;

(ii) fan static pressure against volume flow rate;

(iii) fan static efficiency against volume flow rate.

(b) From the characteristics in part (a), identify the type of impeller, explaining how this identification can be made, and by extrapolation where appropriate estimate the maximum volume the fan could be made to deliver, explaining under what circumstances this will occur.

(CIBSE)

$(1.04 \text{ m}^3/\text{s})$

1.23 (a) Explain with the aid of diagrams the principle of the static regain method of duct sizing, and discuss its implication to air conditioning system design.

(b) Air flows through a duct expander such that its velocity reduces from 15 m/s to 5 m/s.

(i) Plot total, velocity and static pressure graphically to show how these change as the air passes through the expander;

(ii) determine the amount of static regain occurring in this fitting.

Data

Density of air, 1.2 kg/s; pressure loss factor for expander with respect to velocity pressure difference across the fitting, 0.45; upstream total pressure, 125 Pa.

(University of Liverpool)

(55%)

1.24 (a) Explain the principles of the velocity method, constant pressure drop method, and static regain method of air duct sizing. Summarize their advantages and disadvantages.

(b) Calculate the diameters of sections AB, BC and CD of the duct system shown in Fig. 1.42. Assume a velocity of 6 m/s in section AB, and size sections BC and CD using the static regain method. Give your answers to the nearest millimetre. All the ducts are circular. The pressure loss factor for the main at each junction may be taken as 0.1, applied to the velocity pressure in the main downstream of the branch.

(University of Manchester)

(505 mm, 733 mm, 921 mm)

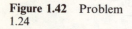 **Figure 1.42** Problem 1.24

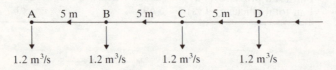

1.25 Two fans, whose characteristics are given below, are used to supply air to a process area. The supply system has a pressure loss of 425 Pa for a flow of 3 m^3/s and the extract system a loss of 350 Pa for a flow of 2.5 m^3/s. The total area through which air may leak away is 0.085 m^2 and can be regarded as a simple orifice with a coefficient of discharge of 0.6. Find the supply and extract volume flows and the pressure, with respect to atmospheric pressure, in the process area.

Supply fan:

Volume flow/(m³/s)	0	1	2	3	4
Fan total pressure/(Pa)	750	755	730	590	275

Extract fan:

Volume flow/(m³/s)	0	0.8	1.6	2.4	3.2
Fan total pressure/(Pa)	600	603	584	472	220

(Polytechnic of the South Bank)

(3.20 m³/s, 2.74 m³/s, 47 Pa)

1.26 (a) Show that for ductwork in parallel, the resistance constant, k, is given by:

$$\frac{1}{\sqrt{k}} = \frac{1}{\sqrt{k_1}} + \frac{1}{\sqrt{k_2}} + \frac{1}{\sqrt{k_3}} + \ldots$$

where $k_1, k_2, k_3 \ldots$ are the resistance constants for the individual ductworks.
(b) A fan supplies air to a room through a length of supply ducting whose resistance results in a total pressure loss of 150 N/m² for a volume flow rate of 4 m³/s. Another fan extracts air from the room through a length of extract ducting which gives a total pressure loss of 50 N/m² for a volume flow rate of 2 m³/s. The room has a leakage path equivalent to a total pressure loss of 25 N/m² for a leakage rate of 0.4 m³/s. The supply and extract fans have characteristic curves given in the tables below. Find the pressure in the room and the volume flow rate in the supply and extract ducting.

Supply fan:

Volume flow/(m³/s)	1	2	3	4	5
Total pressure/(N/m²)	430	435	400	310	160

Extract fan:

Volume flow/(m³/s)	1	2	3	4	5
Total pressure/(N/m²)	340	300	220	130	30

(Newcastle upon Tyne Polytechnic)

(55 N/m²; 4.5 m³/s, 3.9 m³/s)

1.27 (a) Show that the pressure loss in a duct due to frictional resistance to air flow varies directly as the air flow squared. Hence establish that for ductwork in series the resistance constant, k, is given by:

$$k = k_1 + k_2 + k_3 + \ldots$$

where $k_1, k_2, k_3 \ldots$ are the resistance constants for the individual ductworks.
(b) A large process workshop is to be ventilated by a supply system in which the pressure loss is 300 N/m² for a flow of 3 m³/s and an extract system for which the pressure loss is 400 N/m² for a flow of 2.5 m³/s. If there is a loss of air from the space by exfiltration through an aperture, having a pressure loss of 125 N/m² for a flow of 0.5 m³/s, find the pressure in the space, the supply

volume flow and the extract volume flow. Both fans have the following characteristics:

Volume flow/(m³/s)	0	1	2	3	4
Total pressure/(N/m²)	750	755	730	590	275

(Newcastle upon Tyne Polytechnic)

$(70 \text{ N/m}^2; 3.50 \text{ m}^3/\text{s}, 3.15 \text{ m}^3/\text{s})$

1.28 (a) Using the method of Dimensional Analysis, derive the fan similarity laws.

(b) A fan having the characteristics given below delivers air through a supply system for which the pressure loss is 300 N/m² for an air flow of 2.8 m³/s. In parallel with the fan and system is another fan, with the same characteristics, and a system for which the pressure loss is 424 N/m² for an air flow of 2.36 m³/s. Completing a closed circuit back to the fans is a common duct system for which the pressure loss is 300 N/m² for an air flow of 5.2 m³/s.

Determine the operating points of the two fans and the volume flowing through the common branch.

Fan characteristics:

Volume flow/(m³/s)	0	1.0	1.9	2.8	3.8
Total pressure/(N/m²)	748	752	725	580	260

(Newcastle upon Tyne Polytechnic)

$(5.1 \text{ m}^3/\text{s}; \text{flow 1, } 2.8 \text{ m}^3/\text{s}; \text{flow 2, } 2.3 \text{ m}^3/\text{s})$

1.29 A building in a city centre has a ventilation area totalling 1.25 m² at a height of 40 m on the windward face and a ventilation area totalling 0.7 m² at the base on the leeward side.

(a) Exposure to wind effects only:

The wind velocity on a particular day may be assumed to follow the law:

$$u = 1.35z^{0.33} \text{ m/s}$$

(where u is the wind velocity at a height z m). The coefficients of pressure are +0.82 at the windward ventilation opening, and −0.38 at the leeward ventilation opening, both based on the upstream velocity at a height of 40 m.

Calculate:

(i) the overall static pressure difference along the flow path through the ventilation openings;

(ii) the ventilation flow rate along this path.

(b) Stack effect (wind effect ignored):

The same building has an internal temperature of 16 °C and the ambient air temperature is 24 °C.

Calculate:

(i) the effective pressure difference for ventilation;

(ii) the ventilation flow rate.

(c) Combination of wind effect and stack effect:

The same building is now subjected simultaneously to the two effects described in (a) and (b) above.

Calculate:
(i) the combined pressure difference along the flow path;
(ii) the ventilation flow rate.

(Wolverhampton Polytechnic)

(14.98 N/m^2; 1.86 m^3/s; 12.91 N/m^2; 1.73 m^3/s; 27.89 N/m^2; 2.54 m^3/s)

1.30 Re-calculate Problem 1.29 for the case when an additional ventilation area of 1.25 m^2 is added at the height of 40 m on the leeward side, and an additional ventilation area of 0.7 m^2 is added at the base of the building on the windward side.
(14.98 N/m^2; 4.20 m^3/s; 12.91 N/m^2; 3.46 m^3/s; 4.20 m^3/s)

1.31 (a) Sketch a typical set of streamlines and pressure coefficient distribution lines (with approximate values), for flow over:
(i) a conventional pitched roof two-storey house, (elevation);
(ii) a rectangular elevation, i.e. flat roof, building (elevation);
(iii) through a rectangular gap between the end walls of two adjacent co-linear rows of buildings (plan view).
(b) A square-shaped plan building containing a single room has a steady wind blowing normal to side A on which the mean value of the pressure coefficient is +0.8. The mean value of the pressure coefficient on the opposite, i.e. downstream, wall D is −0.2. The values on the other two, parallel, walls B and C are equal at −0.3. Assume that the leakage flow paths through each of the walls are all equal in total area and each has the same coefficient.
Calculate the value of the coefficient of pressure inside the building.
(c) Sketch the variation of height h (on the y-axis) against pressure p (on the x-axis), for a tall, vertical, rectangular plan building, showing:
(i) the effect of a steady wind flowing perpendicular to one face A of the building on the values of p from the windward A and leeward B faces of the building;
(ii) the effect of a temperature difference between the air in the building, t_B, and ambient air, t_o, when $t_B > t_o$;
(iii) the combined effect of (i) and (ii).
The direction of the resultant leakage air flow should be shown. Indicate the neutral pressure level on sketch (iii).

(Wolverhampton Polytechnic)

(−0.154)

1.32 (a) A heat treatment room in a factory liberates a contaminant into a room 16 m × 8 m × 8 m high. Existing ventilation is purely natural, consisting of a roof-top ventilator 2 m × 1 m and two low-level inlets of 0.5 m^2 each. If the mean temperature of the room is 30 °C when the outside temperature is 7 °C and the process is in operation, determine the air change rate which would occur in the absence of wind pressure effects.
(b) It is estimated that the release of contaminant to the space is at a rate of 0.05 m^3/s. If it is necessary to limit this concentration to 1%, determine the ventilation rate necessary to meet these conditions.
(c) Calculate the necessary area of opening at high and low level required to provide for this ventilation rate, assuming the high- and low-level openings will be of equal area and the inside and outside conditions remain as above.

(Newcastle upon Tyne Polytechnic)

(6.79 h^{-1}; 4.95 m^3/s; 1.622 m^2)

1.33 (a) Discuss the factors which you would consider before deciding whether or not to use dilution ventilation to control an environmental hazard.

(b) Dilution ventilation is to be considered for a process in which cyclohexanone is to be released at a rate of 5 litre/h as vapour in a space of volume 280 m^3.

(i) Find the number of air changes per hour required to limit the concentration of cyclohexanone to 0.25 of the Threshold Limiting Value (TLV).

(ii) If after a period of normal running the ventilation is stopped, find the time taken for the concentration to reach the TLV.

Take the value of the TLV for cyclohexanone as 50 parts per million.

Use may be made of the equation:

$$C = \frac{(C_i \dot{V} + \dot{V_c})}{(\dot{V} + \dot{V_c})} + \left\{ C_o - \frac{(C_i \dot{V} + \dot{V_c})}{(\dot{V} + \dot{V_c})} \right\} \exp[-(\dot{V} + \dot{V_c})\tau/V]$$

where the symbols have the meanings as in this chapter.

(CIBSE)

(1.43 h^{-1}; 2.1 h)

1.34 A large paint spraying enclosure with a volume of 17 150 m^3 is constructed within a well-ventilated factory building by tying plastic sheets to a scaffolding framework. Several large doorways are provided for entry and exit purposes, covered by plastic sheets fixed at the sides and top.

Warmed and dried air is supplied at a rate of 2.7 m^3/s. An extract fan removes air at 3 m^3/s.

Paint is sprayed at 60 litre/h and the following information is supplied by the manufacturer.

Paint density, 1.2 kg/litre; solvent content, 30% by mass; drying time, assumed instantaneous; threshold limit value for solvent, 582 mg/m^3; lower explosive limit for solvent in air, 1.8% by volume; specific volume of solvent vapour, 0.4 m^3/kg.

The ventilation equations [1] and [2] describe the build-up and decay respectively of the concentration of a gaseous contaminant released within a ventilated enclosure.

$$C = \frac{\dot{V_c}}{\dot{V}} (1 - \exp[-\dot{V}\tau/V]) \tag{1}$$

$$C = C_o \exp[-\dot{V}\tau/V] \tag{2}$$

where the symbols have the same meanings as in this chapter.

(a) Derive equation [2].

(b) To provide a margin of safety it is required that the concentration of the solvent in the air should not exceed 10% of the lower explosive limit. Should this concentration or the threshold limit value be adopted as the design requirement?

(c) Painting starts at 0900 h and continues till 1200 h when there is a break till 1300 h when painting starts and continues till 1600 h. Plot a graph of the theoretical concentration of contamination over this period.

(d) List the assumptions inherent in the calculations and indicate how the actual concentrations are likely to differ from the theoretical concentrations because of any lack of realism in these assumptions. If the paint does not dry instantaneously but takes 3–4 h, explain how this would affect the solution.

(The Engineering Council)

(TLV = 0.000 23; LEL = 0.0018; 0–0.000 679; 0.000 679–0.000 362; 0.000 362–0.000 734)

REFERENCES

1.1 Eastop T D and McConkey A 1986 *Applied Thermodynamics for Engineering Technologists* 4th edn Longman

1.2 Douglas J F, Gasiorek J M and Swaffield J A 1985 *Fluid Mechanics* 2nd edn Longman

1.3 CIBSE 1986 *Guide to Current Practice* volume A

1.4 Welty J R 1984 *Fundamentals of Momentum, Heat and Mass Transfer* 3rd edn John Wiley

1.5 CIBSE 1986 *Guide to Current Practice* volume C

1.6 BS 3528 1971 *Specification for Convection Type Space Heaters Operating with Steam or Hot Water* BSI London

1.7 BS 4856 1983 *Methods for Testing and Rating Fan Coil Units*: *Unit Heaters and Coolers* BSI London

1.8 Colebrook C F and White C M 1939 Turbulent flow in pipes, with particular reference to the transition region between the smooth and rough pipe laws. *Proc. ICE, II*, 133

1.9 Threlkeld J L 1970 *Thermal Environmental Engineering* Prentice Hall

1.10 Ruskin R E (ed) 1965 *Humidity and Moisture*: *Methods of Measuring Humidity in Gases* Volume 1 Reinhold Publishing Co

1.11 Rogers G F C and Mayhew Y R 1988 *Thermodynamic and Transport Properties of Fluids* 4th edn Basil Blackwell

1.12 McIntyre D A 1980 *Indoor Climate* Applied Science Publishers

1.13 Croome D J and Roberts B M 1981 *Air Conditioning and Ventilating of Buildings* 2nd edn volume 1 Pergamon Press

1.14 BS 5925 1980 *Code of Practice for Design of Buildings*; *Ventilation Principles and Designing for Natural Ventilation* BSI London

1.15 Martin H E 1989 Ventilation for comfort *Building Services* **11** (3)

1.16 CIBSE 1986 *Guide to Current Practice* volumes B2 and B3

2 PRESSURIZED HEATING SYSTEMS

Low-pressure hot water heating systems are covered in Chapter 1. This chapter deals with the more complex systems which use pressurized water or steam. Both systems are frequently employed to provide process as well as space heating loads.

2.1 PRESSURIZED WATER HEATING SYSTEMS

Table 2.1 below is reproduced from Table B1.12 in Ref. 2.1. It classifies different types of water heating systems according to the design flow temperature. Warm and low-temperature hot water (LTHW) heating systems operate with flow temperatures below the saturation temperature of water at atmospheric pressure which is about 100 °C. For a system to operate above 100 °C the pressure must be above atmospheric to prevent boiling.

Table 2.1 Design water temperatures for warm and hot water heating systems

Category	System design water temperatures/(°C)
Warm	40–70
LTHW	70–100
MTHW	100–120
HTHW	Over 120

Note: Account must be taken of the margin necessary between the maximum system operating temperature and saturation temperature at the system operating pressure.

The medium-temperature hot water (MTHW) and the high-temperature hot water (HTHW) systems must be pressurized. Pressure is often used instead of temperature to classify systems so that an HTHW system is also termed a high-pressure hot water system (HPHW). Some of the methods of pressurizing systems are as follows.

Static Head (Gravity) Pressurization

Figure 2.1 shows a system pressurized by means of a static head. The expansion and contraction of the water in the system is accommodated in a high-level expansion and feed tank which is at atmospheric pressure. To increase a static pressure by 1 bar requires a head of about 10 m, which limits the number of sites where such a system could be used and makes it more likely to be used in an MTHW system than in an HTHW system.

Figure 2.1   Static head pressurization

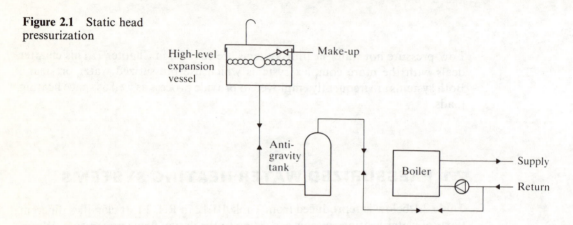

It should be noted that the only difference between this system and many LTHW (LPHW) systems is the introduction of an anti-gravity tank. This prevents hot water from entering the vertical pipe leading to the expansion vessel where the drop of pressure would cause the water to boil. As the system expands during start-up, hot water enters the top of the tank whilst the cooler water from the bottom of the tank is forced into the expansion vessel. The anti-gravity tank and the pipework between the boiler return line and the expansion vessel are left uninsulated to cool the hot water leaving the return line as quickly as possible. One disadvantage of this arrangement is that the water in contact with the atmosphere will absorb oxygen, which can cause corrosion problems when it enters the system. The amount of oxygen absorbed can be reduced by minimizing the surface area of water using a ball blanket (illustrated in Fig. 2.1) or float. It could be eliminated completely by sealing the expansion vessel and filling the top with an inert gas just above atmospheric pressure, but this considerably increases the capital cost.

Sealed Expansion Vessel Pressurization

Figure 2.2 shows a sealed expansion vessel pressurization system. There are several variations which employ the same principle. The simplest system uses a sealed vessel which is initially full of air; as the water warms up it expands compressing the air. More modern systems keep the air and water apart and use a pressurization pump. The one shown in Fig. 2.2 has a diaphragm which is only suitable for very small systems; an alternative for larger systems is a

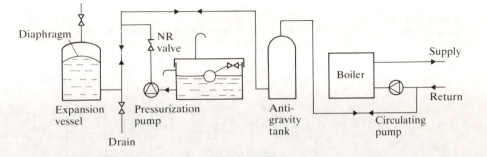

Figure 2.2 Sealed system with pump pressurization

water bag. This consists of a large flexible bag which is connected to the water side on the inside and has the air outside.

With the full expansion accommodated within a pressure vessel, the size of the pressure vessel limits the size of system for which sealed expansion vessels can be used. This arrangement can also be used for LPHW systems where there is no convenient high-level point to install a conventional open expansion tank. The anti-gravity tank in this system is to protect the rubber bags or diaphragms and may be dispensed with for LPHW systems.

Constant-pressure Air Pressurization

An alternative to using a sealed expansion vessel is to maintain a constant pressure using compressed air as shown in Fig. 2.3. As the system expands and the water level rises, the relief valve opens. When the system contracts and the water level falls, air is taken from the receiver to maintain the pressure. To minimize oxygen absorption an alternative to using a float is to heat the water to just below boiling point.

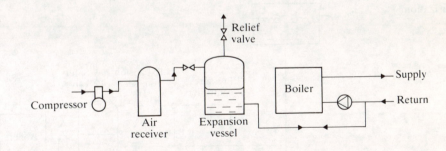

Figure 2.3 Constant-pressure air pressurization

Retention System with Gas Pressurization

A retention system with gas pressurization is shown in Fig. 2.4. It eliminates the problem of oxygen absorption by using nitrogen, which is an inert gas, instead of compressed air. Compared with the previous system an extra receiver and filling bottle are required. When the system expands and the water level rises the relief valve allows nitrogen to pass to the low-pressure receiver. When the system contracts the pressure is maintained by nitrogen from the high-pressure receiver.

Figure 2.4 Retention
system with gas
pressurization

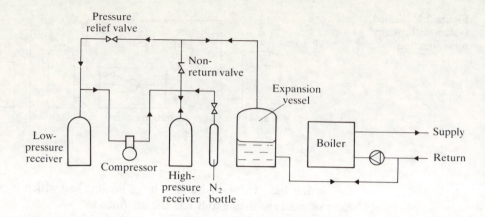

Spill Tank System with Gas Pressurization

With the exception of static head pressurization, all the previous systems have
accommodated the full system expansion in a pressure vessel. A spill tank system
with gas pressurization is shown in Fig. 2.5. In this system some change in
volume can be accommodated in the expansion vessel but the main expansion
between a cold start-up and operating at design temperature takes place in the
spill tank, which is at atmospheric pressure. Pressure controllers actuate either
the spill valve or the pump. When the system contracts after being shut off, the
feed pump takes water from the spill tank to maintain the pressure. When the
system expands on start-up, the spill valve opens to maintain the pressure.

Figure 2.5 Spill tank
system with gas
pressurization

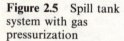

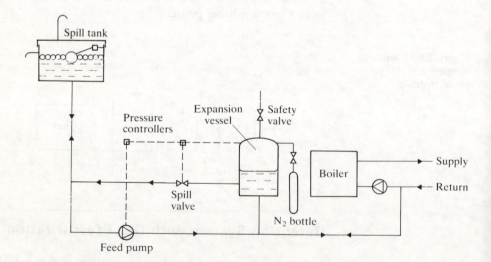

With the simple system shown there may be a problem when the system
expands. Pressurized water above 100 °C can be passed through the spill valve
too close to atmospheric pressure, producing flash steam in the line to the spill
tank. This can be avoided by inserting a heat exchanger upstream of the spill
valve to cool the water to below 100 °C.

Steam Pressurization

In this method steam is generated in conventional boilers but is not delivered to the system. Instead, the saturated water below the liquid level is delivered to the system. In a small system the expansion would take place within the boiler. Figure 2.6 shows a larger system with a number of modular boilers supplying a common expansion vessel (steam drum). The water delivered from the expansion vessel is saturated, so that any reduction in pressure would produce flash steam. The temperature must therefore be reduced as soon as possible to a tolerable value before it enters the feed pump. This is achieved by mixing the saturated water with return water via a by-pass valve.

Figure 2.6 Steam
pressurization

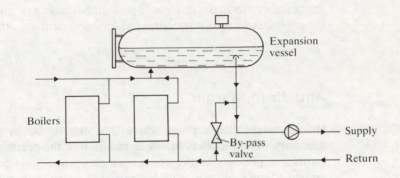

Pressures Around the System

A pressure gauge connected to the side of a pipe at any point in a system would give a reading of static pressure. This reading is made up of three components:

(i) The hydrostatic pressure due to the elevation of the point, which can be found from the following equation:

$$p = \rho g z \qquad [2.1]$$

where p is the pressure in N/m^2, g is the acceleration of gravity in m/s^2 (normally taken as 9.81), ρ is the density in kg/m^3 and z the height in metres relative to some datum.

(ii) The circulating pressure caused by the pump. This will have a maximum value at the pump outlet and a minimum value at the pump inlet. When the pump is stopped it is zero.

(iii) The system pressurization. This is normally by far the highest component but its value is constant throughout the system, whether the pump is running or not.

At any value of fluid temperature, boiling must be avoided: hence it is necessary to find the value of minimum pressure in the system for each possible temperature. If the saturation temperature at a minimum pressure is less than the system temperature, then boiling will not occur. To analyse a system it is therefore necessary to establish points of minimum pressure (there may be more than one at any temperature because the pump may be running or stopped).

Because the system pressurization is constant, points of minimum pressure can be identified by considering only the variations in hydrostatic pressure, circulating pressure, and the pressure resulting from these two acting together. The actual pressures need to be calculated for the critical minimum values only.

Choice of Datum

To calculate the variations in pressure around the system it is necessary to start from a datum. Any point could be selected as a datum but the calculation is simplified if the datum is the point of pressurization. This is the liquid level in the boiler or steam drum for a vapour pressurized system, and the point where the pressurization unit is connected to the system for other units.

Anti-flash Margin

Having established a point where the pressure has a minimum value, it is necessary to add a safety factor to ensure that the pressure is always above the saturation pressure at the system temperature at this point. This is done by adding a temperature margin to the actual water temperature and then looking up the corresponding saturation pressure in tables such as Ref. 2.2 for the higher temperature. The temperature difference which is added is known as an *anti-flash* or *anti-cavitation margin*. (A temperature margin rather than a pressure margin is specified because of the way saturation temperatures and pressures vary. For example, between 80 °C and 100 °C the saturation pressure increases by 0.54 bar but between 160 °C and 180 °C the saturation pressure increases by 4.1 bar. This is also the reason why anti-flash margins decrease as the temperature increases. Typical values are 20 K for LPHW, 15 K for MPHW and 10 K for HPHW.)

To establish the required pressure for a system it is necessary to take into account how the loads are controlled. This is demonstrated in the following examples.

Example 2.1
Figure 2.7(a) shows part of a high-pressure hot water system. The load is a calorifier which is controlled by throttling the hot water. Use the data below to calculate the minimum gas pressure required in the expansion vessel when the pump is running and the decrease in anti-flash margin should the pump stop.

Data

Section	AB	BC	CD	DE	EF	FA
Resistance/(kPa)	5	15	40	5	5	5

Supply temperature 160 °C; return temperature 130 °C; anti-flash margin 10 K.

Figure 2.7 Example 2.1

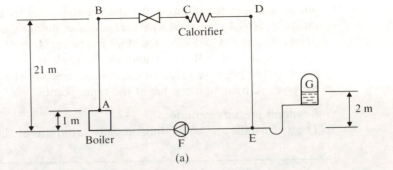

(a)

Solution
Specific volumes can be read from property tables and the densities calculated as follows:

At 160 °C, Density = $1/v_f = 1/0.001\,102 = 907\,\text{kg/m}^3$
At 130 °C, Density = $1/v_f = 1/0.001\,07 = 935\,\text{kg/m}^3$
  Mean density = $(907 + 935)/2 = 921\,\text{kg/m}^3$

The changes in hydrostatic pressure using these values are:

$$\rho gz = 907 \times 9.81 \times z = 8.90z\,\text{kPa}$$
$$\rho gz = 935 \times 9.81 \times z = 9.17z\,\text{kPa}$$
$$\rho gz = 921 \times 9.81 \times z = 9.04z\,\text{kPa}$$

The unit of kPa or kN/m^2 is the most appropriate for this type of calculation.

Datum
In this question the system is connected to the pressurization unit at point E, which is allocated a relative pressure value of 0.

Hydrostatic pressure variations
Point F is on the same level as point E and therefore has a value of zero. Point A is 1 m above F and the water in the boiler has a mean density of 921 kg/m^3. The difference in hydrostatic pressure is therefore $9.04 \times 1 = 9\,\text{kPa}$ to one decimal place, and the value at A is negative with respect to F. Point B is 20 m above A and the density of the water is 907 kg/m^3. The pressure at B is therefore $-9 - (8.9 \times 20) = -187\,\text{kPa}$. Points C and D are on the same level and therefore have the same value. Point E is 21 m below D and the density of the water in the pipe is 935 kg/m^3. The hydrostatic pressure at E is therefore

$$-187 + (9.17 \times 21) = +5.6\,\text{kPa}.$$

The two values of hydrostatic pressure predicted for point E arise because of the difference in density which produce a natural circulation pressure drop of 5.6 kPa around the system.

Circulation pressure variations
The total of the frictional pressure drops (resistances) around the system is $(5 + 15 + 40 + 5 + 5 + 5) = 75\,\text{kPa}$. The natural circulation effect is 5.6 kPa so the pump must supply the balance of 69.4 kPa.
  Starting again at point E: between points E and F there is a resistance or frictional pressure drop of 5 kPa. Since the inlet to the pump, point F_{in}, is downstream of point E it must have a lower pressure, i.e. $-5\,\text{kPa}$. Between the

pump inlet and the pump outlet the pressure rises by 69.4 kPa, resulting in a pressure of $(69.4 - 5) = 64.4$ kPa at the outlet, point F_{out}. The resistance between points F_{out} and A is 5 kPa, resulting in the pressure at A being 59.4 kPa. The pressures at points B, C, D and finally back to E are established in a like manner. The negative value for circulation pressure at point E compensates for the positive hydrostatic value at the same point.

Resultant pressure variations
The hydrostatic and circulating pressure variations are listed in the table below.

Point	Hydrostatic pressure variation (kPa)	Circulating pressure variation (kPa)	Resultant pressure variation (kPa)
E	0	0	0
F_{IN}	0	-5	-5
F_{OUT}	0	$+64.4$	$+64.4$
A	-9	$+59.4$	$+50.4$
B	-187	$+54.4$	-132.6
C	-187	$+39.4$	-147.6
D	-187	-0.6	-187.6
E	5.6	-5.6	0

The resultant pressure variations are found by adding these together taking due account of the algebraic signs. The resultant pressure at point E is zero again after working around the system. These are the variations in pressure with the pump running.

All three pressure variations are shown graphically in Fig. 2.7(b).

Figure 2.7
(*continued*)

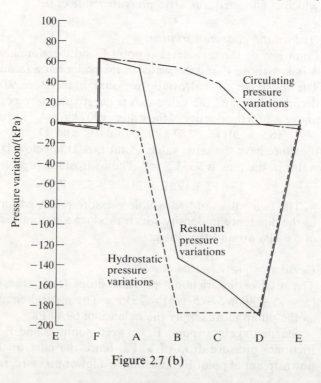

Figure 2.7 (b)

Minimum pressure points
The table or graph is now used to identify any points of minimum pressure which may control the gas pressure.

Point D:
This has the lowest resultant pressure of -187.6 kPa when the pump is running and must be investigated. With the method of control used, the maximum temperature at this point is the return water temperature of 130 °C. Allowing an anti-flash margin of 10 K gives 140 °C, which property tables show has a saturation pressure of 3.614 bar or 361.4 kPa. The absolute pressure at this point must therefore be at least 361.4 kPa.

The expansion vessel is connected to the system at point E and the difference in pressure between point E and point D with the pump running is 187.6 kPa. Since D has a negative value, point E must be at a higher pressure with a value of $(361.4 + 187.6) = 549$ kPa.

Point C:
Of all the points which are exposed to the highest temperature of 160 °C (i.e. A, B and C) the lowest pressure is -147.6 kPa at C. Allowing an anti-flash margin of 10 K, this becomes 170 °C with a corresponding saturation temperature of 7.92 bar or 792 kPa. The pressure required at point E is higher than at C with a value of $(792 + 147.6) = 939.6$ kPa.

Clearly, point C controls the design case with the pump running. Since there is a hydrostatic head between the connection point E and the gas at G, only part of this pressure must be supplied by the gas pressure. When the system has been operating at a steady load for some time, the water in the vertical pipe to the expansion vessel will be close to ambient temperature. However, when the system has just been run up to full load it must be warmer than this; assuming that it is at the return water temperature will introduce a small additional safety factor. Thus,

$$\text{Hydrostatic pressure between E and G} = 9.17 \times 2 = 18.3 \text{ kPa}$$
$$\therefore \quad \text{Pressure in the expansion vessel} = 939.6 - 18.3 = 921.3 \text{ kPa}$$
$$= 9.213 \text{ bar}$$

Pump failure
If the pump fails, the system is subject to the pressure variations due to hydrostatic pressures only. An inspection of the previous table shows that points B and C have the lowest pressure of -187 kPa and, since they also have the highest temperature, this is clearly the worst case.

The pressure at B and C is 187 kPa lower than the pressure at E. The value is therefore $(939.6 - 187) = 752.6$ kPa. Property tables show that this pressure corresponds to a saturation temperature of 167.7 °C. The anti-flash margin is therefore reduced to $(167.7 - 160) = 7.7$ K, which is a decrease of 2.3 K.

Example 2.2
Repeat Example 2.1, using a mean water density of 921 kg/m³ throughout.

Solution
All hydrostatic pressure changes are now calculated using 9.04 z and the pump must overcome the total friction losses of 75 kPa. This results in the table overleaf.

With the pump running, the values of the resultant pressure at the most vulnerable point (i.e. C) is -144.8 kPa and the pressure at G is therefore

Point	Hydrostatic pressure variation (kPa)	Circulating pressure variation (kPa)	Resultant pressure variation (kPa)
E	0	0	0
F_{IN}	0	−5	−5
F_{OUT}	0	+70	+70
A	−9	+65	+56
B	−189.8	+60	−129.8
C	−189.8	+45	−144.8
D	−189.8	+5	−184.8
E	0	0	0

792.4 + 144.8 = 918.5 kPa. If the pump stops, the new pressure at B and C is (792 + 144.8 − 189.8) = 747 kPa. The corresponding saturation temperature is 167.4 °C, which results in a decrease in anti-flash margin of 2.6 K, a change of only 0.3 K to the previous answer.

This example demonstrates that there is little loss in accuracy in using a mean water density. The figures show that it might be worthwhile to carry out the more detailed calculation to justify using a smaller pump. However, this would seldom be advantageous in practice because a larger pump would have to be used with some pressure drop taken up by a regulating valve.

Example 2.3
The primary distribution of a high-pressure hot water system is shown in Fig. 2.8. Using the data provided, draw graphically the total pressure profile for the 'index circuit' and hence determine the minimum safe pressure for the pressurization unit to apply at position 'A' under all operating conditions with the pump running.

Figure 2.8 Example 2.3

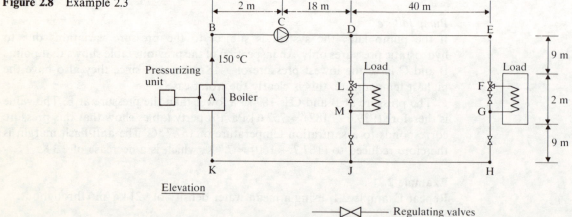

If the arrangement were inverted (turn the diagram upside down) by what amount would the pressure requirement at 'A' be changed? All pipe lengths given are total equivalent lengths.

Data

Pressure head of water at 150 °C	9 kPa/m head
Safety margin	15 K
Pipe losses	500 Pa/m
Control valve authority	0.5
Pressure loss of load (exit of F to G)	50 kPa
Pressure loss of load (exit of L to M)	100 kPa
Boiler loss	negligible
Pump head	300 kPa

(CIBSE)

Solution

To answer this question it is necessary to understand how the system is controlled. The term *valve authority* is defined as follows:

$$\text{Valve authority} = \frac{\Delta p_1}{\Delta p_1 + \Delta p_2}$$

where Δp_1 is the pressure drop across the valve when fully open; Δp_2 is the pressure drop across the remainder of the circuit where the flow is to be controlled. The regulating valves between L and M and between F and G are set to provide the same resistance through the by-pass as through the load and associated pipework. The diverting valve can then divide the flow between two 'equal circuits'.

The regulating valves near J and H balance the total flows to the two loads and take up any excess pressure drop from the pump above that required to satisfy the system resistances.

In this example, with a valve authority of 0.5 the pressure drop across the fully open valve must be the same as across the load. The total pressure drop is therefore twice that of the load. The index circuit can now be identified by working out the pressure drops around the two loads for the section of the system which is unique to each as follows:

$$\text{DEFGHJ pressure drop} = \{(40 + 9)2 \times 0.5\} + (2 \times 50) = 149 \text{ kPa}$$
$$\text{DLMJ} \quad \text{pressure drop} = (9 \times 2 \times 0.5) + (2 \times 100) = 209 \text{ kPa}$$

The index circuit is therefore DLMJ.

The pressurizing unit is on the same level as the boiler and will therefore have the same pressure. Point A is therefore the datum for the pressure variation calculations. The pressure head is given in the question so that any hydrostatic head pressure is given by $9z$ kPa where z is the vertical height in metres. (It must be assumed that the vertical dimensions are both equivalent lengths and heights.)

The calculations for the first part of the question are set out in the table overleaf. The first part of the table works through the three pressure variations, as in the previous examples. This results in a value of $+61$ kPa for the circulation pressure at point A. This is not possible when density differences are not taken into account and this must be the pressure drop across the regulating valve above point J. The bottom part of the table, which is a repeat of points M to A, takes into account the pressure drop across the regulating valve.

The calculations in the table show that the minimum pressure is at point C, which has a value of -96 kPa when the pump is running. It also has the

	Hydrostatic pressure variation (kPa)		Circulation pressure variation (kPa)		Resultant pressure variation (kPa)
Point	Change	Value	Change	Value	Value
A		0		0	0
B	−90	−90	−10 × 0.5	−5	−95
C_{IN}	0	−90	−2 × 0.5	−6	−96
C_{OUT}	0	−90	+300	+294	+204
D	0	−90	−18 × 0.5	+285	+195
L	+81	−9	−9 × 0.5	+280.5	+271.5
M	+18	+9	−2 × 100	+80.5	+89.5
J	+81	+90	−9 × 0.5	+76	+166
K	0	+90	−20 × 0.5	+66	+156
A	−90	0	−10 × 0.5	+61	+61
M	+18	+9	−200	+80.5	+89.5
J	+81	+90	−4.5 − 61	+15	+105
K	0	+90	−10	+5	+95
A	−90	0	−5	0	0

maximum temperature and is therefore the only point which needs to be considered. Adding the anti-flash margin of 15 K to 150 °C gives 165 °C, which property tables show has a corresponding saturation pressure of 7 bar. Since point A has a higher pressure than point C (pressure difference 96 kPa = 0.96 bar), the pressurization unit must be at (7 + 0.96) = 7.96 bar.

If the system is inverted the pressure variations must be re-calculated. This has been done in the table below. The new point of minimum pressure with the pump running is at K with a value of −85 kPa. If the loads had been controlled with throttle valves this point would have only been subjected to return water temperatures. With a diverting valve when there is no load the supply temperature will reach point K. The new pressurization is therefore (7 + 0.85) = 7.85 bar.

	Hydrostatic pressure variation (kPa)		Circulation pressure variation (kPa)		Resultant pressure variation (kPa)
Point	Change	Value	Change	Value	Value
A		0		0	0
B	+90	+90	−5	−5	−85
C_{IN}	0	+90	−1	−6	+84
C_{OUT}	0	+90	+300	+294	+384
D	0	+90	−9	+285	+375
L	−81	+9	−4.5	+280.5	+289.5
M	−18	−9	−200	+80.5	+81.5
J	−81	−90	−4.5 − 61	+15	−75
K	0	−90	−10	+5	−85
A	+90	0	−5	0	0

Figure 2.9 Example 2.4

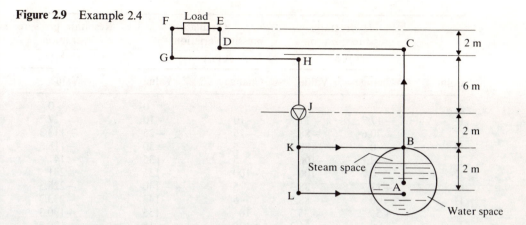

Example 2.4

The following data apply to the high-pressure hot water system shown in Fig. 2.9.

Pipe section	AB	BC	CD	DE	EF	FG	GH	HJ	JK	KL	LA	Total
$\Delta p/(\text{kPa})$	10	15	5	5	100	5	5	10	5	5	5	170
Pipe emission/(kW)	10	40	10	5	50	5	10	30	10	5	5	180

No. of heaters in load	30
Heat output per heater with the fans on	20 kW
Heat output per heater with the fans off	3 kW
Temperature drop across the system at full output	40 K
Boiler steam pressure	800 kPa
Anti-flash margin	10 K
System water density	900 kg/m³
Specific heat of water	4.2 kJ/kg K

Calculate:
(i) the maximum permissible flow temperature at full load;
(ii) the mass flow rate of by-passed water at that condition.

(Dublin Institute of Technology)

Solution

$$\text{Hydrostatic pressure drop} = 900 \times 9.81 \times z = 8.829z \text{ kPa}$$

Ignoring the slight difference in pressure between the liquid surface and point A, the calculations to determine the relevant pressure variations are shown in the table below using point A as the datum.

The mass flow rate around the system can be found from the total emissions from the system as follows:

$$\dot{m} = \frac{\dot{Q}}{c_p \times \Delta t} = \frac{(30 \times 20) + 180}{4.2 \times 40} = 4.64 \text{ kg/s}$$

$$\dot{m} \text{ through each heater} = \frac{4.64}{30} = 0.155 \text{ kg/s}$$

| Point | Hydrostatic pressure variation (kPa) | | Circulation pressure variation (kPa) | | Resultant pressure variation (kPa) |
	Change	Value	Change	Value	Value
A		0		0	0
B	−17.7	−17.7	−10	−10	−27.7
C	−70.6	−88.3	−15	−25	−113.3
D	0	−88.3	−5	−30	−118.3
E	−17.7	−106	−5	−35	−141
F	0	−106	−100	−135	−241
G	+17.7	−88.3	−5	−140	−228.3
H	0	−88.3	−5	−145	−233.3
J_{IN}	+53	−35.3	−10	−155	−190.3
J_{OUT}	0	−35.3	+170	+15	−20.3
K	+17.7	−17.7	−5	+10	−7.7
L	+17.7	0	−5	+5	−5
A		0	−5	0	0

The table above shows that two points need to be investigated.

Point E:

The pressure at point E is $(8 − 1.41) = 6.59$ bar. The corresponding saturation temperature from property tables is $162.5\,°C$. Applying the anti-flash margin of $10\,K$, the maximum allowable temperature at E becomes $152.5\,°C$.

Point F:

The pressure at point F is $(8 − 2.41) = 5.59$ bar. The corresponding saturation temperature from property tables is $156.1\,°C$. Applying the anti-flash margin of $10\,K$, the maximum allowable temperature at F becomes $146.1\,°C$.

If point F was subjected only to the water temperature leaving the heaters at full load, the temperature at inlet to the heaters would be well above $152.5\,°C$ and point E would control. However the question clearly requires that the situation where one fan fails should be taken into account. This would have a small effect on the overall energy balance but a considerable effect on the temperature of the water leaving the particular heater with the fan off.

The temperature drop across a heater with the fan off is

$$\Delta t = \frac{\dot{Q}}{\dot{m}c_p} = \frac{3}{0.155 \times 4.2} = 4.6\ K$$

Assuming that there is a negligible pipe emission loss between point E and the first heater, the maximum allowable temperature at point E is $(146.1 + 4.6) = 150.7\,°C$, which controls.

The emission from the pipes between points E and B is $(5 + 10 + 40) = 55$ kW. The temperature drop along the pipework between points E and B is therefore:

$$\Delta t = \frac{\dot{Q}}{\dot{m}c_p} = \frac{55}{4.64 \times 4.2} = 2.8\ K$$

The flow temperature from B is therefore $(150.7 + 2.8) = 153.5\,°C$ and the return water temperature is at $(153.5 − 40) = 113.5\,°C$. From property tables the saturation temperature at 8 bar is $170.4\,°C$, which will be the temperature of the water leaving the boiler. Y, the mass flow rate through the by-pass KB, can now be found by mixing as follows:

$$Y \times 113.5 + (4.64 - Y)\,170.4 = 4.64 \times 153.5$$
$$Y = 1.38 \text{ kg/s}$$

Expansion Vessel Sizing

The following method is strictly only applicable to simple sealed systems using air pressurization as shown in Fig. 2.10 but the principles developed will apply to any system. The system pressurization takes place in two distinct processes, as follows.

Filling (a)–(b)

This process starts with a sealed vessel with a volume V_1 (above the filling point) filled with air at ambient pressure p_1, and ends when the system has been filled with cold water and the air has been compressed to a pressure p_2 and a volume V_2. The pressure after filling will depend on the elevation of the feed tank and the volume will depend on both the pressure and the process during filling.

Figure 2.10 Expansion vessel sizing

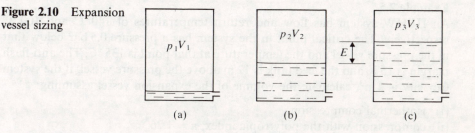

If the filling process took place quickly with the vessel insulated, the process undertaken by the trapped air would be close to adiabatic ($pV^\gamma = $ constant). If it took place very slowly with the vessel uninsulated, the heat produced by compression would be dissipated and the process would be close to isothermal ($pV = $ constant). Expansion vessels are always uninsulated and take a finite time to fill so the process must be polytropic ($pV^n = $ constant), with $\gamma > n > 1$. Reference 2.3 quotes a value for n of 1.26. However, when the system has been running for some time the temperature will fall to close to ambient, which favours the isothermal assumption. For the general case:

$$p_1 V_1^n = p_2 V_2^n \qquad [2.2]$$

Water expansion (b)–(c)

After filling, the system is brought up to temperature, the water expands and there is a further increase in pressure and a decrease in volume. The analysis of the process is the same as for the filling process with the additional

complication that some of the water entering the vessel may be warm until the
system has been running for some time. For the general case,

$$p_2 V_2^n = p_3 V_3^n \qquad [2.3]$$

The water expansion volume, E, is given by

$$E = V_2 - V_3 \qquad [2.4]$$

The water expansion volume can also be found in terms of the system capacity,
C, the cold water density, ρ_c, and the mean density of the hot water in the
system, ρ_H. The system capacity is the internal volume of all the pipework,
emitters, etc., including the cold water in the expansion vessel shown in Fig.
2.10(b).

The mass of water in the system, m, is constant. Therefore,

$$m = \rho_c C = \rho_H(C + E)$$

Rearranging gives

$$E = \frac{(\rho_c - \rho_H)C}{\rho_H} \qquad [2.5]$$

The above equations can be used to calculate the expansion vessel volume V_1.

Example 2.5
An HPHW system has flow and return temperatures of 140 °C and 100 °C
respectively. The critical point in the system has a pressure 0.5 bar below that
in the pressure vessel and the temperature at that point is 135 °C. The anti-flash
margin is 15 K and the feed tank is 12 m above the pressure vessel. If the system
capacity is 2 m³, calculate the volume of the expansion vessel assuming:

(i) isothermal compression;
(ii) compression with the polytropic index, $n = 1.26$.

Solution

Adding the anti-flash margin to the temperature at the critical point in the
system gives 150 °C, which from tables corresponds to a saturation pressure of
4.76 bar. The final pressure in the pressure vessel, p_3, is therefore $(4.76 + 0.5) =$
5.26 bar. At standard atmospheric pressure, $p_1 = 1.013$ bar. Assuming the density
of cold water to be 1000 kg/m³, the hydrostatic pressure from the feed tank is
found from Eqn [2.1]:

$$p = \rho g z = 1000 \times 9.81 \times 12 = 117\,720 \text{ Pa} = 1.18 \text{ bar}$$

Therefore p_2, the absolute pressure after filling, is

$$p_2 = 1.18 + 1.013 = 2.19 \text{ bar}$$

The mean temperature of the hot water is $(100 + 140)/2 = 120$ °C, which from
property tables corresponds to a density of 943.4 kg/m³, and the water expansion
can be found from Eqn [2.5] as follows:

$$E = \frac{(\rho_c - \rho_H)C}{\rho_H} = \frac{(1000 - 943.4)2}{943.4} = 0.12 \text{ m}^3$$

Using Eqn [2.4],

$$E = V_2 - V_3$$

therefore $V_3 = V_2 - 0.12$ [1]

Using Eqns [1] and [2.3] gives:

$$V_3 = V_2 - 0.12 = \left(\frac{p_2}{p_3}\right)^{1/n} V_2$$ [2]

(i) Isothermal compression:
For an isothermal process, $n = 1$. Therefore, from Eqn [2],

$$V_2 - 0.12 = \left(\frac{2.19}{5.26}\right) V_2$$

Hence $V_2 = 0.2056 \text{ m}^3$

From Eqn [2.2],

$$V_1 = V_2\left(\frac{p_2}{p_1}\right) = 0.2056 \times \frac{2.19}{1.013} = 0.4445 \text{ m}^3$$

(ii) Polytropic compression:
Using Eqn [2],

$$V_3 = V_2 - 0.12 = \left(\frac{p_2}{p_3}\right)^{1/n} V_2$$

$$V_2 - 0.12 = \left(\frac{2.19}{5.26}\right)^{1/1.26} V_2$$

Hence $V_2 = 0.2395 \text{ m}^3$

From Eqn [2.2],

$$V_1 = V_2\left(\frac{p_2}{p_1}\right)^{1/n} = 0.2395 \times \left(\frac{2.19}{1.013}\right)^{1/1.26} = 0.4416 \text{ m}^3$$

The isothermal process results in the larger vessel, but the difference is very small.

Example 2.6
The expansion vessel sized in Example 2.5 using the isothermal assumption is cylindrical with a length-to-diameter ratio of 3. Calculate the change in liquid level during filling and when the water expands.

Solution
If the cylinder has a diameter d and L is the length, for any particular volume

$$V_1 = 0.4445 = \frac{\pi}{4} d^2 \times 3d$$

Hence $d = 0.574 \text{ m}; L_1 = 1.721 \text{ m}$

For a fixed diameter the length is proportional to the volume. Therefore:

$$L_2 = 1.721 \times \frac{0.2056}{0.4445} = 0.796 \, \text{m}$$
$$V_3 = 0.2056 - 0.12 = 0.0856 \, \text{m}^3$$
$$L_3 = 1.721 \times \frac{0.0856}{0.4445} = 0.331 \, \text{m}$$

Change in level during filling $= L_1 - L_2 = 1.721 - 0.796 = 0.925 \, \text{m}$
Change in level during expansion $= L_2 - L_3 = 0.796 - 0.331 = 0.465 \, \text{m}$

2.2 STEAM SYSTEMS

Steam is frequently used as the primary medium for distributing heat in factories and large building complexes such as hospitals. When used for heating it is generally the enthalpy of vaporization (h_{fg}) only which is utilized. This decreases as the saturation temperature increases. The decrease in enthalpy of vaporization and the steep increase in pressure as the saturation temperature increases make high-temperature steam an unattractive heat distribution medium. Steam for heat distribution is therefore normally generated at medium or low pressures.

In an ideal system the steam would be generated and delivered to the heaters as a saturated vapour and returned to the boiler as a saturated liquid. In practice:

(i) It is impossible to generate saturated steam without some liquid entrainment.
(ii) There must be friction and heat losses from pipes.
(iii) Some fluid is always lost (boiler blowdown, air purging of heaters, evaporation from hot wells, etc.); therefore treated make-up water is required.
(iv) Condensate is likely to leave the heat exchangers (this includes space heaters, calorifiers, etc.) sub-cooled.
(v) The water must be delivered to the feed pump at a temperature sufficiently below the saturation value to prevent cavitation.
(vi) A means must be found to prevent steam passing through heat exchangers without condensing—steam traps must be used.

Steam Traps

Steam traps allow condensate to pass from the heat exchanger to the hot well without losing any steam. They can operate on a continuous or an intermittent basis. They are classified in groups based on the principle used to open and close the valve. The classification is as follows.

Group 1

These use the difference in density between water and steam. The float trap in Fig. 2.11 is a typical example. If the chamber fills up with steam the liquid level and float fall and the valve closes. When the chamber fills with condensate the valve opens. The air vent at the top right-hand corner operates in the same way

Figure 2.11 Float trap

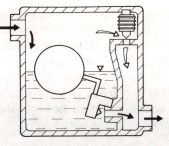

as the thermostatic steam trap described in group 2. It is fully open when cold and therefore allows air to pass to the condensate line during start-up. When the temperature rises, the air vent closes.

The bucket trap is the other main trap in this group. It is so named because the part which operates the valve consists of a cylinder which is open at one end.

Group 1 valves can discharge condensate continuously without the condensate being sub-cooled.

Group 2

These use the difference in temperature between saturated steam and sub-cooled condensate to operate the valve. Because a temperature difference is required to operate the valve, the condensate may back up into the heat exchanger covering some of the heat transfer surface until it has cooled sufficiently to open the valve. The operation is therefore intermittent. Figure 2.12 shows a thermo-static expansion steam trap. The expansion bellows is filled with liquid and as the liquid expands the bellows increases in length and closes the valve. When cold, the valve is fully open, which allows air to be rejected at start-up. It closes slowly as the system is brought up to temperature and should be fully closed at some design condensate temperature. When exposed to steam (or condensate) above this temperature, the valve is closed. It will only open when the temperature falls.

The expansion of a bimetallic strip is an alternative means of providing movement to operate a valve.

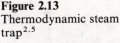

Figure 2.12
Thermostatic trap[2.5]

Group 3

These use the difference in the pressures acting on a disc which cause the disc to rise and fall. A thermodynamic trap is shown in Fig. 2.13(a) and the principle of operation in Figs 2.13(b) and (c). The pressure downstream of the trap will

Figure 2.13
Thermodynamic steam
trap[2.5]

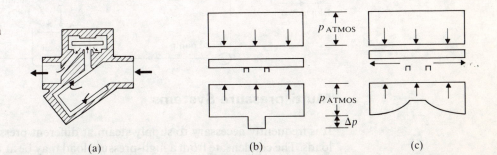

(a)   (b)   (c)

be close to atmospheric and this is the presssure which acts on the top of the disc. Figure 2.13(b) shows the trap in the closed position. Atmospheric pressure is acting on all surfaces of the disc except the central section covering the pipe connected to the heat exchanger. The pressure across the central section will increase until the value of Δp is sufficient to lift the disc. Once the disc rises above its seat a flow will place in an outward radial direction (Fig. 2.13c). The steady-flow energy equation for the flow of incompressible fluids with no heat or mass transfers may be expressed as:

$$p + \tfrac{1}{2}\rho u^2 + \rho g z = \text{constant}$$

The height z is constant in this situation and therefore the pressure must increase as the velocity falls. The area through which the fluid is flowing increases between the centre and the rim of the disc, producing a decrease in velocity and a corresponding increase in pressure. However, the pressure at the rim is atmospheric; therefore the pressure in the centre must be below atmospheric. As soon as a flow is established this decrease in pressure in the centre pulls the disc back down to the closed position and the cycle starts again.

The operation of a thermodynamic steam trap is noisy because cycles may be quite short and the disc will vibrate on its seat. It can also pass steam as well as condensate and is only used where capital costs override energy considerations.

A mathematical treatment of radial flow across flat discs is given by Douglas et al.[2.4]

Sketches of some of the traps not illustrated here together with comments on the relative merits of different types can be found in the CIBSE *Guide to Current Practice*.[2.5]

Single-pressure Systems

A simple sketch of a single-pressure system is shown in Fig. 2.14. The blowdown is to prevent the build-up of dissolved solids in the water in the boiler and may be continuous or intermittent. The condensate from the loads is collected in the hot well, where system losses are made up from a feed tank. The make-up to the feed tank is supplied from a water treatment plant.

Figure 2.14 Single-pressure steam system

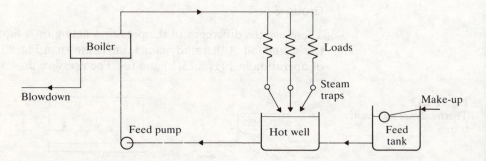

Multi-pressure Systems

It is frequently necessary to supply steam at different pressures to suit different loads. The condensate from a high-pressure load may be at a higher temperature than the saturation temperature of the low-pressure steam. If this is the case, it can be used to provide some of the low-pressure steam by expanding it down to the low pressure which will produce saturated vapour and saturated liquid at the lower pressure. Vapour produced in this way is termed flash steam, and the saturated vapour and saturated liquid are separated in a flash vessel.

Flash steam alone is seldom sufficient to satisfy the low-pressure loads, and additional steam must be provided. This is normally done by passing some of the high-pressure steam through a pressure-reducing valve as shown in Fig. 2.15. The steam will become superheated as it is throttled but in the flash vessel the superheat will be used to produce more low-pressure steam from the condensate.

Figure 2.15 Two-pressure steam system

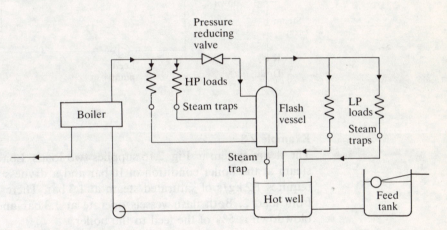

Example 2.7
The high-pressure units on a steam plant operate at 10 bar and the condensate is sub-cooled by 9.5 K. If the mass flow rate of condensate is 10 kg/s, how much saturated steam at 1.5 bar could this condensate produce?

Solution
The relevant enthalpies for the steam and water can be found in property tables such as Ref. 2.2. For the high-pressure condensate it is sufficiently accurate to assume that the specific enthalpy is the same as for a saturated liquid *at the same temperature*. At 10 bar the saturation temperature is 179.9 °C, so allowing for the 9.5 K sub-cooling the temperature is (179.9 − 9.5) = 170.4 °C. At 170.4 °C the specific enthalpy of saturated liquid is 721 kJ/kg K.

After throttling to 1.5 bar the specific enthalpy of the saturated vapour is 2693 kJ/kg K and the specific enthalpy of the saturated liquid is 467 kJ/kg K.

For any adiabatic throttling process with negligible changes in kinetic and potential energy,

$$\text{Enthalpy before throttling} = \text{Enthalpy after throttling}$$
$$10 \times 721 = \dot{m} \times 2693 + (10 - \dot{m})467$$

Flash vapour produced, $\dot{m} = 1.14$ kg/s

Not only is this a useful quantity of low-pressure steam but without this process the high-temperature condensate would have to be cooled to below 100 °C before it could be discharged to a hot well operating at atmospheric pressure.

Figure 2.16 Example
2.8

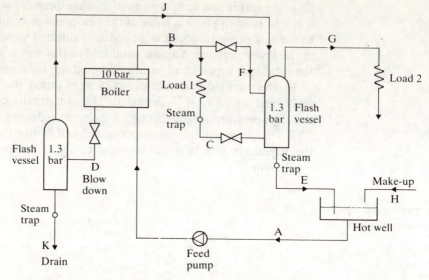

Example 2.8

The steam system in Fig. 2.16 supplies two loads. Load 1 is 3000 kW and uses steam at the boiler condition of 10 bar and a dryness fraction of 0.95. Load 2 requires 1.2 kg/s of saturated steam at 1.3 bar. There is no condensate return from load 2. Both flash vessels operate at 1.3 bar and the continuous boiler blowdown is 5% of the feed to the boiler.

Assuming there is no undercooling of the condensate and negligible pipe losses, calculate:

(i) the mass flow rate through the feed pump;
(ii) the temperature of the feed water when the make-up is at 10 °C.

(Wolverhampton Polytechnic)

Solution
(i) Specific enthalpy values for the two pressure levels and the make-up water temperature are taken from property tables and give the following values:

At 1.3 bar $h_g = 2687$ kJ/kg, $h_{fg} = 2238$ kJ/kg, $h_f = 449$ kJ/kg

At 10 bar $h_g = 2778$ kJ/kg, $h_{fg} = 2015$ kJ/kg, $h_f = 763$ kJ/kg

At 10 °C $h_f = 42$ kJ/kg

It is now necessary to carry out mass and energy balances on each item of plant to establish enough equations to solve the problem.

Boiler:

$$\dot{m}_D = 0.05\dot{m}_A \qquad \text{and} \qquad \dot{m}_B = 0.95\dot{m}_A$$

Load 1:

$$\dot{m}_C = \frac{\dot{Q}_1}{0.95 h_{fg}} = \frac{3000}{0.95 \times 2015} = 1.567 \text{ kg/s}$$

$$\dot{m}_F = (\dot{m}_B - \dot{m}_C) = (0.95\dot{m}_A - 1.567)$$

[1]

Blowdown flash vessel:

$$\dot{m}_K = \dot{m}_D - \dot{m}_J = 0.05\dot{m}_A - \dot{m}_J \qquad [2]$$

Also, $\qquad \dot{m}_D h_D = \dot{m}_J h_J + \dot{m}_K h_G$

Using Eqn [1] to eliminate $\dot{m}_g$,

$$0.05\dot{m}_A \times 763 = \dot{m}_J \times 2687 + (0.05\dot{m}_A - \dot{m}_J) \times 449$$

Hence $\qquad \dot{m}_J = 0.007\dot{m}_A \qquad [3]$

Main flash vessel:

$$\dot{m}_J + \dot{m}_B = \dot{m}_G + \dot{m}_E$$

The value of $\dot{m}_G$ is given in the question as 1.2 kg/s and Eqn [3] can be used to eliminate $\dot{m}_J$:

$$0.007\dot{m}_A + 0.95\dot{m}_A = 1.2 + \dot{m}_E$$
$$\dot{m}_E = (0.957\dot{m}_A - 1.2) \qquad [4]$$

Also, $\qquad \dot{m}_C h_C + \dot{m}_F h_F + \dot{m}_J h_J = \dot{m}_G h_G + \dot{m}_E h_E$

Using Eqns [1] and [4] to eliminate $\dot{m}_F$ and $\dot{m}_E$:

$$(1.567 \times 763) + (0.95\dot{m}_A - 1.567) \times 2778 + 0.007\dot{m}_A \times 2687$$
$$= 1.2 \times 2687 + (0.957\dot{m}_A - 1.2) \times 449$$

Hence, $\qquad \dot{m}_A = 2.62 \text{ kg/s}$

i.e. the mass flow rate through the feed pump is 2.62 kg/s.

(ii) Hot well:
From Eqn [4]

$$\dot{m}_E = (0.957 \times 2.62 - 1.2) = 1.31 \text{ kg/s}$$

Also, $\qquad \dot{m}_H = \dot{m}_A - \dot{m}_E = 2.62 - 1.31 = 1.31 \text{ kg/s}$

and $\qquad m_A h_A = \dot{m}_E h_E + \dot{m}_H h_H$
$$2.62 h_A = 1.31 \times 449 + 1.31 \times 42$$

Hence, $\qquad h_A = 245.5 \text{ kJ/kg and by interpolation, } t_A = 58.7\,°C$

i.e. the temperature of the feed water is 58.7 °C.

Example 2.9

Saturated steam at 8 bar is supplied to the following loads.

(a) A heating calorifier which takes 700 kg/h when the outside temperature is 14 °C; the design load is based on an outside temperature of −1 °C when the inside temperature is 21 °C.
(b) An air heater which continuously heats 540 m³/min of air with a specific heat of 1.34 kJ/m³ K from 12 °C to 30 °C.
(c) A hot water storage calorifier which continuously raises the temperature of 1500 litres/h of water from 10 °C to 50 °C.

It is proposed to use the condensate from all the above loads in a condensate sub-cooler to off-load the hot water storage calorifier. If the steam traps discharge saturated water and the condensate leaving the sub-cooler is at 95 °C, ignore any pipework losses and calculate:

(i)   the maximum condensate flow rate in kg/h;

(ii) the minimum condensate flow rate in kg/h;

(iii) the outside temperature at which the hot water calorifier is completely off-loaded.

(Wolverhampton Polytechnic)

Solution

(i)   Specific enthalpies for the water and steam are taken from property tables and give the following values:

At 8 bar $h_g = 2769 \, \text{kJ/kg}$, $h_{fg} = 2048 \, \text{kJ/kg}$, $h_f = 721 \, \text{kJ/kg}$

At 95 °C $h_f = 398 \, \text{kJ/kg}$

The specific heat of water at the mean hot water temperature of 30 °C is 4.18 kJ/kg.

A sketch of the system is shown in Fig. 2.17. Consider each item of plant in turn.

Figure 2.17   Example 2.9

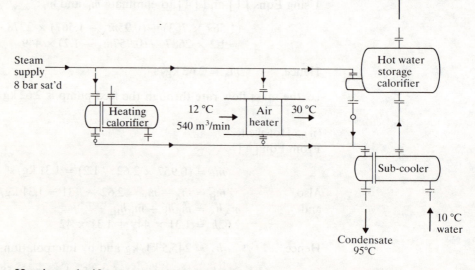

Heating calorifier:

From Eqn [1.1] ($\dot{Q} = UA\Delta t$), the heat load at any time will be directly proportional to the temperature difference between the inside and outside air. Therefore the steam flow rate and temperature difference are related as follows:

$$\frac{\dot{m}_d}{\dot{m}} = \frac{\Delta t_d}{\Delta t} = \frac{\dot{m}_d}{700} = \frac{(21 + 1)}{(21 - 14)} \qquad [1]$$

Hence,   Design mass flow rate, $\dot{m}_d = 2200 \, \text{kg/h}$

Air heater:

$$\dot{Q} = \frac{540}{60} \times 1.34 \times (30 - 12) = 217 \, \text{kW}$$

$$\dot{m} = \frac{\dot{Q}}{h_{fg}} = \frac{217}{2048} \times 3600 = 381.6 \, \text{kg/h}$$

Hot water calorifier:

$$\dot{Q} = \dot{m}c_p \Delta t = \frac{1500}{3600} \times 4.18 \times (50 - 10) = 69.7 \, \text{kW}$$

If a sub-cooler only is used to provide this load, the mass flow rate of condensate required is

$$\dot{m} = \frac{\dot{Q}}{(h_{f@8bar} - h_{f@95°C})} = \frac{69.7}{(721 - 398)} \times 3600 = 776.8 \, \text{kg/h}$$

With the heating calorifier running at design load, giving a condensate flow of 2 200 kg/h, there is clearly plenty of condensate to heat the hot water without any steam being supplied to the hot water storage calorifier. The maximum condensate flow rate is therefore $(2\,200 + 381.6) = 2\,581.6 \, \text{kg/h}$.

(ii)   The minimum condensate flow will occur when there is no space heating requirement and the heating calorifier is off. For this condition there is still the 381.6 kg/h of condensate from the air heater. The part of the load this will satisfy is:

$$\dot{Q} = \dot{m}(h_{f@8bar} - h_{f@95°C}) = \frac{381.6}{3600}(721 - 398) = 34.2 \, \text{kW}$$

The part of the load which must be satisfied by a steam supply is therefore $(69.7 - 34.2) = 35.5 \, \text{kW}$. The mass flow rate of steam required is:

$$\dot{m} = \frac{\dot{Q}}{(h_g - h_{f@95°C})} = \frac{35.5}{(2769 - 398)} \times 3600 = 53.8 \, \text{kg/s}$$

The minimum condensate flow rate is therefore $(381.6 + 53.8) = 435.4 \, \text{kg/h}$.

(iii)   The sub-cooler requires 776.8 kg/h of condensate to completely unload the hot water calorifier, of which the air heater will always supply 381.6 kg/h. This leaves $(776.8 - 381.6) = 395.2 \, \text{kg/h}$ which is required from the heating calorifier. Equation [1] can now be used to calculate the air temperature which corresponds to this flow rate.

$$\frac{\dot{m}_d}{\dot{m}} = \frac{\Delta t_d}{\Delta t} = \frac{2200}{395.2} = \frac{(21 + 1)}{(21 - t_0)}$$

Hence,   Outside air temperature, $t_0 = 17.1 \, °C$

2.3 STEAM BOILER PERFORMANCE

Different types of boilers are comprehensively described in books such as those by Gunn and Horton[2.6] and Babcock and Wilcox.[2.7] They also cover the combustion aspects, as do many textbooks on basic Thermodynamics such as Eastop and McConkey.[2.8] Boilers and combustion in an energy-saving context are covered in Eastop and Croft.[2.9] This section covers the general performance of steam boilers and how they interact with steam systems.

Overall Efficiency

For any boiler the overall thermal efficiency is found from the equation

$$\eta = \frac{\text{Useful heat output} \times 100}{\text{Heat input}}$$

For a steam boiler,

$$\eta = \frac{\dot{m}(h - h_f) \times 100}{\text{Fuel flow rate} \times \text{CV}} \qquad [2.6]$$

where $\dot{m}$ is the mass flow rate of steam, h is the specific enthalpy of the steam, h_f is the specific enthalpy of the feed water, and CV is the calorific value of the fuel. The fuel flow rate and CV may be based on mass or volume depending on whether the fuel used is solid, liquid or gaseous.

For a coal fired boiler a typical energy balance would be:

Energy used to produce steam (overall efficiency)	78%
Energy loss to the exhaust gas	13%
Standing losses (convection and radiation)	4%
Blowdown loss	3%
Energy left in unburnt fuel	2%
Total	100%

However, such a heat balance would only be applicable to a boiler running at its design load. If the load falls, the proportions change. For example, a boiler on standby at the operating temperature which is producing no load will still have the same conduction, convection and radiation losses. Therefore, as the load falls so does the efficiency, and the fall-off in efficiency becomes more pronounced the further the load falls. Since very few boilers run continuously at their design load the efficiency at part-loads is as important as at the design load. The overall performance over a period of time can be specified by a *load*

Figure 2.18 Overall boiler efficiency vs load factor

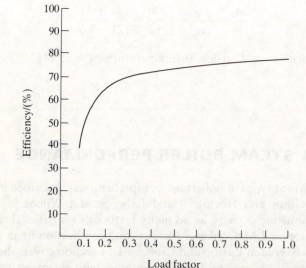

factor. This is defined as follows:

$$\text{Load factor} = \frac{\text{Average load}}{\text{Design load}}$$

If the average load produces steam at the same state from feed water at the same state as in the design case, the load factor may be expressed as:

$$\text{Load factor} = \frac{\text{Average steam mass flow rate}}{\text{Design steam mass flow rate}} \qquad [2.7]$$

Figure 2.18 shows a graph of overall thermal efficiency against load factor for a typical steam boiler.

Combustion Efficiency

This is the ratio of the energy extracted from the products of combustion to the energy input expressed as a percentage:

$$\text{Combustion efficiency} = \frac{\left\{ \begin{array}{l} \text{Useful energy output} + \\ \text{Standing loss} + \text{Blowdown loss} \end{array} \right\}}{\text{Energy input}}$$

The combustion efficiency can be determined from the calorific value of the fuel, the air and exhaust temperatures and an analysis of the products of combustion. A full boiler trial is necessary to determine the standing and blowdown losses. The combustion efficiency is always higher than the overall efficiency and may have an appreciable value when the overall efficiency is very low. When a boiler is on standby the overall efficiency is zero but the combustion efficiency may still be over 70%. In the heat balance discussed earlier the combustion efficiency is $(78 + 4 + 3) = 85\%$.

Equivalent Evaporation

A boiler specified for a particular mass flow rate and state of steam may often have to operate at different conditions. It is also necessary to be able to compare the performance of boilers with different specifications. These comparisons are carried out by means of an equivalent evaporation rate.

The standard chosen is the mass flow rate of saturated steam which would be produced from saturated water at 100 °C. This is termed *from and at 100°C* (F @ A 100 °C).

Equivalent evaporation rates can be found as follows:

$$\text{Heat load} = \dot{m}_{\text{ACTUAL}} \Delta h_{\text{ACTUAL}} = \dot{m}_{\text{F @ A 100°C}} \Delta h_{\text{F @ A 100°C}}$$

$\Delta h_{\text{F @ A 100°C}}$ is h_{fg} at 100 °C, which from property tables is 2256.7 kJ/kg, and h_{ACTUAL} is $(h - h_{\text{f}})$. Therefore rearranging gives:

$$\dot{m}_{\text{F @ A 100°C}} = \dot{m}_{\text{ACTUAL}} \times \frac{(h - h_{\text{f}})}{2256.7} \qquad [2.8]$$

where h and h_{f} are in kJ/kg.

Maximum Continuous Rating (MCR)

Hot water boilers are specified by the design heat load in kilowatts. Steam boilers are specified in terms of the mass flow rate of steam which can be generated continuously. This is known as the maximum continuous rating (MCR). The MCR may be specified at the actual conditions (for example 12 000 kg/h of saturated steam at 8 bar from feed water at 60 °C), or F @ A 100 °C. Because boilers are manufactured in standard sizes which are then adapted to a particular application the $MCR_{F@A100°C}$ is normally given rather than the actual MCR.

Example 2.10

A boiler has an actual MCR of 15 000 kg/h when producing steam at 10 bar with a dryness fraction of 0.98 from feed water at 80 °C.
    Calculate:

(i) the $MCR_{F@A100°C}$.

(ii) the actual MCR when producing steam at 6 bar with a dryness fraction of 0.96 from feed water at 60 °C.

Solution

From property tables:

At 6 bar $h_{fg} = 2087 \text{ kJ/kg}$, $h_f = 670 \text{ kJ/kg}$
At 10 bar $h_{fg} = 2015 \text{ kJ/kg}$, $h_f = 763 \text{ kJ/kg}$
At 80 °C $h_f = 334.9 \text{ kJ/kg}$
At 60 °C $h_f = 251.1 \text{ kJ/kg}$

(i) $h = h_f + x h_{fg} = 763 + 0.98 \times 2015 = 2738 \text{ kJ/kg}$
Using Eqn [2.8]:

$$\dot{m}_{F@A100°C} = \dot{m}_{ACTUAL} \times \frac{(h - h_f)}{2256.7}$$

$$= 15\,000 \times \frac{(2738 - 334.9)}{2256.7}$$

$$= 15\,973 \text{ kg/h}$$

(ii) $h = h_f + x h_{fg} = 670 + 0.96 \times 2087 = 2674 \text{ kJ/kg}$
Using Eqn [2.8]:

$$\dot{m}_{F@A100°C} = \dot{m}_{ACTUAL} \times \frac{(h - h_f)}{2256.7}$$

$$15\,973 = \dot{m}_{ACTUAL} \times \frac{(2674 - 251.1)}{2256.7}$$

$$\dot{m}_{ACTUAL} = 14\,877 \text{ kg/h}$$

Peak Steam Flow Rate

It is possible to have a higher steam flow rate than the MCR value for a short period of time, provided the pressure is allowed to fall. The principle involved is similar to that used to produce flash steam from condensate. However, in this case we are dealing with a non-steady flow situation. A general analysis of the non-steady flow energy equation has been presented by Eastop and McConkey.[2.8]

Figure 2.19 shows sketches of a boiler before, during and after a peak load period. During a peak load period the following assumptions are made.

(i) The changes in density of the water in the boiler are negligible.
(ii) The mass flow rate of feed water is increased to match the new steam mass flow rate which will maintain the liquid level in the boiler so that the boiler water capacity m_c remains constant.

Figure 2.19 Peak load operation: (a) MCR load $\tau = 0$; (b) during peak load; (c) MCR load after τ hours

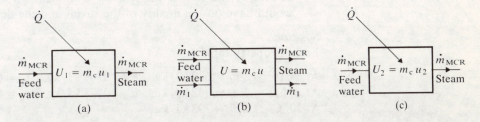

In Fig. 2.19(a) at the start of the peak period the boiler is operating at its MCR with the water capacity, m_c having an internal energy U_1. At the end of the peak period in Fig. 2.19(c) the boiler is again operating at its MCR rating and the water capacity has a new internal energy value U_2. During the peak load period in Fig. 2.19(b) the internal energy is falling, which produces an increase in the steam mass flow rate $\dot{m}_1$ kg/h. Calculations on the performance of a boiler during a peak flow condition are simplified in two ways if a F @ A 100 °C basis is used throughout: firstly, because the steam being generated by the heat supplied (the MCR mass flow rate F @ A 100 °C) remains constant; and secondly, because the specific enthalpy of the additional steam being generated increases by the specific enthalpy of vaporization at 100 °C (2256.7 kJ/kg). The actual MCR and actual increased mass flow rate must of course change as the specific enthalpy of the steam being delivered changes.

The peak mass flow rate F @ A 100 °C can be converted to an average actual mass flow rate using Eqn [2.8] and a mean value for the specific enthalpy of the steam. (The change in specific enthalpy of the steam is quite small.)

Over an interval of time τ hours the increased mass of steam produced is $\dot{m}_1\tau$ and the following energy balance applies:

Decrease in the internal energy of the
water = Increase in the enthalpy of the additional
steam produced
$$U_1 - U_2 = \text{Mass of steam produced} \times h_{fg}$$
$$m_c(u_1 - u_2) = \dot{m}_1\tau \times 2256.7 \qquad [2.9]$$

Peak loads are used when there is a sudden demand for steam for a short period of time, for example when steam is supplied to sterilization units in a hospital. When loads vary in this way it may be economic to increase the water capacity and allow the pressure to fall rather than use a larger boiler with an MCR the same as the peak load. A larger boiler would of course operate at a lower load factor.

With a small fire-tube boiler the feed pump may operate intermittently and the liquid level may fall during a short peak load. This would produce a further increase in the mass flow rate of steam.

Example 2.11

A boiler rated at 7000 kg/h F @ A 100 °C is used to produce steam at 10 bar. If the total boiler water capacity remains constant at 12 tonnes, calculate:

(i)   the peak rate of steam delivery in kg/h F @ A 100 °C that can be achieved over a 15-minute period if the pressure is allowed to drop to 5 bar over the same period;

(ii) the actual average mass flow rate of steam if it remains as a saturated vapour throughout and the feed water has a constant temperature of 70 °C;

(iii) the percentage increase in escape velocity of the steam from the water surface at the end of this period. Comment on the practical effect this would have on the quality of the steam and the design of the system.

Solution

From property tables:

At 5 bar $u_f = 639$ kJ/kg, $h_g = 2749$ kJ/kg, $v_g = 0.3748$ m^3/kg
At 10 bar $u_f = 762$ kJ/kg, $h_g = 2778$ kJ/kg, $v_g = 0.1944$ m^3/kg
At 70 °C $h_f = 293$ kJ/kg.

(i)   Using Eqn [2.9]:

$$m_c(u_1 - u_2) = \dot{m}_1 \tau \times 2256.7$$

$$12\,000(762 - 639) = \dot{m}_1 \times \frac{15}{60} \times 2256.7$$

$$\dot{m}_1 = 2616 \text{ kg/h}$$

The peak mass flow rate F @ A 100 °C is therefore $(7000 + 2616) = 9616$ kg/s.

(ii) Equation [2.8] can be used to find the actual average mass flow rate of steam using a mean value of h_g for the steam as follows:

$$h = \frac{(2778 + 2749)}{2} = 2763.5 \text{ kJ/kg}$$

$$\dot{m}_{F @ A 100°C} = \dot{m}_{ACTUAL} \times \frac{(h - h_f)}{2256.7}$$

$$9616 = \dot{m}_{ACTUAL} \times \frac{(2763.5 - 293)}{2256.7}$$

$$\dot{m}_{ACTUAL} = 8784 \text{ kg/h}$$

(iii) The change in escape velocity is proportional to the change in specific volume of the steam and the change in mass flow rate. Using F @ A 100 °C values for the mass flow rates (the actual mass flow rates must be in proportion) gives:

$$\text{Increase in velocity factor} = \frac{9616}{7000} \times \frac{0.3748}{0.1944} = 2.65$$

In percentage terms this is an increase of 165%.

This large increase in velocity would increase entrainment of water, thus decreasing the dryness fraction of the steam. If this has been taken into account at the design stage it need not be a problem. If it has not been taken into account this increase in velocity may be sufficient to cause priming when large masses of water are carried over from the boiler into the steam system. Whenever a boiler is likely to have to operate at a reduced pressure with an increased mass flow rate, it is essential to design for this as well as the maximum operating pressure. The maximum pressure determines the mechanical strength of components, but the minimum pressure must be used to size valves, pipework, etc.

Example 2.12

A process requires steam for 24 hours per day 365 days per year. Previous steam consumption records indicate the following loads based on a boiler operating pressure of 10 bar with a feed water temperature of 65 °C:

> Annual consumption, 15.78×10^6 kg;
> minimum summer load, 1281 kg/h;
> maximum winter load, 3602 kg/h;
> peak winter load, 5403 kg/h.

Peak loads occur for 10-minute periods once per hour. It is proposed to replace the old boiler with a new one having the following specification:

> MCR evaporation, 4800 kg/h F @ A 100 °C;
> normal working pressure, 10 bar;
> normal water capacity, 8000 litres;
> efficiency at MCR, 80%;
> fuel cost, £4/GJ.

Take the density of the hot water as 900 kg/m³ and, ignoring the mass of steam held in the system, calculate:

(i)   the minimum operating pressure during periods of peak load;
(ii) the time for the working pressure to recover after a peak load period;
(iii) the annual financial saving when using the proposed boiler compared with a larger boiler rated for the peak load at its MCR with the same MCR efficiency.

Use Fig. 2.18 for the variation of overall boiler efficiency with load factor.

(Wolverhampton Polytechnic)

Solution

From property tables:

At 10 bar $u_f = 762$ kJ/kg, $h_g = 2778$ kJ/kg
At 65 °C $h_f = 272$ kJ/kg,

(i)   The maximum and peak loads can be converted to F @ A 100 °C using Eqn [2.8] as follows:

$$\dot{m}_{F@A\,100°C} = \dot{m}_{ACTUAL} \times \frac{(h - h_f)}{2256.7}$$

$$\dot{m}_{max} = 3602 \times \frac{(2778 - 272)}{2256.7}$$

$$\dot{m}_{max} = 4000 \text{ kg/h F @ A 100 °C}$$

$$\text{Peak } \dot{m} = 5403 \times \frac{(2778 - 272)}{2256.7}$$

$$\therefore \quad \text{Peak } \dot{m} = 6000 \text{ kg/h F @ A 100 °C}$$

The increase in mass flow rate required to satisfy the peak load $\dot{m}_1$ is therefore:

$$\dot{m}_1 = \text{Peak } \dot{m} - \text{Boiler MCR}$$
$$\therefore \quad \dot{m}_1 = 6000 - 4800 = 1200 \text{ kg/h}$$

The mass of water in the boiler, m_c, is $8000 \times 900/1000 = 7200$ kg. Equation [2.9] can now be used to determine the specific internal energy of the water at the end of the 10-minute peak period:

$$m_c(u_1 - u_2) = \dot{m}_1\tau \times 2256.7$$

$$7200(762 - u_2) = 1200 \times \frac{10}{60} \times 2256.7$$

Hence $u_2 = 699.3$ kJ/kg

By interpolating from property tables, the pressure is 7.1 bar.

(ii) Recovery time:

$$\text{Spare capacity} = \text{Boiler MCR} - \text{System maximum load}$$
$$= 4800 - 4000$$
$$= 800 \text{ kg/h F @ A } 100\,°C$$

$$\text{Capacity used} = \dot{m}_1\tau = 1200 \times \frac{10}{60} = 200 \text{ kg F @ A } 100\,°C$$

$$\therefore \quad \text{Recovery time} = \frac{\text{Capacity used}}{\text{Spare capacity}} = \frac{200}{800} = 0.25 \text{ h or 15 min}$$

(iii) Annual performance:

The annual steam production is converted to an F @ A $100\,°C$ basis using Eqn [2.8]:

$$\dot{m}_{\text{F @ A }100°C} = \dot{m}_{\text{ACTUAL}} \times \frac{(h - h_f)}{2256.7}$$

$$\dot{m} = 15.78 \times 10^6 \times \frac{(2778 - 272)}{2256.7}$$

$$= 17.52 \times 10^6 \text{ kg/year F @ A } 100\,°C$$

$$\text{Average load} = \frac{\text{Annual production}}{\text{Running time}} = \frac{17.52 \times 10^6}{365 \times 24} = 2000 \text{ kg/h}$$

The alternative to allowing the steam pressure to fall would be to install a boiler which has an MCR equal to the peak load of 6000 kg/h. The load factor can now be worked out for each boiler using Eqn [2.7] as follows:

4800 kg/h boiler:

$$\text{Load factor} = \frac{2000}{4800} = 0.42$$

From Fig. 2.18 the overall efficiency is 73%.

6000 kg/h boiler:

$$\text{Load factor} = \frac{2000}{6000} = 0.33$$

From Fig. 2.18 the overall efficiency is 71%.

The annual energy consumptions can now be worked out for each boiler:

6000 kg/h boiler:

$$\frac{\text{Mass} \times \Delta h}{\eta} = \frac{17.52 \times 10^6 \times 2256.7}{0.71} = 55\,686 \text{ GJ}$$

4800 kg/h boiler:

$$\frac{\text{Mass} \times \Delta h}{\eta} = \frac{17.52 \times 10^6 \times 2256.7}{0.73} = 54\,161\,\text{GJ}$$

$$\text{Saving in energy} = 55\,686 - 54\,161 = 1525\,\text{GJ}$$

$$\therefore \quad \text{Annual financial saving} = 1525 \times £4 = £6100$$

The Effect of Water Capacity on Boiler Response

Different types of boiler with the same MCR can have quite different water capacities. In general, fire-tube boilers (combustion products inside tubes and passages with the water outside), also termed shell boilers, have high water capacities. Water tube boilers (the water inside tubes with the products of combustion outside) have low water capacities. The capacity of a water-tube boiler can be increased by adding an additional steam drum which will act as an accumulator of internal energy.

A boiler with a low water capacity will be able to generate steam at the working pressure from a cold start-up relatively quickly. A boiler with a high water capacity will take longer to reach the working pressure. When the pressure needs to be maintained and the flow rates are below the MCR value, low water capacity boilers respond more quickly to load changes than high water capacity boilers. If the pressure is allowed to fall, boilers with a high water capacity can maintain peak steam flows for a longer time than boilers with low water capacities.

Boiler Blowdown

All water contains some dissolved solids. When water evaporates it leaves most of the dissolved solids behind. If no action is taken the concentration of dissolved solids in the water remaining will increase until solids are deposited on the heat transfer surfaces, which will lead to increased metal temperatures and eventual tube or plate failure. It will also cause carry-over of liquid into the steam system (priming). It is therefore necessary to prevent the build-up of dissolved solids becoming excessive by removing some of the water.

Total dissolved solids (TDS) are measured in *parts per million* (ppm). Tolerable levels vary considerably for different types of boiler: 15000 ppm might be acceptable for the old Lancashire type of boiler, but a modern water-tube boiler could require less than 50 ppm. The Energy Efficiency Office booklet *Boiler Blow-Down*[2.10] quotes 2000–3500 ppm as being typical for modern shell (fire-tube) boilers, which are the ones most commonly used in building services.

Intermittent blowdown

An intermittent blowdown is connected to the lowest point of the boiler. In addition to removing dissolved solids it also removes any sludge which may have accumulated. The valve can be operated on a manual basis of, say, once per shift, or by a time signal. The disadvantage of the latter system is that the blowdown required would be fixed by the design load and for most of the time this would be in excess of what was necessary: any such excess is a waste of energy. With manual operation a sample can of course be taken to determine how much blowdown is necessary but this requires a skilled, conscientious operator. Intermittent blowdown as the sole means of TDS control is most likely to be used on small installations. However, it is always required as a means of removing sludge. Intermittent blowdown should be carried out during periods of light load.

Continuous blowdown

Continuous blowdowns are connected to just below the normal liquid level and only remove dissolved solids. They are situated here because the concentration will be greatest at the point where the water evaporates. The higher the average blowdown required, the more worthwhile it becomes to install one.

For a boiler with a steady load the blowdown required could be calculated and the valve set accordingly. In practice, samples would need to be taken and the valve adjusted from time to time. However, most boilers have fluctuating loads and the blowdown rate should be controlled to maintain the TDS at a satisfactory level.

An automatic TDS level control system is shown in Fig. 2.20. A sample is continuously passed through a water-cooled heat exchanger which reduces the temperature before it enters a monitor. The monitor normally measures thermal conductivity, which varies with the TDS. There are two valves which can discharge the blowdown. One of these is set to the minimum blowdown rate and the other is controlled by a signal from the monitor. The control valve opens only when the TDS level reaches a predetermined limit and it closes again when the TDS content falls.

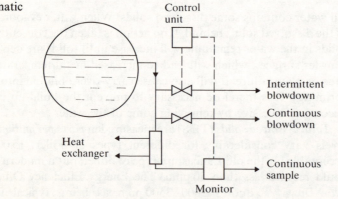

Figure 2.20 Automatic TDS control

Calculation of boiler blowdown flow rate

The TDS level in return condensate is usually quite small and may be neglected. If the TDS of the make-up water is known, the TDS of the boiler feed water can be calculated by mixing.

Considering unit mass flow rate of steam and a mass fraction x of blowdown, the feed water being supplied becomes $(1 + x)$ kg/s. If TDS_F and TDS_B refer to the feed and boiler TDS respectively then assuming no solids in the steam allows the following TDS balance to be carried out:

Mass flow rate of TDS leaving = Mass flow rate of TDS entering

$$x \, TDS_B = (1 + x) \, TDS_F$$

Rearranging,
$$x = \frac{TDS_F}{TDS_B - TDS_F}$$

Expressing the blowdown as a percentage of the steam flow gives:

$$Blowdown = \frac{100 \, TDS_F}{TDS_B - TDS_F} \, \% \qquad [2.10]$$

If unit mass flow rate of feed water is considered, the percentage blowdown expressed as a percentage of the feed water flow rate is:

$$Percentage \; blowdown = \frac{100 \, TDS_F}{TDS_B} \qquad [2.11]$$

Equations [2.10] and [2.11] give the same mass flow rate of blowdown. The Energy Efficiency Office booklet[2.10] recommends the use of Eqn [2.10].

Example 2.13
A boiler produces 20 000 kg/h of saturated steam at 15 bar. The make-up water to the hot well contains 300 ppm TDS. If the TDS in the boiler is not to exceed 2000 ppm, calculate the mass flow rate of blowdown required when:
(i) there is no condensate recovery;
(ii) there is 80% condensate recovery.

Solution
(i) Using Eqn [2.10]:
$$Blowdown = \frac{TDS_F}{TDS_B - TDS_F} \times 100 = \frac{300}{(2000 - 300)} \times 100 = 17.65\%$$

Mass flow rate of blowdown = 20 000 × 0.1765 = 3529 kg/h

(ii) Assuming no dissolved solids in the condensate recovered:

Feed water TDS = 0.2 × 300 + 0.8 × 0 = 60 ppm

Using Eqn [2.10]:

$$Blowdown = \frac{TDS_F}{TDS_B - TDS_F} \times 100 = \frac{60}{(2000 - 60)} \times 100 = 3.1\%$$

Mass flow rate of blowdown = 20 000 × 0.031 = 620 kg/h

This example demonstrates the importance of recovering condensate whenever possible. Blowdown losses are expensive because of the energy consumed and the cost of the water and its treatment. It is invariably worthwhile recovering energy from a continuous blowdown. This is not always true for intermittent blowdowns unless there are a number of boilers with coordinated blowdowns to even out the supply of heat. Some heat recovery schemes have been used in earlier examples.

PROBLEMS

2.1 A high-pressure hot water system shown in Fig. 2.21 is pressurized by nitrogen. Calculate the maximum permissible flow temperature and the anti-flash margin in the worst point in the system if the pump fails.

Figure 2.21 Problem 2.1

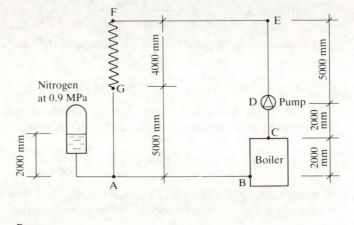

Data

Water density in the nitrogen vessel	1000 kg/m³
Water density in the heating system	900 kg/m³
Anti-flash margin	10 °C

Pipe section	AB	BC	CD	DE	EF	FG	GA	Total
Pressure drop/(kPa)	10	15	5	20	15	60	10	135

(Dublin Institute of Technology)

(163 °C; 9.3 K)

2.2 (a) Critically compare high-pressure hot water and steam for combined process and space heating.

(b) Figure 2.22 shows a nitrogen pressurized high-temperature hot water system. Tabulate the variations in pressure around the system and hence calculate:

(i) the nitrogen pressure required with the pump running;

(ii) the nitrogen pressure required to maintain the same anti-flash margin if the pump fails.

Figure 2.22 Problem 2.2

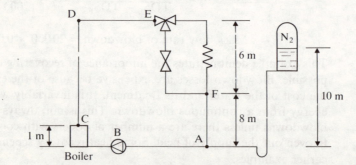

Data

Water temperature leaving the boiler, 180 °C; anti-flash margin 10 K; water density in the system, 900 kg/m³; water density in the expansion pipe, 1000 kg/m³.

Section	AB	BC	CD	DE	EF	FA	Total
Pressure loss/(kPa)	10	10	20	10	90	10	150

(Wolverhampton Polytechnic)

(12.2 bar; 12.8 bar)

2.3 The following data refer to the high-pressure hot water system shown in Fig. 2.23.

Figure 2.23 Problem 2.3

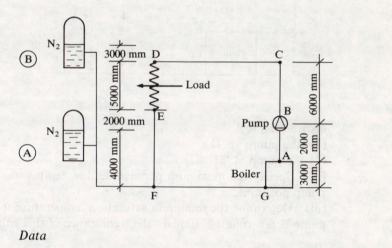

Data

Pipe	AB	BC	CD	DE	EF	FG	GA	Total
p/(kPa)	5	10	20	60	10	20	5	130

Water density in the cold feed, 1000 kg/m³; water density in the system, 900 kg/m³; boiler temperature, 160 °C; anti-flash margin, 10 K.

Calculate:
(i) the nitrogen pressure required when the vessel is at A and B when the pump is running;
(ii) the nitrogen pressure required when the vessel is at A and B to maintain the 10 K anti-flash margin at the weakest point in the system if the pump fails.

(Dublin Institute of Technology)

(8.3, 7.3; 8.5, 7.5)

2.4 (a) Discuss the principal advantages and disadvantages involved with the use of high-pressure, rather than low-pressure, hot water as a heating medium.
(b) A gas-pressurized HPHW heating system is shown in Fig. 2.24. Using the data given below:
(i) Evaluate the pressure at each point in the system and sketch the pressure

Figure 2.24   Problem
2.4

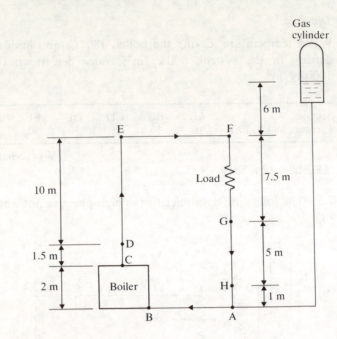

distribution for:

(1)   the pump at 'D';

(2)   the pump at 'H'.

(ii)   Specify the maximum permissible flow temperature for each of the two pump positions.

(iii)   Determine the minimum saturation temperature in the system when the pump is not running. Explain the significance of this value.

Data

Gas pressurization is constant at 9 bar (absolute). Anti-flash margin is 11 K.

Circuit pressure drops are as follows:

Section	AB	BC	CD	DE	EF	FG	GH	HA
Pressure drop/(kPa)	20	7	10	20	20	100	10	10

(CIBSE)

(169 °C at D_{IN}, 163 °C at F; assuming G is exposed only to the return water temperature, 166.6 °C—all based on a water density of 900 kg/m³ throughout).

2.5   Repeat problem 2.4 assuming the water in the expansion vessel and pipe has a density of 1000 kg/m³.

(169.9 °C, 163.9 °C; 167.4 °C)

2.6   The high-pressure hot water heating system shown in Fig. 2.25 serves two process loads via two-position three-way control valves.

Produce a table or graph to show the hydrostatic, circulation and resultant pressure variations for the system (A to J inclusive) and hence determine the minimum pressure required in the pressure vessel in order to maintain a minimum anti-flash margin of 20 K under all operating conditions with the pump running. The characteristic of the direct driven pump is given below.

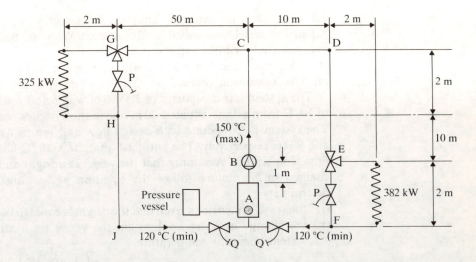

Figure 2.25 Problem 2.6

Assume that all the boiler pressure loss occurs in the section AB, that control valve losses occur in sections EF and GH, and that all thermal losses from pipework, etc., are negligible. The regulating valves marked P are to maintain the pressure differentials from GH and EF at their respective values under both load and no-load conditions and the regulating valves marked Q are to balance the system.

Data

Pressure losses: Boiler, 10 kPa; 382 kW load, 50 kPa; 325 kW load, 30 kPa; three-way valves, 20 kPa; pipework (inc. fittings), 500 Pa/m.

Water density, 918 kg/m³; water specific heat capacity 4.28 kJ/kg K.

Pump characteristics:

Pressure/(kPa)	165	160	150	130	90
Flow rate/(litre/s)	2	4	6	8	10

(CIBSE)

(8.4 bar)

2.7 (a) Describe with the aid of diagrams two methods available for the pressurization of hot water heating systems.

(b) Derive the expression for the sizing of sealed, air-filled expansion vessels for pressurizing heating systems:

$$V_1 = \frac{p_2^{1/n} p_3^{1/n} E}{p_1^{1/n}(p_3^{1/n} - p_2^{1/n})}$$

where
- V_1 = volume of the expansion vessel;
- p_1 = initial pressure (normally atmospheric);
- p_2 = cold fill pressure;
- p_3 = pressure at system operating temperature;
- n = index of compression;
- E = expansion of the water content of the system on heating from cold to the operating temperature.

(Newcastle upon Tyne Polytechnic)

2.8 (a) Describe with sketches the construction of a diaphragm type of expansion vessel for a sealed heating system. What are the principal advantages over a sealed air-filled expansion vessel?

(b) A pressurized hot water heating system is to be installed with a sealed air-filled expansion vessel.

The system has a capacity of 3.5 m³ of water and is filled at a temperature of 10 °C from a tank situated 20 m above the system's sealed expansion vessel. The system is to operate with design flow and return temperatures of 140 °C and 90 °C respectively. The anti-flash margin is 10 °C based on the maximum flow temperature. Assuming that the vessel is upright and cylindrical and that changes in air volume follow the equation $pV^n = $ constant, where $n = 1.26$, determine:

(i) the vessel dimensions given that the height is equal to twice the diameter;
(ii) the change in water level within the vessel on heating from cold to the system operating temperature.

(Newcastle upon Tyne Polytechnic)

(0.99 m, 1.98 m; 262 mm)

2.9 A high-pressure hot water heating system is filled with water at a density of 1000 kg/m³ and operates with a mean water density of 916.6 kg/m³. During the sizing of the expansion vessel, the expansion of the pipework emitters, etc., is not considered. If the coefficient of cubical expansion of the metal is 36×10^{-6} 1/K, calculate the maximum percentage error this could involve.

(5.5%)

2.10 Figure 2.26 shows part of a nitrogen-pressurized hot water heating system to which the following data apply:

Section	AB	BC	CD	DE	EF	FG	GA
Circulation p/(kPa)	10	10	20	90	20	10	10

Figure 2.26 Problem 2.10

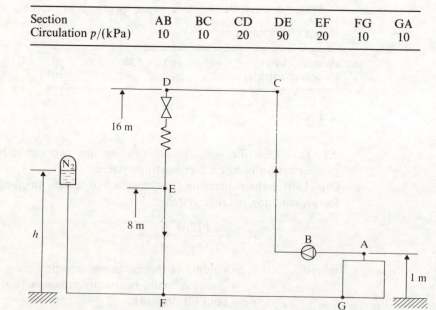

Mean density of the hot water, 900 kg/m³; density of the cold water, 1000 kg/m³; water flow temperature, 180 °C; water return temperature, 140 °C; nitrogen pressure, 12 bar; anti-flash margin with the pump running, 10 K.

Tabulate the relative pressure variations around the system and determine:
(i)   the height h of the expansion vessel;
(ii) the anti-flash margin if the pump should fail.

(Wolverhampton Polytechnic)

(9.56 m; 5.9 K)

2.11   500 kg/h of saturated condensate at 10 bar is passed through a pressure-reducing valve which it leaves at 1.5 bar. Calculate the mass flow rate of saturated steam produced after the valve.
(66.5 kg/h)

2.12   A hot water storage calorifier has a capacity of 2100 litres which can be raised from 12 °C to 60 °C in 2 hours when supplied with saturated steam at 7 bar. Calculate:
(i)   the steam consumption without condensate subcooling;
(ii) the percentage reduction in steam consumption if a condensate sub-cooler is added from which condensate is discharged at 100 °C.
(101.9 kg/h; 11.9%)

2.13   A boiler generates steam at a pressure of 8 bar and a dryness fraction of 0.94. This steam is supplied at 2 kg/s to a process load and the condensate leaves sub-cooled by 5.4 K. The condensate is discharged to a flash vessel which operates at 2 bar and supplies a second load with 0.5 kg/s of steam at a dryness fraction of 0.97. If the condensate from the low-pressure load is sub-cooled by 15.4 K and the condensate from the flash vessel is saturated, calculate:
(i)   the two heat loads;
(ii) the mass flow rate of steam which must be passed from the boiler to the flash vessel through a pressure-reducing valve;
(iii) the mass flow rate of flash steam produced if the two low-pressure condensate streams are passed directly to a hot well operating at atmospheric pressure.
(3898 kW, 1101 kW; 0.322 kg/s; 0.0737 kg/s)

2.14   Figure 2.27 shows a steam system in which the boiler produces steam at 8 bar and a dryness fraction of 0.9. The details of the system loads are as follows:
Load 1—3000 kW using steam at 8 bar and 0.9 dry.

Figure 2.27   Problem 2.14

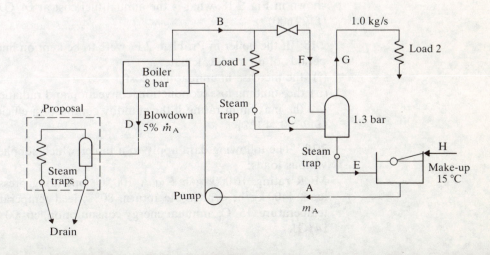

Load 2—1.0 kg/s of saturated steam at 1.3 bar in a direct injection process (no condensate recovery).

It is proposed that a flash vessel, operating at 1.3 bar, is used to recover heat from the boiler blowdown water to heat the fuel oil tanks. Assume that there is no under-cooling at the steam traps and that losses from the condensate pipework and hot well are negligible. Hence determine:

(i) the rate of heat recovery in the blowdown in kW;

(ii) the hot well temperature if the make-up water is at 15 °C.

(Wolverhampton Polytechnic)

(40.8 kW; 67.3 °C)

2.15 (a) Describe with sketches the construction features of a flash steam vessel and show how this vessel may be used in conjunction with high-pressure condensate to provide low-pressure steam for a heating system.

(b) Steam is supplied dry saturated at 10 bar to a process plant, having a duty of 300 kW.

It is proposed to heat a building near the process plant by means of unit heaters, utilizing steam at 1.5 bar, from a flash steam vessel which takes condensate from the process plant. Live steam make-up is also available from the steam main at 10 bar, through a pressure-reducing valve. The space heating load is 85 kW. If the steam available from the flash vessel is dry saturated, calculate:

(i) the quantity of live steam make-up required;

(ii) the condition of the steam as supplied to the unit heaters.

(Newcastle upon Tyne Polytechnic)

(63.7 kg/h; 1.5 bar, 131 °C)

2.16 A boiler is rated at 10 000 kg/h F @ A 100 °C. Calculate:

(i) its MCR steam production when generating saturated steam at 9 bar from feed water at 80 °C;

(ii) the condensate temperature if the system has 80% condensate return with a make-up at 15 °C.

(9253 kg/h; 96.2 °C)

2.17 The average load on a boiler, from log sheets, is 3200 kg/h with an annual production of 9 984 000 kg both at 8 bar saturated from feed water at 80 °C. If the boiler is rated at 6000 kg/h F @ A 100 °C and has the operating characteristics shown in Fig. 2.18, what is the annual fuel cost at £4/GJ?

(£129 600)

2.18 If the boiler in Problem 2.17 were to be kept on line as a standby during periods of no load, calculate:

(i) the increase in annual fuel cost;

(ii) the standing loss (conduction, convection and radiation) as a percentage of the maximum rating if the standby combustion efficiency is 70%.

(£17 700; 4.1%)

2.19 The following data apply to a boiler which supplies saturated steam to various loads:

MCR rating, 16 000 kg/h F @ A 100 °C; operating pressure, 8.5 bar; average load, 4400 kg/h; condensate return, 80%; feed temperature, 75 °C; make-up temperature, 15 °C: annuual energy consumption, 66 000 GJ/annum: fuel cost, £4/GJ.

Due to the implementation of a heat recovery scheme, a constant air heater load of 750 kg/h of steam with 100% condensate return has been eliminated. Assuming that all condensate is returned at the same temperature, calculate:
(i) the new feed water temperature;
(ii) the savings that will result from the heat recovery scheme in £/annum.
Figure 2.18 shows the variation in overall boiler efficiency with average load factor.

(Wolverhampton Polytechnic)

(71.9 °C; £35 000)

2.20 (a) Explain what is meant by the blowdown of boilers and why it is a necessary part of boiler operation. Discuss the advantages of continuous as opposed to intermittent blowdown for large-scale steam-raising plant. How might continuous blowdown be controlled?
(b) A boiler is required to produce 75 000 kg/h of dry saturated steam at 30 bar. The boiler feed water comprises 40% condensate return and 60% make-up water containing 400 ppm of total dissolved solids (TDS). Determine the required rate of continuous blowdown if the boiler TDS is not to exceed 3000 ppm. The blowdown fluid is passed to a flash steam vessel which produces dry saturated steam at 1.5 bar. Determine:
(i) the quantity of flash steam produced;
(ii) the heat recovered if the flash steam is fed to a heat exchanger from which condensate leaves at 70 °C.

(Newcastle upon Tyne Polytechnic)

(6522 kg/h; 1585 kg/h; 1057 kW)

REFERENCES

2.1   CIBSE 1986 *Guide to Current Practice* volume B1, Table B1.12
2.2   Rogers G C F and Mayhew .Y R 1988 *Thermodynamic and Transport Properties of Fluids* 4th edn Basil Blackwell
2.3   Faber O and Kell T J 1989 *Heating and Air Conditioning of Buildings* Architectural Press
2.4   Douglas J F, Gasiorek J M and Swaffield L A 1985 *Fluid Mechanics* 2nd edn Longman
2.5   CIBSE 1986 *Guide to Current Practice* volume B16
2.6   Gunn D and Horton D 1989 *Industrial Boilers* Longman
2.7   Babcock and Wilcox 1972 *Steam: Its Generation and Use* 38th edn
2.8   Eastop T D and McConkey A 1986 *Applied Thermodynamics for Engineers and Technologists* 4th edn Longman
2.9   Eastop T D and Croft D R 1990 *Energy Efficiency* Longman
2.10 *Boiler Blowdown* 1983 Energy Efficiency Office

GENERAL READING

McLaughlin R K, McLean R C and Bonthron W J 1981 *Heating Services Design* Butterworths

3 ENVIRONMENTAL NETWORKS AND THERMAL COMFORT

In Chapter 1 an introduction is given to the transfer of heat to and from buildings in the steady state. This chapter extends the steady-state treatment to include the effects of radiation, both within the building and due to solar radiation falling on the windows and outside walls. Thermal comfort conditions for the building occupants are also considered.

3.1 STEADY-STATE ENERGY NETWORK

The model used in the UK is that recommended by the Chartered Institution of Building Services Engineers.[3.1] An environmental point is introduced, defined as the point at which all the radiant components of the energy input are received.

A body in an enclosed space exchanging heat by radiation with the surfaces of the space will be at an equilibrium temperature given by:

$$(t_{mr} + 273)^4 = \sum \{F_b(t_s + 273)^4\} \qquad [3.1]$$

where t_{mr} is the *mean radiant temperature*; F_b is the geometric factor from the body to any one of the surfaces (for a definition of geometric factor, see Ref. 1.1); t_s is the temperature of any one of the surfaces.

For a body at the centre of a cubical enclosure Eqn [3.1] reduces as an approximation to

$$t_{mr} = t_m = \frac{\sum (A_s t_s)}{\sum (A_s)} \qquad [3.2]$$

where t_m is the mean surface temperature; A_s is the area of any one of the surfaces at a temperature t_s.

Outside Environmental Temperature or sol-air Temperature

An environmental temperature is one which allows for the effects of radiation from and to internal or external surfaces of a building.

For the outside of a building the radiation effect is due entirely to solar radiation. The net radiation per unit area of non-glazed outside surface is given by:

$$\dot{q}_r = (\alpha I - \varepsilon I_L)$$

where α is the absorptivity of the surface; I is the incident solar radiation intensity; ε is the emissivity of the surface; I_L is the long-wave radiation intensity from the surface.

For thermal equilibrium, the net radiation to the surface is equal to the heat convected from the surface through the resistance of the film on the outside surface, R_{so}, plus the heat conducted through the wall to the inside air through the resistance, $(1/U) - R_{so}$. Since $\dot{q} \propto (1/R)$, it follows that the proportion of heat due to the solar radiation which enters the room is given by:

$$\frac{R_{so}}{(1/U) - R_{so} + R_{so}} = UR_{so}$$

i.e.   Heat entering room due to radiation $= UR_{so}(\alpha I - \varepsilon I_L)$

The modified fabric loss, $\dot{Q}_F$, is then given by:

$$\dot{Q}_F = UA(t_{ai} - t_{ao}) - UAR_{so}(\alpha I - \varepsilon I_L)$$

Defining an outside environmental temperature, t_{eo}, as follows:

$$\dot{Q}_F = UA(t_{ai} - t_{eo})$$

we then have

$$UA(t_{ai} - t_{eo}) = UA(t_{ai} - t_{ao}) - UAR_{so}(\alpha I - \varepsilon I_L)$$

i.e.   Outside environmental temperature, $t_{eo} = t_{ao} + R_{so}(\alpha I - \varepsilon I_L)$

$$[3.3]$$

The outside environmental temperature is also known as the *sol-air temperature*.

Values of sol-air temperature are tabulated by CIBSE[3.2] for South-East England for vertical walls of different orientation and for a horizontal surface; a distinction is made between dark and light surfaces which have different values of absorptivity and emissivity. Each table covers a 24-hour period and there are eight tables covering the period from 21 March to 22 October. The variation of outside air temperature is also given.

For example, at 21 June for a south-facing dark wall we have

At 1200 h (sun time): $t_{eo} = 44.5\,°C; t_{ao} = 19.0\,°C$
At 1500 h (sun time): $t_{eo} = 36.5\,°C, t_{ao} = 21.5\,°C$

Note that for a north-facing wall on 21 June at 1200 h (sun time) $t_{eo} = 26.5\,°C$, $t_{ao} = 19.0\,°C$; the outside environmental temperature is still greater than the air temperature even though there is no direct radiation; this is due to diffuse sky radiation.

The values are based on measurements made at the Kew Observatory during the ten-year period 1959–1968 with an average over two days taken for each measurement. The values in the tables may be applied to any region of the UK with only a small error. For other countries the values of sol-air temperature can be calculated using Eqn [3.3].

In the UK in winter, since the sky is overcast for much of the time and the time from sunrise to sunset is comparatively short, it is usual to assume that the sol-air temperature is equal to the outside air temperature when calculating fabric heat losses.

Internal Environmental Temperature

The internal environmental temperature is defined in a similar way to the outside environmental temperature. It is the temperature in the room which, in the absence of incident radiation on the internal walls, would give the same rate of heat transfer through the external fabric as exists with the actual air temperature and the incident radiation on the internal surfaces; that is, the internal environmental temperature, t_{ei}, is given by writing:

$$\text{Fabric loss, } \dot{Q}_F = (t_{ei} - t_{eo})\Sigma(UA_o)$$

where the term $\Sigma(UA_o)$ is the sum of the product of the overall heat transfer coefficient and the area, for each part of the fabric which exchanges heat with the outside of the building.

By making suitable assumptions an approximate expression can be derived relating t_{ei} to the inside air temperature, t_{ai}, and the mean surface temperature inside the room, t_m.

Consider an internal room surface of area A_s at temperature t_s, exchanging heat with the air at t_{ai} and the other surfaces, area $= \Sigma A - A_s$. The surfaces radiating heat to the surface A_s are at a mean temperature $(t_m\Sigma A - t_sA_s)/(\Sigma A - A_s)$, and hence the temperature difference between the remaining surfaces and A_s is given by:

$$\frac{(t_m\Sigma A - t_sA_s)}{\Sigma A - A_s} - t_s = \frac{(t_m - t_s)\Sigma A}{\Sigma A - A_s}$$

$$\text{Heat transfer, } \dot{Q} = \alpha_c A_s(t_{ai} - t_s) + \alpha_r A_s\left(\frac{t_m - t_s}{\Sigma A - A_s}\right)\Sigma A$$

where α_c is the heat transfer coefficient due to convection; α_r is the heat transfer coefficient due to radiation between the surface A_s and the remaining surfaces of area $\Sigma A - A_s$; t_m is the mean temperature of all the internal surfaces as defined by Eqn [3.2].

Then from the definition of t_{ei} we can write:

$$\dot{Q} = (t_{ei} - t_s)\left(\alpha_c + \frac{\alpha_r\Sigma A}{\Sigma A - A_s}\right)A_s$$

Equating this expression for $\dot{Q}$ to that above we have:

$$(t_{ei} - t_s)\left(\alpha_c + \frac{\alpha_r\Sigma A}{\Sigma A - A_s}\right)A_s = \alpha_c A_s(t_{ai} - t_s) + \alpha_r A_s\frac{(t_m - t_s)\Sigma A}{\Sigma A - A_s}$$

Hence,

$$t_{ei} = \frac{\alpha_c t_{ai} + \alpha_r t_m\Sigma A/(\Sigma A - A_s)}{\alpha_c + \alpha_r\Sigma A/(\Sigma A - A_s)} \tag{1}$$

As an approximation, the mean value of α_r can be taken as $5\ \text{W/m}^2\,\text{K}$, and the mean value of α_c as $3\ \text{W/m}^2\,\text{K}$. Also, for a cubical enclosure the area of one of the internal surfaces, A_s, is one-sixth of the total area, ΣA; hence, substituting in Eqn [1]:

$$t_{ei} = t_{ai}\left\{\frac{3}{3 + (5 \times 6/5)}\right\} + t_m\left\{\frac{5 \times 6/5}{3 + (5 \times 6/5)}\right\}$$

i.e.

$$t_{ei} = \frac{1}{3}t_{ai} + \frac{2}{3}t_m \qquad\qquad [3.4]$$

The heat transferred between all the surfaces and the air is given by:

$$(\alpha_c + \alpha_r)\Sigma A(t_m - t_{ei})$$

Then from Eqn [3.4],

$$t_m = \frac{3t_{ei}}{2} - \frac{t_{ai}}{2}$$

Hence the heat transferred between all internal surfaces and the air inside is

$$(\alpha_c + \alpha_r)\Sigma A\left(\frac{3t_{ei}}{2} - \frac{t_{ai}}{2} - t_{ei}\right)$$

$$= (\alpha_c + \alpha_r)\Sigma A\left(\frac{t_{ei} - t_{ai}}{2}\right)$$

$$= \alpha_a \Sigma A(t_{ei} - t_{ai}) \qquad\qquad [2]$$

where α_a is an imaginary heat transfer coefficient linking the air and environmental points and is equal to $(\alpha_c + \alpha_r)/2 = 4.5 \text{ W/m}^2 \text{ K}$.

A steady-state network can then be drawn as shown in Fig. 3.1. The steady-state heat input to the room is $\dot{Q}_i$, which is equal to the sum of the heat inputs at the air point, $\dot{Q}_a$, and environmental point, $\dot{Q}_e$.

Figure 3.1 Steady-state network

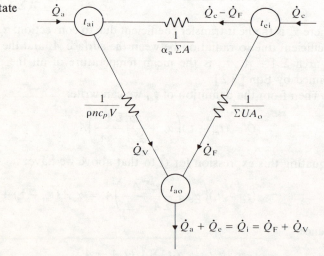

The heat loss due to ventilation (see Eqn [1.7]) is

$$\dot{Q}_V = \rho n c_p V(t_{ai} - t_{ao})$$

giving a resistance for the left-hand arm of the network of $1/\rho n c_p V$.

Note: It is a good approximation in practice to assume a value of ρc_p of 1200 J/m³ K, and hence the expression $\rho n c_p V$ is frequently written as $(nV/3)$ W/K, where n is the air change rate per hour and V is the volume in m³.

In winter it is usual to neglect the effect of solar gains when calculating the fabric loss; this ensures that the heating plant for the building can maintain the required internal temperature on days when solar gains are negligible. Hence the difference between t_{eo} and t_{ao} can be neglected and we have:

$$\text{Fabric loss, } \dot{Q}_F = \Sigma U A_o (t_{ei} - t_{ao})$$

giving a resistance for the right-hand arm of the network of $1/\Sigma U A_o$.

The resistance of the arm at the top of Fig. 3.1 is seen from Eqn [2] to be $1/\alpha_a \Sigma A$.

There are two special cases

(a) $\dot{Q}_a = 0$

This is shown in Fig. 3.2. In this case the heat flow through the top arm is $\dot{Q}_V$, the same as through the left-hand arm, and then

$$\dot{Q}_V \left(\frac{1}{\alpha_a \Sigma A} + \frac{1}{\rho n c_p V} \right) = t_{ei} - t_{ao}$$

Figure 3.2 Steady-state network: (a) $\dot{Q}_a = 0$

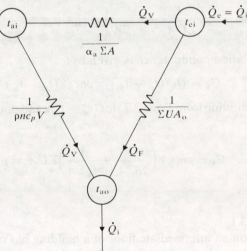

The ventilation loss, $\dot{Q}_V$, is frequently required in terms of the temperature difference, $(t_{ei} - t_{ao})$, and is written as

$$\dot{Q}_V = C_V (t_{ei} - t_{ao}) \qquad [3.5]$$

where C_V is known as the *ventilation conductance*.

It can be seen that, in this case,

$$C_V = \frac{1}{\dfrac{1}{\alpha_a \Sigma A} + \dfrac{1}{\rho n c_p V}} \qquad [3.6]$$

(b) $\dot{Q}_e = 0$

Referring to Fig. 3.3, in this case the heat flow through the top arm is $\dot{Q}_F$, the

Figure 3.3 Steady-state network: (b) $\dot{Q}_e = 0$

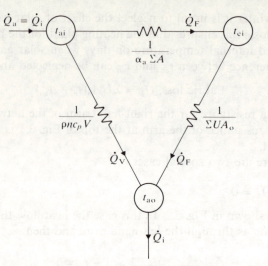

same as through the right-hand arm, and then

$$\dot{Q}_F\left(\frac{1}{\Sigma U A_o} + \frac{1}{\alpha_a \Sigma A}\right) = t_{ai} - t_{ao} \qquad [3.7]$$

The ventilation conductance is given by:

$$C_V = \dot{Q}_V/(t_{ei} - t_{ao}) = \rho n c_p V(t_{ai} - t_{ao})/(t_{ei} - t_{ao})$$

Then substituting from Eqn [3.7] for $(t_{ai} - t_{ao})$, and putting $(t_{ei} - t_{ao}) = \dot{Q}_F/\Sigma U A_o$, we have:

$$C_V = \rho n c_p V\left(\frac{1}{\Sigma U A_o} + \frac{1}{\alpha_a \Sigma A}\right)\Sigma U A_o = \rho n c_p V\left(1 + \frac{\Sigma U A_o}{\alpha_a \Sigma A}\right) \qquad [3.8]$$

Example 3.1

An office on an intermediate floor of a building has one external wall 6 m long, containing windows of total area 3 m²; the depth of the office is 3 m and the ceiling height, 2.5 m. The office is surrounded above, below and on all internal sides by spaces at the same temperature.

(a)   Using the data below calculate for an air change rate of 1 per hour:
(i)   the internal air temperature and the heat input required when all the heat input is at the air point;
(ii)   the internal air temperature and the heat input required when all the heat input is at the environmental point;
(iii) the ventilation conductances for cases (i) and (ii).
(b)   Re-calculate part (a) for the case when the air change rate is 2 per hour.

Data

Room environmental temperature, 20 °C; outside air temperature, −1 °C.
    Thermal transmittances: outside wall, 0.5 W/m² K; window, 4.3 W/m² K.

Solution
(a)(i)   Total area, $\Sigma A = 2\{(6 \times 2.5) + (3 \times 2.5) + (6 \times 3)\} = 81$ m²

For the fabric loss,

$$\Sigma U A_{o} = (3 \times 4.3) + \{(6 \times 2.5) - 3\} \times 0.5 = 18.9 \text{ W/K}$$

For the ventilation loss, taking ρc_p as $1200 \text{ J/m}^3 \text{ K}$.

$$\rho n c_p V = \frac{1 \times 1200}{3600} (6 \times 2.5 \times 3) = 15 \text{ W/K}$$

The network is shown in Fig. 3.4. For the right-hand arm we can write:

$$\dot{Q}_F = 18.9(20 + 1) = 396.9 \text{ W}$$

Figure 3.4 Example 3.1(a): (i)

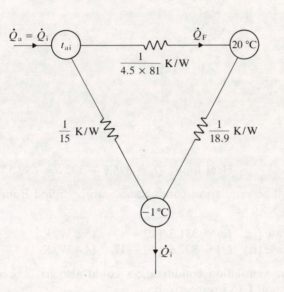

Then,

$$t_{ai} + 1 = \dot{Q}_F \left(\frac{1}{(4.5 \times 81)} + \frac{1}{18.9} \right)$$
$$\therefore \quad \text{Internal air temperature, } t_{ai} = 21.09 \,^{\circ}\text{C}$$

Then, for the left-hand arm,

$$\dot{Q}_V = 15(21.09 + 1) = 331.3 \text{ W}$$

Hence,

$$\text{Heat input, } \dot{Q}_i = \dot{Q}_V + \dot{Q}_F$$
$$= 331.3 + 396.9 = 728.2 \text{ W}$$

(ii) The network is now as shown in Fig. 3.5. As before, $\dot{Q}_F = 396.9 \text{ W}$, and by inspection of the network:

$$\dot{Q}_V \left(\frac{1}{4.5 \times 85} + \frac{1}{15} \right) = 20 + 1$$
$$\therefore \quad \dot{Q}_V = 302.6 \text{ W}$$

Also, from the left-hand arm of the network,

$$\dot{Q}_V = (t_{ai} + 1)15 = 302.6 \text{ W}$$
$$\therefore \quad \text{Internal air temperature, } t_{ai} = 19.17 \,^{\circ}\text{C}$$

Figure 3.5 Example 3.1(a): (ii)

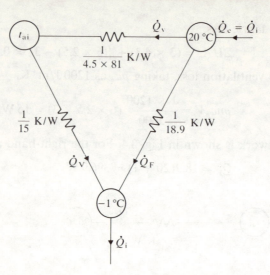

Then,

$$\text{Heat input, } \dot{Q}_i = 396.9 + 302.6 = 699.5 \text{ W}$$

(iii)   The ventilation conductance can be found from the definition given by Eqn [3.5], i.e.

For case (i), $C_V = 331.3/(20 + 1) = 15.8 \text{ W/K}$
For case (ii), $C_V = 302.6/(20 + 1) = 14.4 \text{ W/K}$

The ventilation conductances could also have been calculated using Eqn [3.6] and [3.8] respectively.

It can be seen that the ventilation conductance in both cases is not substantially different from $\rho n c_p V = 15 \text{ W/K}$. Also, the air temperatures, 21.09 °C and 19.17 °C, are very close to the environmental temperature of 20 °C. It can be concluded that for a reasonably well-insulated building with a low ventilation rate it is a good approximation to ignore the difference between t_{ai} and t_{ei} and use the simple procedure to calculate the required heat input (see Example 1.1).

(b)   The ventilation rate is now doubled and hence the resistance of the left-hand arm of the network becomes $1/30 \text{ K/W}$.

For case (i), referring to Fig. 3.4, it can be seen that $\dot{Q}_F$ is unchanged and hence t_{ai} is unchanged at 21.09 °C. With the resistance of the left-hand arm now $1/30$, then

$$\dot{Q}_V = 30(21.09 + 1) = 662.7 \text{ W}$$

Therefore,

$$\text{Heat input, } \dot{Q}_i = 662.7 + 396.9 = 1059.6 \text{ W}$$

The ventilation conductance is now

$$C_V = 662.7/(20 + 1) = 31.6 \text{ W/K}$$

For case (ii), referring to Fig. 3.5 with the new resistance for the left-hand arm of 1/30 K/W, we have,

$$\dot{Q}_V\left(\frac{1}{4.5 \times 81} + \frac{1}{30}\right) = 20 + 1 \qquad \therefore \qquad \dot{Q}_V = 582.1 \text{ W}$$

i.e. $\qquad \dot{Q}_V = 582.1 = 30(t_{ai} + 1)$
$\therefore \qquad t_{ai} = 18.40 \,^{\circ}\text{C}$

Then,

$$\text{Heat input, } \dot{Q}_i = 582.1 + 396.9 = 979.0 \text{ W}$$

The ventilation conductance in this case is given by

$$C_V = 582.1/(20 + 1) = 27.72 \text{ W/K}$$

It can be seen that for the case of the higher ventilation rate with heat input at the environmental point, the internal air temperature is substantially different from the environmental temperature. The network model is therefore very useful when both the internal radiation input and the ventilation rate are high.

Example 3.2

For the office of Example 3.1, assuming an air change rate of 2 per hour and with the other data unchanged, calculate the heat input required when 20% enters at the air point and 80% at the environmental point.

Solution
The resistances are as in part (b) of Example 3.1 and the network is as shown in Fig. 3.6.

Figure 3.6 Example 3.2

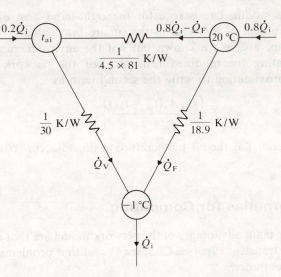

As before,
$$\dot{Q}_F = 18.9(20 + 1) = 396.9 \text{ W}$$

For the top arm of the network:

$$0.8\dot{Q}_i - \dot{Q}_F = 4.5 \times 81(20 - t_{ai})$$
$$\therefore \qquad \dot{Q}_i = 455.63(20 - t_{ai}) + 496.13 \qquad \text{[a]}$$

For the left-hand arm of the network:

$$\dot{Q}_V = \dot{Q}_i - \dot{Q}_F = 30(t_{ai} + 1)$$
$$\therefore \qquad \dot{Q}_i = 30(t_{ai} + 1) + 396.9 \qquad\qquad\qquad\qquad [b]$$

Equating [a] and [b],

$$9112.6 + 496.13 - 455.63t_{ai} = 30 + 396.9 + 30t_{ai}$$

i.e. $\qquad\qquad t_{ai} = 18.91\ ^{\circ}C$

Then,

$$\dot{Q}_V = 30(18.91 + 1) = 597.3\ W$$

i.e. $\qquad\qquad$ Heat input, $\dot{Q}_i = 597.3 + 396.9 = 994.2\ W$

In the above examples it is assumed that all the spaces adjacent to the office are at the same temperature as the office. This may not be accurate if there is a corridor or a store-room, say, at a different temperature, on one side of the room under consideration. The heat loss from the room is then given by:

$$\dot{Q}_F = (t_{ei} - t_{eo})\Sigma UA_o + \Sigma\{U_A A_A(t_{ei} - t_{eA})\}$$

where U_A is the thermal transmittance between the room and an adjacent space at a different environmental temperature, t_{eA}; A_A is the surface area between the room and the adjacent space.

Assuming that $t_{eo} = t_{ao}$, as before, we can write

$$\dot{Q}_F = (t_{ei} - t_{ao})\left[\Sigma UA_o + \Sigma\left\{\frac{U_A A_A(t_{ei} - t_{eA})}{t_{ei} - t_{ao}}\right\}\right]$$

The modified resistance for the right-hand arm of the network is then the reciprocal of the term in the square brackets above. In a case where the air temperatures are known but not the environmental temperatures, when calculating the modified resistance for the network it is a sufficiently good approximation to write the second term as

$$\Sigma\left\{\frac{U_A A_A(t_{ai} - t_{aA})}{t_{ai} - t_{ao}}\right\}$$

where t_{aA} is the air temperature in the adjacent space.

Formulae for Computing

The main advantages of the network model are that it can be extended to cover the transient case (see Chapter 4), and that problems can be easily solved using a computer.

The steady-state network diagram for the general case (Fig. 3.1), is reproduced again as Fig. 3.7; the equations which can be written are as follows:

$$\dot{Q}_F = \Sigma UA_o(t_{ei} - t_{ao}) \qquad\qquad\qquad\qquad [1]$$
$$\dot{Q}_V = \rho n c_p V(t_{ai} - t_{ao}) \qquad\qquad\qquad\qquad [2]$$
$$\dot{Q}_e - \dot{Q}_F = \alpha_a \Sigma A(t_{ei} - t_{ai}) \qquad\qquad\qquad [3]$$
$$\dot{Q}_a = \dot{Q}_V + \dot{Q}_F - \dot{Q}_e \qquad\qquad\qquad\qquad [4]$$

Figure 3.7 Steady-state network

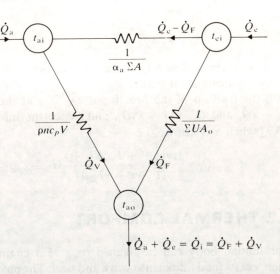

$$\dot{Q}_a + \dot{Q}_e = \dot{Q}_i = \dot{Q}_F + \dot{Q}_V$$

To obtain an expression for $\dot{Q}_a$ and $\dot{Q}_e$ in terms of the temperature difference $(t_{ai} - t_{ao})$, we can eliminate t_{ei}, $\dot{Q}_F$ and $\dot{Q}_V$ from the above equations.

i.e. From Eqn [1], $\quad t_{ei} = t_{ao} + \dot{Q}_F/\Sigma U A_o$

Substituting in Eqn [3],

$$\dot{Q}_e - \dot{Q}_F = \alpha_a \Sigma A t_{ao} + \dot{Q}_F \alpha_a \Sigma A / \Sigma U A_o - \alpha_a \Sigma A t_{ai}$$

i.e. $\quad \dot{Q}_e = \dot{Q}_F\left(1 + \dfrac{\alpha_a \Sigma A}{\Sigma U A_o}\right) - \alpha_a \Sigma A (t_{ai} - t_{ao})$ $\qquad$ [5]

Substituting for $\dot{Q}_V$ from Eqn [2] in Eqn [4],

$$\dot{Q}_a = \rho n c_p V(t_{ai} - t_{ao}) + \dot{Q}_F - \dot{Q}_e$$

i.e. $\quad \dot{Q}_F = \dot{Q}_a + \dot{Q}_e - \rho n c_p V(t_{ai} - t_{ao})$

Substituting in Eqn [5],

$$\dot{Q}_e = \dot{Q}_a\left(1 + \dfrac{\alpha_a \Sigma A}{\Sigma U A_o}\right) + \dot{Q}_e\left(1 + \dfrac{\alpha_a \Sigma A}{\Sigma U A_o}\right)$$
$$- \rho n c_p V\left(1 + \dfrac{\alpha_a \Sigma A}{\Sigma U A_o}\right)(t_{ai} - t_{ao}) - \alpha_a \Sigma A(t_{ai} - t_{ao})$$

Then, dividing through by $(\alpha_a \Sigma A + \Sigma U A_o)/\Sigma U A_o$,

$$\dot{Q}_a + \dot{Q}_e \frac{\alpha_a \Sigma A}{\alpha_a \Sigma A + \Sigma U A_o} = (t_{ai} - t_{ao})\left\{\rho n c_p V + \frac{\alpha_a \Sigma A \Sigma U A_o}{\alpha_a \Sigma A + \Sigma U A_o}\right\}$$

i.e. $\quad \dot{Q}_a + F_{au}\dot{Q}_e = (\rho n c_p V + F_{au}\Sigma U A_o)(t_{ai} - t_{ao})$ $\qquad$ [3.9]

where $\quad F_{au} = \dfrac{\alpha_a \Sigma A}{(\alpha_a \Sigma A + \Sigma U A_o)}$

It can be seen that F_{au} is a dimensionless factor that enables inputs at the environmental point to be referred to the air point.

In a similar way it can be shown that:

$$F_{av}\dot{Q}_a + \dot{Q}_e = (F_{av}\rho n c_p V + \Sigma U A_o)(t_{ei} - t_{ao}) \qquad [3.10]$$

where
$$F_{av} = \frac{\alpha_a \Sigma A}{(\alpha_a \Sigma A + \rho n c_p V)}$$

F_{av} is a dimensionless factor that enables inputs at the air point to be referred to the environmental point.

If the fraction of the heat input, $\dot{Q}_i$, is x at the environmental point, then $\dot{Q}_e = x\dot{Q}_i$ and $\dot{Q}_a = (1-x)\dot{Q}_i$, and hence the input can be found from Eqn [3.9] when x is known, i.e.

$$\dot{Q}_i = \frac{(\rho n c_p V + F_{au}\Sigma U A_o)}{(1 - x + xF_{au})}(t_{ai} - t_{ao}) \qquad [3.11]$$

3.2 THERMAL COMFORT

The human body can be likened to a heat engine, converting the chemical energy of its food intake into work and heat. The physiological system is complex with an exceedingly efficient control mechanism maintaining the body at a mean temperature of about 37 °C. When the body experiences cold air temperatures the blood flow is reduced in the small veins near the surface of the skin. This has two purposes: firstly, the blood flow is increased through the main tissue and organs, thus protecting them; and secondly, the surface tissue acts as a better insulation when the blood flow is reduced, thus preserving the body's energy store. Conversely, when the air temperature is high the blood flow to the surface tissue is increased, giving the blood a greater surface area from which to lose heat. When the body temperature cannot be controlled by the blood flow alone then the sweat glands operate to emit water which evaporates from the surface, thus cooling the body.

About two-thirds of the normal energy intake of the body is required to maintain the functions of the organs and this is known medically as the *basal metabolic rate*; it is the energy intake required when the body is lying down at rest. The absolute value of the metabolic rate varies with body weight, state of health, etc.; an average value is 75 W. As the body is required to perform more and more energetic activities the basic energy production rate increases in an attempt to maintain an energy balance and hence to keep the body temperature at its optimum value.

The basic energy equation is given by:

$$\text{Energy production rate, } \dot{E} = \dot{Q}_c + \dot{Q}_r + \dot{Q}_e \pm \dot{E}_s \qquad [3.12]$$

where $\dot{Q}_c$, $\dot{Q}_r$ and $\dot{Q}_e$ are the rates of heat transfer from the body by convection, radiation and evaporation respectively; E_s is the rate at which energy is stored by the body.

It can be seen that the body is never in a truly steady state; the more irregular the activity of the body the greater will be the value of $\dot{E}_s$. The reader is recommended to consult Threlkeld[3.3] for further discussion on the body's ability to adapt to the external environment. Some representative values of heat emissions from the human body under different activity levels are given by CIBSE.[3.4]

The total heat loss from the body is dependent on the surrounding air temperature and its velocity (influencing convection rate), the temperature of the surrounding surfaces (influencing radiative heat transfer), and on the humidity of the air (influencing evaporation rate from the skin). These factors are therefore important in determining the comfort of an individual. Unfortunately, perceptions of comfort vary from person to person, not simply because of physiological factors such as the variation of metabolic rate between individuals, the levels of clothing worn, or the physical ability to acclimatize, but also owing to complex psychological factors.

Any comfort 'index' must be based on a statistical analysis of people's perceptions under controlled experimental conditions; by this means a comfort range can be established within which a certain percentage of the population will feel comfortable. Many investigators have defined indices of thermal comfort including ASHRAE's *effective temperature*,[3.3,3.5] the CIBSE *dry resultant temperature*[3.1] (see below), Fanger's *Predicted Percentage Dissatisfied*[3.6] (see below), McIntyre's *subjective temperature*[3.7] and Bedford's *calidity*.[3.18]

Dry Resultant Temperature

The traditional instrument for measuring a room temperature which allows for radiation is the *globe thermometer*. This consists of a blackened, hollow copper sphere of 150 mm diameter with a thermometer placed at its centre. The temperature measured is found to be:

$$\text{Globe temperature} = \frac{t_{mr} + 2.35 t_{ai}\sqrt{u}}{1 + 1.35\sqrt{u}}$$

where t_{mr} is the mean radiant temperature (see Eqn 3.1); t_{ai} is the temperature of the air in the room as used previously; u is the mean air velocity in the room.

When a smaller sphere is used (diameter 100 mm), a different temperature is measured called the *resultant temperature*, and this is the criterion chosen by CIBSE for use in the UK. A form of this temperature can be introduced which allows for humidity but it has been shown that in winter the *dry resultant temperature* is an adequate measure of comfort, provided the relative humidity lies between 40 and 70%. It is given by:

$$\text{Dry resultant temperature, } t_c = \frac{t_{mr} + t_{ai}\sqrt{10u}}{1 + \sqrt{10u}} \qquad [3.13]$$

Note that when the mean air velocity is 0.1 m/s, the above expression reduces to

$$t_c = \frac{t_{mr} + t_{ai}}{2}$$

In Section 3.1 the network model is seen to use the approximation of equating mean surface temperature to mean radiant temperature. The CIBSE comfort criterion in the model is therefore defined at the centre of a room as:

$$\text{Dry resultant temperature, } t_c = \frac{t_m + t_{ai}}{2} \qquad [3.14]$$

From Eqn [3.4] the internal environmental temperature is defined as

$$t_{ei} = \frac{1}{3} t_{ai} + \frac{2}{3} t_m$$

Hence substituting for t_m from Eqn [3.4] in Eqn [3.14] we have:

$$t_c = \frac{1}{2}\left(\frac{3}{2}t_{ei} - \frac{1}{2}t_{ai}\right) + \frac{t_{ai}}{2}$$

i.e. $$t_c = \frac{3}{4}t_{ei} + \frac{1}{4}t_{ai}$$ [3.15]

It follows from Eqn [3.15] that the dry resultant temperature can be placed in the network between the air and environmental points (see Fig. 3.8).

Figure 3.8 Steady-state network showing dry resultant temperature

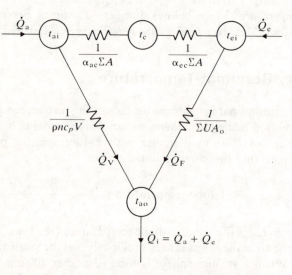

Note also from Eqn [3.15] that

$$t_{ei} - t_c = \frac{1}{4}(t_{ei} - t_{ai})$$

and $$t_c - t_{ai} = \frac{3}{4}(t_{ei} - t_{ai})$$

In the steady state the heat flows are equal through the two resistances connecting the environmental point to the dry resultant point and the air point. It follows that the resistances are directly proportional to the temperature differences. Hence, referring to Fig. 3.8, we have:

$$\frac{1}{\alpha_{ec}\Sigma A} = \frac{1}{4\alpha_a\Sigma A} = \frac{1}{18\Sigma A}$$

and $$\frac{1}{\alpha_{ac}\Sigma A} = \frac{3}{4\alpha_a\Sigma A} = \frac{1}{6\Sigma A}$$

i.e. $\alpha_{ec} = 18 \, \text{W/m}^2\,\text{K}$ and $\alpha_{ac} = 6 \, \text{W/m}^2\,\text{K}$

CIBSE[3.8] gives a table of recommended values of dry resultant temperatures, and an extract of these values is given in Table 3.1 by permission of CIBSE.

Table 3.1 Recommended dry resultant temperatures*

Type of building and usage	Dry resultant temperature, $t_c/(°C)$
Heavy work in factories, warehouse storage	13
Library stores, shop stores, appliance rooms	15
Light work in factories, staircases and corridors	16
Lecture halls, bars, churches, exhibition halls, hospital wards, hotel corridors, police cells, classrooms	18
Museums, banking halls, canteens, offices, laboratories	20
Living rooms, public rooms, sports changing rooms	21
Bathrooms, swimming bath changing rooms, hotel bedrooms	22
Luxury hotel bedrooms	24
Swimming bath hall	26

*Note: When the air velocity exceeds 0.1 m/s the above values should be increased using Fig. 3.9

Figure 3.9 Elevation of dry resultant temperature against air velocity

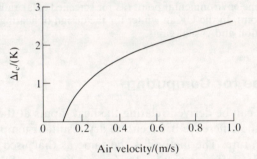

Allowance for air velocities greater than 0.1 m/s is made using a graph of temperature elevation above the value of t_c at 0.1 m/s, against the air velocity. This is reproduced by permission of CIBSE as Fig. 3.9.

Steady-state problems can be solved using a desired value of dry resultant temperature instead of a fixed value of either internal air temperature or internal environmental temperature (see Example 3.3 below). Before giving an example of this use of dry resultant temperature it is necessary to consider the way energy is input to a room in the CIBSE model.

An environmental point is defined as having energy inputs in the ratio of 2 parts radiant to 1 part convective. Therefore in the model the energy input to the environmental point is taken as 1.5 times the direct radiant input with 0.5 times the direct radiant input subtracted at the air point.

Hence for a room where the fraction of the heat input $\dot{Q}_i$ which is due to direct radiation is γ, and the fraction which is due to convection is δ, we can write:

$$\text{Heat input at the environmental point, } \dot{Q}_e = 1.5\gamma\dot{Q}_i$$

and

$$\text{Heat input at the air point, } \dot{Q}_a = \delta\dot{Q}_i - 0.5\gamma\dot{Q}_i$$

Also, in this case, $\gamma + \delta = 1$, and therefore, $\dot{Q}_a = 1 - 1.5\gamma\dot{Q}_i$.

It can be seen that when $\gamma = 2/3$, then $\dot{Q}_e = \dot{Q}_i$, and $\dot{Q}_a = 0$; when γ is larger than $2/3$, then $\dot{Q}_a$ becomes negative.

In general the following values may be taken:

Warm air heating: $\gamma = 0$, $\dot{Q}_e = 0$, $\dot{Q}_a = \dot{Q}_i$
Convectors: $\gamma = 0.1$, $\dot{Q}_e = 0.15\dot{Q}_i$, $\dot{Q}_a = 0.85\dot{Q}_i$
Multi-column radiators: $\gamma = 0.2$, $\dot{Q}_e = 0.30\dot{Q}_i$, $\dot{Q}_a = 0.70\dot{Q}_i$
Double panel radiators: $\gamma = 0.3$, $\dot{Q}_e = 0.45\dot{Q}_i$, $\dot{Q}_a = 0.55\dot{Q}_i$
Single column radiators: $\gamma = 0.5$, $\dot{Q}_e = 0.75\dot{Q}_i$, $\dot{Q}_a = 0.25\dot{Q}_i$
(Floor warming systems and block storage heaters: As for single column radiators)
Vertical and ceiling panels $\gamma = 0.67$, $\dot{Q}_e = \dot{Q}_i$, $\dot{Q}_a = 0$
High temperature systems $\gamma = 0.90$, $\dot{Q}_e = 1.35\dot{Q}_i$, $\dot{Q}_a = -0.35\dot{Q}_i$

Note: Short-wave solar radiation entering a room should be screened from occupants and hence appears as indirect radiation from the room surfaces; when the solar irradiance is multiplied by a suitable solar gain factor (see Chapter 4), the resultant heat input is assumed to be directly at the environmental point. It is also normal practice to assume that casual gains from occupants, machinery and lighting appear at the air point; it may be more accurate in some cases to take a fraction of the casual gains as appearing directly at the environmental point (as for screened solar radiation), and the remainder at the air point. In the UK in winter for the design of heating loads it is usual to ignore solar radiation and casual gains.

Formulae for Computing

A formula can be derived relating the heat inputs at the air and environmental points to the difference between the dry resultant temperature and the outside air temperature. The method is the same as that used earlier (see Eqns [3.9] and [3.10]). Referring to Fig. 3.8, (page 138), two additional equations can be written for the top arm of the network, i.e.

$$\dot{Q}_e - \dot{Q}_F = \alpha_{ec}\Sigma A(t_{ei} - t_c)$$

and

$$\dot{Q}_e - \dot{Q}_F = \alpha_{ac}\Sigma A(t_c - t_{ai})$$

By a similar method to that used to derive Eqn [3.9] we have:

$$F_v\dot{Q}_a + F_u\dot{Q}_e = \{F_v\rho nc_pV + F_u\Sigma UA_o\}(t_c - t_{ao}) \qquad [3.16]$$

where $F_u = \dfrac{\alpha_{ec}\Sigma A}{\alpha_{ec}\Sigma A + \Sigma UA_o}$ and $F_v = \dfrac{\alpha_{ac}\Sigma A}{\alpha_{ac}\Sigma A + \rho nc_pV}$

Then, with $\dot{Q}_e = x\dot{Q}_i$ and $\dot{Q}_a = (1-x)\dot{Q}_i$ as before,

$$\dot{Q}_i = \left\{\frac{F_v\rho nc_pV + F_u\Sigma UA_o}{(1-x)F_v + xF_u}\right\}(t_c - t_{ao}) \qquad [3.17]$$

Another way of expressing the heat input is by writing

$$\dot{Q}_i = \dot{Q}_F + \dot{Q}_v = \Sigma UA_o(t_{ei} - t_{ao}) + \rho nc_pV(t_{ai} - t_{ao})$$
$$= \left\{\Sigma UA_o\frac{(t_{ei} - t_{ao})}{(t_c - t_{ao})} + \rho nc_pV\frac{(t_{ai} - t_{ao})}{(t_c - t_{ao})}\right\}(t_c - t_{ao})$$

Then,

$$\dot{Q}_i = \{F_1\Sigma UA_o + F_2\rho nc_pV\}(t_c - t_{ao}) \qquad [3.18]$$

where $$F_1 = \frac{(t_{ei} - t_{ao})}{(t_c - t_{ao})} \quad \text{and} \quad F_2 = \frac{(t_{ai} - t_{ao})}{(t_c - t_{ao})}$$

By expressing the temperature differences in terms of the network resistances it can be shown[3.1] that F_1 and F_2 can also be expressed as:

$$F_1 = \frac{x(F_v \rho n c_p V + F_u \Sigma U A_o)}{(\Sigma U A_o + \alpha_{ec} \Sigma A)[F_v(1 - x) + x F_u]} + F_u \qquad [3.19]$$

and $$F_2 = \frac{(1 - x)(F_v \rho n c_p V + F_u \Sigma U A_o)}{(\rho n c_p V + \alpha_{ac} \Sigma A)[F_v(1 - x) + x F_u]} + F_v \qquad [3.20]$$

CIBSE[3.9] give tables of values for F_1 and F_2 for various types of radiators for a range of values of $\Sigma U A_o / \Sigma A$ and $\rho n c_p V / \Sigma A$. For the case of warm air heating (i.e. 100% convective), CIBSE[3.9] gives a table in which the values of F_1 and F_2 remain the same for all values of $\rho n c_p V / \Sigma A$. This implies zero infiltration of air into the room. For ducted warm air heating into a well-sealed room this is a good approximation, and the value of F_2 for this special case can be obtained from Eqn [3.20] by putting $x = 0$ and $\rho n c_p V = 0$; for $\rho n c_p V = 0$, $F_v = 1$. For cases of warm air heating where there is also air infiltration, the value of F_2 must be calculated using Eqn [3.20] with $x = 0$, and with the relevant value of $\rho n c_p V$; CIBSE does not give tabulated values for this case.

It should be noted that for the case of ducted warm air heating in a well-sealed building, although there is no air infiltration, there is nevertheless a need for the required volume flow of fresh air to be supplied via the warm air ducting. The energy required to heat the fresh air is given by:

$$\dot{m}_{ao} c_p (t_{ai} - t_{ao}) = \rho n c_p V(t_{ai} - t_{ao})$$

(taking mean values of ρ and c_p as before).

The mass flow rate of fresh air at t_{ao} is equal to the mass flow rate of supply air to the room for the case of zero re-circulation of room air.

Eqn [3.18] applies therefore to the *room* heating load, with the term $\rho n c_p V$ taken as the equivalent air change of the air infiltration into the room; the *plant* load must include the additional energy required to heat the fresh air.

Example 3.3
(a)   The hospital ward shown in Fig. 3.10 is located on an intermediate floor of a multi-storey ward block, and has one external wall. All adjacent rooms can be assumed to be at the same temperature.

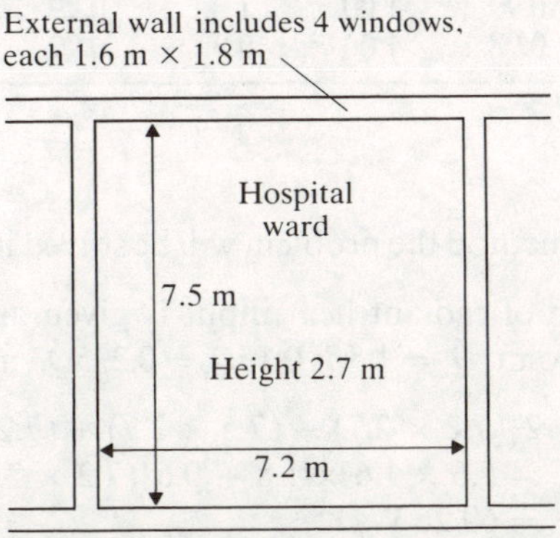

Figure 3.10 Example 3.3: plan of hospital ward

External wall includes 4 windows, each 1.6 m × 1.8 m

Hospital ward

7.5 m

Height 2.7 m

7.2 m

If the ward is heated by steel panel radiators with a radiant component of 15% of the total heat output, for the design conditions stated below, calculate:

(i) the steady-state heat load;

(ii) the ward air temperature.

(b)   If the heating system is controlled by a thermostat that responds to air temperature, how will the room dry resultant temperature vary during the heating season?

Data

Ward dry resultant temperature, 19 °C; outside air temperature, -2 °C; U-value of outside wall, $0.6\,\mathrm{W/m^2\,K}$; window U-value, $5.6\,\mathrm{W/m^2\,K}$; air infiltration rate, 1.0 air change per hour.

(CIBSE)

Solution

Candidates taking the examination were provided with copies of CIBSE Tables A9.1–A9.7, giving values of F_1 and F_2; the relevant extract from these tables is given as Table 3.2.

Table 3.2   Values of F_1 and F_2

(a) 90% convective, 10% radiant (natural convectors and convector radiators):

	$\Sigma UA_o/\Sigma A$							
	0.1		0.2		0.4		0.6	
$\dfrac{\rho n c_p V}{\Sigma A}$	F_1	F_2	F_1	F_2	F_1	F_2	F_1	F_2
0.1	1.00	1.01	0.99	1.03	0.98	1.05	0.97	1.08
0.2	1.00	1.01	0.99	1.02	0.98	1.05	0.97	1.08
0.4	1.00	1.00	0.99	1.02	0.98	1.05	0.98	1.07
0.6	1.00	1.00	1.00	1.01	0.99	1.04	0.98	1.07

(b) 80% convective, 20% radiant (multi-column radiators):

	$\Sigma UA_o/\Sigma A$							
	0.1		0.2		0.4		0.6	
$\dfrac{\rho n c_p V}{\Sigma A}$	F_1	F_2	F_1	F_2	F_1	F_2	F_1	F_2
0.1	1.00	1.01	0.99	1.02	0.99	1.04	0.98	1.06
0.2	1.00	1.00	1.00	1.01	0.99	1.04	0.98	1.06
0.4	1.00	0.99	1.00	1.00	0.99	1.03	0.98	1.05
0.6	1.01	0.98	1.00	0.99	0.99	1.02	0.99	1.04

To illustrate the method the problem will be solved initially from the network.

(a)(i)   The fraction of radiant heat input is given as 0.15; hence the input at the environmental point, $\dot{Q}_e = 1.5 \times 0.15\dot{Q}_i = 0.225\dot{Q}_i$, and therefore $\dot{Q}_a = 0.775\dot{Q}_i$.

$$\Sigma A = 2\{(7.2 \times 2.7) + (7.5 \times 2.7) + (7.2 \times 7.5)\} = 187.38\,\mathrm{m^2}$$
$$\Sigma UA_o = 5.6 \times 1.6 \times 1.8 + 0.6\{(7.2 \times 2.7) - (4 \times 1.6 \times 1.8)\}$$
$$= 69.26\,\mathrm{W/K}$$

$$\rho n c_p V = \frac{1200 \times 1 \times 7.5 \times 7.2 \times 2.7}{3600} = 48.60 \text{ W/K}$$

$$\alpha_{ec} \Sigma A = 18 \times 187.38 = 3372.8 \text{ W/K}$$

$$\alpha_{ac} \Sigma A = 6 \times 187.38 = 1124.3 \text{ W/K}$$

Referring to the network shown in Fig. 3.11, we have

$$\dot{Q}_F = 69.26(t_{ei} + 2) \qquad [1]$$
$$\dot{Q}_V = \dot{Q}_i - \dot{Q}_F = 48.6(t_{ai} + 2) \qquad [2]$$
$$0.225\dot{Q}_i - \dot{Q}_F = 3372.8(t_{ei} - 19) \qquad [3]$$
$$0.225\dot{Q}_i - \dot{Q}_F = 1124.3(19 - t_{ai}) \qquad [4]$$

Figure 3.11 Example 3.3: steady-state network

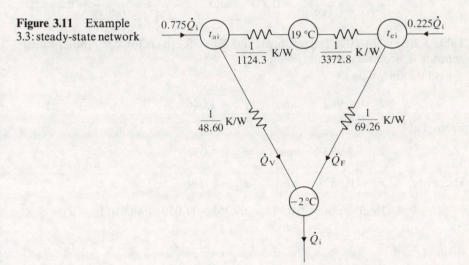

From Eqn [1] in Eqn [3],

$$0.225\dot{Q}_i - 69.26t_{ei} - 138.52 = 3372.8t_{ei} - 64\,084$$

i.e. $$t_{ei} = 18.58 + 0.000\,065\,4\dot{Q}_i \qquad [5]$$

Substituting for t_{ai} from Eqn [2] in Eqn [4],

$$0.225\dot{Q}_i - \dot{Q}_F = 21\,361 - 1124.3\left\{\frac{(\dot{Q}_i - \dot{Q}_F)}{48.6} - 2\right\}$$

i.e. $$23.36\dot{Q}_i = 23\,610 + 24.13\dot{Q}_F$$

Then substituting from Eqn [1],

$$23.36\dot{Q}_i = 23\,610 + 24.13(69.26t_{ei} + 138.52)$$

$$\therefore \quad t_{ei} = 0.014\dot{Q}_i - 16.13 \qquad [6]$$

Then from Eqns [5] and [6],

$$\text{Heat input, } \dot{Q}_i = 2491 \text{ W}$$

(ii)   Substituting in Eqn [6], $t_{ei} = 18.74\,°C$

and in Eqn [1],

$$\dot{Q}_F = 69.26(18.74 + 2) = 1437\,W$$

Therefore, $\dot{Q}_V = 2491 - 1437 = 1054\,W$

Hence from Eqn [2],

$$\text{Ward air temperature, } t_{ai} = \frac{1054}{48.6} - 2 = 19.7\,°C$$

The problem can be solved using Table 3.2. We have

$$\frac{\Sigma U A_o}{\Sigma A} = \frac{69.26}{187.38} = 0.37 \quad \text{and} \quad \frac{\rho n c_p V}{\Sigma A} = \frac{48.6}{187.38} = 0.26$$

Table 3.2(a) is for 10% radiant heat, and Table 3.2(b) is for 20% radiant heat; hence it is necessary to interpolate.

From Table 3.2(a),

$$F_1 = 0.982 \quad \text{and} \quad F_2 = 1.046$$

From Table 3.2(b),

$$F_1 = 0.992 \quad \text{and} \quad F_2 = 1.033$$

Therefore, $F_1 = 0.987 \quad \text{and} \quad F_2 = 1.039$

$\therefore$ Heat input $= \{(0.987 \times 69.26) + (1.039 \times 48.6)\}(19 + 2)$
                    $= 2496\,W$

Also,

$$F_1 = 0.987 = \frac{(t_{ei} + 2)}{(19 + 2)} \qquad \therefore \quad t_{ei} = (0.987 \times 21) - 2 = 18.73\,°C$$

and

$$F_2 = 1.039 = \frac{(t_{ai} + 2)}{(19 + 2)} \qquad \therefore \quad t_{ai} = (1.039 \times 21) - 2 = 19.82\,°C$$

These answers compare closely with the ones obtained from the manual method.

The heat input can also be obtained directly from Eqn [3.17]. It is necessary to evaluate F_u and F_v first, i.e.

$$F_u = \frac{\alpha_{ec}\Sigma A}{\alpha_{ec}\Sigma A + \Sigma U A_o} = \frac{3372.8}{3372.8 + 69.26} = 0.9799$$

and $$F_v = \frac{\alpha_{ac}\Sigma A}{\alpha_{ac}\Sigma A + \rho n c_p V} = \frac{1124.3}{1124.3 + 48.6} = 0.9586$$

Then in Eqn [3.17]

$$\text{Heat input, } \dot{Q}_i = \frac{(0.9799 \times 69.26) + (0.9586 \times 48.6)}{(0.9799 \times 0.225) + (0.9586 \times 0.775)}(19 + 2)$$
$$= 2495\,W$$

Note also that the problem can be solved using Eqns [3.19] and [3.20] to find F_1 and F_2.

$$F_1 = \frac{0.225\{(0.9586 \times 48.6) + (0.9799 \times 69.26)\}}{(69.26 + 3372.8)\{(0.775 \times 0.9586) + (0.225 \times 0.9799)\}} + 0.9799$$

$$= 0.9877$$

$$F_2 = \frac{0.775\{(0.9586 \times 48.6) + (0.9799 \times 69.26)}{(48.6 + 1124.3)\{(0.775 \times 0.9586) + (0.225 \times 0.9799)\}} + 0.9586$$

$$= 1.0371$$

The values found by interpolating from tables were:

$$F_1 = 0.987 \quad \text{and} \quad F_2 = 1.039$$

which agree closely with the exact figures.

(b)   Assuming that the ward inside temperaature, t_{ai}, is maintained constant as the outside temperature, t_{ao}, varies. Since the form of heating is fixed, the proportions of radiant and convective heating will remain the same; only the size of the heat input, $\dot{Q}_i$, will vary as the outside temperature changes. Referring to Eqn [3.20] it can be seen therefore that F_2 remains unchanged, i.e.

$$1.0371 = \text{constant} = \frac{(t_{ai} - t_{ao})}{(t_c - t_{ao})}$$

$$\therefore \quad t_c = t_{ao} + \frac{(t_{ai} - t_{ao})}{1.0371} = 0.036t_{ao} + 0.964t_{ai}$$

Assuming that the ward internal temperature is fixed at the design value of 19.7 °C, then

$$t_c = 0.036t_{ao} + 19.0$$

The variation of t_c can be shown as follows:

$t_{ao}/(°C)$	−2	2	6	10	14
$t_c/(°C)$	18.9	19.1	19.2	19.4	19.5

CIBSE recommend that the dry resultant temperature should not vary by more than ±1.5 K and therefore the variation shown for t_c is within the comfort range. From Table 3.1 and Fig. 3.9, (page 139) a hospital ward with an air velocity of about 0.25 m/s would require a dry resultant temperature of 19 °C.

Fanger's Method

The work of the Danish Engineer P. O. Fanger[3.6] provides a reliable method for determining whether a particular room will be considered comfortable by a sufficiently large number of the occupants.

Fanger's experiments were conducted using 128 college-age Danes and 128 elderly Danes. He compared his results with those of Nevins et al.[3.10], and McNall et al.[3.11] in the USA, and was able to show that there was no statistically significant difference between the results.

His conclusions were:

(a) there is no significant difference in comfort perceptions due to geographical location or season (this even applies to tropical regions as shown by comparison with the work of a large number of investigators);
(b) there is no significant difference due to age;
(c) there is no significant difference due to sex;
(d) there is no significant difference due to body build;
(e) there is no significant difference due to ethnic origin.

Some of these conclusions appear to go against 'perceived wisdom' but scientific explanations can be given to show how variations of physical effects tend to cancel out. For example, older people have a lower metabolic rate but also have a lower insensible perspiration rate, and hence the thermal equilibrium occurs at similar conditions; a similar difference occurs in metabolic rate and insensible perspiration rate between men and women, again leading to thermal equilibrium at the same conditions.

Fanger was able to combine his results with those of Nevins et al.[3.10], and McNall et al.[3.11] to arrive at his comfort index. He expressed the thermal load per unit body surface area in terms of the physical variables, including level of clothing, and defined a thermal sensation factor as a function of the thermal load and the body's internal heat production rate. By using the experimental results he derived a thermal sensation factor which he called *Predicted Mean Vote* (PMV).

The PMV is based on a simple scale in which thermal neutrality is represented by zero, as follows:

+3	+2	+1 Slightly warm	0	−1 Slightly cold	−2	−3
Hot	Warm	warm		cold	Cool	Cold

Fanger tabulated his results for different activity levels and for different clothing levels; an extract from his results is given in Table 3.3. The effect of clothing is introduced using a non-dimensional number given by the thermal resistance from the skin to the outer surface of the clothing divided by the factor $0.155 \ m^2 \ K/W$. The non-dimensional number is called the *clo*; note that it is not a 'unit' although it tends to be used as if it were.

A typical business suit has a clo level of unity; a heavier suit with vest or pullover, woollen socks and cotton underwear has a clo level of about 1.5; lightweight summer clothing such as a summer dress or lightweight trousers with open necked, short-sleeved shirt has a clo level of about 0.5; clothing for normal factory working has a clo level of between 0.7 and 0.8. The values of PMV given in Table 3.3 were calculated by Fanger for conditions where the air temperature is equal to the mean radiant temperature and where the relative humidity is 50%. To allow for variations in mean radiant temperature, t_{mr}, and relative humidity, ϕ, Fanger gives diagrams for three activity levels of (a) $\Delta(PMV)/\Delta t_{mr}$ against clo for various air velocities; (b) $\Delta(PMV)/\Delta\phi$ against clo for various air velocities. These are reproduced as Figs 3.12 and 3.13, (page 148). The corrected PMV is a measure of the general level of discomfort

Table 3.3 Predicted mean vote

(a)   Activity level 58 W/m² (sedentary):

		Predicted Mean Vote, PMV	
Clothing/(clo)	Ambient temp./(°C)	At 0.1 m/s	At 0.2 m/s
1.0	20	−0.87	−1.13
	21	−0.60	−0.84
	22	−0.33	−0.55
	23	−0.07	−0.27
	24	0.20	0.02
	25	0.48	0.31
	26	0.75	0.60
	27	1.02	0.89

(b)   Activity level 116 W/m² (medium):

		Predicted Mean Vote, PMV	
Clothing/(clo)	Ambient temp./(°C)	At 0.2 m/s	At 0.3 m/s
0.75	10	−1.35	−1.54
	12	−1.03	−1.20
	14	−0.70	−0.85
	16	−0.38	−0.51
	18	−0.05	−0.17
	20	0.28	0.18
	22	0.62	0.54
	24	0.97	0.90

(c)   Activity level 174 W/m² (high):

		Predicted Mean Vote, PMV	
Clothing/(clo)	Ambient temp./(°C)	At 0.2 m/s	At 0.3 m/s
0.75	6	−0.75	−0.93
	8	−0.47	−0.64
	10	−0.19	−0.34
	12	0.10	−0.03
	14	0.39	0.27
	16	0.69	0.58
	18	0.98	0.89
	20	1.28	1.20

of the group of volunteers in the experiments. Fanger makes this more meaningful and useful by analysing the preferences of the volunteers to arrive at the *Predicted Percentage Dissatisfied*, PPD. This is expressed diagrammatically in Fig. 3.14, (page 148). As shown by the figure, 5% of the population will be dissatisfied whatever the conditions. It is important to remember that complaints will always be received regardless of how well the environment is designed.

The Fanger results have been adopted as an International Standard, ISO 7730,[3.13] which recommends that for design purposes the PMV should be between +0.5 and −0.5, i.e. corresponding to a PPD of no greater than 10%.

Figure 3.12 $\Delta(\text{PMV})/\Delta t_{mr}$ as a function of the thermal resistance of the clothing with relative velocity as parameter, at three different activity levels. $\Delta(\text{PMV})/\Delta t_{mr}$ is determined for PMV = 0 and indicates the increment of predicted mean vote, when mean radiant temperature is increased by 1 °C (constant vapour pressure)[3.6]

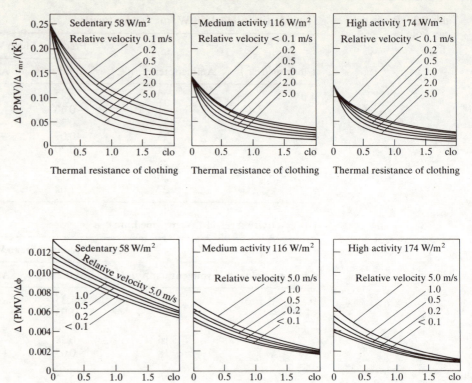

Figure 3.13 $\Delta(\text{PMV})/\Delta\phi$ as a function of the thermal resistance of the clothing with relative velocity as parameter, at three different activity levels. $\Delta(\text{PMV})/\Delta\phi$ is determined for PMV = 0, RH = 50% and indicates the increment of predicted mean vote, when the relative humidity is increased by 1%[3.6]

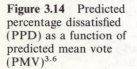

Figure 3.14 Predicted percentage dissatisfied (PPD) as a function of predicted mean vote (PMV)[3.6]

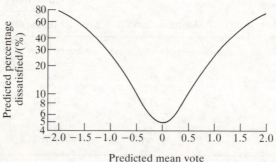

Example 3.4

An office has one external wall, 5 m long, containing a window of area 4 m²; the width of the office from the window to the wall adjacent to the corridor is 4 m, and the ceiling height is 3 m. There are similar offices above, below and on either side. The outside temperature is −1 °C and the dry resultant temperature is to be 20 °C (see Table 3.1).

The thermal transmittances of the external wall and window are 1.0 and 5.6 W/m² K respectively. The air change rate is 1 per hour with an air velocity of 0.1 m/s.

Assuming that all the heat input is at the environmental point and that the level of clothing is 1 clo, calculate:

(i) the required heat input;

(ii) the PMV and the PPD for the office, stating whether conditions are satisfactory by the Fanger comfort criterion.

Assume as a simplification that the mean radiant temperature is equal to the mean surface temperature.

Solution

(i) $\Sigma A = 2\{(5 \times 3) + (4 \times 3) + (5 \times 4)\} = 94 \, \text{m}^2$

$\Sigma U A_o = 1.0 \times \{(5 \times 3) - 4\} + (5.6 \times 4) = 33.4 \, \text{W/K}$

$\rho n c_p V = 1200 \times 5 \times 3 \times 4/3600 = 20 \, \text{W/K}$

Referring to Fig. 3.15, the heat flow through the top arm is equal to the heat flow through the left-hand arm. Therefore,

$$20 + 1 = \dot{Q}_V \left(\frac{1}{6 \times 94} + \frac{1}{20} \right)$$

$$\therefore \quad \dot{Q}_V = 405.62 \, \text{W}$$

Figure 3.15
Example 3.4

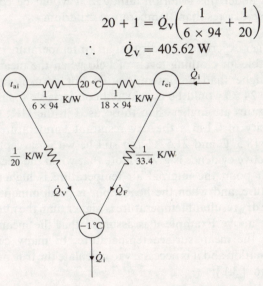

Then,

$$t_{ai} = \frac{405.62}{20} - 1 = 19.28 \, ^\circ\text{C}$$

Also,

$$t_{ei} = 20 + \frac{405.62}{18 \times 94} = 20.24 \, ^\circ\text{C}$$

Hence,

$$\dot{Q}_F = (20.24 + 1) \times 33.4 = 709.42 \, \text{W}$$

Therefore, Heat input $= \dot{Q}_V + \dot{Q}_F = 405.62 + 709.42 = 1115 \, \text{W}$

(ii) Assuming that the mean radiant temperature is equal to the mean surface temperature, from Eqn [3.4] we have

$$t_{ei} = \frac{1}{3} t_{ai} + \frac{2}{3} t_m$$

$$\therefore \quad t_m = \frac{(3 \times 20.24) - 19.28}{2} = 20.72 \, ^\circ\text{C}$$

The activity level in an office is sedentary and hence from Table 3.3, (page 147), extrapolating for an air temperature of 19.28 °C at a velocity of 0.1 m/s, we have:

$$\text{PMV} = -1.06$$

From Fig. 3.12, (page 148), at a velocity of 0.1 m/s, 1.0 clo, we have:

$$\Delta(PMV)/\Delta t_{mr} = 0.12$$
$$\therefore \quad \Delta PMV = 0.12(20.72 - 19.28) = 0.17$$

Therefore the corrected PMV is given by

$$PMV = -1.06 + 0.17 = -0.89$$

Then from Fig. 3.14, (page 148),

$$\text{Predicted Percentage Dissatisfied} = 22\%$$

This is well above the acceptable design figure of 10% and hence the room is too cold under this criterion although it would be considered satisfactory by the simple dry resultant temperature criterion.

The Fanger experiments show that an air temperature of 22–23 °C is considered comfortable for clothing levels of 1 clo when the mean radiant temperature is slightly higher than the air temperature. For warm air heating an air temperature of about 24 °C would be considered comfortable for normal office conditions. These figures are higher than those used in the UK; CIBSE state that for an air velocity of 0.1 m/s the dry resultant temperature for an office can vary between 18.5 °C and 21.5 °C and still be within acceptable comfort limits, but no subjective evidence is given. Note that when the heat input is predominantly at the air point the internal air temperature is higher than the dry resultant temperature, and when the heat input is predominantly at the environmental point the dry resultant temperature is higher than the internal air temperature.

In the above Example it is assumed that the mean radiant temperature is equal to the mean surface temperature. In many cases this is not a good approximation and it is necessary to calculate the true mean radiant temperature using Eqn. [3.1]:

$$(t_{mr} + 273)^4 = \Sigma F_b(t_s + 273)^4$$

When the surface temperatures do not vary substantially the above equation can be reduced to:

$$t_{mr} = \Sigma F_b t_s \qquad [3.21]$$

where F_b is the geometric factor (or angle factor) between a particular room surface at temperature t_s and a point within the room where the mean radiant temperature is to be calculated.

Fanger's method gives better results when it is applied for a symmetrical grid of points within a room at a sitting or standing height and a comfort profile is then plotted. The mean of the values of PMV should be zero. If the mean PMV is subtracted from the PMV at each grid point a new set of PPDs can be found, based on these values. Fanger defines the mean of the new PPD values as the *Lowest Possible Percentage Dissatisfied*, LPPD.

A *figure of merit* can be defined for the room as the LPPD minus the lowest possible PPD, i.e. (LPPD − 5). For comfort conditions the figure of merit should be less than or equal to unity. To simplify calculations for the angle factors, Fanger gives graphs for a seated and a standing person with respect to the walls, floor and ceiling. His figures for a seated person are reproduced as Figs 3.16 and 3.17.

Figure 3.16 Mean value of angle factor between a seated person and a vertical rectangle (above or below his centre) when the person is rotated around a vertical axis. To be used when the location but not the orientation of the person is known[3.6]

Figure 3.17 Mean value of angle factor between a seated person and a horizontal rectangle (on the ceiling or on the floor) when the person is rotated around a vertical axis. To be used when the location but not the orientation of the person is known[3.6]

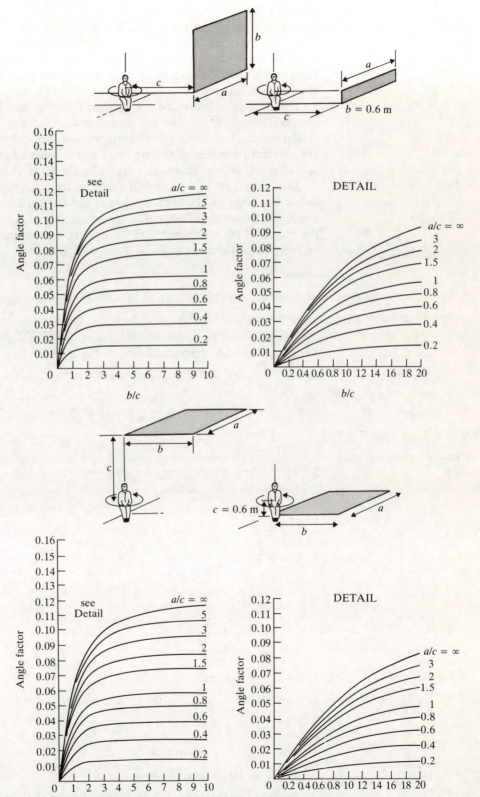

Example 3.5

An office with one external wall, 5 m long, has a window in this wall of size 2 m by 2 m, placed centrally in the wall with the sill 0.6 m from the floor. The width of the office is 4 m and the height 3 m. The office is on an intermediate floor and it may be assumed that the rooms on all five internal sides are at the room temperature.

The heating is provided by a heated ceiling which has a mean surface temperature of 36 °C. The air change rate is 1.5 per hour and the thermal transmittances for the outside wall and window are 0.5 W/m² K and 3.3 W/m² K; the thermal resistance of the air film on the inside surface of the external wall is 0.12 m² K/W; the mean air velocity in the room is 0.1 m/s.

Assuming that the floor is well insulated from the heated ceiling in the room below, and that all the heat input is at the environmental point, calculate the Predicted Percentage Dissatisfied for a seated person with clothing of 1 clo at the centre of the room, for a mean internal air temperature of 20 °C and an external temperature of −2 °C.

Solution

$$\Sigma A = 2\{(5 \times 3) + (4 \times 3) + (5 \times 4)\} = 94 \text{ m}^2$$
$$\Sigma U A_o = 0.5 \times \{(5 \times 3) - 4\} + (3.3 \times 4) = 18.7 \text{ W/K}$$
$$\rho n c_p V = 1.5 \times 1200 \times 5 \times 4 \times 3/3600 = 30 \text{ W/K}$$

From the network diagram shown in Fig. 3.18 we have:

$$\dot{Q}_V = 30(20 + 2) + 660 \text{ W}$$

and

$$t_{ei} = 20 + \frac{660}{4.5 \times 94} = 21.56 \text{ °C}$$

Figure 3.18 Example 3.5

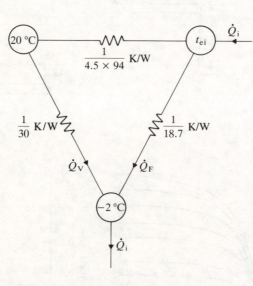

Hence,

$$\dot{Q}_F = 18.7(21.56 + 2) = 440.57 \text{ W}$$

To find the temperature of the inside surface of the window, t_G, the part of the fabric loss through the window is first found by proportion:

$$\dot{Q}_{FG} = \frac{(UA)_G}{\Sigma(UA_o)} \times 440.57 = \frac{4 \times 3.3 \times 440.57}{18.7}$$
$$= 310.99\,W$$

Then,

$$t_G = 21.56 - \frac{(310.99 \times 0.12)}{4} = 12.23\,^{\circ}C$$

The fabric loss through the external walls is given by $(440.57 - 310.99) = 129.58\,W$. Therefore the inside surface temperature of the external wall, t_w, is given by

$$t_w = 21.56 - \frac{(129.58 \times 0.12)}{11} = 20.15\,^{\circ}C$$

Since there is no heat transfer through the internal walls and floor, the surface temperature of the internal walls and the floor can be taken to be equal to the internal environmental temperature of 21.56 °C. This assumes that the adjacent rooms have the same internal radiation as the room itself, and hence are at the same internal environmental temperature.

To find the mean radiant temperature experienced by a seated person at the room centre it is necessary to calculate the angle factors for each individual surface. (*Note*: Using a grid of points would give a better indication of the comfort index of the room; one central point is chosen to illustrate the method.)

The external wall is shown in Fig. 3.19; the wall is symmetrical and the half-wall can be divided into sections A, B, C, D and E. Then, using Fig. 3.16:
For area A, $c = 2$, $a = 1$, $b = 2$; therefore $F_{PA} = 0.025$
Similarly, $F_{PC} = 0.014$

Figure 3.19 External wall, Example 3.5

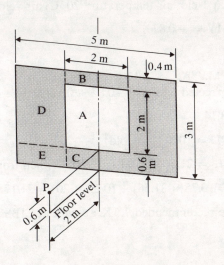
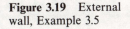

Also taking the relevant values of a, b and c in each case,

$$F_{PB} = F_{PAB} - F_{PA} = 0.029 - 0.025 = 0.004$$
and $$F_{PD} = F_{PABD} - F_{PAB} = 0.054 - 0.029 = 0.025$$
$$F_{PE} = F_{PCE} - F_{PC} = 0.023 - 0.014 = 0.009$$

Therefore, for the whole window,

$$F_{Pw} = 2 \times 0.025 = 0.05$$

and for the external wall,

$$F_{Pe} = 2\{0.004 + 0.025 + 0.009 + 0.014\} = 0.104$$

The ceiling is shown in Fig. 3.20, and from Fig. 3.17,

$$F_{PA} = 0.035$$

Figure 3.20 Ceiling, Example 3.5

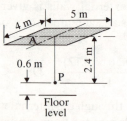

Therefore, for the whole ceiling,

$$F_{Pc} = 4 \times 0.035 = 0.14$$

For the internal walls and floor the angle factor can be found by difference, i.e. $F_{Ps} = 1 - (0.05 + 0.104 + 0.14) = 0.706$. Then using Eqn [3.1],

$$(t_{mr} + 273)^4 = 0.05(12.23 + 273)^4 + 0.104(20.15 + 273)^4$$
$$+ 0.14(36 + 273)^4 + 0.706(21.56 + 273)^4$$

i.e. $t_{mr} = 23.13\,°C$

From Table 3.3 at 1 clo, air temperature 20 °C, air velocity 0.1 m/s, we have

$$PMV = -0.87$$

From Fig. 3.12,

$$\Delta PMV / \Delta t_{mr} = 0.12$$
$$\therefore \quad \text{Corrected PMV} = -0.87 + 0.12(23.13 - 20) = -0.49$$

Then from Fig. 3.14,

$$PPD = 10\% \text{ (too cold)}$$

Note that, since $t_{ei} = 21.56\,°C$ and $t_{ai} = 20\,°C$, then from Eqn [3.15]

$$t_c = (0.75 \times 21.56) + (0.25 \times 20) = 21.17\,°C$$

This is within the recommended 1.5 K above the CIBSE comfort condition of 20 °C.

Local Discomfort

When the Fanger grid method is used, a profile of PPD is obtained for a room from which areas of local discomfort can be identified, even in rooms that may be comfortable on average. The heating system can then be modified to rectify the deficiencies. A person sitting next to a window radiates to the cold window surface and hence may feel uncomfortable; similarly the radiation from a nearby radiator or heating panel may cause local discomfort.

The CIBSE recommends the use of *planar temperature* to check on local discomfort. Planar temperature, t_P, is the radiant temperature on a plane surface as a result of the surfaces on one side of the plane. It can be measured approximately by an instrument consisting of a thermocouple bead embedded flush to the surface of a flat insulated board. The temperature measured will be between the true planar temperature and the air temperature at the bead.

To estimate the planar temperature at a point opposite a flat surface at an absolute temperature T_s, we can write as an approximation:

$$T_P^4 = \Sigma F_P (T_s)^4 + \Sigma F_O (T_O)^4 \qquad [3.22]$$

and $\qquad \Sigma F_P + \Sigma F_O = 0$

where F_P is the form factor for radiation to a parallel plane; F_O is the form factor for radiation to an orthogonal surface; T_s is the temperature of a surface parallel to the plane; T_O is the temperature of a surface orthogonal to the plane.

CIBSE[3.12] gives a graph of form factors for a parallel plane, but not for an orthogonal plane; more accurate tables of factors are given by McLaughlin *et al.*[3.14]

CIBSE recommends that to avoid local discomfort the difference between the local dry resultant temperature and the planar temperature at that point should not exceed 8 K.

As an example, take the case of someone sitting within 1 m of the window in Example 3.5. The factor F_P can be calculated using the tables of McLaughlin *et al.*[3.14] For the external wall the form factor is the factor for the whole wall less that for the window. For the orthogonal surfaces, the ceiling is at 36 °C and the internal walls and floor are at the environmental temperature of 21.56 °C; as an approximation the mean of these surfaces is then

$$T_O = \{(5 \times 36) + (5 + 3 + 3) \times 21.56\}/16 = 26.07 \,°C$$

The point to be considered is shown diagrammatically in Fig. 3.21. To find the form factors the total external area is treated in two symmetrical halves; for the window the form factor is read for a rectangle of 1.0 m × 2.0 m at a

Figure 3.21 External wall for planar temperature

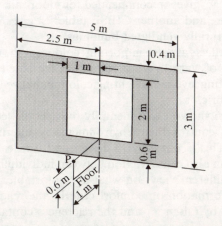

distance of 1 m from the corner and then doubled; for the complete wall the symmetrical half is divided into two rectangles, one of 2.5 m × 2.4 m at a distance 1 m from the corner, and one of 2.5 m × 0.6 m at a distance of 1 m from the corner; these are then added together and doubled. The form factor for the external wall is then the form factor for the total external area less the form factor for the window.

Using this method the form factor for the window is found to be 0.11, and for the external wall 0.26; it follows that the form factor for the orthogonal surfaces is $(1 - 0.37) = 0.63$. That is,

$$(t_P + 273)^4 = 0.11(12.23 + 273)^4 + 0.26(20.15 + 273)^4$$
$$+ 0.63(26.07 + 273)^4$$
$$\therefore \qquad t_P = 23.11\,°C$$

The difference between t_P and t_c is then $(23.11 - 21.17) = 1.94\,K$, which is well within the allowable range of $\pm 8\,K$.

Other Factors

Many factors have been ignored in the above analysis.

In the Fanger method the room is taken to be completely empty of desks, filing cabinets, etc., which would screen the occupants from radiation either from radiators or to windows and other cool surfaces.

The effect of radiation from lighting has been ignored but could be included in the analysis. Similarly, the effect of direct or indirect solar radiation has been neglected.

The allowance for the level of clothing is an oversimplification: the body is not insulated in a uniform way by clothing; also, the clothing must be permeable to perspiration, otherwise discomfort will be experienced. It is unlikely that adding or removing clothing alone will remedy thermal discomfort.

The shape of the vertical thermal gradient within a room is also important. The ideal is for the feet to be a few degrees warmer than the head; it follows that radiant heating from floor level is more effective in sedentary conditions than ceiling heating. The correct number of air changes must be chosen to give the necessary ventilation of odours and tobacco smoke (in areas where smoking is allowed), but draughts of cold air, particularly at foot level, should be avoided. A CIBSE table[3.15] gives recommended outdoor air supply rates for air-conditioned spaces, and another CIBSE table[3.16] gives empirical values of air infiltration for naturally ventilated buildings.

Psychological effects are also important: colours at the red end of the spectrum will give a feeling of warmth while colours at the blue end, or white, will give a cool effect. Lighting levels and lighting colour enhance the effect of the colours of the walls or fabric.

It has also been shown experimentally that people feel more comfortable if they are able to control their environment by altering the heat input or ventilation rate. This is not easy to design and can lead to inefficiencies and waste of fuel; it can also cause friction between individuals working in the same office who have different views on what is comfortable.

For further information on comfort criteria, the reader is recommended to consult the paper by Olsen[3.17] and the references contained therein.

Sick Building Syndrome

Over a period of years starting in the 1960s, a growing number of reports were received of office workers suffering from a variety of poorly defined ailments including headaches, respiratory complaints, sore throats, eye irritation, dry skin, lethargy and general influenza-like symptoms. Initially these were treated with some scepticism by designers and a general view was that most if not all of the complaints had psychological origins. Scientific studies established that there were genuine ailments occurring among staff in certain well designed and well maintained buildings. Genuine symptoms of ill health which appeared when the person was in the building were seen to disappear in the home environment. Buildings which could be shown to be causing this effect were styled 'sick buildings', and the term 'sick building syndrome' was coined. (The term 'tight building' is used in the USA.)

Much research has been undertaken since the term 'sick building' was first used and it is now accepted that there is a serious problem which must be addressed, although the causes of the syndrome and hence possible remedies have not yet been found. Note that the sicknesses occurring due to sick building syndrome are completely different from bacteria-induced illnesses such as Humidifier Fever and Legionnaires' Disease. The bacterium causing the latter illness occurs commonly in rivers, lakes, etc.; the bacteria multiply significantly within the temperature range 25–50 °C and if allowed to enter cooling towers or air conditioning systems can cause a serious respiratory complaint when contaminated water particles become airborne in occupied areas.

Because there are so many likely factors involved in genuine cases of sick building syndrome it is necessary to study a large number of buildings of different designs in order to be able to draw reliable conclusions. It is also necessary to eliminate from any investigation those cases of sickness which may have a psychological origin, and which may induce multiple further cases through mass hysteria.

The World Health Organisation has identified certain common factors in buildings suffering from sick building syndrome:

(a) air-tight buildings with forced ventilation or air conditioning;
(b) a high proportion of re-circulated air;
(c) a high surface-to-volume ratio of surfaces covered in carpets, textiles and plastics.

However, at the present state of knowledge it cannot be proved conclusively whether a building of any particular design will prove to be a sick building or not.

When reports of illness among occupants of a building are received, the greatest difficulty faced by someone responsible for the building is deciding whether there are psychological factors, or whether it is a genuine case of a sick building. When the term 'sick building' becomes implanted in the public consciousness, there is a tendency for people to argue that the building they occupy is a sick building when there may be other reasons for their complaints. It may be that the building has not been adequately designed for thermal comfort, or that there is too high a level of noise, or bad lighting design, or a combination of some or all of these. The problem may be entirely psychological and the possible factors are many and diverse; a good industrial relations policy is essential and may forestall potential cases of pseudo sick building syndrome.

When it is clearly established that the problem is not psychological and is not due to basic design faults or inadequate maintenance, then a proper study can be made of the likely causes.

Scientific measurements are being made of possible air pollutants from man-made materials under normally occurring conditions of temperature and humidity. It is suspected that very small traces of certain gases in combination may be a major contributory factor. The influence of electrostatic fields and ion concentration has been advanced as a possible cause but studies so far have not shown any positive results.

PROBLEMS

3.1   (a)   Define ventilation conductance, C_V, and show that for a space heated entirely by convection:

$$C_V = \rho \dot{V} c_p \left\{ 1 + \frac{\Sigma U A_o}{\alpha_a \Sigma A} \right\}$$

where ρ = air mean density; $\dot{V}$ = volumetric infiltration rate; c_p = air mean specific heat at constant pressure; U = overall heat transfer coefficient for an element conducting heat to the outside; A_o = surface area of an element conducting heat to the outside; α_a = hypothetical heat transfer coefficient between the air and environmental points; A = area of any surface element.

(b)   A naturally ventilated room which is heated entirely by convection has a total steady-state heat input of 1307 W. Using the data below, calculate:
(i)   the inside air temperature;
(ii)   the air infiltration rate.

Data
Total inside surface area, $A = 94 \, \text{m}^2$; $\Sigma U A_o = 38.5 \, \text{W/K}$; $\alpha_a = 4.5 \, \text{W/m}^2 \, \text{K}$; inside environmental temperature = 20 °C; outside air temperature = -2 °C; $\rho c_p = 1200 \, \text{J/m}^3 \, \text{K}$.

                                                 (Wolverhampton Polytechnic)

(22 °C; 0.016 m³/s)

3.2   A cubic-shaped room has one external wall, and each of the room surfaces is of area 20 m². The room is heated to a dry resultant temperature of 19 °C by a ducted warm air heating system. If the air infiltration rate is equivalent to 1.0 air change per hour and the external air temperature is -2 °C, determine:
(i)   the room air temperature;
(ii)   the required heat input to the room;
(iii)   the inside environmental temperature;
(iv)   the average inside surface temperature of the external wall.

Data
For air:
Specific heat capacity, 1.02 kJ/kg K; density, 1.2 kg/m³.
Heat conductances between:
Air and dry resultant point, $6.0 \Sigma A \, \text{W/m}^2 \, \text{K}$; dry resultant and environmental point, $18.0 \Sigma A \, \text{W/m}^2 \, \text{K}$; air and environmental point, $4.5 \Sigma A \, \text{W/m}^2 \, \text{K}$; where

ΣA = area of all room surfaces (m^2). U-value of external wall, 4 W/m^2 K; inside surface resistance of external wall, R_{si}, 0.12 m^2 K/W.
(21.25 °C; 2327 W; 18.25 °C; 8.53 °C)

3.3 A small workshop, 15 m long by 10 m wide by 4.5 m high, has a flat roof and all the walls are external; the window area in the walls is 50 m^2 and there is one door of area 5 m^2. For the data given below calculate the inside air temperature and the total heat input:
(i) for warm air heating;
(ii) for vertical radiant panels (33.33% convective, 66.67% radiant);
(iii) for high-temperature radiant heating (10% convective, 90% radiant).

Data
Thermal transmittances:

Walls, 0.9 W/m^2 K; windows, 5.6 W/m^2 K; door, 3.0 W/m^2 K;
floor, 0.7 W/m^2 K; roof, 0.9 W/m^2 K.
Air change rate, 2 per hour; outside design temperature, -1 °C.

Take the appropriate dry resultant temperature from Table 3.1, assuming light work in the factory. For part (i) assume no air infiltration and 100% fresh air intake to the plant; for parts (ii) and (iii) assume natural ventilation with the required air change achieved by infiltration.
(19.5 °C, 20.1 kW; 13.9 °C, 18.9 kW; 12.1 °C, 18.5 kW)

3.4 It is proposed to heat a flat-roofed factory building, 30 m × 15 m × 5 m high, using forced-convection warm-air heaters. There are 12 windows each of area 4 m^2 in each of the long walls and doors totalling 6 m^2 in each of the end walls. The internal dry resultant temperature is to be 18 °C when the outside temperature is -1 °C.
(a) Using the data below and in Table 3.4, calculate the steady-state design rate of heat loss and the environmental, air and mean surface temperatures that will be achieved.
(b) Comment on the condition achieved, the effects of internal heat gains, and whether any other method of heating should be considered.

Table 3.4 Values of F_1 and F_2 for 100% convective, 0% radiant (forced warm-air heaters)

$\Sigma(UA_o)/\Sigma A$	0.8	1.0	1.5	2.0
F_1	0.96	0.95	0.92	0.90
F_2	1.13	1.16	1.23	1.30

Data
Overall heat transfer coefficient (U-values) in W/m^2 K:
Walls, 0.6; windows, 5.6; roof, 1.0; doors, 2.5; floor, 0.4.

(Engineering Council)

(21.08 kW, 17.1 °C, 20.6 °C, 15.4 °C)

Note: The heat input to the building will be greater than 21.08 kW since some fresh air at -1 °C will be induced and heated in the heaters; the problem does not ask for the heat input to be calculated; to calculate the heat input the percentage of air re-circulated must be known.

3.5 (a) Sketch the energy flow triangles for a room under steady-state conditions. Indicate the energy flows between the internal air, outside air, internal dry resultant and internal environmental temperature points.

(b) Show that in the above case the following heat balance equation applies:

$$F_v Q_a + F_u Q_e = \{F_v \rho n c_p V + F_u \Sigma(UA_o)\}(t_c - t_{ao})$$

where: $$F_v = \frac{\alpha_{ac}\Sigma A}{\rho n c_p V + \alpha_{ac}\Sigma A} \quad \text{and} \quad F_u = \frac{\alpha_{ac}\Sigma A}{\alpha_{ec}\Sigma A + \Sigma UA_o}$$

The symbols have their usual meaning.

(c) A cubical room has one external wall of aggregate U-value 2.1 W/m² K. Each surface has an area of 12 m². The external design air temperature is $-3\,°C$ and the internal design dry resultant temperature is 21 °C. The ventilation loss is 20 W/K. Calculate the heat input required by a heating system which is 30% convective and 70% radiant. Assume: $\alpha_{ac} = 6$ W/m² K; $\alpha_{ec} = 18$ W/m² K.

(Newcastle upon Tyne Polytechnic)

(1.07 kW)

3.6 A room is 5 m × 3 m × 3 m high with an external wall, 5 m long, which contains a window of half the wall area. Using the extract from Fanger's comfort data provided (Table 3.3, page 147, and Figs 3.12 and 3.14) and the data below, calculate:

(i) the mean surface temperature for the room, assuming a mean thermal resistance for the inside surfaces of 0.11 m² K/W, based on the inside environmental temperature;

(ii) the value of the Predicted Percentage Dissatisfied, PPD, for the room, and the significance of this figure for the comfort of the occupants; assume that the mean radiant temperature is approximately equal to the mean surface temperature.

Data
U-values:
External wall, 0.6 W/m² K; window, 5.6 W/m² K.
Heat transfer coefficient for environmental point to air point, 4.5 W/m² K.
Room used for sedentary activity:
Clothing level, 1 clo; air velocity, 0.1 m/s.
Inside environmental temperature, 21 °C; outside air temperature, $-1\,°C$; relative humidity, 50%; environmental temperature of all adjacent rooms, 21 °C.
Room heating entirely by convection.

(Wolverhampton Polytechnic)

(19.64 °C; 7%)

3.7 (a) In the calculation of the steady-state heat loss from a room, clearly explain the use of the following terms:
(i) internal air temperature;
(ii) internal dry resultant temperature;
(iii) internal environmental temperature.
(b) Compare the thermal performance of warm air, underfloor and convector-radiator heating systems with respect to comfort and energy requirements.

(CIBSE)

3.8 (a) State the 'energy (heat) balance equation' for the Human Body and, with reference to the thermo-regulatory mechanisms involved, show how environmental and personal factors will influence the thermal balance.

(b) What is meant by 'Thermal Comfort'? Using selected indices as examples, discuss how this may be assessed.

(Engineering Council)

3.9 The primary aim of most heating, ventilating and air conditioning systems is to provide thermal comfort conditions for the building occupants, throughout the year. By discussing the factors which affect a human being's sensation of thermal comfort, explain the problems faced by the Building Services Engineer in satisfying this aim.

(CIBSE)

3.10 A simple expression for the dry resultant temperature preferred for neutral comfort is given by McIntyre as follows:

$$t_c = 33.5 - 3R_c - (0.08 + 0.05R_c)(M/A)$$

where R_c = thermal resistance of clothing, clo units; M/A = metabolic rate per unit area of body surface, W/m^2.

This expression assumes a low air speed and equal values of room air and surface temperature.

Discuss to what extent this equation may be useful for comfort predictions in moderate thermal environments.

For a man working in an office wearing a standard European business suit, the value of M/A is $70\,W/m^2$. Taking suitable values for R_c, find the preferred dry resultant temperature, (a) if the man is normally dressed, and (b) if he takes off his jacket.

(University of Manchester)

(18.15 °C; 21.40 °C)

3.11 (a) Discuss the major factors influencing a person's sensation of 'thermal comfort', and distinguish between the terms 'thermal balance' and 'thermal comfort'.

(b) Discuss three cases of asymmetric thermal radiation which may lead to discomfort. In each case suggest one measure to remedy the problem.

(CIBSE)

3.12 (a) Discuss the main human physiological mechanisms used to maintain a stable body temperature under varying environmental conditions.

(b) Discuss briefly the following:
(i) comfortable environment;
(ii) tolerable environment;
(iii) heat stress;
(iv) asymmetric radiation.

(CIBSE)

3.13 An office, 5 m long by 4 m wide by 3 m high, is on an intermediate floor of a multi-storey building with similar offices above, below and on either side. The long wall is the only external wall and it contains windows of total area

$4\,m^2$. There is a corridor running alongside the internal 5 m long wall. Using the data below, calculate for the two following cases:
(a) heat input entirely at the air point,
(b) heat input 80% at the environmental point and 20% at the air point,

(i)   the ventilation conductance;
(ii) the heat input required to the office;
(iii) the inside air temperature;
(iv) the dry resultant temperature;
(v)   the PPD for the room, assuming that the mean radiant temperature is equal to the mean surface temperature.

Data
Thermal transmittances:
External wall, $1.0\,W/m^2\,K$; internal walls, $2.7\,W/m^2\,K$; windows, $5.6\,W/m^2\,K$; door, $2.7\,W/m^2\,K$.
Environmental temperature in office, $20\,^{\circ}C$; environmental temperature in corridor, $16\,^{\circ}C$; outside design temperature, $-1\,^{\circ}C$; air infiltration rate, 1 per hour; mean air velocity in office, $0.1\,m/s$; level of clothing of occupants, 1 clo.
((a) $21.94\,W/K$; $1324\,W$; $22.04\,^{\circ}C$; $20.51\,^{\circ}C$; 16%. (b) $19.65\,W/K$; $1276\,W$; $19.63\,^{\circ}C$; $19.91\,^{\circ}C$; 23%)

3.14 (a)   It is proposed to heat a six-bed hospital ward by underfloor heating, giving the floor a mean surface temperature of $25\,^{\circ}C$. The space temperature is to be controlled by an air thermostat set at $18\,^{\circ}C$. Utilizing the following data, determine:
(i) the mean radiant temperature of the space;
(ii) the dry resultant temperature of the space.

Data
Room dimensions: 8 m by 8 m by 3 m high; number of exposed walls, 2; roof exposed.
Average U-values of exposed roof and walls, $0.3\,W/m^2\,K$; thermal resistance of surfaces, $0.123\,m^2\,K/W$; outside temperature, $-4\,^{\circ}C$; air velocity, $0.2\,m/s$.
   Assume no heat flow through the internal walls and that heat transfer through external surfaces is a function of air temperature.
(b)   The activity level and clothing of the occupants of the space are given below. Utilizing the Fanger's tables of PMV, (extracts given here as Table 3.5), and Figs 3.12 and 3.14, determine the likely thermal sensation of the occupants and comment on the suitability of the environment proposed. State any assumptions made.

Occupant	Activity (W/m^2)	Clothing (clo)
Staff working	105	1.0
Patients sitting	70	0.5
Patients in bed	58	1.25

(Newcastle upon Tyne Polytechnic)
($19.6\,^{\circ}C$; $18.8\,^{\circ}C$; 5.5% staff dissatisfied (too warm); 80% seated patients dissatisfied (too cold); 25% patients in bed dissatisfied (too cold))

Table 3.5 Fanger's values of PMV for a range of activity levels

(a) Activity, 105 W/m²; clothing, 1.0 clo; air temperature, 18 °C:

Air velocity/(m/s)	<0.1	0.1	0.2
PMV	0.21	0.20	0.06

(b) Activity, 70 W/m²; clothing, 0.5 clo; air temperature, 18 °C:

Air velocity/(m/s)	<0.1	0.1	0.2
PMV	−2.01	−2.01	−2.38

(c) Activity, 58 W/m²; clothing, 1.25 clo; air temperature, 18 °C:

Air velocity/(m/s)	<0.1	0.1	0.2
PMV	−0.89	−0.91	−1.14

3.15 An office 13 m long by 5 m wide is on an intermediate floor of a building containing similar offices, and occupies a corner position.

One of the 13 m long walls faces North and has three windows equally spaced along its length; each window is 3 m long by 2 m high and the sill is 0.6 m from the floor. The wall opposite the windows is adjacent to a corridor; the 5 m wide outside wall has no windows.

The room is heated by three LPHW radiators, one situated under each window; the length of each radiator is approximately 3 m. Using the data below and making any other assumptions necessary:
(i)   check the 'whole room' comfort using dry resultant temperature and by Fanger's method;
(ii) selecting a suitable-sized grid (say six points), for the seated position (i.e. 0.6 m from the floor), estimate the PPV and PPD for each grid point using the angle factor method, and hence calculate the LPPD and the figure of merit for the office.
(iii) check for local discomfort at a seated location that you consider may be a problem by finding the plane radiant temperature for the point chosen.

Data
Environmental temperature of office, 20 °C; environmental temperature of corridor, 16 °C; outside design temperature, −2 °C; radiator surface temperature, 70 °C; relative humidity, 50%, level of clothing of occupants, 1 clo; mean air velocity in office, 0.1 m/s.
Thermal transmittances:
External walls, 1 W/m² K; windows, 5.6 W/m² K; internal walls, 2.7 W/m² K.
Thermal resistance of all surfaces, 0.123 m² K/W.

Calculations show the internal air temperature of the office to be 23 °C.

(Wolverhampton Polytechnic: set assignment)

Figure 3.22 Grid points for Problem 3.15

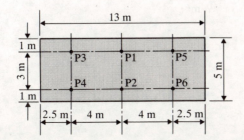

((i) 20.75 °C, 13%; (ii) three grid points chosen say, at 1 m from, and directly opposite each window, with a further three points in the same position lengthwise and 1 m from the corridor wall (see Fig. 3.22): P1 −0.07, 5%; P2 −0.44, 9%; P3 −0.36, 7.6%; P4 −0.47, 9.5%; P5 −0.10, 5.3%; P6 −0.44, 9.0%; LPPD 5.58%, figure of merit 0.58; (iii) P4 19.2 °C).

REFERENCES

3.1 CIBSE 1986 *Guide to Current Practice* volume A5

3.2 CIBSE 1986 *Guide to Current Practice* volume A, Table A2.33

3.3 Threlkeld J L 1970 *Thermal Environmental Engineering* Prentice Hall

3.4 CIBSE 1986 *Guide to Current Practice* volume A1, Table A1.1

3.5 ASHRAE 1976 *Handbook of Fundamentals*

3.6 Fanger P O 1972 *Thermal Comfort* McGraw-Hill (1982 © Robert E Kreiger, Florida)

3.7 McIntyre D A 1973 A guide to thermal comfort. *Applied Ergonomics* **42** (June): 66−72

3.8 CIBSE 1986 *Guide to Current Practice* volume A1, Table A1.3

3.9 CIBSE 1986 *Guide to Current Practice* volume A9, Tables A9.1−7

3.10 Nevins R G, Rholes T H, Springer W and Feyerhem A M 1966 A temperature−humidity chart for thermal comfort of seated persons. *ASHRAE Tr*, **72**(I): 283−91

3.11 McNall J R, Jaax J, Nevins R G and Springer W 1967 Thermal comfort (thermally neutral) conditions for three levels of activity. *ASHRAE Tr.* **73**(I)

3.12 CIBSE 1986 *Guide to Current Practice* volume A1, Fig. A1.17

3.13 ISO 7730 1984 *Moderate Thermal Environments: Determination of the PMV and PPD Indices and Specification of the Conditions for Thermal Comfort* ISO, Geneva

3.14 McLaughlin R K, McLean R C and Bonthron W J 1981 *Heating Services Design* Butterworth

3.15 CIBSE 1986 *Guide to Current Practice* volume A1, Fig. A1.8 and Table A1.5

3.16 CIBSE 1986 *Guide to Current Practice* volume A4, Table A4.12

3.17 Olsen B W 1987 Recent developments in design criteria for human comfort. *CIBSE Conference Proceedings, Advances in Air Conditioning* pp 3−14

3.18 Bedford T 1964 *Basic Principles of Ventilation and Heating* H K Lewis and Co Ltd, London

4 TRANSIENT HEAT FLOW IN BUILDINGS

In real buildings the steady state is never reached, so that the theory outlined in Chapter 3 is limited and gives only an overall mean value of possible thermal loads, or enables a load to be calculated for a limited pseudo-steady-state period. The normal daily cycle causes a basic variation in external conditions; in addition, weather changes can cause fluctuations from day to day or even within a shorter time period.

In winter most buildings are heated for the period of occupancy only, with perhaps a pre-heat period just prior to occupancy; thermostatic control during occupancy allows for variations in external conditions and changes in internal gains. In summer the greatest fluctuation in conditions is due to the daily solar cycle and in fully air-conditioned buildings the air conditioning system is usually activated only during occupancy, again with thermostatic control.

The thermal analysis of a building is therefore exceedingly complex and difficult, and many different models for computer simulation have been used. The book by Clarke[4.1] is recommended for a review of possible mathematical models; articles by Clarke *et al.*[4.2] give a concise overview. In this chapter the basic theory of transient heat transfer is covered briefly and the mathematical model recommended by CIBSE is illustrated through worked examples.

4.1 TRANSIENT HEAT TRANSFER

Basic transient heat transfer theory is covered by Eastop and McConkey.[4.3] For example, for a one-dimensional body with temperature changing with time we have:

$$\frac{\partial^2 t}{\partial x^2} = \frac{1}{\kappa} \frac{\partial t}{\partial \tau} \qquad [4.1]$$

where t is the temperature at a point in the body at any distance x from one surface; τ is time; κ is the thermal diffusivity, $\lambda/\rho c$; λ is the thermal conductivity of the body; ρ is the density of the body; c is the specific heat of the body. A wall of a building in which the height and width of the wall are large compared with the thickness approximates to Eqn [4.1].

Equation [4.1] can be solved by the separation of the variables method, or by the Laplace transformation method,[4.1] or by a numerical method using finite differences,[4.1,4.3] to give the temperature distribution within the wall at any time τ with given boundary conditions at $x = 0$ and $\tau = 0$.

It can be seen that the thermal diffusivity $(\lambda/\rho c)$ determines the rate at which a body responds to changes in the external conditions; a body with a high thermal diffusivity heats up and cools down much faster than a body with a low thermal diffusivity.

Equation [4.1] is for the heat transfer within the body but the heat transfer from the surfaces of the body to the surroundings is also required. At the surface at any time, τ, we have: $\dot{q}_c + \dot{q}_r = \lambda(\partial t/\partial x)_s$, where $\dot{q}_c$ is the heat flux at the surface at any time τ due to convection; $\dot{q}_r$ is the net heat flux on the surface at any time τ due to short- and long-wave radiation; $(\partial t/\partial x)_s$ is the temperature gradient at the wall surface at any time τ. Using a numerical method on a computer the boundary values of fluid temperature and radiation can be input on an hourly basis with each layer of the wall subdivided into a number of space intervals.[4.1]

Approximation for a Complete Building

When the thermal resistance of the fluid film on the surface is high compared with the internal thermal resistance, the heat transfer from the body to its surroundings is controlled by the surface heat transfer. Under these conditions the temperature variation within the body is negligible compared with the variation of the mean temperature of the body with time. This is known as Newtonian cooling.

We have:

$$\alpha A(t - t_F) = -\rho Vc \frac{dt}{d\tau}$$

where α is the heat transfer coefficient from the surface of the body to the fluid at a constant temperature, t_F; V is the volume of the body of surface area A.

This equation can be solved between initial and final conditions 1 and 2 at $\tau = 0$ and $\tau = \tau$, as:

$$\int_1^2 \frac{dt}{(t - t_F)} = -\frac{\alpha A}{\rho Vc} \int_0^\tau d\tau$$

i.e.

$$\frac{t_2 - t_F}{t_1 - t_F} = e^{-(\alpha A/\rho Vc)\tau} \tag{4.2}$$

The term $\rho Vc/\alpha A$ is the time constant, τ_c; a large time constant therefore implies a longer time for the body to cool down or heat up between specific temperatures. Figure 4.1 shows Eqn [4.2] plotted for two time constants. A body with a large time constant has a thermal capacity (ρVc) which is much greater than the heat transfer rate per unit temperature difference between the body and its surroundings (αA). The body can be considered to have a certain 'thermal inertia'; when the time constant is large the body takes longer to respond to a cooling or heating process and hence has a high 'thermal inertia'.

The simplified theory above can be applied to a complete building. The heat loss from a building is in two parts: the fabric heat loss and the ventilation heat loss, both of which can be written in terms of a factor times a temperature

Figure 4.1 Cooling of a body

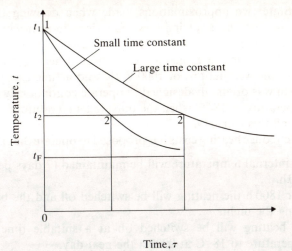

difference between inside and outside. Taking the simple model, neglecting radiation internally, an overall heat loss coefficient can be defined as:

$$\text{Total rate of heat loss from building} = H(t_{ai} - t_{ao})$$

where H is the heat loss coefficient.

The heat loss coefficient is also given by:

$$H = \Sigma U A_o + \rho n c_p V$$

The term H can be substituted in Eqn [4.2] for the term (αA); note also that the thermal capacity of the complete building must be substituted for the term $(\rho c V)$ for a solid body.

The Example below illustrates the application of this simple theory to a building.

Example 4.1

(a) Show that the cooling of a body over a time interval τ_1 to τ_2 can be given by Eqn [1]. Subsequent re-heating over a period τ_2 to τ_3 can be represented by Eqn [2].

$$\tau_2 - \tau_1 = -\tau_c \times \ln\{(t_2 - t_0)/(t_1 - t_0)\} \qquad [1]$$

$$\tau_3 - \tau_2 = -\tau_c \times \ln\left\{\frac{\dot{Q} - H(t_3 - t_0)}{\dot{Q} - H(t_2 - t_0)}\right\} \qquad [2]$$

where t_0 is the temperature of the environment; t_1 is the initial temperature of the body; t_2 is the temperature of the body at time τ_2, i.e. just as re-heating commences; t_3 is the temperature reached by the body at time τ_3; H is the heat loss coefficient of the body; $\dot{Q}$ is the rate at which heat is supplied during the period τ_2 to τ_3; τ_c is the time constant of the body. Hence show that if t_1 and t_3 are equal, then Eqns [1] and [2] can be combined to give the result

$$\tau_3 - \tau_2 = -\tau_c \times \ln\left\{\frac{1 - (H/\dot{Q})(t_3 - t_0)}{1 - (H/\dot{Q})(t_3 - t_0)\exp[-(\tau_2 - \tau_1)/\tau_c]}\right\} \qquad [3]$$

(i) State two approximations made when deriving the time constant for a building and applying these equations to transient heating.

(ii) Discuss briefly the role of plant oversize in saving energy used for heating.

(b) A building in England has an overall heat loss coefficient (i.e. including fabric and ventilation) of 1000 W/K and a time constant of 40 h. Its heating system was originally designed to operate continuously and provide an internal temperature of 18 °C when the outside temperature is 0 °C, with no oversizing beyond this capacity.

To reduce heating costs it is proposed to operate the existing system as follows:

The internal temperature will be maintained (7 days per week) from 0900 h to 1800 h.

After 1800 h the heating will be switched off and the building allowed to cool during the night.

The heating will be switched on at a suitable time to bring the internal temperature to 18 °C at 0900 h the next day.

Calculate, graphically or otherwise, the reduction in annual heating energy to be expected from these measures by considering a representative 24 h period. You can assume that the outside temperature stays constant throughout but you should explain briefly why you have chosen the value which has been used in the calculation.

(Engineering Council)

Solution

(a) Substituting, $\tau_c = \rho c V / \alpha A$, in Eqn [4.2] we can derive Eqn [1].

i.e.
$$\tau_2 - \tau_1 = -\tau_c \times \ln\{(t_2 - t_0)/(t_1 - t_0)\}$$

The derivation of Eqn [2] is not required but it will be given to illustrate the procedure.

For the period from τ_2 to τ_3 we have at any time τ when the temperature of the body is t:

$$\dot{Q} - H(t - t_0) = \rho V c \frac{dt}{d\tau}$$

i.e.
$$\int_2^3 \frac{dt}{\dot{Q} - H(t - t_0)} = \frac{1}{\rho V c} \int_\tau^{\tau_3} d\tau$$

$$\therefore \quad -\frac{1}{H} \times \ln\left\{\frac{\dot{Q} - H(t_3 - t_0)}{\dot{Q} - H(t_2 - t_0)}\right\} = \frac{1}{\rho V c}(\tau_3 - \tau_2)$$

Then, writing $\rho c V / H = \rho c V / \alpha A = \tau_c$, Eqn [3] is derived, i.e.

$$\tau_3 - \tau_2 = -\tau_c \times \ln\left\{\frac{\dot{Q} - H(t_3 - t_0)}{\dot{Q} - H(t_2 - t_0)}\right\}$$

From Eqn [1]:

$$t_2 - t_0 = (t_1 - t_0)\exp[-(\tau_2 - \tau_1)/\tau_c]$$

Substituting in Eqn [2], and putting $t_1 = t_3$ (given), then Eqn [3] is obtained:

i.e.
$$\tau_3 - \tau_2 = -\tau_c \times \ln\left\{\frac{1 - (H/\dot{Q})(t_3 - t_0)}{1 - (H/\dot{Q})(t_3 - t_0)\exp[-(\tau_2 - \tau_1)/\tau_c]}\right\}$$

(i)   In the simplified approach above it is assumed that there is no temperature variation within the body itself; the model is known as 'lumped parameter', or Newtonian cooling (see earlier). For a building this implies that the structure is at a mean temperature equal to the internal air temperature with no significant temperature variations within the structure as cooling or heating occurs. A suitable mean value of density and specific heat for the whole building must also be chosen. The volume refers to the volume of the material the structure is made of, since the air within the building is assumed to be of negligible thermal capacity compared with the solid structure. It is also assumed in the derivation of the equations above that the outside temperature, t_0, remains constant during the heating or cooling process.

As stated earlier, the heat loss coefficient depends on the fabric heat loss plus the ventilation heat loss.

The major assumption in this simplified approach is that the building can be taken to act as a single unit when cooling down or heating up; in practice there is an interaction between the elements of the building structure and the air within the building, as well as a temperature variation within the structure itself.

(ii)   A building may be heated continuously over each 24-hour period or it may be heated intermittently, i.e. during occupancy only. When the heating system for a building is switched off at night during the unoccupied period, the building cools down. The heating system must therefore be switched on some hours before occupancy the following day to heat the building back up to the comfort condition. Depending on the building time constant, the time taken for the heating-up process may be too long, and in such a case the plant must be oversized.

Overload capacity is the capability of the system, i.e. the heat emitters as well as the heating plant, to provide a heat output greater than the calculated steady-state requirement based on certain design temperatures. The outside design temperatures chosen are based on prevailing weather conditions and are usually specified for buildings of either low or high thermal inertia, and assume an overload capacity of about 20%.

For example, for a building in London the recommended outside design temperatures are:

High thermal inertia $-2.0\,°C$
Low thermal inertia $-3.0\,°C$

Values for outside design temperature on this basis are given for eight locations in the UK in Ref. 4.4. Statistically, a temperature below these values will not occur on more than one occasion during the heating season.

To calculate the actual energy used during a typical heating season it is necessary to know the daily mean seasonal outside temperature. For example, for London from October to April inclusive the seasonal mean outside temperature is approximately $7\,°C$.

When intermittent heating is used, then a saving in fuel costs occurs; the full load is applied during the pre-heat period and the normal heating period, but the system is shut down during the remainder of the unoccupied period, thus giving a net reduction in energy used. It can be shown that as the overload capacity is increased the necessary pre-heat period becomes shorter and the fuel saving increases. The decrease in fuel costs must be high enough to compensate for the increased capital cost of the oversized system so that the system pays for itself within an acceptable time period. It should also be noted that an oversized system will tend to operate at a lower efficiency when running on part-load, and this must be taken into account in the calculations.

In an extreme case it may be found that continuous heating with a smaller heating system is a better financial proposition. Also, a system designed with no overload capacity will still have excess heating capacity on days when the outside temperature is above the design value.

(b)   The heating capacity can be found as follows:

$$Q = H(18 - 0) = 1000 \times 18 = 18\,000\text{ W} = 18\,\text{kW}$$

Part (a) of the problem asks for the derivation of Eqn [3] which may be taken to imply that Eqn [3] is useful for solving part (b), but the problem is best solved by adding Eqn [1] to Eqn [2], to give

$$\tau_3 - \tau_1 = -\tau_c \times \ln\left\{\frac{(t_2 - t_0)[\dot{Q} - H(t_3 - t_0)]}{(t_1 - t_0)[\dot{Q} - H(t_2 - t_0)]}\right\} \qquad [4]$$

Then, with $\tau_c = 40$ h, $t_1 = t_3 = 18\,°\text{C}$, $H = 1\,\text{kW/K}$, and $\dot{Q} = 18\,\text{kW}$, and taking a mean seasonal outside temperature of $7\,°\text{C}$, say,

$$\tau_3 - \tau_1 = -40 \times \ln\left\{\frac{(t_2 - 7)[18 - (18 - 7)]}{(18 - 7)[18 - (t_2 - 7)]}\right\}$$

The variation of temperature with time is shown diagrammatically in Fig. 4.2; the time τ_1 is put equal to zero at 1800 h when the heating is switched off. That is,

$$\tau_3 = 15 = -40 \times \ln\left\{\frac{7(t_2 - 7)}{11(25 - t_2)}\right\} \qquad [a]$$

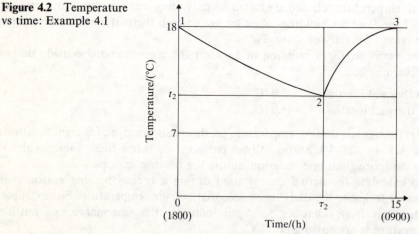

Figure 4.2 Temperature vs time: Example 4.1

Also, from Eqn [1],

$$\tau_2 = -40 \times \ln\left\{\frac{(t_2 - 7)}{11}\right\} \qquad [b]$$

From Eqn [a],

$$\frac{7(t_2 - 7)}{11(25 - t_2)} = e^{-15/40} = 0.687$$

$$\therefore \quad t_2 = 16.35\,°C$$

and substituting in Eqn [b],

$$\tau_2 = 6.5\,h,\ \text{i.e. pre-heating starts at } 0030\,h$$

This gives a pre-heating period of 8.5 h, which is excessively long. With the information as given, the length of the pre-heating period can be reduced by choosing a higher mean seasonal outside temperature. For example, with a mean seasonal outside temperature of 10 °C the time τ_2 is 9.0 h (i.e. pre-heating starts at 0300 h), giving a pre-heating period of 6 h, still long. Also, there is no part of the British Isles with as high a mean seasonal outside temperature as 10 °C (the highest value is 9 °C for the Scilly Isles and the Channel Islands).

Continuing with the problem and assuming a mean seasonal outside temperature of 7 °C as above (the value for London), we have:

$$\text{Energy for continuous heating} = \frac{1 \times (18 - 7) \times 24 \times 3600}{1000}$$

$$= 950.4\,\text{MJ/day}$$

Energy for intermittent heating

$$= \frac{1 \times (18 - 7) \times 9 \times 3600}{1000} + \frac{1 \times (18 - 0) \times 8.5 \times 3600}{1000}$$

$$= 356.4 + 550.8 = 907.2\,\text{MJ/day}$$

i.e.   Reduction in heating energy = 43.2 MJ/day

Consider introducing an overload capacity. Take a 20% increase over the initial capacity, i.e. $\dot{Q} = 1.2 \times 18 = 21.6\,\text{kW}$. Then, from Eqn [4],

$$15 = -40 \times \ln\left\{\frac{(t_2 - 7)[21.6 - (18 - 7/]}{(18 - 7)[21.6 - (t_2 - 7)]}\right\}$$

i.e. $t_2 = 15.99\,°C$

and using Eqn [b],

$$\tau_2 = 8.06\,h$$

i.e. pre-heating starts at 0204 h.

This gives a pre-heat period of 6.94 h, and a reduction in energy usage of

$$\text{Reduction} = 950.4 - \{356.4 + (21.6 \times 6.94 \times 3.6)\}$$

$$= 54.4\,\text{MJ/day}$$

Assuming it is desired to have a pre-heat period of 4 hours, say, then $\tau_2 = 15 - 4 = 11\,h$.

Therefore, from Eqn [1],

$$11 = -40 \times \ln\left\{\frac{t_2 - 7}{11}\right\}$$

i.e. $t_2 = 15.36\,°C$

and in Eqn [2],

$$4 = -40 \times \ln\left\{\frac{[\dot{Q} - (18 - 7)]}{[\dot{Q} - (15.36 - 7)]}\right\}$$

i.e. $\dot{Q} = 36.1 \, \text{kW}$

This represents a 100% overload, which is excessively high.

The reduction in energy use is given by

$$\text{Reduction} = 950.4 - \{356.4 + (36.1 \times 4 \times 3.6)\}$$
$$= 74.2 \, \text{MJ/day}$$

This is to be compared with the figure of 43.2 MJ/day with no overload, and 54.4 MJ/day with 20% overload; the very large increase in capital cost for a 100% overload capacity is likely to rule out the scheme on financial grounds. An overload of more than about 25% is normally financially non-viable.

4.2 TRANSIENT ENERGY TRANSFER EQUATIONS

When the Laplace transformation method is used to solve Eqn [4.1] there are two basic ways of solving the resulting transmission matrix. One method is concerned with the response of the system to time-series temperature pulses and is known as the time-domain or *response function method*; it is the method used by ASHRAE and summarized by Clarke.[4.1] The analysis produces a certain number of response factors which allow the equations to be solved quickly using a computer. The *ASHRAE Handbook* quotes response factors; hourly weather data are also required.

The second method of solution of the matrix is to represent temperature changes by a series of periodic cycles using Fourier series; the principle of superposition is then used to sum the effects of the individual harmonics. This method is known as the frequency-domain or *harmonic method*; it is the method used by CIBSE. It was originally developed as a manual method but is now available for computer use with commercial software. The remainder of this chapter will discuss the basis of the method.

The model adopted is that using environmental temperature, described in Chapter 3, with steady cyclic variations related to the 24-hour weather cycle. It is frequently described as a 'means and swings' method. Best results are obtained when the weather conditions are stable over a period of several days.

Thermal response factors related to the 24-hour frequency are applied to the actual temperature and heat flux excitations; the three factors used are as follows.

Admittance, *Y*

In ac electrical networks with a resistance and capacitance the current has a phase lead over the voltage. The electrical term for the combination of resistance and capacitance governing the current–voltage relationship is the impedance; the reciprocal of impedance is the admittance.

In transient heat flow when the temperature varies cyclically (strictly speaking, sinusoidally), there is an analogy to the ac circuit. Each building element has a thermal resistance and because of its thermal capacity it also has the equivalent of an electrical capacitance. Hence a thermal admittance, Y, can be defined as the energy absorbed by unit area of a surface for each degree of temperature swing about the mean environmental temperature. It represents the response of the fabric to the cyclic load on it, and is the cyclic equivalent of the thermal transmittance, U, with the same units. The swing in the environmental temperature due to the energy absorbed occurs at a time τ_a in advance of the energy absorption. Values of Y and τ_a for various materials and structures are given by CIBSE.[4.5] (Note that CIBSE uses the symbol ω instead of τ_a.)

A low value of Y denotes a material giving a large temperature variation, i.e. a lightweight material, and vice versa. For example, for a 100 mm lightweight concrete block, $Y = 2.0 \text{ W/m}^2\text{ K}$; for a 100 mm heavyweight concrete block, $Y = 5.7 \text{ W/m}^2\text{ K}$. It can be shown that the variation in temperature is rapidly damped in a thick material and hence the admittance is constant for a material above a certain thickness. For example, for external brickwork of 105 mm $Y = 4.2 \text{ W/m}^2\text{ K}$, for 220 mm thickness $Y = 4.6 \text{ W/m}^2\text{ K}$, and for 335 mm thickness $Y = 4.7 \text{ W/m}^2\text{ K}$.

For a very thin layer of material with a very low thermal capacity the admittance tends to the value of the transmittance and the heat flow is in phase with the temperature. For example, for single glazing, $Y = U = 5.7 \text{ W/m}^2\text{ K}$.

Decrement Factor, *f*

This is defined as the ratio of the cyclic energy fluctuation within a structure due to the variation in external conditions (such as external air temperature and solar radiation), to the steady-state heat transfer through the structure. The effect of the energy fluctuation does not appear at the inside of the building until a time τ_f after the change in the external conditions causing the fluctuation.

Surface Factor, *s*

This is defined as the ratio of the energy re-admitted to a room from a surface to the energy absorbed by the surface, with reference to the variation of the cyclic radiant energy about its mean value incident on the surface.

Due to the thermal capacity of the structure, the effect of the re-admitted energy is not apparent in the room until a time, τ_s, after the energy strikes the surface. The surface factor decreases as the thermal capacity of the room surfaces increases. The radiant energy striking an internal surface may be from solar radiation passing through a window, or from the radiant component of equipment and lighting.

The Admittance Method

The model described in Chapter 3 can be extended to fluctuating energy transfers.

Note: Mean values are now given the overbar ' ¯ ' to distinguish them from fluctuating values, denoted by a tilde ' ~ ' and instantaneous values, indicated by a prime ' ′ '.

For the steady state:

$$\bar{Q}_i = \bar{Q}_e + \bar{Q}_a = \bar{Q}_F + \bar{Q}_v = \{\Sigma(UA_o) + C_v\}(\bar{t}_{ei} - \bar{t}_{ao}) \qquad [4.3]$$

where C_v is the ventilation conductance.

Alternatively, using the factor defined in Chapter 3 (Eqn [3.10]) as

$$F_{av} = \alpha_a \Sigma A / \{(\alpha_a \Sigma A) + \rho n c_p V\}$$

we multiply the air-related terms by this factor to refer all terms to the environmental point, i.e.

$$F_{av}\bar{Q}_a + \bar{Q}_e = \{F_{av}\rho n c_p V + \Sigma(UA_o)\}(\bar{t}_{ei} - \bar{t}_{ao}) \qquad [4.4]$$

Similarly, using the factor defined (Eqn [3.9]) as

$$F_{au} = \alpha_a \Sigma A / \{(\alpha_a \Sigma A) + \Sigma(UA_o)\}$$

we have

$$\bar{Q}_a + F_{au}\bar{Q}_e = \{\rho n c_p V + F_{au}\Sigma(UA_o)\}(\bar{t}_{ai} - \bar{t}_{ao}) \qquad [4.5]$$

When the dry resultant temperature t_c is used, then a further two factors are introduced, defined (Eqn [3.16]) as

$$F_u = \alpha_{ec} \Sigma A / \{(\alpha_{ec} \Sigma A) + \Sigma(UA_o)\}$$

and

$$F_v = \alpha_{ac} \Sigma A / \{(\alpha_{ac} \Sigma A) + \rho n c_p V\}$$

We then have

$$F_v\bar{Q}_a + F_u\bar{Q}_e = \{F_v\rho n c_p V + F_u\Sigma(UA_o)\}(\bar{t}_c - \bar{t}_{ao}) \qquad [4.6]$$

For the cyclic case there are cyclic inputs at the air and environmental points, $\tilde{Q}_a$ and $\tilde{Q}_e$, and temperature fluctuations at the air and environmental points of $\tilde{t}_{ai}$ and $\tilde{t}_{ei}$. Also, the external air temperature fluctuation is $\tilde{t}_{ao}$. Then, assuming that $\tilde{Q}_a$ and $\tilde{Q}_e$ are due to energy and temperature fluctuations from outside the building, the combined effect of these is to vary the inside temperature, causing a fluctuating ventilation rate, $\rho n c_p V \tilde{t}_{ai}$, and a thermal capacity effect, $\Sigma(YA)\tilde{t}_{ei}$. Substituting cyclic terms for steady-state terms in Eqn [4.4], we have:

$$F_{av}\tilde{Q}_a + \tilde{Q}_e = \{F_{av}\rho n c_p V + \Sigma(YA)\}\tilde{t}_{ei} \qquad [4.7]$$

To refer the terms to the fluctuating inside air temperature, a new dimensionless room factor, F_{ay}, is required, given by

$$F_{ay} = \alpha_a \Sigma A / \{\alpha_a \Sigma A + \Sigma(YA)\}$$

Then we have an alternative equation,

$$\tilde{Q}_a + F_{ay}\tilde{Q}_e = \{\rho n c_p V + F_{ay}\Sigma(YA)\}\tilde{t}_{ai} \qquad [4.8]$$

Introducing the dry resultant temperature as before, the equation can be expressed in terms of $\tilde{t}_c$ by introducing the factor F_v as before, and the new factor F_y, given by

$$F_y = \alpha_{ec} \Sigma A / \{\alpha_{ec} \Sigma A + \Sigma(YA)\}$$

Then,

$$F_v\tilde{Q}_a + F_y\tilde{Q}_e = \{F_v\rho n c_p V + F_y\Sigma(YA)\}\tilde{t}_c \qquad [4.9]$$

The fluctuating inputs at the air and environmental points are due to separate effects which can be added together, taking into account any time lags.

Let us now consider the various inputs in turn.

(i) Opaque surfaces solar gain

The fluctuating solar intensity causes a thermal energy wave to be conducted through the wall, arriving at a time τ_f after the initial external fluctuation. The combined effects of the fluctuating air temperature and the fluctuating solar intensity can be combined using the outside environmental temperature. Then the fluctuating energy gain, $\tilde{Q}_f$, at time, τ, at the environmental point due to these effects is,

$$\tilde{Q}_f = \{\Sigma f(UA_o)\}\tilde{t}_{eo(\tau - \tau_f)} \qquad [4.10]$$

where $\qquad \tilde{t}_{eo}(\tau - \tau_f) = t'_{eo(\tau - \tau_f)} - \bar{t}_{eo}$

and the decrement factor, f, and the time lag, τ_f, are as defined previously.

The time $(\tau - \tau_f)$ is the time when the energy entered the wall. For example, if the energy entering a room at 1200 h is required and the time lag for the wall is 9 hours, then $(\tau - \tau_f)$ is at 0300 h and the external conditions at this time must be used in the calculations.

(ii) Transparent surfaces solar gain

Firstly consider the steady-state solar gain through glass, Q_G, defined in two parts as

$$\bar{Q}_{Ga} = \Sigma(A_G \bar{S}_a \bar{I}_t) \qquad \text{and} \qquad \bar{Q}_{Ge} = \Sigma(A_G \bar{S}_e \bar{I}_t) \qquad [4.11]$$

where $\bar{S}_a$ is a solar gain factor for the air point, defined as

$$\bar{S}_a = \frac{\text{Mean solar gain at the air point}}{\text{Mean solar intensity on facade}}$$

and $\bar{S}_e$ is a solar gain factor for the environmental point, defined as

$$\bar{S}_e = \frac{\text{Mean solar gain at the environmental point}}{\text{Mean solar intensity on the facade}}$$

$\bar{I}_t$ is the total solar irradiance on the surface.

The first term in Eqn [4.11] is an input at the air point, and the second term is an input at the environmental point. Solar gain at the air point occurs when there is convective heating from an internal blind which has been heated by solar radiation to a temperature above the room temperature. The factors $\bar{S}_a$ and $\bar{S}_e$ depend on the incidence angles of the solar radiation, and the arrangement of the window/blind; values for S are given by CIBSE.[4.6]

For the fluctuating case, the fluctuating component of the total solar irradiance, $\tilde{I}_t$, is given by

$$\tilde{I}_t = I'_t - \bar{I}_t$$

where I'_t is the instantaneous solar irradiance.

Then the fluctuating energy, $\tilde{Q}_G$, is given in two parts by

$$\tilde{Q}_{Ga} = \Sigma(A_G \tilde{S}_a \tilde{I}_t) \qquad \text{and} \qquad \tilde{Q}_{Ge} = \Sigma(A_G \tilde{S}_e \tilde{I}_{t(\tau - \phi_s)}) \qquad [4.12]$$

where $\tilde{S}_a$ and $\tilde{S}_e$ are the alternating solar gain factors defined as the instantaneous cyclic solar gain divided by the instantaneous solar intensity on the façade.

The factor $\tilde{S}_e$ incorporates the surface factor for the inside surface receiving the radiation, and the energy input at the environmental point is at a time τ_s after the instantaneous solar gain on the façade.

For glazing the time lag varies between 0 and 2 hours. Very high internal surface factors, i.e. lightweight walls of low admittance, give a negligible time lag; moderately high internal surface factors, i.e. mediumweight walls with moderately high admittance, give rise to a time lag of about one hour; low internal surface factors, i.e. heavyweight walls with high admittance, give rise to a time lag of about two hours.

(iii) Gain due to fluctuation in outside air temperature

The fluctuating gain conducted through the opaque surfaces due to the fluctuations in the outside air temperature is included in Eqn [4.10] since the fluctuating outside environmental temperature is used in that equation.

The fluctuating gain conducted through the glass due to fluctuations in outside air temperature, $\tilde{Q}_{Gc}$, is given by,

$$\tilde{Q}_{Gc} = \tilde{t}_{ao}\Sigma(UA_G)\tag{4.13}$$

It is assumed that there is no time delay for this term; the input is at the environmental point, as with other fabric heat conduction terms.

(iv) Cyclic ventilation gain

The fluctuation in the outside air temperature causes a fluctuating ventilation component, $\tilde{Q}_V$, at the air point, given by

$$\tilde{Q}_V = \rho n c_p V \tilde{t}_{ao}\tag{4.14}$$

Note: It is assumed in Eqn [4.14] that the ventilation rate, $\rho n V$, remains constant.

(v) Fluctuating casual gain

The changes in casual gains due to people, lighting and equipment usually have a square profile, but the harmonic method assumes a sine wave. The fluctuating casual gain, $\tilde{Q}_C$, is therefore given by

$$\tilde{Q}_C = \dot{Q}'_C - \bar{Q}_C$$

In many cases the casual gains have separate convective and radiant components and hence a proportion will be input at the air point and the remainder at the environmental point. The part which is input at the environmental point must be multiplied by the surface factor, s, of the internal surfaces and will have a time lag of τ_s. It is normal practice to neglect this effect and to assume that the total casual gain is at the air point.

The total fluctuating gains at the air and environmental points are then given by:

$$\tilde{Q}_a = \tilde{Q}_{Ga} + \tilde{Q}_V + \tilde{Q}_C\tag{4.15}$$

and $\qquad \tilde{Q}_e = \tilde{Q}_f + \tilde{Q}_{Ge} + \tilde{Q}_{Gc}\tag{4.16}$

The fluctuating gains calculated from Eqns [4.15] and [4.16] can then be substituted in one of the equations [4.7], [4.8] or [4.9].

The General Case

Take the general case where a plant load, Q_P, provides either heating or cooling to the room. For the steady state based on dry resultant temperature, say, we have, from Eqn [4.6],

$$F_v \bar{Q}_a + F_u \bar{Q}_e = \{F_v \rho n c_p V + F_u \Sigma(UA_o)\}(\bar{t}_c - \bar{t}_{ao})$$

and for the fluctuating inputs we have, from Eqn [4.9],

$$F_v \bar{Q}_a + F_y \tilde{Q}_e = \{F_v \rho n c_p V + F_y \Sigma(YA)\}\tilde{t}_c$$

These equations can now be added to give an equation for the general case. The terms $\bar{Q}_a$ and $\bar{Q}_e$ can be assumed to include the input from the plant, i.e.

$$\bar{Q}_a = \bar{Q}_{Ga} + \bar{Q}_C + (1-x)\bar{Q}_P \quad \text{and} \quad \bar{Q}_e = \bar{Q}_{Ge} + x\bar{Q}_P$$

where $\bar{Q}_{Ga}$ and $\bar{Q}_{Ge}$ are defined by Eqn [4.11]; $\bar{Q}_C$ is the mean casual gain from people, lighting and equipment, assumed to be input at the air point; x is the fraction of the plant input which appears at the environmental point.

The terms $\tilde{Q}_a$ and $\tilde{Q}$ are given by Eqns [4.15] and [4.16]. In the case of air conditioning an equation in terms of the internal air temperature is frequently used; in that case Eqns [4.5] and [4.8] are used in the above in place of Eqns [4.6] and [4.9]. Also, the plant load is assumed to be entirely at the air point for air conditioning and hence $x = 0$ in the above.

Response Factor

It was shown earlier that the steady-state input is given by Eqn [4.3]:

$$\bar{Q}_i = \{C_v + \Sigma(UA_o)\}(\bar{t}_{ei} - \bar{t}_{eo})$$

Billington[4.7] assumed an equivalent equation for the fluctuating energy component given by,

$$\tilde{Q}_i = \{C_v + \Sigma(YA)\}\tilde{t}_{ei}$$

He then defined a thermal damping factor as

$$\frac{\{C_v + \Sigma(UA_o)\}}{\{C_v + \Sigma(YA)\}} = \frac{\bar{Q}_i}{\tilde{Q}_i} \frac{\tilde{t}_{ei}}{(\bar{t}_{ei} - \bar{t}_{eo})}$$

CIBSE[4.5] define a response factor, f_r, given by

$$\frac{\{\rho n c_p V + \Sigma(YA)\}}{\{\rho n c_p V + \Sigma(UA_o)\}} \qquad [4.17]$$

which is approximately the reciprocal of Billington's thermal damping factor.

The response factor can be used to define whether a building is classified as *heavyweight* or *lightweight*. A heavyweight building would normally have a response factor greater than 6, and modern highly insulated buildings may have response factors of the order of 10.

4.3 HEATING LOADS

The plant load required to heat a building to a certain dry resultant temperature can be found using the equations discussed at the end of Section 4.2. In the UK a simpler procedure is used, as shown by calculations given in Chapter 2; the solar and casual gains in winter are usually neglected and the thermal inertia of the building is allowed for by the choice of the mean outside design temperature. CIBSE[4.4] give different outside design temperatures for a series of locations for low thermal inertia, i.e. lightweight buildings, and high thermal inertia, i.e. heavyweight buildings.

Intermittent Heating

The case of intermittent heating is considered earlier in the chapter using the concept of building time constant (see Example 4.1). The equations developed in Section 4.2 for the admittance method can be applied using, in the first instance, a simple rectangular temperature–time response as shown in Fig. 4.3. The plant load, $\dot{Q}_P$, is applied for N hours in every 24-hour period, giving a mean heat input, $\bar{\dot{Q}}_i = N\dot{Q}_P/24$; then the fluctuating component,

$$\tilde{\dot{Q}}_i = \dot{Q}_P - \bar{\dot{Q}}_i = \dot{Q}_P(1 - N/24) \qquad [4.18]$$

Figure 4.3 Intermittent heating: idealized response

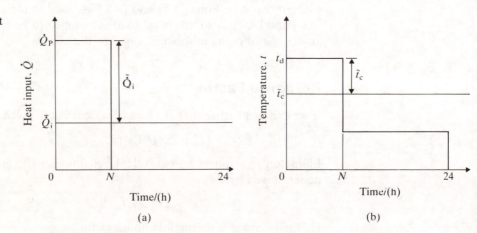

In the simplified case shown in Fig. 4.3, the inside design dry resultant temperature, t_d, is maintained for the period during which the plant load is applied; the 24-hour mean inside temperature is $\bar{t}_c$, and the fluctuating component is given by $\tilde{t}_c = t_d - \bar{t}_c$. The steady-state plant load, $\dot{Q}_{PS}$, required for the case of continuous heating with the design inside temperature, t_d, maintained for the 24-hour period, is less than $\dot{Q}_P$.

With intermittent heating, the inside temperature will take a finite time to reach the design value after the heat input is switched on. A more realistic approximation to the temperature changes is shown in Fig. 4.4. As discussed earlier (see Example 4.1), the period of heating can be extended to allow for pre-heating the building before occupancy so that the inside temperature is

HEATING LOADS

Figure 4.4 Intermittent heating: real response

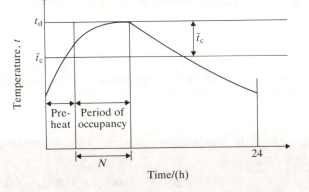

approximately equal to the design temperature when the occupants arrive. For example, the period of occupancy might be as shown on Fig. 4.4. Also, the heat input can be increased during the pre-heating period so that the rate of increase of temperature is greater and the period consequently shorter. CIBSE[4.8] show that using the admittance method to calculate $\dot{Q}_P$ allows for a pre-heat period of about four hours; for pre-heat periods shorter than 4 h the calculated value of $\dot{Q}_P$ should be multiplied by a factor which rises to 1.9 at zero pre-heat, and falls to 0.7 for an eight-hour pre-heat period.

Example 4.2
The data below refer to a building which is occupied from 8 am to 6 pm for a heating season of 230 days.

Using the data given, neglecting solar and casual gains, calculate:
(i) the plant load for continuous heating over a 24-hour period;
(ii) the plant load assuming a four-hour pre-heating period;
(iii) the mean daily saving in energy during the heating season by intermittent heating.

Data
$\Sigma A = 5000 \, \text{m}^2$; $\Sigma(UA_o) = 4 \, \text{kW/K}$; $\Sigma(YA) = 15 \, \text{kW/K}$; $\rho nc_p V = 1.5 \, \text{kW/K}$; inside design dry resultant temperature, 20 °C; outside design temperature, −3 °C; seasonal mean outside temperature, 6 °C; the heat input is 50% at the environmental point and 50% at the air point.

Solution
(i) The continuous heating or steady-state case can be found using the network method of Chapter 3 which, for the case of a known dry resultant temperature, can be found from Eqn [4.6], i.e.

$$F_v \bar{Q}_a + F_u \bar{Q}_e = \{F_v \rho nc_p V + F_u \Sigma(UA_o)\}(\bar{t}_c - \bar{t}_{ao})$$

where
$$F_v = \alpha_{ac} \Sigma A / (\alpha_{ac} \Sigma A + \rho nc_p V)$$
$$= 6 \times 5000 / \{6 \times 5000) + 1500\} = 0.9524$$

and
$$F_u = \alpha_{ec} \Sigma A / \{\alpha_{ec} \Sigma A + \Sigma(UA_o)\}$$
$$= 18 \times 5000 / \{(18 \times 5000) + 4000\} = 0.9575$$

Also, $\qquad \bar{Q}_a = \bar{Q}_e = 0.5\dot{Q}_{PS}$

i.e. $\qquad (0.9524 + 0.9575) \times 0.5\dot{Q}_{PS} = \{(0.9524 \times 1.5)$
$$+ (0.9575 \times 4.0)\}(20 + 3)$$
$$\therefore \quad \dot{Q}_{PS} = 126.7\,\text{kW}$$

(ii) For the case of intermittent heating, $N = 10\,\text{h}$ (8 am–6 pm), and therefore, referring to Fig. 4.3(a), (page 178),

$$\bar{Q}_i = 10\dot{Q}_P/24 = 0.4167\dot{Q}_P$$

Then using the steady-state equation [4.6], the mean temperature, $\bar{t}_c$, can be found, i.e.

$$(0.9524 + 0.9575) \times 0.5 \times 0.4167\dot{Q}_P = \{(0.9524 \times 1.5)$$
$$+ (0.9575 \times 4.0)\}(\bar{t}_c + 3)$$
$$\therefore \quad \dot{Q}_P = 13.2(\bar{t}_c + 3) \qquad [1]$$

For the fluctuating case, Eqn [4.9] applies, i.e.

$$F_v\tilde{Q}_a + F_y\tilde{Q}_e = \{F_v\rho n c_p V + F_y\Sigma(YA)\}\tilde{t}_c$$

where $\qquad F_y = \alpha_{ec}\Sigma A/\{\alpha_{ec}\Sigma A + (YA)\}$
$$= 18 \times 5000/\{(18 \times 5000) + 15\,000\} = 0.8571$$

Assuming that the fluctuating inputs are also 50% at the environmental and air points, then:

$$(0.9524 + 0.8571) \times 0.5\tilde{Q}_i = \{(0.9524 \times 1.5) + (0.8571 \times 15)\}\tilde{t}_c$$
$$\therefore \quad \tilde{Q}_i = 15.8\tilde{t}_c \qquad [2]$$

Also, from Eqn [4.18]:

$$\tilde{Q}_i = \dot{Q}_P\{1 - (N/24)\} = 0.583Q_P \qquad [3]$$

Substituting for $\dot{Q}_P$ from Eqn [1],

$$\tilde{Q}_i = 13.2(\bar{t}_c + 3) \times 0.583 = 7.7(\bar{t}_c + 3)$$

Substituting for $\tilde{Q}_i$ from Eqn [2],

$$15.8\tilde{t}_c = 7.7(\bar{t}_c + 3)$$
$$\therefore \quad \tilde{t}_c = 0.487(\bar{t}_c + 3) \qquad [4]$$

Referring to Fig. 4.4, (page 179),

$$t_d = \bar{t}_c + \tilde{t}_c$$

i.e. $\qquad 20 = \bar{t}_c + \tilde{t}_c$

Substituting in Eqn [4],

$$\tilde{t}_c = 0.487(20 - \tilde{t}_c + 3)$$
$$\therefore \quad \tilde{t}_c = 7.5\,\text{K}$$

and $\qquad \bar{t}_c = 20 - 7.5 = 12.5\,°\text{C}$

Then, substituting in Eqn [1],

$$\dot{Q}_P = 13.2(12.5 + 3) = 204.6\,\text{kW}$$

For a four-hour pre-heat period the recommended correction factor is unity.[4.8] Therefore the plant load for intermittent heating is 204.6 kW; this compares with the steady-state plant load for continuous heating of 126.7 kW calculated in (i) above, giving a plant operation ratio of $204.6/126.7 = 1.62$.

Note: CIBSE[4.6] derive a formula for the plant operation ratio and using this formula gives an identical answer; CIBSE[4.9] give an empirical expression for plant operation ratio allowing for pre-heat time given by

$$1.2 \left\{ \frac{(24 - N)(f_r - 1)}{24 + N(f_r - 1)} \right\} + 1$$

where N is the number of hours' operation including pre-heat.

The response factor, f_r, can be found from Eqn [4.17] as

$$\frac{1.5 + 15}{1.5 + 4} = 3$$

Then the plant operation ratio for 14 hours' operation (including four hours' pre-heat) is given by,

$$1.2 \left\{ \frac{(24 - 14)(3 - 1)}{24 + 14(3 - 1)} \right\} + 1 = 1.46$$

This gives a lower figure than the ratio of 1.62 calculated by the method above.

(iii)   The mean seasonal outside temperature is given as 6 °C over the 230-day heating season, therefore the annual energy saving by intermittent heating is given by:

$$\{(126.7 \times 24) - (204.6 \times 14)\} \times \frac{(20 - 6)}{(20 + 3)} \times 230 = 24\,696 \text{ kW h}$$
$$= 88.9 \text{ GJ}$$

4.4 AIR CONDITIONING LOADS

Due to the temperate climate in the UK, the growth in fully air-conditioned buildings has been slow. For many buildings it has been considered sufficient to ensure adequate screening of the sun's rays on days in summer when the weather is fine. The normal maximum air temperature of about 25 °C occurs in late afternoon on sunny days in July, with a highest sol-air temperature for horizontal surfaces of about 52 °C at solar noon also in July. The admittance method can be used to estimate the maximum temperature likely to occur inside a given naturally ventilated building or room on a particular day in summer, as in the following example.

Example 4.3
The non-air-conditioned room shown in Fig. 4.5 is situated on the intermediate floor of an office block in London. It has three occupants from 0800 h to 1700 h sun time. The windows are of single clear glass with internal white venetian blinds. If the room is ventilated continuously at a rate of three air changes per hour, determine the peak internal environmental temperature during a sunny period in July. State all assumptions made and make no more than one prediction.

Figure 4.5 Plan of room: Example 4.3

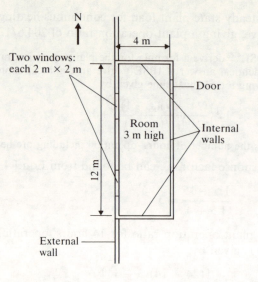

Data

External wall: U-value, 0.4 W/m^2 K; Y-value, 3.5 W/m^2 K; decrement factor, 0.4; time lag, 5 h.

Internal wall: Y-value, 2.0 W/m^2 K.

Floor: Y-value, 2.9 W/m^2 K.

Ceiling: Y-value, 2.1 W/m^2 K.

Window: U-value, 5.6 W/m^2 K; Y-value, 5.6 W/m^2 K.

Occupant sensible gain, 90 W/person.

Use CIBSE Tables A8.1, A8.3(e), A8.2 and A8.6 (given here as Tables 4.1, 4.2, 4.3 and 4.4 by permission of CIBSE).

Assume the following formulae:

$$\frac{1}{C_V} = \frac{3}{nV} + \frac{1}{4.5\Sigma A}$$

and

$$f_r = \frac{\Sigma(YA) + C_V}{\Sigma(UA_o) + C_V}$$

(CIBSE)

Solution

$$\Sigma A = (2 \times 4 \times 3) + (2 \times 12 \times 3) + (2 \times 12 \times 4) = 192 \, \text{m}^2$$

(neglecting the area of the window recess).

$$\begin{aligned}
\Sigma(UA_o) &= (2 \times 2 \times 2 \times 5.6) + \{(12 \times 3) - (2 \times 2 \times 2)\} \times 0.4 \\
&= 56 \, \text{W/K} \\
\Sigma(YA) &= (2 \times 4 \times 3 \times 2) + (3 \times 12 \times 2) + (2 \times 2 \times 2 \times 5.6) \\
&\quad + 3.5 \times \{(12 \times 3) - (2 \times 2 \times 2)\} + (2.1 \times 12 \times 4) \\
&\quad + (2.9 \times 12 \times 4) \\
&= 502.8 \, \text{W/K}
\end{aligned}$$

(assuming that the door has the same U-value and Y-value as the internal walls, and that the U-value and Y-value given for the windows includes the window frames).

Table 4.1 Total solar irradiance (W/m²) on vertical and horizontal surfaces for South-East England (approximately correct for UK)

51.7°N

Date	Orientation	Daily mean	04	05	06	07	08	09	10	11	12	13	14	15	16	17	18	19	20
June 21	N	90	35	155	165	105	100	125	145	160	165	160	145	125	100	105	165	155	35
	NE	135	60	305	445	465	420	320	195	170	175	170	155	130	105	75	50	20	5
	E	185	50	295	495	600	625	585	490	355	185	180	165	140	110	80	50	25	5
	SE	180	10	120	280	420	525	580	585	430	290	165	140	110	80	50	25	5	
	S	155	5	25	50	80	185	320	435	510	535	510	435	320	185	80	50	25	5
	SW	180	5	25	50	80	110	140	165	290	430	535	585	580	525	420	280	120	10
	W	185	5	25	50	80	100	140	165	180	185	355	490	585	625	600	495	295	50
	NW	135	5	20	50	75	105	130	155	170	175	170	195	320	420	465	445	305	60
	H	315	5	80	210	360	505	640	750	825	850	825	750	640	505	360	210	80	5
July 23 and May 22	N	85		100	130	90	110	140	165	180	185	180	165	140	110	90	130	100	
	NE	125		210	370	410	380	300	195	195	200	195	175	145	115	80	50	20	
	E	175		210	420	540	580	555	480	360	215	205	185	155	125	90	55	20	
	SE	180		95	255	400	510	570	580	540	450	320	185	155	125	90	55	20	
	S	165		20	55	90	210	340	450	525	550	525	450	340	210	90	55	20	
	SW	180		20	55	90	125	155	185	320	450	540	580	570	510	400	255	95	
	W	175		20	55	90	125	155	185	205	215	360	480	555	580	540	420	210	
	NW	125		20	50	80	115	145	175	195	200	195	195	300	380	410	370	210	
	H	295		55	180	325	470	605	715	785	815	785	715	605	470	325	180	55	
August 22 and April 22	N	60		5	60	60	90	120	145	160	165	160	145	120	90	60	60	5	
	NE	95		10	240	330	315	240	155	170	180	170	155	125	95	65	30	0	
	E	145		10	295	475	545	535	460	340	190	185	165	135	105	65	30	0	
	SE	175		5	190	380	510	585	600	565	490	340	185	135	105	65	30	0	
	S	175		0	30	95	235	375	490	565	590	565	490	375	235	95	30	0	
	SW	175		0	30	65	105	135	185	340	470	565	600	585	510	380	190	5	
	W	145		0	30	65	105	135	165	185	190	340	460	535	545	475	295	10	
	NW	95		0	30	65	95	125	155	170	180	170	155	240	315	330	240	10	
	H	240		0	90	225	375	510	620	690	715	690	620	510	375	225	90	0	
Sept 22 and March 21	N	40			0	30	60	90	115	130	135	130	115	90	60	30	0		
	NE	60			0	205	235	165	125	140	145	140	125	100	65	30	0		
	E	120			0	340	515	535	465	330	155	150	130	105	70	35	0		
	SE	175			0	295	525	650	685	665	545	395	215	105	70	35	0		
	S	200			0	95	270	445	585	675	710	675	585	445	270	95	0		
	SW	175			0	35	70	105	215	395	545	665	685	650	525	295	0		
	W	120			0	35	70	105	130	150	155	330	465	535	515	340	0		
	NW	60			0	30	65	100	125	140	145	140	125	165	235	205	0		
	H	180			0	95	245	395	515	595	625	595	515	395	245	95	0		
Oct 22 and Feb 21	N	30				0	30	60	90	105	110	105	90	60	30	0			
	NE	35				10	105	80	95	110	120	110	95	65	30	0			
	E	75				25	290	395	370	265	125	120	100	70	35	0			
	SE	135				20	320	515	590	580	500	370	215	70	35	0			
	S	170				10	185	380	530	620	655	620	530	380	185	10			
	SW	135				0	35	70	215	370	500	580	590	515	320	20			
	W	75				0	35	70	100	120	125	265	370	395	290	25			
	NW	35				0	30	65	95	110	120	110	95	165	105	10			
	H	110				5	100	230	340	415	440	415	340	230	100	5			
Nov 22 and Jan 21	N	15					5	30	55	70	75	70	55	30	5				
	NE	20					15	35	60	75	80	75	60	35	5				
	E	40					45	230	265	200	90	80	65	35	5				
	SE	100					55	325	470	495	440	325	180	50	5				
	S	130					35	250	435	545	585	545	435	250	35				
	SW	100					5	50	180	325	440	495	470	325	55				
	W	40					5	35	65	80	90	200	265	230	45				
	NW	20					5	35	60	75	80	75	60	35	15				
	H	60					10	95	190	260	285	260	190	95	10				
Dec 21	N	10						15	40	50	55	50	40	15					
	NE	10						20	40	55	60	55	40	20					
	E	30						155	225	175	65	60	45	20					
	SE	85						230	410	460	410	300	155	35					
	S	115						180	385	510	555	510	385	180					
	SW	85						35	155	300	410	460	410	230					
	W	30						20	45	60	65	175	225	155					
	NW	10						20	40	55	60	55	40	20					
	H	40						50	130	195	220	195	130	50					

Notes:

1. N, NE, E etc: total irradiance on vertical surfaces

   H: total irradiance on horizontal surface

2. This table is based on horizontal surface measurements at Kew for the period 1959–1968: weather for two consecutive days averaged.

   See Section A2, Table A2.27 CIBSE

Table 4.2

July 23

Sun Time	Air Temp. t_{ao} (°C)	Horizontal Dark	Horizontal Light	North Dark	North Light	North-East Dark	North-East Light	East Dark	East Light	South-East Dark	South-East Light	South Dark	South Light	South-West Dark	South-West Light	West Dark	West Light	North-West Dark	North-West Light
00	16.0	12.5	12.5	15.0	15.0	15.0	15.0	15.0	15.0	15.0	15.0	15.0	15.0	15.0	15.0	15.0	15.0	15.0	15.0
01	14.5	11.5	11.5	14.0	14.0	14.0	14.0	14.0	14.0	14.0	14.0	14.0	14.0	14.0	14.0	14.0	14.0	14.0	14.0
02	13.5	10.5	10.5	13.0	13.0	13.0	13.0	13.0	13.0	13.0	13.0	13.0	13.0	13.0	13.0	13.0	13.0	13.0	13.0
03	13.0	10.0	10.0	12.0	12.0	12.0	12.0	12.0	12.0	12.0	12.0	12.0	12.0	12.0	12.0	12.0	12.0	12.0	12.0
04	13.0	9.5	9.5	12.0	12.0	12.0	12.0	12.0	12.0	12.0	12.0	12.0	12.0	12.0	12.0	12.0	12.0	12.0	12.0
05	13.0	12.0	11.0	17.5	15.0	22.5	18.0	22.5	18.0	17.0	15.0	13.5	13.0	13.5	13.0	13.5	13.0	13.0	13.0
06	13.5	17.5	14.5	19.5	16.5	31.0	23.0	33.5	24.5	25.5	20.0	15.5	14.5	15.5	14.5	15.5	14.5	15.5	14.0
07	14.5	24.5	18.5	18.0	16.0	34.0	25.0	40.5	28.5	33.5	24.5	18.0	16.0	18.0	16.0	18.0	16.0	18.0	16.0
08	16.0	31.5	23.0	20.5	18.0	33.5	25.5	43.5	31.0	40.0	29.0	25.5	20.5	21.0	18.5	21.0	18.5	20.5	18.0
09	17.5	38.5	27.5	23.0	20.0	31.0	24.5	44.0	31.5	44.5	32.0	33.5	26.0	24.0	20.5	24.0	20.5	23.5	20.5
10	19.0	44.5	31.5	26.0	22.5	27.5	23.5	41.5	31.0	46.5	34.0	40.0	30.5	27.0	23.0	27.0	23.0	26.5	22.5
11	20.5	49.0	34.5	28.5	24.5	29.0	24.5	37.5	29.5	46.0	34.5	45.5	34.0	35.0	28.0	29.5	25.0	29.0	24.5
12	21.5	51.5	36.5	30.0	26.0	30.5	26.5	31.5	26.5	43.0	33.0	48.0	36.0	43.0	33.0	31.5	26.5	30.5	26.5
13	23.0	51.5	37.5	31.0	27.0	31.5	27.5	32.0	27.5	38.0	31.0	48.0	36.5	49.0	37.0	40.0	32.0	31.5	27.5
14	24.0	49.5	36.5	31.0	27.5	31.5	28.0	32.0	28.0	32.0	28.0	45.0	35.5	51.5	39.0	46.5	36.0	32.5	28.5
15	24.5	45.5	35.0	30.5	27.5	31.0	27.5	31.5	28.0	31.5	28.0	40.5	33.0	51.5	39.0	51.0	39.0	38.5	32.0
16	24.5	40.5	32.0	29.0	26.5	29.5	27.0	30.0	27.0	30.0	27.0	34.0	29.5	49.0	37.5	52.5	39.5	42.5	34.0
17	24.5	34.5	28.5	28.0	26.0	27.5	26.0	28.0	26.0	28.0	26.0	28.0	26.0	43.5	34.5	50.5	38.5	45.0	35.0
18	24.0	28.0	24.5	29.5	26.5	25.5	24.5	25.5	24.5	25.5	24.5	25.5	24.5	35.5	30.0	44.0	34.5	41.5	33.0
19	23.0	22.0	21.0	27.0	25.0	23.0	22.5	23.0	22.5	23.0	22.5	23.0	22.5	26.5	24.5	32.5	27.5	32.5	28.0
20	21.5	18.5	18.5	21.0	21.0	21.0	21.0	21.0	21.0	21.0	21.0	21.0	21.0	21.0	21.0	21.0	21.0	21.0	21.0
21	20.5	17.0	17.0	19.5	19.5	19.5	19.5	19.5	19.5	19.5	19.5	19.5	19.5	19.5	19.5	19.5	19.5	19.5	19.5
22	19.0	15.5	18.0	15.5	18.0	18.0	18.0	18.0	18.0	18.0	18.0	18.0	18.0	18.0	18.0	18.0	18.0	18.0	18.0
23	17.5	14.0	14.0	16.5	16.5	16.5	16.5	16.5	16.5	16.5	16.5	16.5	16.5	16.5	16.5	16.5	16.5	16.5	16.5
Mean	19.0	27.5	22.0	22.0	20.0	24.0	21.5	26.5	22.5	27.0	23.0	26.0	22.5	27.0	23.0	26.5	22.5	24.0	21.5

Note: I_1 (horizontal surface) = 79 W/m^2, I_1 (vertical surface) = 18 W/m^2, where I_1 = net long-wave radiation loss

Table 4.3 Mean solar gain factors, $\bar{S}_e$, for various types of glazing and shading (strictly accurate for UK only, approximately correct worldwide)

Position of shading and type of sun protection		Mean solar gain factors*, $\bar{S}_e$, for stated window type	
Shading	Type of sun protection	Single	Double
None	None	0.76	0.64
	Lightly heat absorbing glass	0.51	0.38
	Densely heat absorbing glass	0.39	0.25
	Lacquer coated glass, grey	0.56	—
	Heat reflecting glass, gold (sealed unit when double)	0.26	0.25
Internal	Dark green open weave plastic blind	0.62	0.56
	White venetian blind	0.46	0.46
	White cotton curtain	0.41	0.40
	Cream holland linen blind	0.30	0.33
Mid-pane	White venetian blind	—	0.28
External	Dark green open weave plastic blind	0.22	0.17
	Canvas roller blind	0.14	0.11
	White louvred sunbreaker, blades at 45°	0.14	0.11
	Dark green miniature louvred blind	0.13	0.10

Notes: *All glazing clear except where stated otherwise. Factors are typical values only and variations will occur due to density of blind weave, reflectivity and cleanliness of protection

Table 4.4 Alternating solar gain factors, $\tilde{S}_e$, for various types of glazing and shading, lightweight and heavyweight structures (strictly accurate for UK only—SW façade)

Position of shading and type of sun protection		Alternating solar gain factors*, $\tilde{S}_e$, for the following building and window types			
		Heavyweight building		Lightweight building	
Shading	Type of sun protection	Single	Double	Single	Double
None	None	0.42	0.39	0.65	0.56
	Lightly heat absorbing glass	0.36	0.27	0.47	0.35
	Densely heat absorbing glass	0.32	0.21	0.37	0.24
	Lacquer coated glass, grey	0.37	—	0.50	—
	Heat reflecting glass, gold (sealed unit when double)	0.21	0.14	0.25	0.20
Internal	Dark green open weave plastic blind	0.55	0.53	0.61	0.57
	White venetian blind	0.42	0.44	0.45	0.46
	White cotton curtain	0.27	0.31	0.35	0.37
	Cream holland linen blind	0.24	0.30	0.27	0.32
Mid-pane	White venetian blind	—	0.24	—	0.27
External	Dark green open weave plastic blind	0.16	0.13	0.22	0.17
	Canvas roller blind	0.10	0.08	0.13	0.10
	White louvred sunbreaker, blades at 45°	0.08	0.06	0.11	0.08
	Dark green miniature louvred blind	0.08	0.06	0.10	0.07

Notes: *All glazing clear except where stated otherwise. Factors are typical values only and variations will occur due to density of blind weave, reflectivity and cleanliness of protection

$$\text{Volume, } V = 12 \times 4 \times 3 = 144 \, \text{m}^3$$

Then,

$$\frac{1}{C_V} = \frac{3}{3 \times 144} + \frac{1}{4.5 \times 192} = 0.0081$$

$$\therefore \quad C_V = 123.43 \, \text{W/K}$$

and

$$f_r = \frac{502.8 + 123.43}{56 + 123.43} = 3.49$$

Therefore the building can be classed as lightweight. The external wall faces due West and for 23 July we have:

From Table 4.1, $\bar{I}_t = 175 \, \text{W/m}^2$

The peak irradiance of $I'_t = 580 \, \text{W/m}^2$ occurs at 1600 h and since the building is lightweight no time lag needs to be taken.

From Table 4.2, $\bar{t}_{eo} = 26.5 \, °\text{C}$

(assuming a dark surface, which is the worst case).

For conduction of heat through the external wall a time lag of 5 h is given; hence,

From Table 4.2, $t'_{eo} = 29.5 \, °\text{C}$ at 1100 h

Also,

From Table 4.2, $t'_{ao} = 24.5 \, °\text{C}$ at 1600 h

and $\bar{t}_{ao} = 19 \, °\text{C}$

The steady-state mean solar gain factor can be read from Table 4.3 for single clear glass with internal white venetian blinds, i.e.

$$\bar{S}_e = 0.46$$

Similarly, from Table 4.4, fluctuating mean solar gain factor is

$$\tilde{S}_e = 0.45$$

The equation given for the ventilation conductance, C_V, is Eqn [3.7] (i,e, the case when $\dot{Q}_a$ is zero), and hence as an approximation it can be assumed that there is a negligible difference between the internal air temperature and the internal environmental temperature. The mean temperature in the room at 1600 h can then be found by evaluating the steady-state heat gains and then using the steady-state equation,

$$\bar{Q}_i = \bar{Q}_V + \bar{Q}_F = C_V(t_{ei} - t_{ao}) + \Sigma\{(UA_o)(\bar{t}_{ei} - \bar{t}_{eo})\} \qquad [1]$$

with C_V given approximately by the equation given in the Example.

For $\bar{Q}_i$ we have

$$\bar{Q}_i = \bar{Q}_C + \bar{Q}_G$$

where $\bar{Q}_C = 3 \times 90 \times 9/24 = 101.25 \, \text{W}$

and $\bar{Q}_G = \bar{S}_e \bar{I}_t A_G = 0.46 \times 175 \times (2 \times 2 \times 2) = 644 \, \text{W}$

i.e. $\bar{Q}_i = 101.25 + 644 = 745.25 \, \text{W}$

Then, substituting in Eqn [1],

$$745.25 = 123.43(\bar{t}_{ei} - 19) + \{0.4 \times 28(\bar{t}_{ei} - 26.5)\} + \{5.6 \times 8(\bar{t}_{ei} - 19)\}$$

i.e. $\quad \bar{t}_{ei} = 23.6\,°C$

Note: The fabric loss is considered in two parts: the external wall, $\Sigma(UA_o) = 0.4 \times 28$, loses heat from the internal environmental temperature to the mean outside environmental temperature, 26.5 °C; the glass, $\Sigma(UA_o) = 5.6 \times 8$, loses heat from the internal environmental temperature to the mean air temperature, 19 °C. Note also that the ventilation loss is defined as $C_V(\bar{t}_{ei} - \bar{t}_{ao})$ (see Eqn [3.5]).

Similarly, the fluctuating heat input is given by

$$\tilde{Q}_i = \{C_V + \Sigma(YA)\}\tilde{t}_{ei} \qquad [2]$$

The fluctuating components are as follows:

$$\tilde{Q}_i = \tilde{Q}_C + \tilde{Q}_G + \tilde{Q}_f + \tilde{Q}_{Gc} + \tilde{Q}_v$$

where

$$\tilde{Q}_C = (3 \times 90) - 101.25 = 168.75\,W$$
$$\tilde{Q}_G = \tilde{S}_e A_G(I'_t) = 0.45 \times 8 \times (580 - 175) = 1458\,W$$
$$\tilde{Q}_f = f\Sigma(UA_o)_{wall}\{t'_{eo(\tau - 5)} - \bar{t}_{eo}\}$$
$$\qquad = 0.4 \times 0.4 \times 28 \times (29.5 - 26.5) = 13.4\,W$$
$$\tilde{Q}_{Gc} = \tilde{t}_{ao}\Sigma(UA)_G = (24.5 - 19) \times 5.6 \times 8 = 246.4\,W$$
$$\tilde{Q}_v = \rho n c_p V\tilde{t}_{ao} = 0.33 \times 3 \times 144(24.5 - 19) = 784.1\,W$$

i.e. $\quad \tilde{Q}_i = 168.75 + 1458 + 13.4 + 246.4 + 784.1 = 2670.7\,W$

Then substituting in Eqn [2],

$$2670.7 = (123.43 + 502.8)t_{ei}$$
$$\therefore \quad \tilde{t}_{ei} = 4.3\,K$$

Therefore,

$$\text{Room temperature at 1600 h} = \bar{t}_{ei} + \tilde{t}_{ei} = 23.6 + 4.3 = 27.9\,°C$$

A more accurate method is to take separate air and environmental point inputs and to use Eqns [4.4] and [4.7], (page 174). In that case the solar gain factors are now given by CIBSE,[4.6] reproduced here as Tables 4.5 and 4.6.

$$\bar{S}_a = 0.16, \bar{S}_e = 0.31, \tilde{S}_a = 0.17, \tilde{S}_e = 0.28$$

The factor in Eqns [4.4] and [4.7] is as follows:

$$F_{av} = 4.5 \times 192/\{(4.5 \times 192) + 144\} = 0.857$$

Separating the air and environmental inputs we have, for the steady state,

$$\bar{Q}_a = \bar{Q}_C + \bar{Q}_{Ga} = 101.25 + \bar{S}_a A_G \bar{I}_t = 101.25 + (0.16 \times 8 \times 175)$$
$$\qquad = 325.25\,W$$
$$\bar{Q}_e = \bar{Q}_{Ge} = 0.31 \times 8 \times 175 = 434\,W$$

Then substituting in Eqn [4.4],

$$0.857 \times 325.25 + 434 = \{0.857(0.33 \times 3 \times 144) + 44.8\}(\bar{t}_{ei} - 19)$$
$$+ 11.2(\bar{t}_{ei} - 26.5)$$

$$\therefore \quad t_{ei} = 23.4\,°C$$

Table 4.5 Solar gain factors with no internal shading

Description	$\bar{S}_e$	$\tilde{S}_e$	
		Light	Heavy
Single glazing			
Clear 6 mm	0.76	0.64	0.47
Surface tinted (STG) 6 mm	0.60	0.53	0.41
Body tinted (BTG) 6 mm	0.52	0.47	0.38
Body tinted (BTG) 10 mm	0.42	0.39	0.34
Clear with reflecting film (metallic)	0.32	0.29	0.23
Clear with strongly reflecting film (metallic)	0.21	0.19	0.16
Clear with reflecting film (tinted)	0.28	0.26	0.23
Reflecting	0.36	0.33	0.27
Strongly reflecting	0.18	0.17	0.15
Double glazing (outer pane first)			
Clear 6 mm + clear 6 mm	0.64	0.56	0.42
STG + clear 6 mm	0.48	0.43	0.34
BTG 6 mm + clear 6 mm	0.40	0.37	0.30
BTG 10 mm + clear 6 mm	0.30	0.28	0.24
Reflecting + clear 6 mm	0.28	0.25	0.21
Strongly reflecting + clear 6 mm	0.13	0.12	0.10
Lightly reflecting sealed double unit	0.32	0.29	0.21
Strongly reflecting sealed double unit	0.15	0.14	0.11
Single glazing with external shade			
Clear 6 mm + light horizontal slats	0.16	0.11	0.09
Clear 6 mm + light vertical slats	0.18	0.13	0.10
Clear 6 mm + dark horizontal slats	0.13	0.09	0.08
Clear 6 mm + holland blind	0.13	0.10	0.08
Clear 6 mm + miniature louvres (1)	0.16	0.10	0.09
Clear 6 mm + miniature louvres (2)	0.12	0.09	0.09
BTG 6 mm + light horizontal slats	0.13	0.09	0.08
BTG 6 mm + light vertical slats	0.14	0.12	0.09
Double glazing + external shade			
Clear 6 mm + clear 6 mm + light horizontal slats	0.13	0.09	0.07
Clear 6 mm + clear 6 mm + light vertical slats	0.15	0.10	0.08
Clear 6 mm + clear 6 mm + light roller blind	0.10	0.09	0.07
Clear 6 mm + clear 6 mm + miniature louvres (1)	0.12	0.07	0.06
Clear 6 mm + clear 6 mm + miniature louvres (2)	0.09	0.06	0.06
Clear 6 mm + clear 6 mm + dark horizontal slats	0.10	0.06	0.06
Other			
Triple glazing clear 6 mm + clear 6 mm + clear 6 mm	0.55	0.50	0.39
Clear 6 mm + clear 6 mm + mid pane light slats	0.28	0.26	0.24

TYPE OF SHADE

Light slatted blind (vertical or horizontal) width/spacing ratio	1.2
blade angle (downward tilt for horizontal)	45°
blade absorptance	0.4
Dark slatted blind (vertical or horizontal) width/spacing ratio	1.2
blade angle (downward tilt for horizontal)	45°
blade absorptance	0.8
Miniature fixed louvres (1)	
number of slats per inch	17
width/spacing ratio	0.85
blade angle (downward tilt)	20°
blade absorptance	0.96

Table 4.5 *cont.*

Type of shade

Miniature fixed louvres (2)	
number of slats per inch	23
width/spacing ratio	1.15
blade angle (downward tilt)	20°
blade absorptance	0.98
Linen roller blind	
transmittance	0.13
absorptance	0.20
reflectance	0.67

Table 4.6 Solar gan factors
with internal shading

Description	$\bar{S}_e$	$\bar{S}_a$	Lightweight		Heavyweight	
			$\tilde{S}_e$	$\tilde{S}_a$	$\tilde{S}_e$	$\tilde{S}_a$
Single glazing + internal shade						
Clear 6 mm　　　　 + light horizontal slats	0.31	0.16	0.28	0.17	0.24	0.17
Clear 6 mm　　　　 + light vertical slats	0.32	0.16	0.30	0.18	0.24	0.18
Clear 6 mm　　　　 + dark horizontal slats	0.35	0.23	0.36	0.26	0.34	0.26
Clear 6 mm　　　　 + linen blinds	0.20	0.11	0.18	0.11	0.14	0.11
BTG 6 mm　　　　 + light slatted blinds	0.19	0.20	0.18	0.22	0.17	0.22
BTG 10 mm　　　 + light slatted blinds	0.14	0.21	0.14	0.22	0.13	0.22
Reflecting　　　　 + light slatted blinds	0.14	0.15	0.14	0.16	0.12	0.16
Strongly reflecting + light slatted blinds	0.06	0.10	0.06	0.10	0.06	0.10
Double glazing + internal shades						
Clear 6 mm + clear 6 mm + light slatted blinds	0.26	0.19	0.25	0.21	0.21	0.21
Clear 6 mm + clear 6 mm + dark slatted blinds	0.30	0.26	0.31	0.29	0.30	0.29
BTG 6 mm + clear 6 mm + light slatted blinds	0.15	0.15	0.14	0.16	0.13	0.16
BTG 10 mm + clear 6 mm + light slatted blinds	0.10	0.13	0.10	0.14	0.09	0.14
Reflecting + clear 6 mm + light slatted blinds	0.11	0.10	0.10	0.11	0.09	0.11
Strongly reflecting + clear 6 mm + light slatted blinds	0.04	0.06	0.04	0.06	0.04	0.06

Note: In this case the outside environmental temperature is not equal to the air
temperature for the external wall; Eqn [4.4] was derived with winter conditions in mind
where solar gains are neglected

The fluctuating components are given by:

$$\tilde{Q}_a = \tilde{Q}_C + \tilde{Q}_{Ga} + \tilde{Q}_V = 168.75 + \{0.17 \times 8 \times (580 - 175)\} + 784.1$$

$$= 1503.7 \, \text{W}$$

$$\tilde{Q}_e = \tilde{Q}_f + \tilde{Q}_{Ge} + \tilde{Q}_{Gc} = 13.4 + \{0.28 \times 8 \times (580 - 175)\} + 246.4$$

$$= 1167.0 \, \text{W}$$

Then substituting in Eqn [4.7], (page 174),

$$(0.857 \times 1503.7) + 1167 = \{(0.857 \times 144) + 502.8)\}\tilde{t}_{ei}$$

i.e. $\tilde{t}_{ei} = 3.9\,K$

(compare with the previous answer of 4.3 K).
Then,

$$\text{Room temperature at } 1600\,h = 23.4 + 3.9 = 27.3\,°C$$

(compare with the previous answer of 27.9 °C).

The Example asks for only one prediction of peak internal environmental temperature. The value calculated is not necessarily the maximum temperature attained on that particular day. The mean value, $\bar{t}_{ei}$, is always the same regardless of the time chosen (in this case 1600 h), since it is calculated from the daily mean values. For the fluctuating values, that transferred through the external wall is very small compared with the fluctuating gains through the glass or by ventilation. In this case the maximum solar irradiation coincides with the maximum sol-air temperature and the maximum outside air temperature, and therefore the peak does occur at 1600 h. A glance at Tables 4.1 and 4.2 will show that for a wall facing due South, for example, this is not the case and a trial and error solution is required to determine the time of the peak internal temperature.

The above example is for a naturally ventilated building. When a building is fully air-conditioned, then it can be assumed that the air conditioning load appears as a negative input at the air point with the method otherwise the same as above.

Example 4.4

The room in Fig. 4.6 is situated on the intermediate floor of an office block in London. It has four occupants and is to be air-conditioned during 10 hours of plant operation. The window is of single reflecting glass with no blind and the external wall is dark-coloured. The infiltration rate is one air change per hour. Calculate the sensible cooling load to be extracted at the air point at 1300 h (sun time) in July if the room is conditioned to an air temperature of 22 °C.

Figure 4.6   Plan of room: Example 4.4

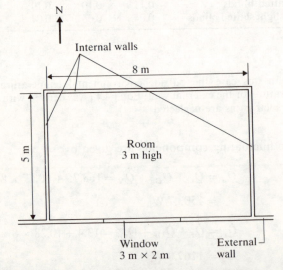

Data

External wall: U-value, $0.5\,\text{W}/\text{m}^2\,\text{K}$; Y-value, $3.7\,\text{W}/\text{m}^2\,\text{K}$; decrement factor, 0.4; time lag, 6 h.

Internal wall: Y-value, $2.2\,\text{W}/\text{m}^2\,\text{K}$.

Floor: Y-value, $3.7\,\text{W}/\text{m}^2\,\text{K}$.

Ceiling: Y-value, $2.3\,\text{W}/\text{m}^2\,\text{K}$,

Window: U-value, $5.7\,\text{W}/\text{m}^2\,\text{K}$; Y-value, $5.7\,\text{W}/\text{m}^2\,\text{K}$.

Occupants sensible gain, 90 W per person.

Use CIBSE Table A9.14, given here as Table 4.7, and also Table A2.33 given previously as 4.2 (page 184) also referred to as CIBSE Table A8.3(e).

(CIBSE)

Solution

$$\Sigma A = (2 \times 5 \times 3) + (2 \times 8 \times 3) + (2 \times 8 \times 5) = 158\,\text{m}^2$$
$$\Sigma(UA_o) = (3 \times 2 \times 5.7) + \{(8 \times 3) - (3 \times 2)\} \times 0.5 = 34.2 + 9.0$$
$$= 43.2\,\text{W}/\text{K}$$
$$\Sigma(YA) = 2.2 \times \{(2 \times 5 \times 3) + (8 \times 3)\} + (3.7 \times 8 \times 5)$$
$$+ (2.3 \times 8 \times 5)$$
$$+ (5.7 \times 3 \times 2) + 3.7 \times \{(8 \times 3) - (3 \times 2)\}$$
$$\Sigma(YA) = 459.6\,\text{W}/\text{K}$$
$$\rho n c_p V = 0.333 \times 1 \times (8 \times 5 \times 3) = 40\,\text{W}/\text{K}$$

From Eqn [4.17] the response factor is

$$f_r = \frac{\Sigma(YA) + \rho n c_p V}{\Sigma(UA_o) + \rho n c_p V} = \frac{459.6 + 40}{43.2 + 40} = 6$$

The building may be therefore taken as mediumweight. From Table 4.7 the cooling load (uncorrected) due to solar gain can be read at 1200 h (i.e. allowing a one hour time lag for a mediumweight building) as $361\,\text{W}/\text{m}^2$.

The correction factor for single glazed reflecting glass is given in Table 4.7 as 0.42, and the correction factor to refer to the internal air temperature instead of the dry resultant temperature is given in the table as 0.73. That is,

$$\text{Cooling load for solar gain} = 361 \times 0.42 \times 0.73 \times (3 \times 2) = 664.1\,\text{W}$$

Note: In Table 4.7 the solar gain is taken to be entirely at the air point, and includes both steady and fluctuating gains.

From Table 4.2,

$$\bar{t}_{eo} = 26\,°\text{C}; \ \bar{t}_{ao} = 19\,°\text{C}$$

At 0700 h, $t'_{eo} = 18\,°\text{C}$ (allowing 6 h time lag as given)

At 1300 h, $t'_{ao} = 23\,°\text{C}$

Also, $\quad \bar{t}_{ai} = 22\,°\text{C}$ (given)

The steady-state inputs at the air point are:

$$\bar{Q}_a = \bar{Q}_c + \bar{Q}_{\text{solar}}$$
$$= (4 \times 90 \times 10/24) + 664.1$$
$$= 150 + 664.1$$
$$= 814.1\,\text{W}$$

Table 4.7 Cooling load/(W/m³) due to solar gain through vertical glazing (10 h plant operation) for constant dry resultant temperature

Lightweight building

							Sun time						
Date	Climatic constants	Orientation	0800	0900	1000	1100	1200	1300	1400	1500	1600	1700	1800	Orientation
21 June	$I = 0.66$	N	89	104	122	137	147	150	146	135	119	103	92	N
		NE	358	307	220	159	165	169	164	153	138	120	102	NE
	$k_c = 1.96$	E	465	477	438	355	241	179	175	164	149	131	112	E
		SE	322	392	425	419	372	286	188	149	134	116	98	SE
	$k_r = 0.20$	S	89	127	208	289	342	359	337	280	196	116	83	S
		SW	102	121	138	158	202	300	381	423	424	386	313	SW
	$C = 0.14$	W	116	135	152	167	177	189	257	366	444	477	460	W
		NW	104	123	140	155	165	169	166	166	232	313	357	NW
23 July and 22 May	$I = 0.89$	N	85	107	128	145	157	161	156	143	125	104	86	N
		NE	314	276	205	162	173	177	172	159	141	120	98	NE
	$k_c = 1.33$	E	419	440	412	343	244	188	183	170	152	131	109	E
		SE	309	381	418	419	380	304	209	162	144	123	101	SE
	$k_r = 0.20$	S	97	146	231	309	361	377	356	301	219	134	89	S
		SW	106	128	149	171	224	317	388	421	416	375	300	SW
	$C = 0.18$	W	114	136	156	174	186	197	258	353	417	439	413	W
		NW	102	124	144	162	173	177	173	168	214	281	312	NW
24 August and 20 April	$I = 0.93$	N	59	82	102	119	131	134	129	117	100	79	58	N
		NE	243	214	150	130	142	145	140	128	111	90	67	NE
	$k_c = 1.34$	E	364	403	385	317	216	158	153	141	123	102	80	E
		SE	296	383	430	435	399	323	219	144	126	105	82	SE
	$k_r = 0.20$	S	90	167	264	346	398	415	394	338	253	154	80	S
		SW	88	110	131	156	235	335	406	437	427	376	285	SW
	$C = 0.23$	W	86	108	129	146	157	168	233	329	390	402	355	W
		NW	72	95	115	132	143	147	142	135	160	220	240	NW
22 September and 22 March	$I = 0.97$	N	32	55	76	93	104	107	102	90	73	51	29	N
		NE	143	142	91	96	107	110	106	94	76	54	32	NE
	$k_c = 1.61$	E	263	371	375	306	191	122	118	106	88	67	44	E
		SE	245	396	474	494	459	376	254	136	101	79	56	SE
	$k_r = 0.20$	S	97	208	329	428	490	510	485	418	315	192	84	S
		SW	62	85	109	153	272	388	466	495	469	383	225	SW
	$C = 0.27$	W	50	73	94	111	122	134	209	318	377	364	243	W
		NW	36	59	80	97	108	111	107	98	97	146	133	NW
23 October and 20 February	$I = 1.18$	N	10	32	53	70	81	85	80	68	50	28	8	N
		NE	19	61	54	71	82	85	81	69	51	29	9	NE
	$k_c = 1.31$	E	43	215	272	237	149	94	89	77	59	37	17	E
		SE	57	254	379	424	409	345	242	128	71	50	30	SE
	$k_r = 0.20$	S	51	162	289	390	452	472	447	380	275	145	42	S
		SW	32	54	81	143	255	353	412	420	367	233	43	SW
	$C = 0.31$	W	19	40	62	79	90	101	161	243	268	198	31	W
		NW	10	32	54	71	82	86	81	69	55	58	14	NW
21 November and 21 January	$I = 1.29$	N	4	9	28	44	55	58	54	42	25	7	4	N
		NE	5	11	28	44	55	58	54	42	25	7	4	NE
	$k_c = 1.34$	E	10	47	160	168	106	63	58	47	30	11	9	E
		SE	23	68	246	338	349	301	212	108	43	24	21	SE
	$k_r = 0.20$	S	31	60	201	322	396	419	390	310	184	49	29	S
		SW	21	26	51	121	224	308	349	329	228	54	21	SW
	$C = 0.35$	W	9	13	32	49	59	69	115	169	149	36	9	W
		NW	4	9	28	44	55	58	54	42	25	9	4	NW
22 December	$I = 1.18$	N	3	4	16	31	40	43	39	29	14	3	3	N
		NE	3	4	16	31	46	43	39	29	14	3	3	NE
	$k_c = 1.70$	E	6	13	111	138	87	47	43	32	18	6	6	E
		SE	18	27	180	297	324	283	196	94	71	18	18	SE
	$k_r = 0.20$	S	25	33	152	287	372	398	365	273	136	25	25	S
		SW	18	20	39	107	207	289	322	286	162	18	18	SW
	$C = 0.40$	W	6	7	20	34	44	52	95	137	100	6	9	W
		NW	3	4	16	31	40	43	39	29	14	3	3	NW

Correction factors for tabulated values

Type of glass	Single glazing		Double glazing with clear 6 mm glass inside		Single glazing and external light slatted blinds used intermittently	
	Lightweight	Heavyweight	Lightweight	Heavyweight	Lightweight	Heavyweight
Clear 6 mm	1.00	0.85	0.85	0.70	0.19	0.19
BTG 6 mm	0.70	0.60	0.53	0.45	0.16	0.16
BTG 10 mm	0.58	0.49	0.40	0.34	0.14	0.14
Reflecting	0.49	0.42	0.37	0.31	0.13	0.13
Strongly reflecting	0.25	0.21	0.16	0.14	0.09	0.10
Additional factor for air point control	0.76	0.73	0.76	0.73	0.76	0.73

Since all the solar gain is taken to be at the air point, there is no gain at the environmental point in this case, i.e. $\tilde{Q}_e = 0$.

The fluctuating components at the air point are:

$$\tilde{Q}_a = \tilde{Q}_v + \tilde{Q}_c$$
$$= 40(23 - 19) + \{(4 \times 90) - 150\}$$
$$= 160 + 210 = 370 \, W$$

The fluctuating components at the environmental point are:

$$\tilde{Q}_e = \tilde{Q}_f + \tilde{Q}_{Gc}$$
$$= \{0.4 \times 9 \times (18 - 26)\} + \{3 \times 2 \times 5.7 \times (23 - 19)\}$$
$$= -28.8 + 136.8 = 108 \, W$$

Then using Eqns [4.5] and [4.8] we require the following factors:

$$F_{au} = 4.5 \times 158/\{(4.5 \times 158) + 43.2\} = 0.9427$$
$$F_{ay} = 4.5 \times 158/\{(4.5 \times 158) + 459.6\} = 0.6074$$

Adding Eqns [4.5] and [4.8] to $\dot{Q}_P$, the required plant load at the air point,

$$\dot{Q}_P + \bar{Q}_a + F_{au}\bar{Q}_e + \tilde{Q}_a F_{ay}\tilde{Q}_e$$
$$= \{\rho n c_p V + F_{au}\Sigma(UA_o)\}(\bar{t}_{ai} - \bar{t}_{ao}) + \{\rho n c_p V + F_{ay}\Sigma(YA)\}\tilde{t}_{ai}$$

Substituting the values calculated above, noting that $\bar{Q}_e = 0$, and since $\bar{t}_{ai}$ is controlled at 22 °C, then $\tilde{t}_{ai} = 0$, and we have:

$$\dot{Q}_P + 814.1 + 0 + 370 + (0.6074 \times 108)$$
$$= \{40 + (0.9427 \times 43.2)\}(22 - 19) + 0$$

i.e. $$\dot{Q}_P = -1008 \, W$$

This represents the fraction of the plant load required for one room. To calculate the plant load for the building it is necessary to plot the variation in plant load through the middle of the day for each façade since the value of the solar gain is different according to the direction of the sun; the time at which the total value is a maximum will give the necessary plant load.

The solution using Table 4.7 can be compared with that obtained using Tables 4.1 and 4.5. From Table 4.1 for a south-facing wall in July at 1200 h (allowing 1 h time lag), the solar irradiation $I'_t = 550 \, W/m^2$, and the mean value $\bar{I}_t = 165 \, W/m^2$. From Table 4.5, for single reflecting glass with no blind we have

$$\bar{S}_e = 0.36, \tilde{S}_e = 0.27$$

Note that all the solar gain appears at the environmental point since it passes directly into the room; with shading some of the solar gain appears as convection at the air point.

Hence

$$\bar{Q}_{Ge} = 0.36 \times 3 \times 2 \times 165 = 356.4 \, W$$
$$\tilde{Q}_{Ge} = 0.27 \times 3 \times 2 \times (550 - 165) = 623.7 \, W$$

Then the solar gain referred to the air point is given by

$$F_{au}\bar{Q}_{Ge} + F_{ay}\bar{Q}_{Ge}$$
$$= (356.4 \times 9427) + (623.7 \times 0.6074)$$
$$= 714.8 \, W$$

This compares with the previous value of 664.1 W. The reason for the difference is that the value of 664.1 W allows for 10-hour plant operation whereas the value of 714.8 W represents the solar gain for the full day. The basis of the calculation of the values in Table 4.7 is not given by CIBSE.

When the air conditioning plant is operated intermittently, then the plant load must be made larger to compensate for the load incurred in the off-period. Harrington-Lynn[4.10] has shown that for cases of low or zero infiltration rates the plant load for intermittent operation is greater than the continuous value for a constant internal air temperature by an amount

$$\dot{Q}_P = \frac{24\{F_{ay}\Sigma(YA) - F_{au}\Sigma(UA_o)\}\bar{Q}_{off}}{(24 - N)F_{au}\Sigma(UA_o) + NF_{ay}\Sigma(YA) + 24\rho nc_p V} \qquad [4.19]$$

where $\bar{Q}_{off}$ is the 24-hour mean of the continuous operating load which would otherwise have occurred in the off-period; N is the number of hours of plant operation.

Example 4.5

A small well-sealed laboratory is continuously air-conditioned at a constant internal temperature of 20 °C. It has one external wall and the spaces above, below and adjacent to all the internal walls are at the same air temperature. The laboratory is occupied for eight hours per day by five people and there is equipment with a total heat gain of 1500 W which is operated for five of the eight hours of occupancy.

Using the data below, assuming that the peak air conditioning load occurs at 1800 h, calculate:

(i) the required air conditioning plant load for continuous air conditioning;
(ii) the plant load required if the plant is in operation for only 8 h.

Data

Mean total solar irradiance, 185 W/m²; total solar irradiance at 1600 h, 625 W/m²; total solar irradiance at 1700 h, 600 W/m².

Mean sol-air temperature, 24.5 °C; sol-air temperature at 0900 h, 21.0 °C; external air temperature at 1800 h, 21.0 °C; mean external air temperature, 16.5 °C.

Decrement factor for external wall, 0.28; decrement factor for glazing, 1.0; time lag for external wall, 9 h; time lag for glazing, 0 h.

Mean solar gain factor at air point, 0.19; mean solar gain factor at environmental point, 0.26; alternating solar gain factor at air point, 0.21; alternating solar gain factor at environmental point, 0.21.

Energy output per person, 130 W.

Total internal surface area, ΣA, 190 m²; window area, 20 m²; external wall area, 10 m²; U-value for windows, 3.3 W/m² K; U-value for external walls, 0.6 W/m² K; $\Sigma(YA) = 760$ W/K.

The 24-hour mean of the continuous operating load which would otherwise have occurred in the off-period may be taken as 1000 W.

Solution

(i) Steady-state gains at the air point:

$$\bar{Q}_{Ga} = \bar{S}_a A_G \bar{I}_t = 0.19 \times 20 \times 185 = 703 \text{ W}$$
$$\bar{Q}_C = (5 \times 130 \times 8/24) + (1500 \times 5/24) = 529.2 \text{ W}$$

i.e. $\qquad \bar{Q}_a = 703 + 529.2 = 1232.2 \text{ W}$

Steady-state gains at the environmental point:

$$\bar{Q}_e = \bar{Q}_{Ge} = \bar{S}_e A_G \bar{I}_t = 0.26 \times 20 \times 185 = 962 \text{ W}$$

For heat transfer through the fabric,

$$\Sigma(UA_o) = (3.3 \times 20) + (0.6 \times 10) = 72 \text{ W/K}$$

From Eqn [4.17], with $\rho n c_p V = 0$ for a well-sealed building we have:

$$f_r = \frac{\rho n c_p V + \Sigma(YA)}{\rho n c_p V + \Sigma(UA_o)} = \frac{0 + 760}{0 + 72} = 10.56$$

Therefore the building is heavyweight and the solar gain must be taken at 1600 h (i.e. a two-hour time lag). Fluctuating gains at the air point:

$$\tilde{Q}_{Ga} = \bar{S}_a A_G(I_t' - \bar{I}_t) = 0.21 \times 20 \times (625 - 185) = 1848 \text{ W}$$
$$\tilde{Q}_C = \{(5 \times 130) - (5 \times 130 \times 8/24)\} + \{1500 - (1500 \times 5/24)\}$$
$$= 1620.8 \text{ W}$$

From Eqn [4.14],

$$\tilde{Q}_V = \rho n c_p V \tilde{t}_{ao} = 0$$

since $\rho n c_p V = 0$ for a well-sealed building.

$$\therefore \qquad \tilde{Q}_a = 1848 + 1620.8 + 0 = 3468.8 \text{ W}$$

Fluctuating gains at the environmental point:

$$\tilde{Q}_f = 0.28 \times 10 \times (21.0 - 24.5) = -9.8 \text{ W}$$
$$\tilde{Q}_{Ge} = 0.21 \times 20 \times (625 - 185) = 1848 \text{ W}$$
$$\tilde{Q}_{Gc} = 3.3 \times 20 \times (21.0 - 16.5) = 297 \text{ W}$$

i.e. $\qquad \tilde{Q}_e = 2135.2 \text{ W}$

Then introducing the required plant load, $\dot{Q}_P$, and adding Eqns [4.5] and [4.8], we have:

$$\dot{Q}_P + \bar{Q}_a + \tilde{Q}_a + F_{au}\bar{Q}_e + F_{ay}\tilde{Q}_e$$
$$= \{\rho n c_p V + F_{au}\Sigma(UA_o)\}(\bar{t}_{ai} - \bar{t}_{ao}) + \{\rho n c_p V + F_{ay}\Sigma YA\}\tilde{t}_{ai}$$

where $\qquad F_{au} = (4.5 \times 190)/\{(4.5 \times 190) + 72\} = 0.9223$

and $\qquad F_{ay} = (4.5 \times 190)/\{(4.5 \times 190) + 760\} = 0.5294$

Substituting the calculated values (noting that $\rho n c_p V = 0$ and $\tilde{t}_{ai} = 0$):

$$\dot{Q}_P + 1232.2 + 3468.8 + (0.9223 \times 962) + (0.5294 \times 2135.2)$$
$$= \{0 + (0.9223 \times 72)\}(20 - 16.5) + 0$$

i.e. Plant load $= -6310 \text{ W}$

(ii) 24-hour mean load in off-period $= 1000 \text{ W}$ (given). Then using Eqn [4.19],

$$\Delta\dot{Q}_P = \frac{24\{(0.5294 \times 760) - (0.9223 \times 72)\} \times 1000}{(24 - 8)(0.9223 \times 72) + (8 \times 0.5294 \times 760) + 0}$$
$$= 1883 \text{ W}$$

Therefore the intermittent air conditioning plant load is

$$6310 + 1883 = 8193 \text{ W}$$

Note: CIBSE practice has been followed in using the term 'plant load'; the load calculated is in fact the room sensible cooling load which is used as the basis of the calculation for the plant load when the complete system is designed as explained in Chapter 8.

A summary of the equations and room factors for the steady and fluctuating states is given in Table 4.8.

Table 4.8 Equations and room factors for steady and fluctuating heat transfer

---

Steady state

$$F_{av}\bar{Q}_a + \bar{Q}_e = \{F_{av}\rho nc_p V + \Sigma(UA_o)\}(\bar{t}_{ei} - \bar{t}_{ao})$$
$$\bar{Q}_a + F_{au}\bar{Q}_e = \{\rho nc_p V + F_{au}\Sigma(UA_o)\}(\bar{t}_{ai} - \bar{t}_{ao})$$
$$F_v\bar{Q}_a + F_u\bar{Q}_e = \{F_v\rho nc_p V + F_u\Sigma(UA_o)\}(\bar{t}_c - t_{ao})$$

Fluctuating state

$$F_{av}\tilde{Q}_a + \tilde{Q}_e = \{F_{av}\rho nc_p V + \Sigma(YA)\}\tilde{t}_{ei}$$
$$\tilde{Q}_a + F_{ay}\tilde{Q}_e = \{\rho nc_p V + F_{ay}\Sigma(YA)\}\tilde{t}_{ai}$$
$$F_v\tilde{Q}_a + F_y\tilde{Q}_e = \{F_v\rho nc_p V + F_y\Sigma(YA)\}\tilde{t}_c$$

Dimensionless room factors
Air point to environmental point:
    Steady and fluctuating states, $F_{av} = \alpha_a\Sigma A/\{\alpha_a\Sigma A + \rho nc_p V\}$
Environmental point to air point:
    Steady state, $F_{au} = \alpha_a\Sigma A/\{\alpha_a\Sigma A + \Sigma(UA_o)\}$
    Fluctuating state, $F_{ay} = \alpha_a\Sigma A/\{\alpha_a\Sigma A + \Sigma(YA)\}$
Air point to dry resultant point:
    Steady and fluctuating states, $F_v = \alpha_{ac}\Sigma A/\{\alpha_{ac}\Sigma A + \rho nc_p V\}$
Environmental point to dry resultant point:
    Steady state, $F_u = \alpha_{ec}\Sigma A/\{\alpha_{ec}\Sigma A + \Sigma(UA_o)\}$
    Fluctuating state, $F_y = \alpha_{ec}\Sigma A/\{\alpha_{ec}\Sigma A + \Sigma(YA)\}$

Normal values
$\alpha_a = 4.5 \text{ W/m}^2\text{ K}$; $\alpha_{ac} = 6.0 \text{ W/m}^2\text{ K}$; $\alpha_{ec} = 18.0 \text{ W/m}^2\text{ K}$

PROBLEMS

4.1   A building is heated to a constant internal temperature from 8 am until 8 pm, seven days per week, throughout a heating season. The heating system is switched off at 8 pm each day and the building is allowed to cool. At a pre-determined time, the system is switched on again at a rate which brings the internal temperature to the required day-time temperature at 8 am. Select a range of possible rates of re-heating and draw a graph showing the variation of the efficiency of the intermittent operation (compared with continuous heating) with the rate of re-heating. Discuss other design factors which would bear on the selection of a maximum installed heating capacity and the re-heating rate.

Data
Building fabric heat loss coefficient, 100 kW/K; building time constant (for 8 pm to 8 am), 36 h; building ventilation rate (8 am to 8 pm), 20 m³/s; building ventilation rate, (8 pm to 8 am), 2 m³/s; volumetric specific heat of air,

1200 J/m³ K; building internal temperature (8 am to 8 pm), 18 °C; outside air temperature (seasonal mean), 6 °C; winter external design temperature, −2 °C.

(Polytechnic of the South Bank)

(Percentage saving against plant load: 4.07% at 2700 kW; 4.28% at 3000 kW; 4.74% at 3300 kW)

4.2 A single-storey office block has dimensions and thermal properties as detailed below.

Element	Area (m²)	U-value (W/m² K)	Y-value (W/m² K)
External wall	290	0.4	3.1
Windows	45	5.7	5.7
Floor	600	0.2	4.0
Roof	600	0.3	2.1
Internal wall (both sides)	1100	—	2.2

(a) By determining the building thermal response factor indicate whether the building is thermally heavyweight, mediumweight, or lightweight. Comment on the effect of each of the building elements on the overall thermal weight of the building.

(b) If the building is intermittently heated and naturally ventilated, briefly discuss the effect of the building thermal weight on the operation of the heating system and on the thermal performance of the building during winter and summer.

Data

Volume of office block, 1800 m³; winter ventilation rate, 1 air change per hour; summer ventilation rate, 4 air changes per hour.

(CIBSE)

(Heavyweight in winter; mediumweight in summer)

4.3 A building with the shape and dimensions shown in Fig. 4.7 is attached to another building with the same internal temperature, along one of its shorter walls; the other walls are external and have windows; there is no glazing in the roof. The building is heated intermittently with an eight-hour operating period and one-hour pre-heat period.

Figure 4.7 Building of Problem 4.3

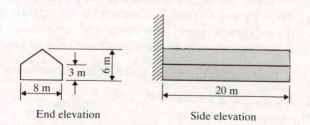

End elevation   Side elevation

Using the data below, calculate:
(i) the 24-hour average dry resultant temperature;
(ii) the required plant load.

Data

Design dry resultant temperature, 16 °C; design outside air temperature, -1 °C. Window area, 80 m²; U-value of windows, 5.3 W/m² K; Y-value of windows, 5.3 W/m² K; U-value of external walls, 0.73 W/m² K; Y-value of external walls, 3.6 W/m² K; U-value of pitched roof, 6.5 W/m² K; Y-value of roof, 6.5 W/m² K; U-value of floor, 0.45 W/m² K; Y-value of floor, 2.9 W/m² K; Y-value of internal walls, 3.6 W/m² K.

Air infiltration rate, 0.75 changes per hour.

Correction factor for one-hour pre-heating, 1.4.

(Wolverhampton Polytechnic)

(5.65 °C; 51 kW)

4.4 An intermediate-floor room in a building, 8 m by 5 m by 3 m high with one external wall on the long axis, has two alternative methods of construction proposed as detailed below. In addition it is proposed to consider two alternative modes of heating, i.e. continuous or intermittent.

Construction Method A:

Description	ΣUA_o (W/K)	ΣYA (W/K)	C_v (W/K)
External walls: brick/block/cavity fill and plaster Internal walls: lightweight block/plaster Floors/ceiling: cast concrete with carpet Glazing: single, 30%	65.1	726.6	30

Construction Method B:

Description	ΣUA_o (W/K)	ΣYA (W/K)	C_v (W/K)
As Method A, except Internal walls: lightweight partition Floors/ceiling: cast concrete with plasterboard ceiling and plenum	65.1	411.4	30

(a) Determine for *each* method of construction, for *both* continuous and intermittent heating:

(i)   the thermal weight ratio of the room;
(ii) the mean internal environmental temperature over 24 hours;
(iii) the minimum internal environmental temperature over 24 hours;
(iv) the maximum internal environmental temperature over 24 hours;
(v)   the energy input from the heat source over 24 hours.

(b) Tabulate your results to make brief comments on them in relation to:

(i)   the design of a suitable system;
(ii) energy consumption;
(iii) possible damage to the building fabric and contents.

Data

Design internal environmental temperature, 21 °C; design external air temperature, 0 °C.

Intermittent heating: 12 hours on and 12 hours off.

Assume that the resulting internal environmental temperature for intermittent heating follows a sinusoidal pattern, and that for continuous heating steady-state conditions prevail.

(Newcastle upon Tyne Polytechnic)

((i) 7.96, 4.64; (ii) 21 °C, 18.66 °C, 17.28 °C; (iii) 21 °C, 16.31 °C, 13.56 °C; (iv) 2 kW, 3.55 kW, 3.28 kW; (v) 48 kW h, 42.6 kW h, 39.4 kW h)

4.5 (a)   Explain the effect ventilation may have on the peak temperature reached in a room in a UK building with a south-facing wall of open aspect and a large glazed area.

(b)   Estimate

(i) the mean temperature, and

(ii) the peak temperature,

that might be reached in a room, given the following information:

Room volume, 40 m^3; infiltration rate, 3 air changes per hour; volumetric specific heat, $1200 \text{ J/m}^3 \text{ K}$; window area in outer wall, 3 m^2; U-value of window, $3.0 \text{ W/m}^2 \text{ K}$; solar gain factors for the window, $\bar{S}_e = 0.7$, $\tilde{S}_e = 0.5$; opaque area of outer wall, 7 m^2; U-value of outer wall, $1.0 \text{ W/m}^2 \text{ K}$; Y-value of outer wall, $2.0 \text{ W/m}^2 \text{ K}$; floor area, 16 m^2; Y-value of floor, $4.0 \text{ W/m}^2 \text{ K}$; Y-value of remaining internal surfaces, $2.5 \text{ W/m}^2 \text{ K}$; daily mean ambient temperature, 14 °C; swing in ambient temperature, $+6 \text{ K}$; radiation incident daily on the outer wall, $14 \times 10^6 \text{ J/m}^2$; maximum incident intensity of solar radiation, 540 W/m^2.

(University of Liverpool)

(19.9 °C; 23.9 °C)

4.6   The daily mean inside environmental temperature in an office is related to the mean outside air temperature by the relation:

$$\bar{t}_{ei} - \bar{t}_{ao} = \frac{\bar{Q}_c + \bar{S} A_G \bar{I}_t}{U_G A_G + (A_o - A_G) U_w + C_V}$$

where $\bar{Q}_c$ denotes mean casual gains; A_G the window area; A_o the total façade area in which the window is built; $\bar{I}_t$ the daily mean solar intensity falling on the façade; U_G the U-value for the windows; U_w the U-value for the external wall; C_V the ventilation conductance.

(i)   Find the criterion that the temperature elevation $(\bar{t}_{ei} - \bar{t}_{ao})$ should increase as the window area A_G increases.

(ii) The elevation can be restrained if the ventilation rate is increased; write down an explicit expression for C_V in terms of the other quantities.

(iii) Hence find the value for C_V and the associated air change rate in an office, given the following information:

Room width, 5 m; room depth, 3 m; room height, 2.5 m; $U_G = 3.0 \text{ W/m}^2 \text{ K}$; $U_w = 0.6 \text{ W/m}^2 \text{ K}$; $\bar{Q}_c = 400 \text{ W}$; daily incident solar radiation on a south elevation, 12 MJ/m^2; solar gain factor, 0.6; 40% of façade glazed; $\bar{t}_{ei} = 22 \text{ °C}$; $\bar{t}_{ao} = 16 \text{ °C}$.

(iv) Calculate the peak temperature reached, assuming that it is due to daily variation in solar radiation and ambient air temperature:

Maximum solar intensity on façade, 500 W/m²; alternating solar gain factor, 0.45; admittance of all opaque surfaces, 3.0 W/m² K; variation of ambient temperature, +6 K.

(v)   Comment critically on the basis on which the calculation of the peak temperature rests.

(University of Liverpool)

(116.6 W/K; 9.33; 27 °C)

4.7 (a)   Estimate the daily mean and the peak-to-mean environmental temperature in a room which has one exposed wall and five internal walls, given the following information:

Room dimensions: height, 2.5 m; width, 3.2 m; depth, 4.0 m; window area, 3.6 m²; steady ventilation rate, 2 air changes per hour; U-values: window, 5.7 W/m² K; opaque wall, 0.8 W/m² K; Y-values: all vertical walls, 3.0 W/m² K; floor, 4.0 W/m² K; ceiling, 3.0 W/m² K; daily incidence of solar radiation on outer wall, 12×10^6 J/m²; maximum intensity of solar radiation, 510 W/m²; solar gain factor for the window, 0.7; dynamic solar gain factor, 0.5; ambient temperature: daily mean, 14 °C; peak-to-mean daily variation, 6 K.

Assume that there are no further heat gains to the room and ignore the heat conducted through the opaque part of the wall.
(b)   Explain how the solar gain factors are related to the physical parameters, i.e. dimensions and materials, of the room.
(c)   The estimates of daily variation in room temperature as found using CIBSE *Guide* methods are easy to obtain but they are very coarse. Explain why this is the case.

(University of Liverpool)

(16.2 °C; 1.76 K)

4.8 (a)   A south-facing room has one external wall with a window. Using the data given below, construct a mathematical model to predict the peak summertime temperature as a function of window area.
(b)   Use the model to calculate the window area to minimize the peak summertime temperature.

Data
Total room surface area, 300 m²; area of external wall plus window, 30 m²; room volume, 270 m³; U-value of external wall, 0.6 W/m² K; U-value of window, 3.6 W/m² K; Y-value of all room surfaces, 4.5 W/m² K; people load: mean, 300 W; peak, 900 W; solar gain through glass to air point: mean, 16 W/m²; peak, 41 W/m²; solar gain through glass to environmental point: mean, 31 W/m²; peak, 59 W/m²; mean outside air temperature, 14 °C; peak outside air temperature, 19 °C; number of air changes per hour, 1; peak lighting load, $80(30 - A)$ W, where A is the area of the window; mean lighting load, $\frac{1}{3}$ of the peak lighting load.

(Polytechnic of the South Bank)

(18.1 m²; 37.9 °C)

4.9 (a)   Explain the terms
(i)   decrement factor,
(ii) admittance,
(iii) lag,
as they are used in the admittance technique for estimation of building thermal performance.
(b)   Detail the difference between *environmental temperature* and *dry resultant temperature*; explain the usage of each, and their relationship with air temperature and room surface temperatures.

Figure 4.8   Plan of room: Problem 4.9

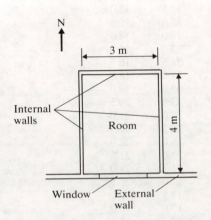

(c)   The heat balance of a room may be stated by the steady-state equation [1] and the cyclic equation [2]. Considering only the external wall and window of the room shown in Fig. 4.8 (i.e. neglecting the partitions, floor and ceiling and any infiltration or internal heat gains) and using the data given in the table below, and the table overleaf, calculate, for the location and date to which the tables are relevant, the necessary 'air point' cooling load at 1500 hours to maintain the internal dry resultant temperature constant at 21 °C.

Element	Area (m²)	Y (W/m² K)	U (W/m² K)	Decrement factor f	Decrement factor Lag (h)	Surface factor s	Surface factor Lag (h)
External wall	9	2.4	0.46	0.43	8	0.79	1
Window	3	3.0	3.00	1.00	0	—	—

$$F_v \bar{Q}_a + F_u \bar{Q}_e = F_v \rho n c_p V(\bar{t}_c - \bar{t}_{ao}) + F_u \Sigma(UA_o)(\bar{t}_c - \bar{t}_{eo}) \qquad [1]$$

$$F_v \tilde{Q}_a + F_y \tilde{Q}_e = \{F_v \rho n c_p V + F_y \Sigma(YA)\}\tilde{t}_c \qquad [2]$$

(Symbols are as defined previously, and therefore the listing as given in the original problem is omitted.)
Solar gain factors: $\bar{S}_e = 0.26$; $\bar{S}_a = 0.19$; $\tilde{S}_e = 0.25$; $\tilde{S}_a = 0.21$.
Room factors: $F_u = 0.989$; $F_v = 1.00$; $F_y = 0.962$.

(The Engineering Council)

(275.9 W)

Vertical south-facing surface:

Time (h)	I (W/m^2)	t_{eo} (°C)	t_{ao} (°C)
0	0	13.0	14.0
1	0	11.5	12.5
2	0	11.0	12.0
3	0	10.5	11.0
4	0	10.0	11.0
5	15	11.0	11.0
6	50	12.5	12.0
7	70	14.0	12.5
8	85	18.0	14.0
9	185	23.0	15.0
10	285	27.5	16.5
11	345	31.0	18.0
12	370	33.0	19.0
13	345	33.5	20.5
14	285	32.0	21.0
15	190	29.5	21.5
16	85	26.0	22.0
17	70	23.0	21.5
18	50	21.5	21.0
19	15	20.0	20.5
20	0	18.5	19.0
21	0	17.0	18.0
22	0	15.5	16.5
23	0	14.0	15.0
Average	101.87	19.85	16.45

4.10 The opaque fabric of the room shown in Fig. 4.9 can be assumed to be perfectly insulated on its external faces.

(i) Calculate the dry resultant temperature at 3 pm.

(ii) Identify the steps in the calculation where you take account of the mass of the structure and explain briefly the difference between the step (or steps) which use the admittance value in some form and the one that does not.

(iii) Draw the conductance diagram for direct and cyclic heat flows, inserting the conductances and temperatures derived in part (i).

(iv) Calculate the air temperature at 3 pm making use of these diagrams.

Figure 4.9 Room for Problem 4.9

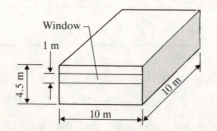

Window

1 m

4.5 m

10 m

10 m

Data

Glazing: single clear glass; U-value 5.6 W/m^2 K.

Average admittance of room surfaces, 5 W/m^2 K; ventilation, 10 air changes per hour; gains from occupants and lighting, 0.

Climatic data:

Time	Solar intensity	Outside air temperature
(h)	(W/m^2)	(°C)
1200	430	19.2
1300	535	20.3
1400	585	21.2
1500	580	21.7
24-hour average	180	16.5

(Polytechnic of the South Bank)

$(18.4 + 3.3, 21.7\,°C; 17.6 + 4.2, 21.8\,°C)$

4.11 A 7 m^2 by 4 m high west-facing laboratory is continuously air-conditioned at a temperature of 24 °C. Using the data given below calculate the percentage increase in peak load and the percentage saving in room energy consumption if the air conditioning system is switched off between 8 pm and 8 am.

Data
One external wall: U-value, 0.6 W/m^2 K; one window: size, 7 m by 2 m; U-value of window, 3.21 W/m^2 K; average admittance of walls, 2.0 W/m^2 K; average admittance of floor and ceiling, 4.0 W/m^2 K.

Air conditioning load:

Time/h	1	2	3	4	5	6	7	8	9	10	11	12	13
Load (W)	976	924	895	875	865	941	1052	1604	2155	2261	2362	2443	2500

Time/h	14	15	16	17	18	19	20	21	22	23	24
Load (W)	2889	3207	3415	3487	2952	2228	1698	1167	1130	1085	789

(Polytechnic of the South Bank)

(22.6%; 6.61%)

REFERENCES

4.1 Clarke J A 1985 *Energy Simulation in Building Design* Adam Hilger Ltd

4.2 Clarke J A, Holmes M J, Bowman N and Lomas K 1986 Does simulation work? *Building Services* **8** (3)

4.3 Eastop T D and McConkey A 1986 *Applied Thermodynamics for Engineering Technologists* 4th edn Longman

4.4 CIBSE 1986 *Guide to Current Practice* volume A2, Table A2.2

4.5 CIBSE 1986 *Guide to Current Practice* volume A3

4.6 CIBSE 1986 *Guide to Current Practice* volume A5, Tables A5.3 and A5.4

4.7 Billington N S 1976 Thermal insulation and thermal capacity of buildings. *Building Services Engineer* **43**: 226–33

4.8   CIBSE 1986 *Guide to Current Practice* volume A5, Table A5.6

4.9   CIBSE 1986 *Guide to Current Practice* volume A9, Appendix 9.1

4.10 Harrington-Lynn J 1974 The admittance procedure: intermittent plant operation. *Building Services Engineer* **42**: 219–21

5 MOISTURE TRANSFER

Moisture will be transferred between a room and adjoining areas and across any permeable structure whenever there is a difference in the partial pressure of the water vapour in the air. At any point in time in a typical building, moisture will be flowing in many different directions: within rooms, between rooms, across the building fabric, to and from the outside air through air gaps and ventilation openings.

An understanding of moisture transfer is essential for predicting whether condensation will occur on surfaces or within structures. Condensation will take place whenever water vapour is in contact with a surface at a lower temperature than its dew point. The first step in any condensation problem is therefore to establish the temperature at any point in the structure being investigated.

5.1 SURFACE AND INTERFACE TEMPERATURES

The equations governing the heat transfer through structures with plane boundaries are as follows:

Heat flow rate

$$Q_F = UA(t_i - t_o) \qquad [1.1]$$

where

$$\frac{1}{U} = \frac{1}{\alpha_i} + \sum \frac{\Delta x}{k} + \frac{1}{\alpha_o} \qquad [1.2]$$

and

$$\frac{1}{U} = R_{TT} = R_{si} + \sum R_T + R_{so} \qquad [5.1]$$

R_{TT} is the total thermal resistance between the inside and outside air and R_T the thermal resistance of individual sections of the structure including any air space. The surface resistances R_{si} and R_{so} are equal to the reciprocal of their respective combined (convection and radiation) surface heat transfer coefficients α_i and α_o.

A thermal resistance results in a temperature drop in the same way as an electrical resistance results in a voltage drop. The share of the total temperature difference between the inside and outside air which occurs across any section of a structure is proportional to the thermal resistance of that section. For any section x:

$$\Delta t_x = R_{Tx} \frac{(t_i - t_o)}{R_{TT}}$$ [5.2]

Example 5.1

A window is constructed of 6 mm glass with a thermal conductivity of 1 W/m K. The internal air temperature is 20 °C and the external air temperature 2 °C. If the internal and external surface heat transfer coefficients are 8 W/m² K and 20 W/m² K respectively, calculate the surface temperatures.

Solution

The system is shown in Fig. 5.1. Using Eqn [5.1]:

$$\frac{1}{U} = R_{TT} = \frac{1}{\alpha_i} + \frac{\Delta x}{k} + \frac{1}{\alpha_o} = \frac{1}{8} + \frac{0.006}{1} + \frac{1}{20}$$

$$\therefore \quad R_{TT} = 0.125 + 0.006 + 0.05 = 0.181 \text{ m}^2 \text{ K/W}$$

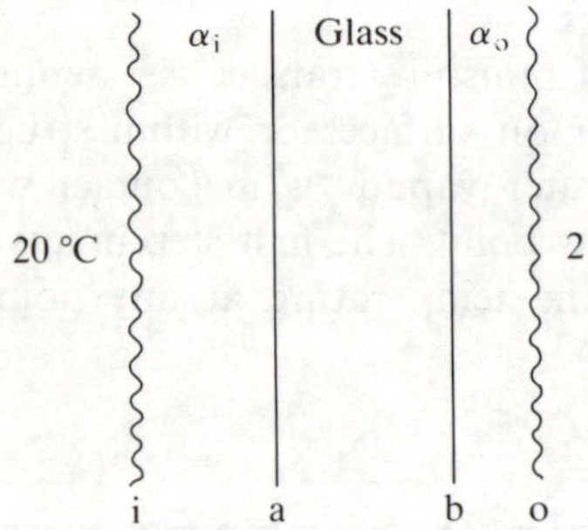

Figure 5.1 Example 5.1

20 °C α_i Glass α_o 2 °C

i   a   b   o

Using Eqn [5.2],

$$t_a = t_i - \Delta t_{i-a} = 20 - 0.125 \frac{(20 - 2)}{0.181} = 7.57 \,^\circ C$$

$$t_b = t_a - \Delta_{ta-b} = 7.57 - (0.006 \times 99.45) = 6.97 \,^\circ C$$

Check:

$$t_o = t_b - \Delta t_{b-o} = 6.97 - (0.05 \times 99.45) = 2 \,^\circ C$$

The term $(t_i - t_o)/R_{TT}$ which in this example is 99.45 is repeated in each calculation for structures with any number of sections.

When information is given in terms of a U-value and a surface coefficient there is no need to use thermal resistances to establish a surface temperature. In this example the U-value is $1/0.181 = 5.525$ W/m² K and the inside surface temperature could have been calculated by substituting U for $1/R_{TT}$ and α_i for $1/R_{si}$ in Eqn [5.2] as follows:

$$t_a = t_i - \frac{U}{\alpha_i}(t_i - t_o)$$ [5.3]

$$= 20 - \frac{5.525}{8}(20 - 2) = 7.57 \,^\circ C$$

5.2 MOISTURE TRANSFER THROUGH STRUCTURES

Fick's Law

This law was originally proposed for the diffusion of a vapour into a gas without appreciable displacement of the gas and is as follows:

$$\text{mass flow rate of vapour } \dot{m}_s = -DA\frac{d\rho_s}{dx} \qquad [5.4]$$

In this equation D is a diffusion coefficient, A the area perpendicular to the direction of vapour transfer, ρ_s the density of water vapour and x the distance in the direction of the flow of vapour. This equation can also be applied to the transfer of vapour through a solid structure but the form of the equation is not convenient for Building Services work and it is therefore modified as follows.

Using the characteristic gas equation $p_s V_s = m_s R_s T$ and $\rho_s = m_s/V_s$, Eqn [5.4] becomes:

$$\dot{m}_s = -\frac{DA}{R_s T}\frac{dp_s}{dx}$$

Integrating,

$$\dot{m}_s = -\frac{DA(p_{s2} - p_{s1})}{R_s T(x_2 - x_1)}$$
$$= \frac{DA(p_{s1} - p_{s2})}{R_s T(x_2 - x_1)}$$

This equation can be conveniently written as:

$$\dot{m}_s = \frac{DA}{R_s T}\frac{\Delta p_s}{\Delta x}$$

Substituting μ for $D/R_s T$,

$$\dot{m}_s = \mu A\frac{\Delta p_s}{\Delta x} \qquad [5.5]$$

This form of Fick's law is the basic equation for calculating the mass flow rate of moisture through building structures.

Permeability μ

This is defined as the mass flow rate of vapour per unit area per unit pressure difference through unit thickness of material. In the SI system the units become kg/s m Pa or kg m/N s. This results in very small numbers for both the permeability and mass flow rate, which is why the units of g/h m bar were once widely used.

The above derivation shows that permeability is temperature-dependent. In most Building Services calculations the temperature range is relatively small and the changes in permeability with temperature are negligible. This is not always the case for relative humidity values.

There are two relatively simple experimental means of determining permeability values; the wet cup method and the dry cup method. In the former the specimen seals the top of a cup which is partly filled with water. In the latter the specimen seals a cup containing an absorbent. In either case the cup is then placed in a controlled atmosphere maintained at 50% relative humidity and a constant temperature. The cups are then periodically weighted to determine the loss or gain of water.

Unfortunately the results from wet and dry cup tests can be quite different, particularly for materials such as wood. ASHRAE[5.1] states that wet cup values can be three times higher than dry cup values. The differences are because the relative humidity range is 0–50% for the dry cup test and 50%–100% for the wet cup test. If a large number of tests are carried out over more limited ranges of humidity, the results can be used to draw a graph of permeability against relative humidity at a fixed temperature. Figure 5.2 shows a plot of permeability against relative humidity with the standard dry and wet cup readings being the mean values for the lower and upper range of humidity. Most permeability data currently available to engineers have been obtained from standard wet and dry cup tests and any data source should state which test has been used. All data for any calculation should be from one type of test.

Figure 5.2

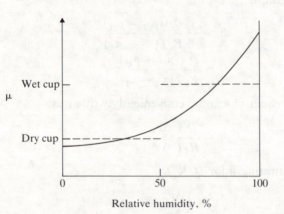

Analogies

Flows of electrical current, heat and water vapour are analogous to each other. They each require a driving force (potential difference, temperature difference and pressure difference) and the quantity transferred in each case is limited by a resistance (electrical, thermal or vapour). Vapour resistances can be treated in a similar way to electrical and thermal resistances. Equation [5.5] can therefore be modified as follows:

$$\dot{m}_s = \mu A \frac{\Delta p_s}{\Delta x}$$

$$\dot{m}_s = A \frac{\Delta p_s}{R_v} \qquad [5.6]$$

where $R_v = \Delta x / \mu$ and is the vapour resistance with units of N s/kg. When a composite structure forms a number of resistances in series and the area is constant throughout, Eqn [5.6] becomes:

$$\dot{m}_s = A \frac{(p_{si} - p_{so})}{R_{TV}} \qquad [5.7]$$

where R_{TV} is the total vapour resistance for the structure (comparable with total thermal resistance), and $(p_{si} - p_{so})$ is the total vapour pressure difference (comparable with total temperature difference). R_{TV} can be found from an equivalent equation to [1.2], namely:

$$R_{TV} = \sum \frac{\Delta x}{\mu} \qquad [5.8]$$

The only difference in form between Eqns [1.2] and [5.8] is the omission of surface vapour resistances in Eqn [5.8]. These are so small that they may be neglected. The share of the total vapour pressure difference between the inside and outside air which occurs across any section of a structure is proportional to the vapour resistance of that section. This leads to an equation which is equivalent to Eqn [5.2]: for any section Δx,

$$\Delta p_x = R_{Vx} \frac{(p_i - p_o)}{R_{TV}} \qquad [5.9]$$

Vapour Resistivity r

This is defined as the vapour resistance of unit thickness of material. Hence:

$$R_V = r \Delta x \qquad [5.10]$$

where Δx is the thickness of the material and r has the units N s/kg m. Resistivity is related to permeability by the following equation:

$$r = \frac{1}{\mu} \qquad [5.11]$$

Vapour resistivity is therefore used as an alternative to permeability and is currently recommended by the CIBSE.[5.2] Using vapour resistivity, Eqn [5.8] becomes:

$$R_{TV} = \sum r \Delta x \qquad [5.12]$$

In the SI system, resistivity values are very large numbers which are usually expressed as giga quantities (10^9), whereas permeability values are very small numbers in the order of 10^{-11}.

Present permeability and vapour resistivity data are rather crude when compared with, say, thermal conductivity data. CIBSE[5.2] (Table A10.4), which uses resistivity rather than permeability, gives minimum and typical values of 25 GN s/kg m and 40 GN s/kg m for brickwork and 21 GN s/kg m and 80 GN s/kg m for foamed phenolic. If maximum values had been included the spread would have been much greater. Clearly there is a need for more detailed data before too much reliance can be placed on vapour transfer calculations.

Diffusion Resistance Factor

This is defined as the ratio of the permeability of still air to the permeability of a material.

$$\text{Diffusion resistance factor} = \frac{\mu_a}{\mu} \tag{5.13}$$

The permeability of air, μ_a, is quoted as approximately 200×10^{-12} kg/s m Pa in the CIBSE guide. Resistance factors can be used as an alternative to permeability or resistivity data.

Permeance P

The permeance or permeance coefficient is defined as the mass flow rate of vapour per unit area per unit of pressure difference. The equation corresponding to Eqn [5.6] is:

$$\dot{m}_s = P A \Delta p_s \tag{5.14}$$

Clearly $P = 1/R_v$ (the reciprocal of the vapour resistance) and the units are kg/N s. When applied to a complete structure, permeance is the equivalent of thermal transmittance in heat transfer. The term is more commonly used for vapour barriers, which can have a dramatic effect on vapour transfer but a negligible effect on heat transfer.

Vapour Conductance

This is basically just another name for permeance and is normally applied only to a section of a structure. It is the equivalent of thermal conductance in heat transfer.

Example 5.2
(a) State the conditions necessary for the occurrence of:
(i) surface condensation;
(ii) interstitial condensation.
(b) A wall has the construction detailed in Fig 5.3 and the table of data below. Determine whether or not there is a risk of interstitial condensation, and if so within which section it could occur.

Figure 5.3   Example 5.2

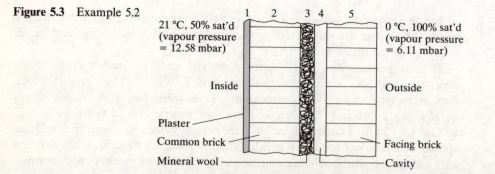

21 °C, 50% sat'd
(vapour pressure
= 12.58 mbar)

0 °C, 100% sat'd
(vapour pressure
= 6.11 mbar)

Inside

Outside

Plaster

Common brick

Mineral wool

Facing brick

Cavity

(c)   Assuming the room temperature remains constant, at what level of internal percentage saturation would condensation on the wall surface become a problem?

Section	Thickness (mm)	Thermal conductivity (W/m K)	Thermal conductance (W/m² K)	Vapour permeability (10^{-11} kg m/N s)
si			8.3	
1	15	0.16		2.0
2	100	0.71		1.5
3	25	0.04		18.0
4	25		5.5	19.2
5	100	0.96		0.6
so			16.7	

(Engineering Council)

Solution
(a)   (i)   Surface condensation will occur whenever the vapour pressure in the air is above the saturation vapour pressure corresponding to the surface temperature.

An alternative statement is that surface condensation will occur whenever the surface temperature is below the dew point of the air.
(ii)   Interstitial condensation will occur whenever the predicted vapour pressure is above the saturation vapour pressure at any point in a structure.

An alternative statement is that interstitial condensation will occur whenever the temperature is below the predicted vapour dew point at any point in a structure.

The two different approaches to answering part (a) lead to three different methods of solving part (b).

Method 1:
The first step is to calculate the thermal resistances of each section and of the whole structure. It should be noted that the term 'thermal conductance' is synonymous with surface heat transfer coefficient for the outer surfaces and the cavity conductance may be treated in the same way as a heat transfer coefficient. With a large number of resistances there is less likelihood of a mistake if a tabular procedure is used as follows.

Section	Thermal resistance (m² K/W)
si	$1/\alpha_i = 1/8.3$ $= 0.1205$
1	$\Delta x/k = 0.015/0.16 = 0.938$
2	$\Delta x/k = 0.100/0.71 = 0.1408$
3	$\Delta x/k = 0.025/0.04 = 0.625$
4	$1/C = 1/5.5$ $= 0.1818$
5	$\Delta x/k = 0.1/0.96$ $= 0.1042$
so	$1/\alpha_o = 1/16.7$ $= 0.0599$
Total	R_{TT} $= 1.3259$

A similar procedure can now be followed to determine the vapour resistances using Eqn [5.8].

Section	Vapour resistance $(10^{11}\,kg/Ns)$	
1	$\Delta x/\mu = 0.015/2$	$= 0.0075$
2	$\Delta x/\mu = 0.1/1.5$	$= 0.0667$
3	$\Delta x/\mu = 0.025/18$	$= 0.0014$
4	$\Delta x/\mu = 0.025/19.2$	$= 0.0013$
5	$\Delta x/\mu = 0.1/0.6$	$= 0.1667$
Total	R_{TV}	$= 0.2436$

Equation [5.2] can be used to calculate the temperature drop across each thermal resistance and Eqn [5.9] to calculate the predicted vapour pressure drop across each vapour resistance as follows:

$$\Delta t_x = R_{Tx}\frac{(t_i - t_o)}{R_{TT}} = R_{Tx}\frac{(21 - 0)}{1.3259} = R_{Tx}15.84\ K$$

$$\Delta p_x = R_{Vx}\frac{(p_i - p_o)}{R_{TV}} = R_{Vx}\frac{(12.58 - 6.11)}{0.2436} = R_{Vx}26.57\ mbar$$

A tabular procedure, as shown in the table below, can now be used to calculate the temperature, saturation vapour pressure p_g (by interpolation from property tables such as Ref. 5.3) and predicted vapour pressure p_s, for each plane. The predicted vapour pressures and vapour pressure drops are the values for vapour flow continuity, i.e. no condensation. Figure 5.4 shows how planes have been identified by letters to avoid confusion with sections.

Plane	Temp (°C)		p_g (mbar)	p_s (mbar)
i		21.0	24.86	12.58
a 21 −	$0.1205 \times 15.84 =$	19.09	22.09	12.58
b 19.09 −	$0.0938 \times 15.84 =$	17.61	20.24	$12.56 - 0.0075 \times 26.57 = 12.38$
c 17.61 −	$0.1408 \times 15.84 =$	15.38	17.47	$12.38 - 0.0667 \times 26.57 = 10.61$
d 15.38 −	$0.625 \times 15.84 =$	5.48	9.02	$10.61 - 0.0014 \times 26.57 = 10.57$
e 5.48 −	$0.1818 \times 15.84 =$	2.6	7.37	$10.57 - 0.0013 \times 26.57 = 10.54$
				Check:
f 2.6 −	$0.1042 \times 15.84 =$	0.95	6.55	$10.54 - 0.1667 \times 26.57 = 6.11$
Check:				
o 0.95 −	$0.0599 \times 15.84 =$	0	6.11	6.11

The predicted vapour pressure is above the saturation vapour pressure at planes d and e; therefore condensation must be taking place. The condensation will be at plane e, the outer face of the cavity (section 4) because this is where p_s exceeds p_g by the greatest amount.

A graphical presentation is useful to understand why condensation occurs at e and not d. The lines of saturation vapour pressure and predicted vapour

Figure 5.4 Example 5.2: vapour pressure curves

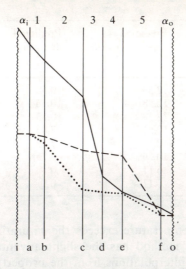

——— Saturation vapour pressure
– – – Predicted vapour pressure
········· Actual vapour pressure

pressure have been drawn on Fig. 5.4 and can be seen to cross. This cannot be correct and the predicted vapour pressure line is the line which must be wrong. The actual vapour pressure line can touch but never cross the saturation line. The point where it will touch the saturation line is where there was the greatest overlap between the predicted and saturation pressure lines.

If the actual vapour pressure line had been required, the values at planes i, a, f and o are the same as for the predicted line and the value at e is the saturation value of 7.37 mbar. The values at points b, c, and d can be calculated by considering sections 1, 2, 3 and 4 as a sub-structure with a total vapour resistance R_{TV} of $(0.0075 + 0.0667 + 0.0014 + 0.0013) \times 10^{11} = 0.0769 \times 10^{11}$ kg/N s and a total vapour pressure difference of $(12.58 - 7.37) = 5.21$ mbar. Equation [5.9] can now be used as follows:

$$p_{sb} = p_{sa} - R_{V1}\frac{(p_{sa} - p_{ge})}{R_{TV}} = 12.58 - 0.0075 \times \frac{5.21}{0.0769} = 12.07 \text{ mbar}$$

$$p_{sc} = p_{sb} - R_{V2}\frac{(p_{sa} - p_{ge})}{R_{TV}} = 12.07 - 0.0667 \times 67.75 = 7.55 \text{ mbar}$$

$$p_{sd} = p_{sc} - R_{V3}\frac{(p_{sa} - p_{ge})}{R_{TV}} = 7.55 - 0.0014 \times 67.75 = 7.46 \text{ mbar}$$

Check: $p_{se} = p_{sd} - R_{V4}\dfrac{(p_{sa} - p_{ge})}{R_{TV}} = 7.46 - 0.0013 \times 67.75 = 7.37 \text{ mbar}$

The vertical scale of Fig. 5.4 is not consistent to enable the three vapour pressure lines to be clearly distinguished.

Method 2:
The temperature and predicted vapour pressure profiles are calculated as in Method 1 but there is no need to look up corresponding saturation temperatures. Instead the dew point temperatures corresponding to the predicted vapour pressure values are found in property tables. This results in the table below.

With this method the temperature and predicted dew point are compared and at planes d and e the temperature is below that of the predicted dew point, showing that condensation is taking place. The condensation is at e, where the

Plane	Temperature (°C)	Predicted vapour pressure (mbar)	Predicted dew point (°C)
i	20	12.58	10.36
a	19.09	12.58	10.36
b	17.61	12.38	10.13
c	15.38	10.61	7.85
d	5.48	10.57	7.79
e	2.6	10.54	7.75
f	0.95	6.11	0
o	0	6.11	0

predicted dew point temperature exceeds the saturation temperature by the greatest amount. This method has the slight advantage of requiring fewer (in this case one less) interpolations from the property tables. However, this advantage is lost if a further calculation is required to determine the rate of condensation, the maximum room humidity for no condensation, or the minimum permeance of a vapour barrier to prevent condensation.

Method 3:

This method, usually referred to as the Glazer method after its originator, uses graphs to eliminate some of the calculations. Another example using this method can be found in Ref. 5.4.

The thermal and vapour resistances of each section are calculated as for the previous methods. The cumulative thermal and vapour resistances through the structure are then calculated as shown in the table below.

Section	Thermal resistance, R_T (m^2 K/W)		Vapour resistance, R_V (10^{11} kg/N s)	
	Section	Cumulative	Section	Cumulative
α_i	0.1205	0.1205	0	0
1	0.0938	0.2143	0.0075	0.0075
2	0.1408	0.3551	0.0667	0.0742
3	0.6250	0.9801	0.0014	0.0756
4	0.1818	1.1619	0.0013	0.0769
5	0.1042	1.2661	0.1667	0.2436
α_o	0.0599	1.326	0	0.2436
Total	1.3260		0.2436	

The cumulative thermal resistances are then used to plot temperature against thermal resistance through the structure. Since temperature changes are proportional to thermal resistances this only requires the two end points and results in the straight line graph shown in Fig. 5.5. The plane between each section is then located by using the cumulative thermal resistance up to that plane. For example, plane d is where the cumulative thermal resistance is 0.9801 m^2 K/W. The temperature at each plane can then be read off from the point where the plane intersects the temperature profile. For example, at plane (d) the temperature is 5.4 °C.

Figure 5.5 Example
5.2: temperature vs
thermal resistance

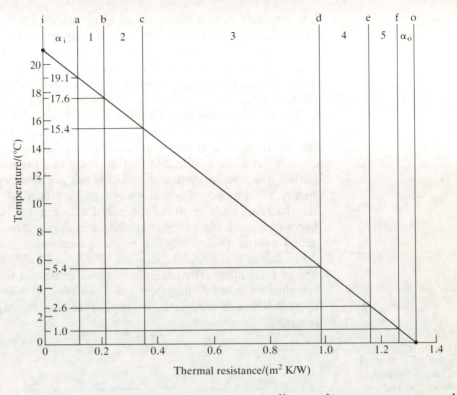

The saturation vapour pressure corresponding to the temperature at each
plane is then interpolated from property tables as for Method 1. Graphs are
plotted of vapour pressures against vapour resistance as shown in Fig. 5.6. Each
plane in this graph is identified by the cumulative vapour resistance. For example,
up to plane e there is a cumulative vapour resistance of 0.0769×10^{11} kg/N s.

Figure 5.6 Example
5.2: vapour pressure vs
vapour resistance

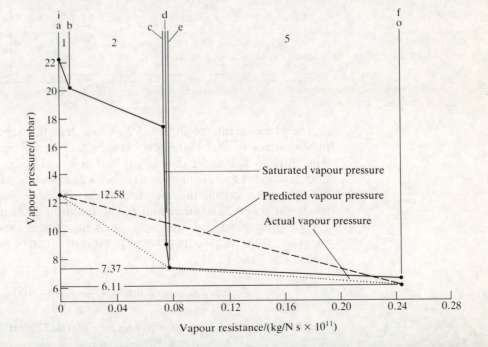

The saturated vapour pressure values are then used to draw a saturated pressure line. Since vapour pressure changes are proportional to vapour resistances, the predicted vapour pressure follows a straight line from the internal and external values of vapour pressure (12.58 mbar and 6.11 mbar respectively). As with Method 1 the overlap between the two lines shows that condensation will take place.

The actual vapour pressure line is found by drawing tangents from the actual vapour pressure at the two end planes (12.58 mbar and 6.11 mbar) just to touch the saturated vapour pressure line. Vapour barriers are covered later in this chapter but if it is required to find the minimum vapour resistance of a vapour barrier which will prevent condensation this can also be carried out graphically. Figure 5.7 is a plot of actual vapour pressure against vapour resistance with the resistance scale extended compared with Fig. 5.6. A line from the external vapour pressure of 6.11 mbar at plane o is drawn through the point where the actual vapour pressure line touches the saturated vapour pressure line, i.e. 7.37 mbar at plane e, and extended until it meets the internal vapour pressure line of 12.58 mbar. The minimum vapour resistance for a vapour barrier R_{VB} can then be read off the graph. An alternative to drawing Fig. 5.7 to scale is to sketch it and calculate the vapour barrier resistance R_{VB} using similar triangles.

Figure 5.7 Example 5.2: actual vapour pressure vs vapour resistance

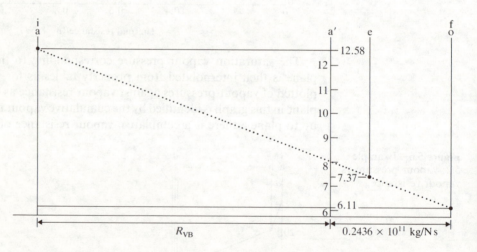

The planes identified in Fig. 5.7 assume that the vapour barrier is on the inside surface so that plane a is between the vapour barrier and the plaster. Alternative positions for the vapour barrier would be between the plaster and brick (sections 1 and 2) or between the brick and insulation (sections 2 and 3).

(c) From the calculations in part (b) the saturation vapour pressure at the inside surface is 22.09 mbar and of the room air 24.86 mbar. The saturation vapour pressure at the surface becomes the critical vapour pressure for the room air. The critical room percentage saturation can be found as follows, using Eqns [1.22] and [1.24]:

$$\text{Percentage saturation} = \frac{p_s(p - p_g)}{p_g(p - p_s)} \times 100$$

Assuming standard atmospheric pressure of 1.01325 bar.

$$\text{Percentage saturation} = \frac{22.09\,(1.01325 - 0.02486)}{22.86\,(1.02325 - 0.02209)} \times 100$$
$$= 88.6\%$$

Surface condensation will occur only if the room percentage saturation rises above this value. It should also be noted that in practice by this point there would have been so much interstitial condensation that the thermal conductivity values and hence the surface temperature would have changed. This would result in surface condensation at a lower room percentage saturation.

Example 5.3

An existing 343 mm thick solid brick wall is to be internally lined with 50 mm of insulating board. The wall is to be exposed to indoor conditions of 19 °C and 50% saturation and outdoor conditions of 2 °C and 92% saturation.

(a) Using the data given below and Sections C1 and C2 of the CIBSE Guide, determine the position in the wall at which interstitial condensation will occur.

(b) Determine the rate at which condensation will occur at the interface between the insulating board and the brick.

(c) List the principal means of preventing interstitial and surface condensation in a wall.

Data

Thermal resistance of inside surface	0.12 m² K/W
Thermal conductivity of insulating board	0.03 W/m K
Thermal conductivity of brick	0.75 W/m K
Thermal conductivity of outside surface	0.06 m² K/W
Vapour resistivity of brick	50 GN s/kg m
Vapour resistivity of insulating board	200 GN s/kg m
	(CIBSE)

Solution

(a) CIBSE tables[5.5] list properties of humid air from which the vapour pressures for the inside and outside air can be read off directly as 1.11 and 0.6494 kPa respectively. The most convenient unit of pressure for this type of calculation is the millibar so the values become 11.1 mbar and 6.494 mbar. CIBSE tables[5.6] list properties of water at saturation which can be found in many other property tables (e.g. Ref. 5.3).

With only four thermal and two vapour resistances, a tabular calculation is unnecessary. Using Eqns [5.1] and [5.12] the thermal and vapour resistances can be calculated as follows:

$$R_{TT} = R_{si} + \sum \frac{\Delta x}{k} + R_{so}$$
$$= 0.12 + \frac{0.05}{0.03} + \frac{0.343}{0.75} + 0.06$$
$$= 0.12 + 1.667 + 0.457 + 0.06$$
$$= 2.304 \text{ m}^2 \text{ K/W}$$
$$R_{TV} = \sum r\Delta x$$
$$= 200 \times 0.05 + 50 \times 0.343$$
$$= 10 + 17.15$$
$$= 27.15 \text{ GN s/kg}$$

Equations [5.2] and [5.9] can now be used to calculate the temperature and pressure drops across each section as follows:

$$\Delta t_x = R_{Tx} \frac{(t_i - t_o)}{R_{TT}} = R_{Tx} \frac{(19 - 2)}{2.304} = R_{Tx}\, 7.38 \text{ K}$$

$$\Delta p_x = R_{Vx} \frac{(p_i - p_o)}{R_{TV}} = R_{Vx} \frac{(11.1 - 6.494)}{27.15} = R_{Vx}\, 0.17 \text{ mbar}$$

The saturation vapour pressure p_g, and the predicted vapour pressure, p_s, can now be calculated as shown in the table below.

Plane	Temp (°C)		p_g (mbar)		p_s (mbar)
i		19	21.96		11.1
a	$19 \;\; - 0.12 \;\times 7.38 =$ 18.11		20.75		11.1
b	$18.11 - 1.667 \times 7.38 =$ 5.81		9.23	$11.1 - 10 \times 0.17 \quad= 9.4$	9.4
				Check:	
c	$5.81 - 0.457 \times 7.38 =$ 2.44		7.28	$9.4 - 17.15 \times 0.17 = 6.5$	6.5
	Check:				
d	$2.44 - 0.06 \;\; \times 7.38 =$ 2.0		7.05		6.5

The only plane where p_s exceeds p_g is b, which is where condensation will occur (the second part of the question confirms this).

(b)   Figure 5.8 shows plots of the saturation line, the predicted vapour pressure line and the actual vapour pressure line. The latter just touches the saturation line at plane b. The mass flow rate of condensate is the difference between the vapour flow rate into plane b, through section 1, and the vapour flow rate from plane b, through section 2. The mass flow rates of vapour can be found using Eqn [5.6] as follows:

$$\dot{m}_{cond} = \dot{m}_{s1} - \dot{m}_{s2}$$
$$= \frac{\Delta p_{s1}}{R_{V1}} - \frac{\Delta p_{s2}}{R_{V2}}$$
$$= \left\{ \frac{(11.1 - 9.23)}{10} - \frac{(9.23 - 6.5)}{17.15} \right\} \frac{10^2}{10^9}$$
$$= 2.78 \times 10^{-9} \text{ kg/s m}^2$$

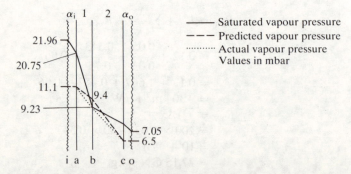

Figure 5.8   Example 5.3

(c) Surface condensation can be prevented by increasing the ventilation rate, which will lower the room humidity, or by improving the insulation, which will increase the surface temperature. The latter method on its own can lead to interstitial condensation, which part (b) demonstrates. Increasing the ventilation rate can also prevent interstitial condensation. The other measure is to install a vapour barrier as close to the source of moisture as possible.

Vapour Barriers

A vapour barrier is a thin layer of material which has a high vapour resistance (low permeance). It will normally have a negligible effect on the thermal resistance of the structure.

Vapour barriers may be classified as follows:

(a) structural, e.g. aluminium, reinforced plastic;
(b) membrane, e.g. aluminium foil, treated papers, polythene sheets;
(c) coatings, e.g. asphalt, resin or polymer-based paints.

Doubling the thickness of a vapour barrier does not necessarily double the vapour resistance, so it cannot be treated as a homogeneous material. For example, according to CIBSE,[5.2] doubling the thickness of a polythene sheet only increases the resistance by 60%, whereas ASHRAE[5.1] quotes an example where doubling the thickness of a mastic barrier increased the resistance by 2800%. Vapour barriers are therefore designated in terms of a vapour resistance or permeance value for a specified barrier.

One problem with vapour barriers, particularly the membrane type, is that laboratory data can never be reproduced in a practical building because of the effect of joints and fastenings. A carefully installed barrier may have a vapour resistance of half the laboratory figure and a badly installed one very much less. Vapour barriers must always be placed as close to the source of moisture as possible. A wrongly placed barrier will increase rather than decrease condensation.

Figure 5.9 shows the effect of correctly and incorrectly placed vapour barriers. Graphs (a), (b) and (c) show structures with the vapour barrier in a correct position and graphs (d), (e) and (f) the same structures with the vapour barrier in an incorrect position. In case (c) the vapour barrier would have also been satisfactory at the interface between the brick and insulation. With the vapour barrier in the correct position the predicted and actual vapour pressure lines follow the same path and never cross the saturation line.

For the brick structure with no insulation and the vapour barrier on the outside face, case (d), condensation will take place at the interface between the brick and vapour barrier. Condensation will take place at the same position when the insulation is on the outside of the brickwork, i.e. in case (f).

When the insulation is on the inside of the brickwork, case (e), some condensation will take place at the interface between the insulation and brickwork, and vapour will continue to condense throughout the brickwork. This is shown by the actual vapour pressure line coinciding with the saturation line through this section.

In each case in Fig. 5.9 it has been assumed that the vapour barrier has a sufficiently high vapour resistance for the other vapour resistances in the

Figure 5.9 Vapour barriers

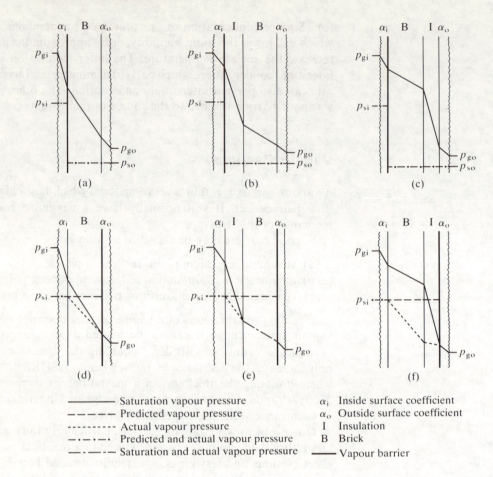

	Saturation vapour pressure	α_i	Inside surface coefficient
-----	Predicted vapour pressure	α_o	Outside surface coefficient
---------	Actual vapour pressure	I	Insulation
-·-·-·	Predicted and actual vapour pressure	B	Brick
-·--·--	Saturation and actual vapour pressure	———	Vapour barrier

structure to be negligible in comparison. A partial vapour barrier such as a coat of paint would not have such a dramatic effect. It should also be noted that the vapour pressure of the outside air has been omitted from the cases where condensation occurs. This is because it is irrelevant to interstitial condensation in these examples. A fully effective vapour barrier divides a structure into two sections as far as vapour transfer is concerned, and each side of the vapour barrier must be investigated as a separate problem. If the vapour pressure of the outside air was above the saturation vapour pressure at the external surface, then condensation would take place at both sides of the vapour barrier.

The above analysis has been for winter conditions in a temperate climate. In tropical climates and in air-conditioned buildings in summer in temperate climates, the water vapour pressure is higher outside the building than inside. The direction of vapour flow is therefore from outside to inside and vapour barriers should be on the outside of the structure. In temperate climates the winter case is normally the most important one.

Example 5.4

Re-calculate the mass flow rate of condensate in Example 5.3 if an effective vapour barrier is applied to the outside surface. (It does happen!)

Solution
The vapour pressure lines will now follow the paths shown in Fig. 5.9(e). The vapour pressure at the inside surface is 11.1 mbar and at the interface between the insulation and brick 9.23 mbar. Using Eqn [5.6], the mass flow rate of vapour through the insulation is:

$$\dot{m}_s = \frac{\Delta p_s}{R_V}$$
$$= \frac{(11.1 - 9.23)}{10 \times 10^9} \times 10^2$$
$$= 18.7 \times 10^{-9} \text{ kg/s m}^2$$

With an effective vapour barrier all this vapour will condense. The amount of condensation is now 6.8 times higher than without the vapour barrier. Some of this condensation will take place at the interface and the rest within the brickwork. (Note that the rate of condensation was unaffected by the vapour pressure at the other side of the vapour barrier.)

Example 5.5
A wall of a building consists of two layers of 115 mm brick, separated by an air gap, with a 19 mm layer of plaster on the inside surface. The surface resistances and the air gap offer a negligible resistance to vapour flow. Using the data below, tabulate and sketch the predicted vapour pressure and saturation vapour pressure lines to determine whether condensation will occur, and if so, where.

It is proposed to prevent condensation by coating the surface of the plaster with a low-permeability paint. Assuming that the paint offers a negligible resistance to heat transfer, calculate its minimum vapour resistance in N s/kg to prevent condensation.

Data
Plaster: $k = 0.432$ W/m K $\mu = 2 \times 10^{-11}$ kg m/N s
Brick: $k = 1.3$ W/m K $\mu = 4 \times 10^{-11}$ kg m/N s
Air gap thermal resistance, 0.176 m^2 K/W
$\alpha_i = 8$ W/m^2 K $\alpha_o = 25$ W/m^2 K
Room state: 23 °C and 50% relative humidity.
Outside air 0.01 °C saturated.

(Wolverhampton Polytechnic)

Solution
From property tables the saturation vapour pressure is 28.08 mbar at 23 °C and 6.11 mbar at 0.01 °C. The vapour pressure of the inside air can be found from Eqn [1.22] as follows:

$$p_s = \phi p_g = 0.5 \times 28.08 = 14.04 \text{ mbar}$$

The sections and planes of the structure are identified in Fig. 5.10 and the thermal resistances can be found using Eqn [5.1] as shown in the table overleaf. For the vapour resistances,

$$R_{TV} = \sum \frac{\Delta x}{\mu} = \left(\frac{0.019}{2} + \frac{0.115}{4} + \frac{0.115}{4} \right) \times 10^{11}$$
$$= (0.95 + 2.875 + 2.875) \times 10^9$$
$$= 6.7 \times 10^9 \text{ N s/kg}$$

MOISTURE TRANSFER

Section	Thermal resistance $(m^2\,K/W)$	
si	$1/\alpha_i = 1/8$	$= 0.1250$
1	$\Delta x/k = 0.019/0.432$	$= 0.0440$
2	$\Delta x/k = 0.115/1.3$	$= 0.0885$
3		0.1760
5	$\Delta x/k = 0.115/1.3$	$= 0.0885$
so	$1/\alpha_o = 1/25$	$= 0.0400$
Total	R_{TT}	$= 0.5620$

Figure 5.10 Example 5.5

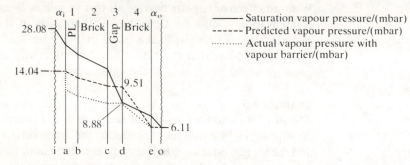

Equation [5.2] can be used to calculate the temperature drop across each thermal resistance and Eqn [5.9] to calculate the predicted vapour pressure drop across each vapour resistance as follows:

$$\Delta t_x = R_{Tx}\frac{(t_i - t_o)}{R_{TT}} = R_{Tx}\frac{(23 - 0)}{0.562} = R_{Tx}40.93\ K$$

$$\Delta p_x = R_{Vx}\frac{(p_i - p_o)}{R_{TV}} = R_{Vx}\frac{(14.04 - 6.11)}{6.7} = R_{Vx}1.184\ mbar$$

A tabular procedure, as shown in the table below, can now be used to calculate the temperatures, saturation vapour pressures and predicted vapour pressures.

Plane		Temp (°C)	p_g (mbar)		p_s (mbar)
i		23.0	28.08		14.04
a	$23 - 0.125 \times 40.93 =$ 17.88	17.88	20.45		14.04
b	$17.88 - 0.044 \times 40.93 =$ 16.08	16.08	18.26	$14.04 - 0.95 \times 1.184 =$ 12.92	12.92
c	$16.08 - 0.0885 \times 40.93 =$ 12.46	12.46	14.45	$12.92 - 2.875 \times 1.184 =$ 9.51	9.51
d	$12.46 - 0.176 \times 40.93 =$ 5.26	5.26	8.88	Check:	9.51
e	$5.26 - 0.0885 \times 40.93 =$ 1.63	1.63	6.87	$9.51 - 2.875 \times 1.184 =$ 6.11	6.11
o	Check: $1.63 - 0.04 \times 40.93 =$ 0	0	6.11		6.11

The predicted vapour pressure is above the saturation vapour pressure at d; therefore condensation must be taking place at this plane.

To eliminate condensation the actual vapour pressure at d must not be greater than 8.88 mbar. The drop in vapour pressure across the outer layer of brick (section 4) is therefore $(8.88 - 6.11) = 2.77$ mbar. Across the plaster, the inner layer of brick (sections 1 and 2) and the vapour barrier, the pressure drop must be $(14.04 - 8.88) = 5.16$ mbar. Knowing that vapour pressure drops are proportional to vapour resistances, the minimum resistance of the vapour barrier R_{VB} can be found as follows.

$$\frac{R_{V1} + R_{V2} + R_{VB}}{R_{V3}} = \frac{\Delta p_{i-c}}{\Delta p_{c-d}}$$

$$\frac{0.95 + 2.875 + R_{VB}}{2.875} = \frac{5.16}{2.77}$$

Hence $$R_{VB} = 1.53 \times 10^9 \text{ N s/kg}$$

This is quite a modest value of vapour resistance which can easily be achieved with paint. The actual vapour pressure line with the vapour barrier installed is shown in Fig. 5.10.

Limitations

Comments have already been made on the unreliability of available permeability data and on the effect of condensation on thermal conductivity values. There are two further limitations to calculations on vapour transfer through structures. The first is that, although Chapter 3 showed that heat transfer is governed by the environmental temperatures of the air on either side of a structure, in this chapter the difference between environmental and air temperature has been ignored. This difference is fortunately quite small for most applications. If it is not, it should be taken into account. The second limitation is that when condensation occurred it was assumed that the temperature and hence the saturation vapour pressure profile remained constant. In practice, once condensation starts the temperature must rise due to the release of enthalpy of condensation. For condensation within walls this effect is very small and if it had been taken into account in Example 5.2 it would have changed the saturation temperature at point e by less than 0.1 °C. The only situation where this effect is appreciable is where there is a very low resistance to vapour flow, such as condensation on a window. This is demonstrated in the numerical part of the following example.

Example 5.6
(a) Problems due to excessive moisture have occurred fairly frequently in recent years in certain modern dwellings. Suggest the changes in the life style of the occupant and other causes that have led to these problems.
(b) How can such problems be eliminated by good design?
(c) How can they be reduced or eliminated by appropriate behaviour of the occupant?
(d) Estimate the rate of condensation (in kg/h) upon a kitchen window area of 3 m², given the following data: air temperature in kitchen, 18 °C; relative humidity, 80%; outside air temperature, 0 °C; enthalpy of condensation, 2.48×10^6 J/kg; inside and outside thermal film conductances, 8 and 16 W/m² K respectively; inside vapour film conductance (transport of water vapour per unit pressure difference), 18×10^{-9} kg/m² s Pa.

Also calculate the glass temperatures in the absence and presence of condensation and the saturated vapour pressure of the condensate.

The calculation can be done on the psychrometric chart supplied.

(University of Liverpool)

Solution to part (d)

Thermal and vapour film conductances are alternative terms for heat transfer coefficient and permeance respectively. The chart provided was not the CIBSE chart. It used relative humidity rather than percentage saturation and also incorporated lines of constant vapour pressure. This enabled vapour pressures to be read off directly without calculation. The problem can also be solved using data from the property tables.

Making the very reasonable assumption that the temperature drop across the glass is negligible, the system is as shown in Fig. 5.11. With no condensation the temperature of the glass can be found using Eqns [1.2] and [5.3] as follows:

$$\frac{1}{U} = \frac{1}{\alpha_i} + \frac{1}{\alpha_o} = \frac{1}{8} + \frac{1}{16}$$

Hence $\quad U = 5.33 \text{ W/m}^2 \text{ K}$

$$\therefore \quad t_g = t_i - \frac{U}{\alpha_i}(t_i - t_o) = 18 - \frac{5.33}{8}(18 - 0) = 6 \,^\circ\text{C}$$

Figure 5.11 Example 5.6

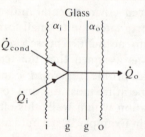

From property tables the saturation vapour pressure at 18 °C is 0.020 63 bar. The vapour pressure of the inside air can now be found as follows:

$$p_{si} = \phi p_{gi} = 0.8 \times 0.020\,63 = 0.0165 \text{ bar}$$

For equilibrium vapour transfer the glass provides a perfect vapour barrier, and all the vapour passing through the vapour film will condense on the inside surface of the glass. The equation $Q = \alpha A \Delta t$ and Eqn [5.14] can now be used in an energy balance as follows:

$$\dot{Q}_{cond} + \dot{Q}_i = \dot{Q}_o$$
$$\dot{m}_s h_{fg} + \alpha_i A(t_i - t_g) = \alpha_o A(t_g - t_o)$$
$$PA(p_{si} - p_g)h_{fg} + \alpha_i A(t_i - t_g) = \alpha_o A(t_g - t_o)$$
$$18 \times 10^{-9} \times 3(0.0165 - p_g)10^5 \times 2.48 \times 10^6 + 8 \times 3(18 - t_g)$$
$$= 16 \times 3(t_g - 0)$$
$$221 - 13392p_g + 432 - 24t_g = 48t_g$$
$$9.07 - 186p_g = t_g$$

Since p_g depends on t_g, this equation can be solved by trial and error. The temperature t_g must be above the temperature of 6 °C when no condensation occurs.

Try $t_g = 8$ °C: $p_g = 0.010\,72$ bar   LHS = 7.08 °C
Try $t_g = 7$ °C: $p_g = 0.010\,01$ bar   LHS = 7.2 °C
Try $t_g = 7.2$ °C: $p_g = 0.010\,15$ bar   LHS = 7.18 °C

The last line is sufficiently accurate. Equation [5.14] can now be used to calculate the mass flow rate of vapour which all condenses.

$$\begin{aligned}
\dot{m}_s &= PA\Delta p_s \\
&= 18 \times 10^{-9} \times 3 \times (0.0165 - 0.010\,15) \times 10^5 \\
&= 3.429 \times 10^{-5} \text{ kg/s} \\
&= 0.123 \text{ kg/h}
\end{aligned}$$

In this example the increase in glass temperature when condensation occurred was 1.2 K, which corresponds to an increase of saturation vapour pressure of 0.000\,804 bar.

5.3 TRANSIENT MOISTURE TRANSFER IN ROOMS

The mass flow rates of water vapour through structures are normally quite small compared with the mass flow rates of vapour between and within spaces. The specific humidity in non-air-conditioned rooms will vary depending on the ventilation rate, the occupancy and the specific humidity of the outside air. To investigate the possibility of temporary condensation on surfaces, it is necessary to determine how the room air humidity changes with time.

Nomenclature

τ	time
V_R	room volume
$\dot{V}_R$	volume flow rate of extract air
m_R	mass of air in the room
$\dot{m}$	mass flow rate of air to and from the room
n	air changes per unit time
$\dot{m}_{sp}$	moisture production rate within the room
ω_τ	specific humidity of the room air at any time τ
ω_a	specific humidity of the ambient (outside) air
ω_o	specific humidity of the room air at time = 0

During an interval of time $d\tau$:

Mass of moisture produced $\quad\quad\quad = \dot{m}_{sp}\,d\tau$
Mass of moisture gained from outside $= \dot{m}\omega_a\,d\tau$
Mass of moisture lost to outside $\quad\quad = \dot{m}\omega_\tau\,d\tau$

Net gain of moisture $= (\dot{m}_{sp} + \dot{m}\omega_a - \dot{m}\omega_\tau)\,d\tau$

Change in ω_τ, $\quad d\omega_\tau = \dfrac{1}{m_R}(\dot{m}_{sp} + \dot{m}\omega_a - \dot{m}\omega_\tau)\,d\tau$

Rearranging and multiplying both sides by $(-\dot{m})$ gives:

$$\frac{-\dot{m}\,d\omega_\tau}{(\dot{m}_{sp} + \dot{m}\omega_a - \dot{m}\omega_\tau)} = \frac{-\dot{m}\,d\tau}{m_R}$$

On integration this becomes:

$$\ln(\dot{m}_{sp} + \dot{m}\omega_a - \dot{m}\omega_\tau) = -\frac{\dot{m}\tau}{m_R} + C$$

and in exponential form,

$$(\dot{m}_{sp} + \dot{m}\omega_a - \dot{m}\omega_\tau) = e^{\frac{-\dot{m}\tau}{m_R} + C}$$

To evaluate C,

At $\tau = 0$, $\qquad \omega_\tau = \omega_o$,

$$e^C = (\dot{m}_{sp} + \dot{m}\omega_a - \dot{m}\omega_o)$$
$$\therefore \qquad (\dot{m}_{sp} + \dot{m}\omega_a - \dot{m}\omega_\tau) = (\dot{m}_{sp} + \dot{m}\omega_a - \dot{m}\omega_o)e^{-\dot{m}\tau/m_R}$$

Rearranging,

$$\dot{m}\omega_\tau = (1 - e^{-\dot{m}\tau/m_R})\dot{m}_{sp} + (1 - e^{-\dot{m}\tau/m_R})\dot{m}\omega_a + \dot{m}\omega_o e^{-\dot{m}\tau/m_R}$$

Dividing by $\dot{m}$,

$$\omega_\tau = (1 - e^{-\dot{m}\tau/m_R})\frac{\dot{m}_{sp}}{\dot{m}} + (1 - e^{-\dot{m}\tau/m_R})\omega_a + \omega_o e^{-\dot{m}\tau/m_R}$$

Subtracting ω_o from both sides of the equation,

$$\omega_\tau - \omega_o = (1 - e^{-\dot{m}\tau/m_R})\frac{\dot{m}_{sp}}{\dot{m}} + (1 - e^{-\dot{m}\tau/m_R})\omega_a$$
$$+ (e^{-\dot{m}\tau/m_R} - 1)\omega_o$$

Rearranging,

$$\omega_\tau - \omega_o = \left(\frac{\dot{m}_{sp}}{\dot{m}} + \omega_a - \omega_o\right)(1 - e^{-\dot{m}\tau/m_R}) \qquad [5.15]$$

From the continuity equation, $\dot{m} = \rho_a \dot{V}_a = \rho_R \dot{V}_R$. Consider the volume flow rate from the room, which would be the volume flow rate handled by an extract fan. Eliminating $\dot{m}$, Eqn [5.15] becomes:

$$\omega_\tau - \omega_o = \left(\frac{\dot{m}_{sp}}{\rho_R \dot{V}_R} + \omega_a - \omega_o\right)(1 - e^{-\dot{V}_R\tau/V_R}) \qquad [5.16]$$

The term $\dot{V}_R/V_R$ is the air change rate, n,

i.e. $\qquad \omega_\tau - \omega_o = \left(\dfrac{\dot{m}_{sp}}{\rho_R \dot{V}_R} + \omega_a - \omega_o\right)(1 - e^{-n\tau}) \qquad [5.17]$

ρ_R has a typical value of $1.2\,kg/m^3$.

These equations can be used to predict the specific humidity after any interval of time. This can then be used to determine when surface condensation will occur, or the minimum ventilation rate to prevent condensation.

Example 5.7

A room with a volume of $100\,m^3$ is maintained at a temperature of $21\,°C$.

Initially the room has been unoccupied for some time and the specific humidity is the same as that of the outside air. Twenty people then enter the room and start a meeting. Use the data below to calculate how long after the meeting starts condensation will begin to form on the windows.

Data

Specific humidity of the outside air	0.008
Temperature of the outside air	14 °C
Moisture output per person	0.05 kg/h
Thermal transmittance of glass	5 W/m² K
Inside surface film coefficient	7 W/m² K
Atmospheric pressure	1.013 bar
Density of room air	1.2 kg/m³
Air change rate	0.5 h⁻¹

Solution

The inside surface temperature of the glass, t_s, can be found using Eqn [5.3] as follows:

$$t_s = t_i - \frac{U}{\alpha_i}(t_i - t_o) = 21 - \frac{5}{7}(21 - 14) = 16\,°C$$

This is the air dew point temperature below which condensation will form on the glass. It corresponds to a saturation pressure of 0.018 17 bar. The specific humidity when this vapour pressure is reached, ω_τ, is then calculated as follows:

$$\omega_\tau = 0.622\frac{p_s}{(p - p_s)} = 0.622\frac{0.018\ 17}{(1.013 - 0.018\ 17)} = 0.011\ 36$$

The mass flow rate of moisture produced within the room, $\dot{m}_{sp}$, and the volume flow rate of room air, $\dot{V}_R$, can now be calculated as follows:

$$\dot{V}_R = n V_R = 0.5 \times 100 = 50\ m^3/h$$
$$\dot{m}_{sp} = 20 \times 0.05 = 1\ kg/h$$

Values can now be inserted in Eqn [5.17], leaving the time in hours as the only unknown.

$$\omega_\tau - \omega_o = \left(\frac{\dot{m}_{sp}}{\rho_R \dot{V}_R} + \omega_a - \omega_o\right)(1 - e^{-n\tau})$$

$$0.01136 - 0.008 = \left(\frac{1}{1.2 \times 50} + 0.008 - 0.008\right)(1 - e^{-0.5\tau})$$

$$0.2016 = 1 - e^{-0.5\tau}$$

$$-\ln 0.7984 = 0.5\tau$$

$$\therefore \qquad \tau = 0.45\ \text{hours}$$

Example 5.8

A bathroom of volume 20 m³ is maintained at a temperature of 22 °C. When unoccupied the relative humidity is 40%. The ambient air is at −2 °C. The windows are double glazed with a U-value of 3 W/m² K and an internal surface resistance of 0.125 m² K/W. The main moisture load is when people take showers. It is estimated that a shower takes five minutes, during which time water evaporates at the rate of 1.27 kg/h.

To avoid condensation on the inside of the windows it is proposed to fit an extract fan which will discharge air to ambient while drawing air from other rooms which are at 18 °C and 50% relative humidity. If the pressure thoughout is 1 bar, calculate the minimum capacity of the extract fan.

Any formula may be used without proof and an initial estimate of 0.05 m³/s taken for the trial and error solution. The density of the room air may be taken as 1.2 kg/m³.

(Wolverhampton Polytechnic)

Solution
The window surface temperature and the relevant specific humidities are calculated as in Example 5.7 as follows:

$$\alpha_i = \frac{1}{R_{si}} = \frac{1}{0.125} = 8 \text{ W/m}^2 \text{ K}$$

$$t_s = t_i - \frac{U}{\alpha_i}(t_i - t_o) = 22 - \frac{3}{8}(22 + 2) = 13 \text{ °C}$$

Critical value:

$$\omega_\tau = 0.622 \frac{p_s}{(p - p_s)} = 0.622 \frac{0.014\,97}{(1 - 0.014\,97)} = 0.009\,45$$

Initial room air at $\tau = 0$:

$$p_s = \phi p_g = 0.4 \times 0.026\,42 = 0.010\,59$$

$$\omega_o = 0.622 \times \frac{p_s}{(p - p_s)} = 0.622 \times \frac{0.010\,59}{(1 - 0.010\,59)} = 0.006\,64$$

Supply air:

$$p_s = \phi p_g = 0.5 \times 0.020\,63 = 0.010\,315$$

$$\omega_a = 0.622 \times \frac{p_s}{(p - p_s)} = 0.622 \times \frac{0.010\,315}{(1 - 0.010\,315)} = 0.006\,48$$

Inserting values in Eqn [5.16] gives:

$$\omega_\tau - \omega_o = \left(\frac{\dot{m}_{sp}}{\rho_R \dot{V}_R} + \omega_a - \omega_o\right)(1 - e^{-\dot{V}_R\tau/V_R})$$

$$0.009\,45 - 0.006\,64 = \left(\frac{1.27}{3600 \times 1.2\dot{V}_R} + 0.006\,48 - 0.006\,64\right)(1 - e^{-\dot{V}_R \times 300/20})$$

$$0.002\,81 = \left(\frac{0.000\,294}{\dot{V}_R} - 0.000\,16\right)(1 - e^{-15\dot{V}_R})$$

For the trial and error solution,

Try $\dot{V}_R = 0.05$ m³/s: RHS = 0.003 018
Try $\dot{V}_R = 0.06$ m³/s: RHS = 0.002 813

This is sufficiently close and the answer is 0.06 m³/s.

Example 5.9

An occupied building has an internal temperature of 20 °C and a relative humidity of 53%. The occupants then leave the building and the heating system is switched off. This results in the internal temperature falling by 1 K every half-hour. When the building is unoccupied the ventilation rate falls to 0.1 air changes per hour.

The thermal transmittance of the windows is 5 W/m² K and the internal heat transfer coefficient 8 W/m² K. The pressure remains constant at 1 bar. If the ambient air remains at 5 °C saturated, how long will it take for condensation to start to form on the windows?

(Wolverhamption Polytechnic)

Solution

The inside surface temperature of the glass, t_s, for any room temperature, t_R, can be found using Eqn [5.3] as follows:

$$t_s = t_i - \frac{U}{\alpha_i}(t_i - t_o) = t_R - \frac{5}{8}(t_R - 5)$$ [a]

The saturation pressure, p_g, corresponding to t_s is the critical room vapour pressure, p_s, for condensation. This can be used to calculate the critical (dew point) specific humidity ω_D from the equation:

$$\omega_D = 0.622 \frac{p_s}{(p - p_s)} = 0.622 \frac{p_s}{(1 - p_s)}$$ [b]

The initial room specific humidity and the specific humidity of the ambient air can now be calculated as follows.

Initial room air at $\tau = 0$:

$$p_s = \phi p_g = 0.53 \times 0.023\,37 = 0.012\,39 \text{ bar}$$
$$\omega_o = 0.622 \frac{p_s}{(p - p_s)} = 0.622 \frac{0.012\,39}{(1 - 0.012\,39)} = 0.0078$$

Ambient air:

$$p_s = p_g = 0.008\,719 \text{ bar}$$
$$\omega_a = 0.622 \frac{p_s}{(p - p_s)} = 0.622 \frac{0.008\,719}{(1 - 0.008\,719)} = 0.005\,47$$

These values can now be inserted in Eqn [5.17].

$$\omega_\tau - \omega_o = \left(\frac{\dot{m}_{sp}}{\rho_R \dot{V}_R} + \omega_a - \omega_o\right)(1 - e^{-n\tau})$$
$$\omega_\tau - 0.0078 = (0 + 0.005\,47 - 0.0078)(1 - e^{-0.1\tau})$$
$$\omega_\tau = 0.005\,47 + 0.002\,33\,e^{-0.1\tau}$$ [c]

Equations [a], [b] and [c] can now be used to establish the dew point specific humidity, ω_D, and the room specific humidity, ω_τ, after any time τ hours. A trial and error solution is now required. The table below is a summary of these calculations. It can be seen from the table that the answer is about 1.5 hours.

τ (h)	t_R (°C)	t_s (°C)	p_g (bar)	ω_D	ω_τ
0.5	19	10.25	0.01248	0.00786	0.00769
1.0	18	9.88	0.01217	0.00766	0.00758
1.5	17	9.50	0.01187	0.00747	0.00748
2.0	16	9.15	0.01159	0.00729	0.00738

Figure 5.12 Example 5.9

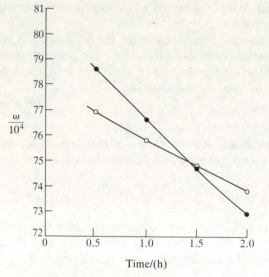

Time/(h)

It is normally necessary to plot ω_D and ω_r against time, which is illustrated in Fig. 5.12.

The example demonstrates the temporary condensation problems which can arise when well-sealed, mechanically ventilated buildings are closed down at night.

PROBLEMS

5.1   A 200 mm brick wall covers an area of 20 m². The brick has a resistivity of 30 GN s/kg m. The inside air is at 21 °C and 50% relative humidity and the external air at 2 °C saturated.

Calculate the mass flow rate of water vapour through the wall in kg/h.

$(6.45 \times 10^{-3}$ kg/h$)$

5.2   (a)   List four factors in modern buildings which tend to increase condensation risk.

(b)   The following table lists the thermal and vapour properties of a wall. Carry out a steady-state calculation to determine the risk of interstitial condensation

| | | | Resistivity | |
Layer	Thickness (mm)	Thermal (m K/W)		Vapour (GN s/kg m)
R_{si}	—	(0.12 m² K/W)		0
Timber	13	7.0		45
Expanded polystyrene	38	28.0		100
Cavity	50	(0.18 m² K/W)		0
Brickwork	110	1.4		100
Render	13	0.8		100
R_{so}	—	(0.06 m² K/W)		0

in the wall under the following conditions:

Inside air: Dry bulb temp.   22 °C, RH 80%
Outside air: Dry bulb temp.   −3 °C, RH 100%

A psychrometric chart is provided.

(University of Manchester)

Note: The problem can be solved using property tables instead of a chart.

(Condensation occurs on the outside face of the cavity where the saturation vapour pressure is 6.43 m bar and the predicted vapour pressure 16.92 m bar.)

5.3   The *U*-value of a traditional cavity brick wall is to be upgraded by the addition to the inside of 50 mm fibre glass between studding and finishing with 12 mm plasterboard, as shown in Fig. 5.13. The wall is exposed to indoor conditions of 20 °C, 46% saturation and outdoor conditions of 0 °C, 92% saturation.

(a)   Using the data below and Sections C1 and 2 of the CIBSE *Guide*,[5.5,5.6] show that interstitial condensation will occur in the upgraded structure.

(b)   Explain how the risk of interstitial condensation may be significantly reduced.

Figure 5.13 Example 5.3

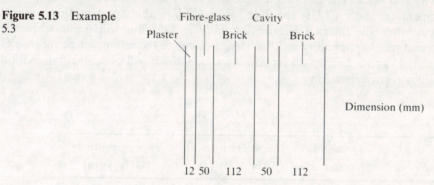

Plaster   Fibre-glass   Cavity
                Brick   Brick

Dimension (mm)

12 50   112   50   112

Data
Permeance of plasterboard, 150×10^{-11} kg/N s; permeability of fibre glass, 20×10^{-11} kg m/N s; permeability of brick, 5×10^{-11} kg m/N s; permeability of air gap, 20×10^{-11} kg m/N s; thermal resistance of inside surface, 0.12 m² K/ W; thermal conductivity of plasterboard, 0.16 W/m K; thermal conductivity of fibre glass, 0.04 W/m K; thermal conductivity of brick, 0.70 W/m K; thermal resistance of cavity, 0.18 m² K/W; thermal resistance of external surface, 0.06 m² K/W.

(CIBSE)

Note: From the CIBSE *Guide*, at 20 °C and 46% saturation the vapour pressure is 1.089 kPa and at 0 °C and 92% saturation the vapour pressure is 0.5622 kPa.

(Condensation occurs on the outside face of the cavity where the saturation vapour pressure is 7.15 mbar and the predicted vapour pressure 7.71 mbar.)

5.4   The walls of a building are constructed of 105 mm brickwork, 50 mm foamed urea–formaldehyde (UF foam), 100 mm medium concrete block and 13 mm lightweight plaster.

The inside air temperature is 19 °C and the conditions outside are 1 °C, 100% relative humidity.

Using the data given, calculate the lowest inside relative humidity which will lead to interstitial condensation within the wall.

Data

Section	Thermal conductivity (W/m K)	Permeability $(10^{-11} \, \text{kg m/N s})$
Brick	0.73	2.1
UF foam	0.04	2.5
Concrete	1.45	2.0
Plaster	0.48	2.6

Heat transfer coefficients: outside, 20 W/m^2 K; inside, 8 W/m^2 K.

(Wolverhampton Polytechnic)

(42%)

5.5 The construction of a wall, from inside to outside, is concrete, insulation, cavity and brick. The external design state is -3 °C saturated and the internal temperature is 18 °C. Use the data below to show that condensation will not occur when the internal relative humidity is 40% but will occur when the internal relative humidity is 70%. Hence calculate the minimum permeance of a partial vapour barrier on the inside surface which will prevent condensation at 70% relative humidity.

Data

Layer	Thickness (mm)	Thermal resistance (m^2 K/W)	Vapour resistivity (GN s/kg m)
R_{si}	—	0.12	0
Concrete	200	0.53	50
Insulation	80	1.90	200
Cavity	50	0.18	5
Brickwork	102	0.12	35
R_{so}	—	0.14	0

$(4.4 \times 10^{-11} \, \text{kg/N s})$

5.6 (a) Explain the principles of moisture movement in walls.

(b) What forms of construction and environmental circumstances are likely to lead to interstitial condensation problems?

(c) Explain in general terms the reasons for (i) temporary condensation, (ii) permanent condensation in a building.

(d) Estimate the rate of condensation upon a factory canteen window given the following information:

Inside vapour film conductance, 18×10^{-9} s/m (kg/m^2 s transport of water vapour per N/m^2 pressure difference); inside film conductance, 8 W/m^2 K;

outside film conductance, $16 \text{ W/m}^2 \text{ K}$; specific enthalpy of condensation, $2.48 \times 10^6 \text{ J/kg}$; air temperature in canteen, $18 \,°\text{C}$; relative humidity, 85%; outside air temperature, $2 \,°\text{C}$.

(University of Liverpool)

$(0.0417 \text{ kg/h m}^2)$

5.7 At the start of a meeting of 50 people in a room of volume 250 m^3 the temperature and relative humidity are $19 \,°\text{C}$ and 50%, when the outside conditions are $5 \,°\text{C}$, relative humidity 80%. The natural ventilation rate is 0.3 air changes per hour.

Using the data below calculate the time taken for condensation to begin to form on the window surfaces, assuming that the inside temperature is maintained at $19 \,°\text{C}$ throughout and that the outside conditions of temperature and relative humidity do not change with time.

Work from first principles throughout.

Data
Thermal transmittance of windows, $5 \text{ W/m}^2 \text{ K}$; resistance of the inside surface, $0.12 \text{ m}^2 \text{ K/W}$; moisture output per person, 0.05 kg/h; atmospheric pressure, 1.013 bar; density of inside air, 1.2 kg/m^3.

(Wolverhampton Polytechnic)

(9.1 minutes)

5.8 A room of volume $V \text{ m}^3$ is ventilated with outside air at a rate of n air changes per hour. The outside air moisture content is $\omega_o \text{ kg/kg}$. Water vapour is generated inside the room at a rate $\dot{m} \text{ kg/h}$. Write down a mass balance on the water vapour, and hence obtain a simple differential equation which includes ω_i, the inside air moisture content. From this, show that the steady state value of ω_i is given by:

$$\omega_i = \omega_o + \frac{\dot{m}}{\rho V n}$$

where ρ is the inside air density, assumed to be a constant at 1.2 kg/m^3.

In a room of volume 70 m^3 the mean rate of vapour generation is 0.45 kg/h. Find the necessary minimum value of ventilation rate required to maintain the inside air relative humidity below 70% when the air temperature is steady at $22 \,°\text{C}$. The mean value of ω_o is 0.008 kg/kg.

(University of Manchester)

(1.5 per hour)

5.9 A meeting room has a volume of 100 m^3 and is maintained at a temperature of $21 \,°\text{C}$. The ambient air is at $14 \,°\text{C}$ with a specific humidity of 0.009. When the room is unoccupied the specific humidity in the room is the same as that of the outside air. A meeting then starts which is scheduled to last for three hours. During the meeting the occupants produce moisture at the rate of 1.44 kg/h. Calculate the minimum rate at which air must be extracted to prevent condensation on the windows.

Work from first principles and use an initial estimate of $0.15 \text{ m}^3/\text{s}$ for the 'trial and error' solution.

Data
Thermal transmittance of windows, 5 W/m² K; internal heat transfer coefficient,
7 W/m² K; atmospheric pressure, 1.01 bar; density of inside air, 1.2 kg/m³.

(Wolverhampton Polytechnic)

(0.14 m³/s)

REFERENCES

5.1   ASHRE 1976 *Handbook of Fundamentals* chapter 20
5.2   CIBSE 1986 *Guide to Current Practice* volume A10
5.3   Rogers G C F and Mayhew Y R 1988 *Thermodynamic and Transport Properties of Fluids* 4th edn Basil Blackwell
5.4   Silcock G W H and Shields T J 1985 Predicting condensation—an alternative approach *Energy in Buildings* (January)
5.5   CIBSE 1986 *Guide to Current Practice* volumes C1.6 to C1.66
5.6   CIBSE 1986 *Guide to Current Practice* volumes C2.2 to C2.7

6 MASS TRANSFER PLANT

Moisture in atmospheric air is necessary for human comfort and its control is a central part of most air conditioning systems. Air can be de-humidified by cooling it below the dew point and humidified by allowing water to evaporate into it; in both cases the mechanism is one of combined heat and mass transfer. In air conditioning plant water cooling is frequently required and a cooling tower is an effective way of doing this; the water on evaporating into an air stream is cooled as the air stream becomes more humid, another example of combined heat and mass transfer.

In Chapter 5, Fick's law is introduced and applied to moisture transfer through permeable structures. This chapter considers convective mass transfer, the analogy with heat transfer, and simple applications to plant in air conditioning systems.

6.1 MOLECULAR DIFFUSION

Molecular diffusion of one gas into another or one liquid into another is governed by Fick's law, which can be expressed as:

$$\dot{n} = -D \frac{dc}{dy}$$

where $\dot{n}$ is the number of kmol per unit area per unit time of one of the constituents diffusing in the direction y due to a difference in concentration, c, measured by the number of kmol per unit volume; D is the diffusion coefficient with units of m^2/s.

In the case of water vapour, assuming it acts as a perfect gas, the concentration can be expressed as

$$c = n_s = p_s / \tilde{R} T$$

where n_s is the number of kmol of water vapour per unit volume; p_s is the partial pressure of the water vapour; $\tilde{R}$ is the molar gas constant; T is the absolute temperature of the mixture.

$$\therefore \qquad \dot{n}_s = -\frac{D}{\tilde{R} T} \frac{dp_s}{dy} \qquad\qquad [6.1]$$

Evaporation of water into stagnant air at the same temperature is one example of molecular diffusion. The simple law of Eqn [6.1] applies to both the water vapour and the bulk of the air when the air is completely stagnant and at the same temperature as the evaporating water; as we shall see in later sections, the law also applies very close to a water surface in the case of an air stream crossing the surface.

Consider evaporation of water at absolute temperature T into stagnant air at temperature T, as shown in Fig. 6.1. The total pressure of the mixture is p, assumed constant over the distance y; the partial pressures of the water vapour and air at any distance y from the surface are p_s and p_a, i.e.

$$p = p_s + p_a \quad \text{and} \quad \frac{dp}{dy} = \frac{dp_s}{dy} + \frac{dp_a}{dy} = 0$$

$$\therefore \quad \frac{dp_s}{dy} = -\frac{dp_a}{dy}$$

Figure 6.1 Diffusion of water vapour into air

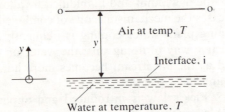

At any section distance y from the water surface, the diffusion of water vapour and of air are both governed by Eqn [6.1]. In the case of air diffusing towards the water surface, since it has no exit through the surface it forms a re-circulatory system at a mean velocity u, say.

We then have, for the air:

$$\dot{n}_a = -\frac{D}{\tilde{R}T}\frac{dp_a}{dy} + un_a = 0$$

where n_a is the number of kmol of air per unit volume at any distance y from the interface.

$$\therefore \quad u = \frac{D}{\tilde{R}Tn_a}\frac{dp_a}{dy}$$

Also, since $p_a = p - p_s$, then $dp_a/dy = -dp_s/dy$. Therefore,

$$u = -\frac{D}{\tilde{R}Tn_a}\frac{dp_s}{dy} \tag{1}$$

For the water vapour, the circulation of air causes an additional transfer of water vapour of un_s, where n_s is the number of kmol of water vapour per unit volume at any distance y. That is,

$$\dot{n}_s = -\frac{D}{\tilde{R}T}\frac{dp_s}{dy} + un_s$$

and substituting for u from Eqn [1],

$$\dot{n}_s = -\frac{D}{\tilde{R}T}\left\{1 + \frac{n_s}{n_a}\right\}\frac{dp_s}{dy}$$

Then, since $p_s/p_a = n_s/n_a$, we have

$$\dot{n}_s = -\frac{D}{\tilde{R}T}\left\{1 + \frac{p_s}{p_a}\right\}\frac{dp_s}{dy}$$

i.e.
$$\dot{n}_s = -\frac{Dp}{\tilde{R}Tp_a}\frac{dp_s}{dy} \qquad [6.2]$$

This equation is known as *Stefan's law*.

The transfer of water vapour between the surface and a section at any distance y from the surface is given by separating the variables in Eqn [6.2] and integrating, i.e.

$$\dot{n}_s\frac{\tilde{R}T}{Dp}\int_0^y dy = -\int_{p_{si}}^{p_{so}}\frac{dp_s}{(p - p_s)}$$

$$\therefore \quad \dot{n}_s = \frac{Dp}{\tilde{R}Ty}\ln\left\{\frac{p - p_{so}}{p - p_{si}}\right\} = \frac{Dp}{\tilde{R}Ty}\ln\left\{\frac{p_{ao}}{p_{ai}}\right\}$$

where p_{so} and p_{ao} are the partial pressures of the water vapour and air at the section o–o, and p_{si} and p_{ai} are the partial pressures at the air–water interface.

Defining a logarithmic mean air pressure as

$$\bar{p}_{ln} = (p_{ao} - p_{ai})/\ln(p_{ao}/p_{ai})$$
$$= (p_{si} - p_{so})/\ln(p_{ao}/p_{ai})$$

we then have

$$\dot{n}_s = \frac{D}{\tilde{R}T}\frac{p}{\bar{p}_{ln}}\frac{(p_{si} - p_{so})}{y} \qquad [6.3]$$

The mass transfer rate for a surface area A is then

$$\dot{m}_s = \dot{n}_s\tilde{m}_s A = \frac{DA}{R_s T}\frac{p}{\bar{p}_{ln}}\frac{(p_{si} - p_{so})}{y} \qquad [6.4]$$

In the case of mass transfer of water vapour in air, the velocity of re-circulation is very small and it is usually possible to assume that the ratio $p/\bar{p}_{ln}$ is unity; the numerical values in Example 6.1 illustrate this. That is,

$$\dot{n}_s = -\frac{D}{\tilde{R}T}\frac{dp_s}{dy}$$

and

$$\dot{m}_s = -\frac{DA}{R_s T}\frac{dp_s}{dy} \qquad [6.5]$$

Example 6.1

The water storage tank shown in Fig. 6.2 has its water surface 1 m below its top. On the outside of the tank, dry air at low velocity passes across the open end of the ventilation pipe which is 0.5 m long and 25 mm inside diameter. The air inside the tank is saturated with water vapour.

Figure 6.2   Water
storage tank of Example
6.1

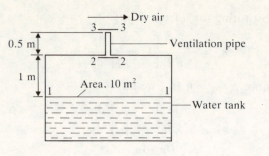

Taking the system pressure and temperature to be 1.0132 bar and 27 °C, calculate:

(i)   the mass flow rate of water vapour diffusing through the ventilation pipe into the dry air stream;

(ii) the rate of heat loss due to diffusion.

The diffusion coefficient for water vapour diffusing into air is 2.56×10^{-5} m^2/s.

(Polytechnic of the South Bank)

Solution

(i)   The transfer of water vapour from the tank is entirely by molecular diffusion due to the difference in partial pressures between the saturated air in the tank and the dry air outside. There is no temperature difference and hence no convection current, although there will be a small velocity of air due to the concentration difference of the dry air causing a circulatory movement in the stand pipe as in the theory section above. The tank surface area of 10 m^2 is very large compared with the stand pipe cross-sectional area and hence the difference in partial pressures between the water surface, section 1–1, and the entry to the stand pipe, section 2–2, is negligible. It can therefore be assumed that the partial pressure of the water vapour inside the tank is the saturation value at 27 °C, i.e.

From tables, $p_{s1} = p_{s2} = 0.035\,64$ bar

Also, outside the tank, at section 3–3, $p_{s3} = 0$.

The total pressure throughout is given as 1.0132 bar; therefore the air partial pressures are

$$p_{a1} = p_{a2} = 1.0132 - 0.035\,64 = 0.977\,56 \text{ bar}$$

and $p_{a3} = 1.0132$ bar

The logarithmic air partial pressure, $p_{\ln}$, is then

$$\frac{(p_{a3} - p_{a2})}{\ln (p_{a3}/p_{a2})} = \frac{1.0132 - 0.977\,56}{\ln (1.0132/0.977\,56)} = 0.995\,27 \text{ bar}$$

Then applying Eqn [6.4] between sections 2–2 and 3–3 we have

$$
\begin{aligned}
\dot{m}_s &= \frac{DA}{R_s T} \frac{p}{\bar{p}_{\ln}} \frac{(p_{s2} - p_{s3})}{y} \\
&= \frac{2.56 \times 10^{-5} \times (\pi \times 0.025^2/4) \times 1.0132 \times 0.035\,64 \times 10^5}{(8314.5/18) \times 300 \times 0.995\,27 \times 0.5} \\
&= 6.58 \times 10^{-10} \text{ kg/s}
\end{aligned}
$$

(The accuracy of neglecting the difference in partial pressures between sections

1–1 and 2–2 can be shown by equating the mass flow from 1–1 to 2–2, to that from 2–2 to 3–3, using Eqn [6.4], i.e.

$$\frac{Dp \times 10(p_{s1} - p_{s2})}{R_s T \bar{p}_{ln1}} \frac{1}{1} = \frac{Dp(\pi \times 0.025^2/4)(p_{s2} - p_{s3})}{R_s T \bar{p}_{ln2}} \frac{1}{0.5}$$

The ratio of the two values of $\bar{p}_{ln}$ is very close to unity and hence

$$(p_{s1} - p_{s2}) = 0.000\,098(p_{s2} - p_{s3}) = 0.000\,003\,5 \text{ bar}$$

Note also that if the ratio $p/\bar{p}_{ln}$ is neglected in Eqn [6.4], then the mass transfer is 6.46×10^{-10} kg/s, an error of about 1.8%.)

(ii)   The rate of heat loss from the water in the tank is given by the product of the rate of mass transfer and the specific enthalpy of vaporization at the temperature of the system, i.e.

$$\text{Rate of heat loss} = 6.58 \times 10^{-10} \times 2437.2 \times 10^3$$
$$= 0.0016 \text{ W}$$

It can be seen that the rate of heat loss is so small that the initial assumption of constant temperature is justified.

6.2 CONVECTIVE MASS TRANSFER

In the previous section it was shown that mass transfer by diffusion occurs in completely stagnant conditions; the only movement within the fluids is due to the re-circulation when one of the fluids is prevented from passing through a solid or liquid surface. As shown in Section 6.1, the effect of this re-circulatory movement is very small in the case of water vapour transfer in air.

Convective mass transfer occurs when there is a forced flow of fluid across the surface. The mechanism of molecular diffusion is still present but the effect is small and is confined to the thin layer of fluid at the surface. This is similar to the convection of heat from a hot surface where conduction of heat is significant only at the surface layer of fluid.

Considering the mass transfer of water vapour in air, the driving force for convective mass transfer is the difference in concentration of the water vapour between that at the liquid surface and that at the free stream of the fluid flowing across the surface.

By analogy with heat transfer we can then define a molar transfer coefficient, β, as,

$$\dot{n}_s = \beta(c_i - c_o)$$

where subscripts i and o refer to the interface and free stream condition.

Then, since the concentration is $n_s = p_s/\tilde{R}T$, we can write

$$\dot{n}_s = \frac{\beta}{\tilde{R}T}(p_{si} - p_{so}) \qquad [6.6]$$

The mass transfer rate, $\dot{m}_s$, is

$$\dot{n}_s \tilde{m}_s A = \dot{n}_s A \tilde{R}/R_s$$

i.e. $$\dot{m}_s = \frac{\beta A}{R_s T}(p_{si} - p_{so}) \qquad [6.7]$$

The units of β in Eqns [6.6] and [6.7] are m/s.

Sometimes a mass transfer coefficient is defined as

$$\beta_G = \beta/\tilde{R}T$$

so that

$$\dot{n}_s = \beta_G(p_{si} - p_{so})$$

and

$$\dot{m}_s = \beta_G \tilde{m}_s A(p_{si} - p_{so}) \qquad [6.8]$$

The units of β_G are kmol/m^2 s bar.

Alternatively, in terms of mass flow rate, from Eqn [6.7] we have

$$\beta_P = \beta/R_s T$$

so that

$$\dot{m}_s = \beta_P A(p_{si} - p_{so}) \qquad [6.9]$$

The units of β_P are then kg/m^2 s bar, or kg/N s.

For the particular case of water vapour and air mixtures encountered in air conditioning, the partial pressure of the vapour is always very small compared with that of the air. Re-writing Eqn [6.7] as

$$\dot{m}_s = \beta A \left(\frac{p_{si}}{R_s T} - \frac{p_{so}}{R_s T} \right)$$

$$= \beta A \left(\frac{m_{si}}{V} - \frac{m_{so}}{V} \right) = \frac{\beta A m_{ao}}{V} \left(\frac{m_{si}}{m_{ao}} - \frac{m_{so}}{m_{ao}} \right)$$

$$= \frac{\beta A m_{ao}}{V} \left(\frac{m_{si}}{m_{ai}} \frac{m_{ai}}{m_{ao}} - \frac{m_{so}}{m_{ao}} \right)$$

Then, taking the approximation $m_{ai} = m_{ao}$, and the definition of specific humidity, $\omega = m_s/m_a$, we have

$$\dot{m}_s = \beta A \rho_{ao}(\omega_i - \omega_o)$$

where ρ_{ao} is the density of the dry air in the free stream.

Then introducing a new mass transfer coefficient, β_ω,

$$m_s = \beta_\omega A(\omega_i - \omega_o) \qquad [6.10]$$

where $\beta_\omega = \beta\rho_{ao}$, and has units of kg/m^2 s.

The use of different mass transfer coefficients is confusing and in teaching the subject it is better to use only one definition throughout. In this text several definitions are introduced as above, since different authors use different definitions; also, some of the Examples and Problems in this text are taken from sources which use different definitions. Table 6.1 gives a summary of the expressions for mass transfer and the various mass transfer definitions.

A non-dimensional mass transfer number analogous to the Nusselt number, Nu, in heat transfer can be introduced, called the Sherwood number, Sh.

The Nusselt number can be considered as the ratio of the heat transferred by convection to that by conduction through a typical dimension, L, of the system, i.e. $Nu = \alpha L/\lambda$, where α is the heat transfer coefficient and λ is the thermal conductivity. In a similar way the Sherwood number can be considered

Table 6.1 Mass transfer rates and mass transfer coefficients

Expressions for mass transfer, $\dot{m}_s$ (usual units, kg/s)	Equation in text
$\dfrac{\beta A}{R_s T}(p_{si} - p_{so})$	[6.7]
$\beta_G \tilde{m}_s A (p_{si} - p_{so})$	[6.8]
$\beta_P A (p_{si} - p_{so})$	[6.9]
$\beta_\omega A (\omega_i - \omega_o)$	[6.10]

Relationship between mass transfer coefficients	Normal units
β	m/s
$\beta_G = \beta/\tilde{R}T$	$\text{kmol/m}^2 \text{ s bar}$
$\beta_P = \beta/R_s T$	$\text{kg/m}^2 \text{ s bar}$
$\beta_\omega = \beta\rho_{ao}$	$\text{kg/m}^2 \text{ s}$

Nomenclature
A = liquid–air interface surface area; T = absolute temperature of the air–vapour mixture; R_s = specific gas constant for water vapour; $\tilde{m}_s$ = molar mass of water vapour; $\tilde{R}$ = molar gas constant; p_{si} = partial pressure of water vapour at the liquid–air interface; p_{so} = partial pressure of water vapour in the free stream; ω_i = specific humidity of air at the liquid–air interface; ω_o = specific humidity of air in the free stream; ρ_{ao} = partial density of the dry air in the free stream

as the ratio of the mass transfer by convection, β, to the mass transferred by molecular diffusion through a typical dimension, L.

From Eqn [6.4], the mass transfer by molecular diffusion is

$$\frac{DA}{R_s T \bar{p}_{ln}} \frac{p}{} \frac{(p_{si} - p_{so})}{L}$$

and from Eqn [6.7] the mass transfer by convection is

$$\frac{\beta A}{R_s T}(p_{si} - p_{so})$$

Then from the definition of the Sherwood number we have

$$Sh = \frac{\beta A}{R_s T} \frac{R_s T \bar{p}_{ln} L}{DA} \frac{}{p} = \frac{\beta L}{D}\left\{\frac{\bar{p}_{ln}}{p}\right\} \qquad [6.11]$$

The Sherwood number can be written approximately as $\beta L/D$ for most cases of water vapour in air since $\bar{p}_{ln}$ is approximately equal to p, as explained earlier.

Empirical relationships expressed in terms of non-dimensional numbers can be found for mass transfer and these bear a close resemblance to the empirical expressions for heat transfer. In general, the expression takes the form

$$Sh = (\text{constant}) \times Re^n \times Sc^m$$

where Re is the Reynolds number, uL/v; u is the velocity of the fluid; L is a typical linear dimension; v is the kinematic viscosity of the fluid, η/ρ; Sc is the Schmidt number, v/D; D is the diffusion coefficient; η is the dynamic viscosity of the fluid; n and m are indices to be determined by experiment.

It can be seen that the Schmidt number plays the same part in mass transfer as the Prandtl number, $c_p\eta/\lambda = v/\kappa$, in heat transfer ($\kappa$ is the thermal diffusivity, $\lambda/\rho c_p$, where λ is the thermal conductivity of the fluid, ρ is the fluid density, and c_p is the specific heat of the fluid at constant pressure).

Example 6.2

Water is evaporated by allowing it to run down as an annular film on the inside surface of a vertical tube while air is blown through the tube. Using the data below and the empirical relationship given, calculate:
(i) the mass transfer coefficient;
(ii) the mass transfer per unit length of pipe.

Data

System pressure and temperature, 1.013 25 bar and 27 °C; pipe inside diameter, 50 mm; velocity of air in pipe, 2 m/s; diffusion coefficient, 2.5×10^{-5} m^2/s; mean value of the relative humidity of the bulk of air in the pipe, 50%. Take properties of air at the system conditions, and properties of saturated steam, from the tables of Rogers and Mayhew.[6.1]

For mass transfer from the inside of a pipe, take the following expression, where the Sherwood number is as defined by Eqn [6.11], and the linear dimension in the Sherwood and Reynolds numbers is the pipe inside diameter:

$$Sh = 0.023 Re^{0.83} Sc^{0.44}$$

Solution

From tables,[6.1] at 1.013 25 bar and 27 °C, $v = 1.568 \times 10^{-5}$ m^2/s. Therefore,

$$\text{Schmidt number, } Sc = \frac{v}{D} = \frac{1.568 \times 10^{-5}}{2.5 \times 10^{-5}} = 0.627$$

Also,

$$Re = \frac{ud}{v} = \frac{2 \times 0.05}{1.568 \times 10^{-5}} = 6377.6$$

Then,

$$Sh = 0.023(6377.6)^{0.83}(0.627)^{0.44} = 26.94$$

Substituting in Eqn [6.11],

$$26.94 = \frac{\beta L \bar{p}_{\text{ln}}}{Dp} \tag{1}$$

(where, in this case, L is the inside diameter of the pipe)

(i)   To find the mass transfer coefficient, β, it is necessary to calculate first the value of $\bar{p}_{\text{ln}}$.

From tables, p_g at 27 °C is 0.035 64 bar; therefore the partial pressure of the water vapour in the bulk of air in the pipe is $0.5 \times 0.035 64 = 0.017 82$ bar. The partial pressure of the air at the interface, p_{ai}, is

$$(1.013 25 - 0.035 64) = 0.977 61 \text{ bar}$$

and the partial pressure of the bulk of the air, p_{ao}, is

$$(1.013 25 - 0.017 82) = 0.995 43 \text{ bar}$$

Then,

$$\bar{p}_{\text{ln}} = \frac{0.995 43 - 0.977 61}{\ln(0.995 43/0.977 61)} = 0.9865$$

Therefore, from Eqn [1], $\beta = 26.94 Dp/L\bar{p}_{\text{ln}}$, we have

$$\beta = \frac{26.94 \times 2.5 \times 10^{-5} \times 1.01325}{0.05 \times 0.9865} = 0.0138 \text{ m/s}$$

The mass transfer coefficient could also be expressed as

$$\beta_{\text{G}} = \beta/\tilde{R}T = 0.0138/8314.5 \times 300$$
$$= 0.000553 \times 10^{-5} \text{ kmol/m}^2 \text{ s Pa}$$
$$= 0.000553 \text{ kmol/m}^2 \text{ s bar}$$

or $\qquad \beta_{\text{P}} = 0.000553 \times 18 = 0.00996 \text{ kg/m}^2 \text{ s bar}$

(ii) The mass transfer per unit length can then be found using Eqn [6.9]:

$$\dot{m}_{\text{s}} = \beta_{\text{P}} A (p_{\text{si}} - p_{\text{so}})$$
$$= 0.00996 \times (\pi \times 0.05) \times (0.03564 - 0.01782)$$
$$= 27.88 \times 10^{-6} \text{ kg/s per metre length}$$

6.3 ANALOGIES BETWEEN HEAT, MASS AND MOMENTUM TRANSFER

Reynolds Analogy

Reynolds showed that when a fluid flows across a surface there is an analogy between the transfer of momentum and the transfer of heat. In its simplest form the analogy can be expressed as:

$$St = f/2 \qquad\qquad\qquad [6.12]$$

St is the Stanton number, $Nu/RePr$, and f is the friction factor, defined as $\tau/(\rho u^2/2)$, where τ is the shear stress in the fluid at the surface, ρ is the fluid density, and u is the fluid velocity across the surface.

The Stanton number is useful because it can be evaluated entirely from experimental measurements, as shown below:

$$St = \frac{Nu}{RePr} = \frac{\alpha L}{\lambda} \frac{\eta}{\rho u L} \frac{\lambda}{c_p \eta} = \frac{\alpha}{\rho u c_p} \qquad\qquad [6.13]$$

For the general case where a fluid approaches a surface at a temperature t_1 and leaves it at t_2, the heat transferred from the surface, area A, is equal to the enthalpy increase of the fluid crossing the surface, i.e.

$$\alpha A \Delta t = \rho u A_{\text{c}} c_p (t_2 - t_1)$$

where α is the heat transfer coefficient; Δt is the temperature difference between the surface and the bulk of the fluid; A_{c} is the cross-sectional area of the flow. Therefore,

$$St = \frac{\alpha}{\rho u c_p} = \frac{A_{\text{c}}}{A} \frac{(t_2 - t_1)}{\Delta t}$$

It can be seen that for a given geometry the Stanton number can be found by measuring the fluid and surface temperatures.

Prandtl–Taylor Analogy

The simple analogy given by Eqn [6.12] can be shown to be exactly true when $Pr = 1$. For air–water vapour mixtures at the pressure and temperatures encountered in air conditioning, the Prandtl number is about 0.7.

Various investigators have improved the Reynolds analogy by considering the boundary layer of fluid on the surface. The Prandtl–Taylor analogy considers a laminar sub-layer on the surface; a simple derivation is given in most heat transfer textbooks, see for example the one by Welty.[6.2] For heat transfer for flow in smooth pipes we have:

$$St = \frac{f/2}{1 + 5(Pr - 1)\sqrt{(f/2)}}$$

(*Note*: This reduces to the simple Reynolds analogy when $Pr = 1$.)

The above expression gives good accuracy for a range of Prandtl numbers from 0.5 to 2.0.

Colburn *j*-Factor

Colburn[6.3] found experimentally that replacing the denominator of the Prandtl–Taylor expression with $Pr^{2/3}$ gave good results over a wide range of Prandtl numbers, i.e.

$$St\, Pr^{2/3} = j_H = f/2 \qquad [6.14]$$

The expression $\{St\, Pr^{2/3}\}$ is known as the Colburn *j*-factor, j_H. The Colburn expression is exactly true for heat transfer in laminar flow over a flat surface, as shown by the following analysis.

The exact mathematical solution gives the following expression for heat transfer in the laminar boundary layer of a flat plate:

$$Nu = 0.332 Re^{1/2} Pr^{1/3}$$

and the friction factor is

$$f = 0.664/Re^{1/2}$$

Hence,

$$Nu = (f/2) Re\, Pr^{1/3}$$

and

$$\frac{Nu}{Re\, Pr} Pr^{2/3} = \frac{f}{2}$$

i.e.

$$St\, Pr^{2/3} = j_H = f/2$$

Colburn plotted j_H against Re on a logarithmic basis and obtained a straight line, as shown for flow in a pipe in Fig. 6.3(b); a plot of friction factor, f, against Re is given in Fig. 6.3(a) and the similarity can be clearly seen. Transition to turbulence can be caused by any disturbance in the stream and the value of the Reynolds number above which turbulent flow prevails can vary considerably. However, it is found that below a Reynolds number of about 2 000 the viscous forces in the fluid are able to damp out any disturbance and the flow is always laminar.

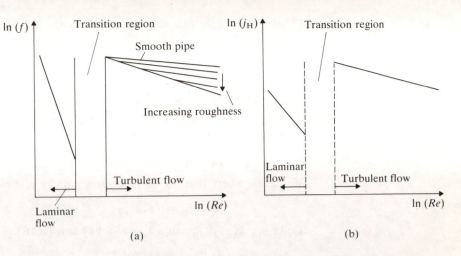

Figure 6.3 Friction factor and Colburn *j*-factor for heat transfer plotted on a log–log basis vs Reynolds number for flow in a pipe

For flow across a curved body, or an inclined flat surface, the boundary layer separates from the surface because of the adverse pressure gradient. This causes a turbulent wake which leads to an additional pressure loss known as *form drag*; the friction loss due to the boundary layer is added to the form drag to give the overall drag. In such cases the heat transfer is increased because of the turbulence of the separated boundary layer and analogies of the form of Eqns [6.12] and [6.14] no longer apply. The graph of j_H against Reynolds number is still linear on a logarithmic basis but the graph of drag coefficient, C_d, against Reynolds number on a logarithmic basis takes the form of that shown in Fig. 6.4. The drag coefficient is defined as the pressure drop across the body divided by $(\rho u^2/2)$, where u is the free stream velocity of the fluid.

Figure 6.4 Drag coefficient vs Reynolds number for a cylinder and a sphere

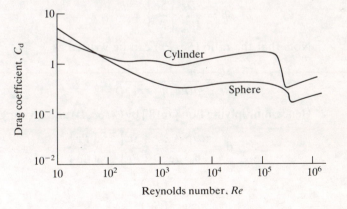

Mass Transfer

It can be shown that a similar analogy holds true for mass transfer. Chilton and Colburn[6.4] defined a mass transfer Stanton number as

$$St_M = Sh/Re\,Sc = \frac{(\beta L \bar{p}_{ln})}{(Dp)} \frac{(v)}{(uL)} \frac{(D)}{(v)}$$

$$= \frac{\beta \bar{p}_{ln}}{up} \qquad\qquad [6.15]$$

(This expression can be compared with that for the Stanton number, $St = Nu/Re\,Pr$; the Schmidt number, $Sc = v/D = \eta/\rho D$, plays the same role as the Prandtl number, $Pr = v/\kappa = c_p\eta/\lambda$.)

A j-factor for mass transfer, j_M, is then defined as

$$j_M = St_M Sc^{2/3} \qquad\qquad [6.16]$$

and it can be shown experimentally that over a range of Prandtl numbers from 0.5 to 50, for both internal and external flows, we can write

$$j_M = j_H$$

For cases where there is no form drag we can also write

$$j_M = j_H = f/2 \qquad\qquad [6.17]$$

Equating j_M and j_H using Eqns [6.14] and [6.16] gives

$$St\,Pr^{2/3} = St_M Sc^{2/3}$$

i.e.
$$\frac{Nu}{Re\,Pr}Pr^{2/3} = \frac{Sh}{Re\,Sc}Sc^{2/3}$$

$$\frac{Nu}{Re\,Pr^{1/3}} = \frac{Sh}{Re\,Sc^{1/3}}$$

i.e.
$$Sh = Nu\left\{\frac{Sc}{Pr}\right\}^{1/3}$$

Using Eqn [6.11] for the Sherwood number,

$$\frac{\beta L \bar{p}_{\ln}}{Dp} = \frac{\alpha L}{\lambda}\left\{\frac{Sc}{Pr}\right\}^{1/3}$$

$$\therefore \quad \beta = \frac{\alpha Dp}{\lambda \bar{p}_{\ln}}\left\{\frac{Sc}{Pr}\right\}^{1/3} \qquad\qquad [6.18]$$

Note that the ratio (Sc/Pr) can be written as

$$\frac{Sc}{Pr} = \frac{\eta}{\rho D}\frac{\lambda}{c_p\eta} = \frac{\lambda}{\rho c_p D}$$

Hence, multiplying Eqn [6.18] by $(\lambda/\rho c_p D)$ and dividing by (Sc/Pr), we have

$$\beta = \frac{\alpha Dp}{\lambda \bar{p}_{\ln}}\frac{\lambda}{\rho c_p D}\left\{\frac{Sc}{Pr}\right\}^{1/3}\left\{\frac{Pr}{Sc}\right\}$$

i.e.
$$\beta = \frac{\alpha p}{\rho c_p \bar{p}_{\ln}}\left\{\frac{Pr}{Sc}\right\}^{2/3} \qquad\qquad [6.19]$$

Hence in cases of mass transfer where the heat transfer coefficient of an equivalent geometry is known, the mass transfer coefficient can be found; note that both the Schmidt and Prandtl numbers are properties of the fluid.

As stated earlier, for most cases of water vapour–air mixtures, it is a good appoximation to assume that $\bar{p}_{\ln} = p$. Therefore, from Eqn [6.19] we have

$$\frac{\alpha}{\beta \rho c_p} = \left\{\frac{Sc}{Pr}\right\}^{2/3} = Le \qquad\qquad [6.20]$$

where Le is the Lewis number.

From Eqn [6.10] we can introduce the mass transfer coefficient, $\beta_\omega = \beta\rho_{ao}$ (where ρ_{ao} is the density of the dry air in the free stream), and therefore

$$Le = \frac{\alpha\rho_{ao}}{\beta_\omega\rho c_p} = \frac{\alpha}{\beta_\omega c_{pma}} \qquad [6.21]$$

where $c_{pma} = \rho c_p/\rho_{ao}$ is the specific heat of the mixture at constant pressure per unit mass of dry air; ρ is the density of the mixture; c_p is the specific heat of the mixture at constant pressure.

Example 6.3

(a)   On their work on fluid flow around curved bodies, Chilton and Colburn have produced results that indicate similarity between the phenomena of heat and mass transfer. Discuss these results and their relationship with skin friction and form drag. Illustrate your answer with sketches of the graphs of Chilton and Colburn j-factors and $(\tau/\rho u^2)$ against Re.

(b)   The mean heat transfer coefficient, α_c, for equipment used for vaporizing water into dry air at a temperature of 52 °C and a pressure of 1.0132 bar is 2.5 kW/m² K. Use the Colburn analogy to predict the mass transfer coefficient, β_G, for the equipment at the same air and liquid flow rates. For this water vapour–air mixture the diffusion coefficient, D, is 0.3×10^{-4} m²/s; take all other properties from tables.[6.1] You may assume that $p = \bar{p}_{ln}$, but explain why you have done so.

(Polytechnic of the South Bank)

Solution

(a)   When a fluid flows round a convex curved surface (see Fig. 6.5), a boundary layer forms and gradually thickens in the same way as for a flat surface, but because of the curvature of the surface the free stream is accelerated and this acceleration is accompanied by a corresponding fall in pressure round the surface. As the fluid reaches the turning point of the curve, the fluid begins to decelerate with a corresponding rise in pressure. The boundary layer is therefore made to flow against an adverse pressure gradient and it separates from the surface as shown in the figure, causing a turbulent wake. The overall resistance to the flow is made up of the frictional resistance of the boundary layer plus the resistance due to the separated boundary layer or wake, called the form drag.

Figure 6.5 Flow round a convex surface

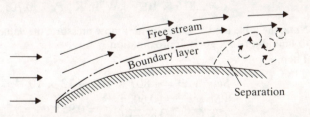

Free stream

Boundary layer

Separation

For very low Reynolds numbers the flow remains laminar over the surface and the boundary layer stays attached to the surface; for example, for flow across a cylinder this is true for $Re < 0.5$. As the Reynolds number increases, the point of separation moves further and further upstream with the boundary layer still remaining laminar up to the point of separation. Figure 6.4 in Section 6.3 (page 245) shows the variation in total drag against Reynolds number for flow across a cylinder and for flow across a sphere. The sharp drop in drag at a high Reynolds number is due to transition to turbulence occurring in the boundary layer before it separates from the surface. When this occurs, the point of separation moves to a point well round the rear surface of the body; for a cylinder this drop in drag occurs at a Reynolds number of about 2×10^5. A fuller explanation of the various flow regimes is given by Douglas et al.[6.5]

For a wide range of Reynolds numbers similar to those encountered in practice, the form drag on a cylinder accounts for most of the total drag (e.g. in the range $3 \times 10^4 < Re < 2 \times 10^5$). The analogies between heat and momentum transfer, and between mass and momentum transfer, are therefore not reliable for flow across bodies such as cylinders and spheres unless the Reynolds number is very low. It can be shown experimentally that there is still a close analogy between heat transfer and mass transfer for such bodies. Provided the same range of Reynolds numbers is considered, corresponding to one of the flow regimes described above, we can write:

$$j_M = j_H \neq \tau / \rho u^2$$

The term $(\tau / \rho u^2)$ corresponds to $f/2$, where f is the friction factor, and τ is the shear stress in the fluid at the surface.

(b)   The problem is solved by starting from the relationship $j_M = j_H$, and proceeding as shown earlier in the derivation of the expression for mass transfer coefficient, Eqn [6.18],

i.e. $$\beta = \frac{\alpha D p}{\lambda \bar{p}_{ln}} \left\{ \frac{Sc}{Pr} \right\}^{1/3}$$

In this case the ratio $p/\bar{p}_{ln}$ can be assumed equal to unity since the partial pressures of the water vapour are small compared with the total pressure.

The Example requires the mass transfer coefficient, β_G, to be calculated. We have

$$\beta_G = \frac{\beta}{\tilde{R}T} = \frac{\alpha D}{\lambda \tilde{R}T} \left\{ \frac{Sc}{Pr} \right\}^{1/3} \qquad\qquad [1]$$

From tables for dry air at 325 K:

$$\lambda = 2.816 \times 10^{-5} \, kW/m \, K; \quad Pr = 0.701; \quad \nu = 1.807 \times 10^{-5} \, m^2/s$$

Note: the table used[6.1] is for dry air at low pressure; the values may be taken as applying to humid air to a good approximation.

Then,

$$Sc = \nu / D = 1.807 \times 10^{-5} / 0.3 \times 10^{-4} = 0.602$$
$$\left\{ \frac{Sc}{Pr} \right\}^{1/3} = \left\{ \frac{0.602}{0.701} \right\}^{1/3} = 0.951$$

Substituting in Eqn [1],

$$\beta_G = \frac{2.5 \times 0.3 \times 10^{-4} \times 0.951 \times 10^5}{2.816 \times 10^{-5} \times 8314.5 \times 325}$$
$$= 0.0937 \, kmol/m^2 \, s \, bar$$

or $$\beta_P = 0.0937 \times 18 = 1.686 \, kg/m^2 \, bar$$

The mass transfer coefficient can also be expressed as

$$\beta = \beta_G \tilde{R} T = 0.0937 \times 10^{-5} \times 8314.5 \times 325 = 2.532 \text{ m/s}$$

and as

$$\beta_\omega = \beta \rho_{ao} = 2.532 \times p_{ao}/R_a T = 2.532 \times 1.0132 \times 10^5/287 \times 325$$
$$\therefore \qquad \beta_\omega = 2.749 \text{ kg/m}^2 \text{ s}$$

(The equations for mass transfer using various definitions of mass transfer coefficient are given in Table 6.1, page 241.)

The Example does not ask for the rate of evaporation but this can be calculated using any of the mass transfer equations. For example, using Eqn [6.9]:

$$\dot{m}_s/A = \beta_P(p_{si} - p_{so}) = 1.686(0.1369 - 0)$$
$$= 0.231 \text{ kg/m}^2 \text{ s}$$

where $p_{si} = p_g$ at $52\,°C = 0.1369$ bar; $p_{so} = 0$ since the air is dry in the free stream.

Example 6.4

(a) The heat and mass transport across a wet bulb thermometer offers an interesting example of the use of the Colburn analogy. From a consideration of the wet bulb thermometer and the principles involved, starting with the thermal balance equation,

$$(\alpha_c + \alpha_r)A(t_{db} - t_{wb}) = \dot{m}_s h_{fg} \qquad [1]$$

show that,

$$\frac{\alpha_c}{\beta_G} = \frac{h_{fg} p \tilde{m}_a}{\{1 + (\alpha_r/\alpha_c)\}} \frac{(\omega_i - \omega_o)}{(t_{db} - t_{wb})} \qquad [2]$$

Then at this stage of the development introduce the Colburn analogy, $j_H = j_M = St_M Sc^{2/3}$ to show, in combination with Eqn [2], that

$$\frac{\omega_i - \omega_o}{t_{db} - t_{wb}} = \frac{c_{pma}}{h_{fg}} \left\{\frac{Sc}{Pr}\right\}^{2/3}$$

where A is the interface area; α_c is the convective heat transfer coefficient; α_r is the radiative heat transfer coefficient; ω_i is the specific humidity of the air at the interface; ω_o is the specific humidity of the air in the free stream; $\dot{m}_s$ is the mass transfer rate of water from the wet bulb; p is the total pressure of the system; $\tilde{m}_a$ is the molar mass of dry air; t_{db} is the dry bulb temperature of the air stream; t_{wb} is the temperature reading of the wet bulb thermometer; c_{pma} is the specific heat at constant pressure of the air in the free stream per unit mass of dry air.

Note: It may be assumed that the air–wick interface temperature is at the wet bulb temperature and that $p = \bar{p}_{ln}$.

(b) The wet and dry bulb temperatures of moist air are 11 °C, (sling), and 18 °C. Calculate its specific humidity if its pressure is 1.013 bar. Take $Pr = 0.71$ and $Sc = 0.60$, and neglect heat transfer by radiation.

(Polytechnic of the South Bank)

Solution

$$\omega_i - \omega_o = \frac{m_{si}}{m_{ai}} - \frac{m_{so}}{m_{ao}} = \frac{p_{si} R_a}{p_{ai} R_s} - \frac{p_{so} R_a}{p_{ao} R_s}$$

$$= \frac{R_a}{R_s} \left\{ \frac{p_{si}}{(p - p_{si})} - \frac{p_{so}}{(p - p_{so})} \right\}$$

Since the partial pressure of the water vapour is very small compared with the total pressure, we have

$$(p - p_{si}) = (p - p_{so}) = p_a$$

$$\therefore \qquad \omega_i - \omega_o = \frac{\tilde{m}_s}{p_a \tilde{m}_a} (p_{si} - p_{so}) \qquad [3]$$

By definition of mass transfer coefficient, β_G, (Eqn [6.8]), we have

$$\dot{m}_s = \beta_G \tilde{m}_s A (p_{si} - p_{so})$$

Substituting for $(p_{si} - p_{so})$ from Eqn [3],

$$\dot{m}_s = \beta_G \tilde{m}_s A (\omega_i - \omega_o) \frac{p_a \tilde{m}_a}{\tilde{m}_s}$$

$$= \beta_G A p_a \tilde{m}_a (\omega_i - \omega_o)$$

Then substituting in Eqn [1] given in the Example,

$$(\alpha_c + \alpha_r) A (t_{db} - t_{wb}) = \beta_G A p_a \tilde{m}_a (\omega_i - \omega_o) h_{fg}$$

i.e. $\qquad \dfrac{\alpha_c}{\beta_G} = \dfrac{h_{fg} p_a \tilde{m}_a}{\{1 + (\alpha_r/\alpha_c)\}} \dfrac{(\omega_i - \omega_o)}{(t_{db} - t_{wb})}$

Then, neglecting radiation effects,

$$\frac{(\omega_i - \omega_o)}{(t_{db} - t_{wb})} = \frac{\alpha_c}{\beta_G h_{fg} p_a \tilde{m}_a} \qquad [4]$$

From the Colburn analogy,

$$St\, Pr^{2/3} = St_M\, Sc^{2/3}$$

Substituting for St from Eqn [6.13], and for St_M from Eqn [6.15],

$$\frac{\alpha_c}{\rho c_p u} Pr^{2/3} = \frac{\beta \bar{p}_{ln}}{u\, p} Sc^{2/3}$$

Then writing $\beta = \beta_G \tilde{R} T$, we have:

$$\alpha_c = \frac{\beta_G \tilde{R} T \bar{p}_{ln} \rho c_p}{p} \left\{ \frac{Sc}{Pr} \right\}^{2/3}$$

Assuming that $p = \bar{p}_{ln}$, and substituting in Eqn [4],

$$\frac{(\omega_i - \omega_o)}{(t_{db} - t_{wb})} = \frac{\tilde{R} T \rho c_p}{h_{fg} p_a \tilde{m}_a} \left\{ \frac{Sc}{Pr} \right\}^{2/3}$$

The term $(p_a \tilde{m}_a / \tilde{R} T)$ is the density of the dry air in the mixture, ρ_{ao}, and $c_{pma} = \rho c_p / \rho_{ao}$ is the specific heat of the mixture at constant pressure per unit mass of dry air, i.e.

$$\frac{(\omega_i - \omega_o)}{(t_{db} - t_{wb})} = \frac{c_{pma}}{h_{fg}} \left\{ \frac{Sc}{Pr} \right\}^{2/3}$$

(c)   From the above equation,

$$(\omega_i - \omega_o) = (18 - 11) \frac{c_{pma}}{h_{fg}} \left\{ \frac{0.60}{0.71} \right\}^{2/3} \qquad [5]$$

From tables,[6.1] at 11 °C p_g = 0.013 12 bar and h_{fg} = 2474.9 kJ/kg.
  Then using Eqn [1.23], $\omega = 0.622 p_s/(p - p_s)$, we have

$$\omega_i = 0.622 \times 0.013\,12/(1.013 - 0.013\,12)$$
$$= 0.008\,16$$

Also, using Eqn [1.31],

$$c_{pma} = 1.005 + 1.88\omega \text{ kJ/kg K}$$

Substituting in Eqn [5],

$$0.008\,16 - \omega_o = (18 - 11) \frac{(1.005 + 1.88\omega_o)}{2474.9} \times (0.60/0.71)^{2/3}$$

Therefore the specific humidity of the air in the free stream, ω_o, is 0.005 59.

6.4 DIRECT CONTACT PROCESSES

A direct contact process is defined as one in which a moving air stream comes
in contact with particles of water. Apparatus to humidify or to de-humidify air
by spraying water into an air stream are examples of direct contact processes; a
further example is the cooling tower in which water is cooled by coming in direct
contact with an air stream.

Cooling Tower

Figure 6.6 is a diagrammatic representation of a cooling tower or a humidifier
using hot water. It is assumed that the temperatures of the air and water at their
respective inlets and outlets remain constant with time. The case of counter-flow

Figure 6.6 Direct
contact cooling tower or
heated water humidifier

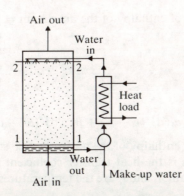

is shown in Fig. 6.6 but the analysis which follows can be applied in a similar way to the case of a parallel flow humidifier using hot water. Note that the analysis does not apply to a humidifier with pumped re-circulation without heating; in that case all the water emerging is evaporated and the process occurs at constant wet bulb temperature (see page 33).

The temperature variations with volume of space are as shown in Fig. 6.7. At any section the specific humidity of the air stream is ω, and the specific humidity is increased by $d\omega$ in the increased space, dV, as the water temperature increases by dt_w in the direction of the air stream.

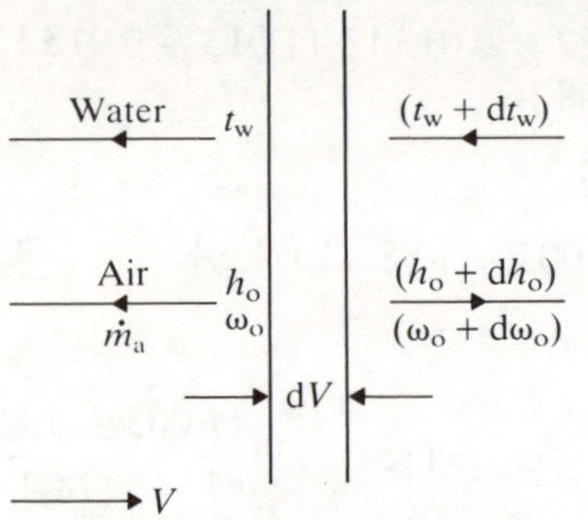

Figure 6.7 Property variations across an infinitely small volume change

Applying Eqn [6.10] to the general cross-section:

$$d\dot{m}_s = \beta_\omega dA(\omega_i - \omega_o)$$

or

$$\dot{m}_a d\omega_o = \beta_\omega a \, dV(\omega_i - \omega_o)$$

where a is the interfacial water surface area of the droplets per unit volume of space.

$$\therefore \qquad \frac{d\omega_o}{\omega_i - \omega_o} = \frac{\beta_\omega a \, dV}{\dot{m}_a} \qquad\qquad [6.22]$$

Then, integrating between sections 1 and 2, we have

$$\int_1^2 \frac{d\omega_o}{\omega_i - \omega_o} = \frac{\beta_\omega aV}{\dot{m}_a} \qquad\qquad [6.23]$$

Considering the total energy transfer during the process at any cross-section, we have

Increase of enthalpy of the air stream = (Heat transfer by convection from the droplets to the air) + (Heat transfer due to the evaporation from the water droplets to the air)

i.e. $\dot{m}_a dh_o = \alpha a \, dV(t_i - t_o) + \beta_\omega a \, dV(\omega_i - \omega_o)h_{fg}$

where h_o is the specific enthalpy of the air in the free stream at any section per unit mass of dry air; α is the heat transfer coefficient for convection from the interface to the air stream; t_i and t_o are the temperatures of the interface and free stream respectively; h_{fg} is the specific enthalpy of vaporization at the interface temperature.

Then, introducing the Lewis number from Eqn [6.21], $Le = \alpha/\beta_\omega c_{pma}$, we have:

$$\dot{m}_a dh_o = \beta_\omega c_{pma} Le\, a\, dV (t_i - t_o) + \beta_\omega a\, dV (\omega_i - \omega_o) h_{fg}$$

$$= \beta_\omega a\, dV \{ Le\, c_{pma}(t_i - t_o) + (\omega_i - \omega_o) h_{fg} \} \qquad [1]$$

Using Eqn [1.30] for the specific enthalpy of the air per unit mass of dry air, we have

$$h = c_{pma} t + 2500\omega$$

where h is in kJ/kg, c_{pma} is in kJ/kg K, and t is in °C.

Then, in Eqn [1],

$$\dot{m}_a dh_o = \beta_\omega a\, dV \{ Le(h_i - h_o) - 2500 Le(\omega_i - \omega_o) + (\omega_i - \omega_o) h_{fg} \}$$

i.e.

$$\dot{m}_a dh_o = \beta_\omega a\, dV Le \left\{ (h_i - h_o) - (\omega_i - \omega_o)\left(2500 - \frac{h_{fg}}{Le} \right) \right\}$$

For the conditions appertaining to air conditioning equipment the Lewis number is approximately unity, and h_{fg} is in the range 2400–2500. Therefore the expression $(\omega_i - \omega_o)(2500 - h_{fg}/Le)$ in the above equation is always very small compared with the specific enthalpy difference and can be neglected. Then,

$$\dot{m}_a dh_o = \beta_\omega a\, dV (h_i - h_o) \qquad [6.24]$$

(F. Merkel was the first to derive this relationship in the 1920s and it is sometimes referred to as the Merkel enthalpy potential.)

$$\therefore \quad \int_1^2 \frac{dh_o}{h_i - h_o} = \int_0^V \frac{\beta_\omega a\, dV}{\dot{m}_a}$$

and

$$\int_1^2 \frac{dh_o}{h_i - h_o} = \frac{\beta_\omega a V}{\dot{m}_a} \qquad [6.25]$$

The term on the left-hand side of Eqn [6.25] is known as the Number of Transfer Units, N_{tu}. The N_{tu} for a cooling tower or humidifier can be defined as the change of specific enthalpy of the air divided by the mean specific enthalpy driving force between the air and the air–water interface.

Combining Eqn [6.23] and [6.26], we have

$$\frac{\beta_\omega a\, dV}{\dot{m}_a} = \frac{d\omega_o}{\omega_i - \omega_o} = \frac{dh_o}{h_i - h_o}$$

$$\therefore \quad \frac{dh_o}{d\omega_o} = \frac{(h_i - h_o)}{(\omega_i - \omega_o)} \qquad [6.26]$$

Equation [6.26] gives the slope of the process line at any position on the psychrometric chart.

Also, for the heat transfer to the air stream,

$$\dot{m}_a c_{pma} dt_o = \alpha a\, dV (t_i - t_o)$$

$$\therefore \quad \frac{dt_o}{(t_i - t_o)} = \frac{\alpha a\, dV}{\dot{m}_a c_{pma}}$$

and introducing the Lewis number, $\alpha/\beta_\omega c_{pma}$,

$$\frac{dt_o}{(t_i - t_o)} = \frac{\beta_\omega Le\,a\,dV}{\dot{m}_a}$$

Then putting $Le = 1$,

$$\frac{dt_o}{(t_i - t_o)} = \frac{\beta_\omega a\,dV}{\dot{m}_a}$$

This equation can now be compared with Eqn [6.25],

$$\frac{dh_o}{(h_i - h_o)} = \frac{\beta_\omega a\,dV}{\dot{m}_a}$$

Then combining the two equations:

$$\frac{dh_o}{dt_o} = \frac{h_i - h_o}{t_i - t_o} \qquad\qquad [6.27]$$

Equation [6.27] gives the slope of the process line at any position on the enthalpy–temperature graph.

At any cross-section the mass flow rate of water decreases as the water evaporates into the air stream, but it is a good approximation to assume that the decrease in mass flow rate is very small compared with the mass flow rate itself. Therefore we can write,

$$\dot{m}_a\,dh_o = \dot{m}_w c_w\,dt_w$$

where c_w is a mean specific heat for water,

$$\therefore \qquad \frac{dh_o}{c_w\,dt_w} = \frac{\dot{m}_w}{\dot{m}_a} \qquad\qquad [6.28]$$

or

$$\frac{h_{o2} - h_{o1}}{t_{w2} - t_{w1}} = \frac{c_w \dot{m}_w}{\dot{m}_a} \qquad\qquad [6.29]$$

If the properties of humid air are plotted as specific enthalpy against temperature, as shown in Fig. 6.8, the line, AB, given by Eqn [6.29] can be plotted on the diagram and is known as the operating line or energy balance line. Figure 6.8(a) shows the case for counter-flow, and Fig. 6.8(b) shows the case for parallel-flow; note that for parallel flow Eqns [6.28] and [6.29] have a minus sign and hence the slope of AB is negative as shown.

Figure 6.8 Specific enthalpy vs temperature for counter-flow and parallel-flow humidification

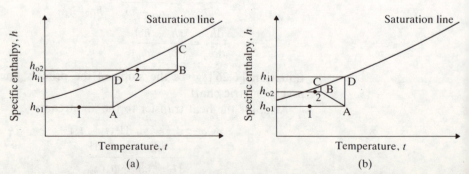

(a)   (b)

Operating line method

Initially, as an approximation, the temperature of the water droplets at any cross-section, t_w, will be assumed to be equal to the interface temperature, t_i, at that cross-section. This implies that the resistance to heat transfer of the air film at the air–water interface is very large compared with the resistance to heat transfer from the bulk of the water to the water surface.

Point D on Fig. 6.8 therefore represents the air–water interface condition at the air inlet, and the specific enthalpy difference at section 1, $(h_{i1} - h_{o1})$ is given by the vertical line AD; similarly the line BC represents $(h_{i2} - h_{o2})$.

From Eqn [6.25] it can be seen that the number of transfer units, $\beta_\omega a V / \dot{m}_a$, is given by the integral of $dh_o/(h_i - h_o)$, and hence the volume V required for any particular inlet and outlet conditions can be obtained by plotting values of the reciprocals of the vertical lines, i.e. $1/(h_i - h_o)$, against h_o and evaluating the area under the line. The shaded area in Fig. 6.9 gives the number of transfer units, $\int_1^2 dh_o/(h_i - h_o) = \beta_\omega a V / \dot{m}_a$. The value of the product $\beta_\omega a$ is found by experiment and is known for any given humidifier or cooling tower. The required volume of the tower can then be found for a known mass flow rate of air, m_a.

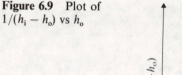

Figure 6.9 Plot of $1/(h_i - h_o)$ vs h_o

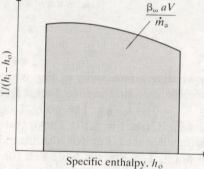

Example 6.5

A vertical induced-draught cooling tower is to be used to cool water from a temperature of 50 °C to 26 °C. The water flow rate is 0.9 kg/s and the air approach condition is 20 °C dry bulb and 11.1 °C (sling).

Using a water flow rate of 0.3 kg/m² s and an air/water ratio 1.4 times the minimum to obtain a recommended value of the product βa of 0.24 s^{-1}, determine:

(i) the air condition curve, using the operating line method, on the specific enthalpy–temperature chart provided (see Fig. 6.10);

(ii) the necessary packed height of the tower, using a suitable numerical integration method, and its area of cross-section;

(iii) the temperature of the air at exit from the tower, and the temperature of the air at a section where the water temperature is 30 °C.

Note: The resistance to heat and mass transfer in the liquid phase may be considered negligible compared with that in the gas phase.

(Polytechnic of the South Bank)

Figure 6.10 Specific enthalpy vs temperature chart for humid air

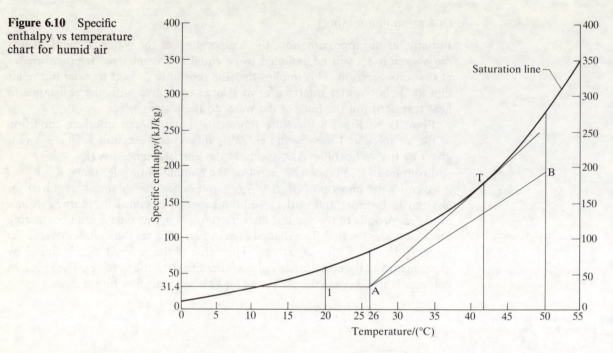

Solution

The process is shown diagrammatically in Fig. 6.11.

The air inlet specific enthalpy can be found from tables[6.6] at 20 °C and 11.1 °C (sling), as $h_{o1} = 31.4$ kJ/kg. The point 1 can then be fixed at $h_{o1} = 31.4$ kJ/kg and $t_{a1} = 20$ °C (see Fig. 6.10). The operating line starts from point A and, for the ideal case when the air at exit is saturated, will touch the saturation line at the tangent point, T, as shown. Note that for this ideal case the specific enthalpy difference, $h_i - h_o$, is zero at the air exit, implying an infinitely high tower. The line AT represents the minimum air/water ratio.

From the specific enthalpy–temperature chart, the specific enthalpy at point T is found to be 180 kJ/kg at a water temperature of 41.5 °C. Therefore from

Figure 6.11 Diagram of cooling tower for Example 6.5

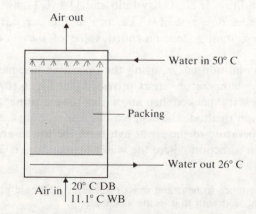

Eqn [6.29] for the slope of the operating line, we have

$$\frac{180 - 31.4}{41.5 - 26} = \frac{c_w \dot{m}_w}{\dot{m}_a} = 9.587 \text{ kJ/kg K}$$

Taking a mean specific heat of water at $(50 + 26)/2 = 38\,°C$ as 4.179 kJ/kg K, we have

$$\dot{m}_a = \frac{4.179 \times 0.3}{9.587} = 0.1308 \text{ kg/m}^2 \text{ s}$$

The actual air/water ratio is 1.4 times the minimum value; hence the actual slope of the operating line is

$$\frac{c_w \dot{m}_w}{\dot{m}_a} = \frac{9.587}{1.4} = 6.85 \text{ kJ/kg K}$$

The actual air mass flow rate per unit area is then

$$0.1308 \times 1.4 = 0.1831 \text{ kg/m}^2 \text{ s}$$

The actual operating line can now be drawn from point A to point B, where $h_{oB} = 31.4 + 6.85(50 - 26) = 195.8$ kJ/kg.

(i) The air condition curve on the specific enthalpy–temperature chart can be plotted taking the slope of the line at any point from Eqn [6.27], i.e.

$$\frac{dh_o}{dh_o} = \frac{h_i - h_o}{t_i - t_o}$$

At point 1 the specific enthalpy at the air/water interface, h_i, is equal to the specific enthalpy of saturated air per unit mass of dry air at $t_i = 26\,°C$, so the slope at point 1 can be found by joining point 1 to point C on Fig. 6.12, i.e. slope is CA/1A. Taking a point D on the line 1C, draw a horizontal line to cut the operating line at point E; then draw a vertical line through E to cut the saturation line at point F. The slope of the condition curve at point D is then given by the line DF, i.e. slope is FE/DE. Taking a point G on line DF, draw a horizontal line to cut the operating line at point H; then draw a vertical line cutting the saturation line at point I. The slope of the condition line at

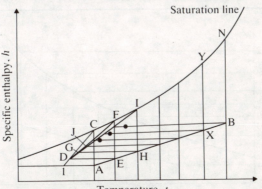

Figure 6.12 Air condition line plotted on specific enthalpy vs temperature chart for Example 6.5

point G is then given by the line GI. The method is then continued until a horizontal line is drawn through the penultimate point to cut the operating line at X; the final part of the condition line is then drawn from the penultimate point to Y and where the line cuts the horizontal line through B, point 2 is fixed. Since a numerical method is necessary to evaluate $N_{tu} = \int_1^2 dh_o/(h_i - h_o)$ (see below), it is convenient to divide the operating line into a number of equal increments as shown in Fig. 6.12.

This method gives a series of small straight lines, the locus of which is the condition line; clearly, the smaller the increments chosen, the greater the accuracy. A similar method using Eqn [6.26] gives the condition line on the psychrometric chart.

(ii)   The size of the tower can be found from Eqn [6.25] using numerical integration to evaluate the integral. Simpson's rule can be used to evaluate the area under the plot of $1/(h_i - h_o)$ against h_o. For an integral of the form $\int y \, dx$ the range $(b-a)$ is divided into an *even* number of intervals, n. Taking the ordinates of the curve at the points chosen as $y_0, y_1, y_2, \ldots y_n$, the integral is given by:

$$\frac{(b - a)}{3n} \{y_0 + y_n + 4(y_1 + y_3 + \ldots + y_{n-1})$$
$$+ 2(y_2 + y_4 + \ldots + y_{n-2})\} \qquad [6.30]$$

For this example, divide the temperature range, $(50-26)$ K, into six intervals, say, of 4 K. Then from Fig. 6.10 the values of $(h_i - h_o)$ can be found by measuring the vertical distances from the operating line to the saturation line. The table below gives the measured values, with the numbers referring to a sketch of the chart given in Fig. 6.13.

Point	t_i (°C)	h_i (kJ/kg)	h_o (kJ/kg)	$h_i - h_o$ (kJ/kg)	$1/(h_i - h_o)$ (kg/kJ)
0	26	80.8	31.4	49.5	0.020
1	30	100.0	58.8	41.2	0.024
2	34	123.0	86.2	36.8	0.027
3	38	150.7	113.6	37.1	0.027
4	42	184.2	141.0	43.2	0.023
5	46	225.0	168.4	56.6	0.018
6	50	275.2	195.8	79.4	0.013

Figure 6.13 Sketch of specific enthalpy vs temperature chart for Example 6.5

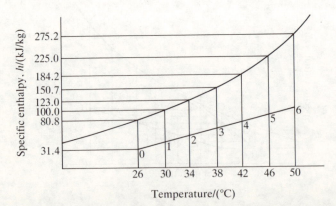

The temperature interval of $(50 - 26)/6 = 4$ K is equivalent to an interval in h_o of $(4 \times 6.85) = 27.4$ kJ/kg, i.e. $(b - a)/n = 27.4$ kJ/kg.

Applying Simpson's rule, Eqn [6.30], the integral $dh_o/(h_i - h_o)$ is given by

$$\frac{27.4}{3}\{0.020 + 0.013 + 4(0.024 + 0.027 + 0.018)$$
$$+ 2(0.027 + 0.023)\}$$

i.e.

$$\int_1^2 \frac{dh_o}{h_i - h_o} = 3.74$$

Note: The specific enthalpies for the table above can be read from the chart (or values of h_i read from tables[6.6], and values of h_o taken as $\{31.4 + 6.85(t_i - 26)\}$, where the operating line slope is 6.85 kJ/kg K). Alternatively the vertical distances from the operating line to the saturation line can be measured with a ruler, the reciprocal taken and the area evaluated. If the interval, $(b - a)/n$, is also measured with a ruler, and then multiplied by the operating line slope of 6.85, then the area will be given in the correct non-dimensional form.

Using Eqn [6.25],

$$3.74 = \frac{\beta_\omega aV}{\dot{m}_a}$$

The value of βa is given as 0.24 s^{-1}; hence the product $\beta_\omega a$ is equal to $0.24 \times 1.2 = 0.288 \text{ kg/m}^3$ s, where the density, ρ_{ao}, is taken as approximately 1.2 kg/m^3. The volume of the tower is equal to the height, z, times the cross-sectional area, i.e. $V = zA$. Therefore we have

$$z = \frac{3.74 \times 0.1831}{0.288} = 2.38 \text{ m}$$

The mass flow rate of water is given as 0.9 kg/s; therefore the cross-sectional area, A, of the tower is $0.9/0.3 = 3 \text{ m}^2$.

(iii)   From the condition line drawn on the specific enthalpy–temperature chart the temperature of the air at exit is found to be 43.5 °C. The air temperature when the water temperature is 30 °C is found by locating the point on the operating line at 30 °C, then drawing a horizontal line through it to cut the condition line. The air temperature is found by this means to be 23.5 °C.

Stevens diagram method

Carey and Williamson[6.7] first used a Stevens diagram to find the value of $\int_1^2 dh_o/(h_i - h_o)$, based on the assumption that the values of $(h_i - h_o)$ vary in a parabolic manner. A Stevens diagram is a graphical construction of three known values of the difference function, $y = h_i - h_o$, at the initial (y_1), final (y_2), and mean value (y_m), of h_o. The diagram is a plot of (y_m/y_1) against (y_m/y_2) for various values of a factor, f, where:

$$\int_1^2 \frac{dh_o}{h_i - h_o} = \frac{h_{o2} - h_{o1}}{f y_m} \qquad [6.31]$$

A Stevens diagram is shown in Fig. 6.14.

Figure 6.14 Stevens diagram

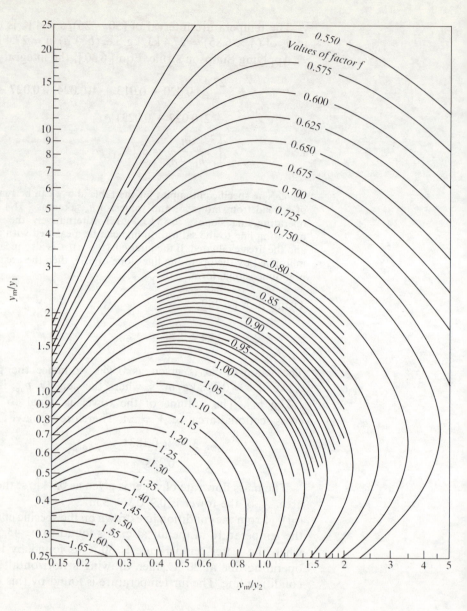

Referring back to Fig. 6.13, the three difference values required in order to apply the Stevens diagram would be at point 0 for y_1, at point 6 for y_2, and at point 3 for y_m. Taking the data from Example 6.5, we have:

$$y_1 = 49.5 \text{ kJ/kg}; \quad y_2 = 79.4 \text{ kJ/kg}; \quad y_m = 37.1 \text{ kJ/kg}$$

i.e. $\quad y_m/y_1 = 37.1/49.5 = 0.75, \quad$ and $\quad y_m/y_2 = 37.1/79.4 = 0.47$

Then, from the Stevens diagram, $f = 1.17$, and therefore using Eqn [6.31],

$$\int_1^2 \frac{dh_o}{h_i - h_o} = \frac{h_{o2} - h_{o1}}{f y_m} = \frac{195.8 - 31.4}{1.17 \times 37.1} = 3.79$$

This gives a tower height of

$$(3.79 \times 0.1831)/0.288 = 2.41 \text{ m}$$

which compares with the value of 2.38 m found using Simpson's rule.

Thermal resistance from a water droplet to the water–air interface

In the above analysis it is assumed that the water droplet is at the same temperature as the water–air interface; a more accurate solution is obtained if the thermal resistance for heat transfer from the bulk of the water to the interface is taken into account. The real temperature variation at any section of the tower is as shown with reference to a water droplet in Fig. 6.15. For the rate of heat transfer, $\dot{q}$, from the water droplet to its interface with the air we can write:

$$\dot{q} = \frac{(t_w - t_i)}{R_w}$$

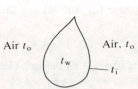
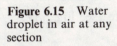

Air t_o Air, t_o

t_w

t_i

Water droplet

Figure 6.15 Water droplet in air at any section

where R_w is the thermal resistance for heat transfer from the bulk of the water droplet to its interface per unit interface area.

In the steady state the rate of heat transfer, $\dot{q}$, is equal to the rate of increase in specific enthalpy of the air for unit interface area, given by Eqn [6.24], i.e.

$$\frac{(t_w - t_i)}{R_w} = \beta_\omega (h_i - h_o)$$

$$\therefore \quad \frac{h_i - h_o}{t_i - t_w} = -\frac{1}{R_w \beta_\omega} \qquad [6.32]$$

In the case of a cooling tower or humidifier using hot water, the water temperature throughout is above the air temperature and hence the water temperature, t_w, is always greater than the interface temperature. Figure 6.16 shows a representation of the process on the specific enthalpy–temperature chart when the thermal resistance of the water droplet is taken into consideration. For any point P within the tower or humidifier, the specific enthalpy and temperature of the air are h_o and t_o, the temperature of the water is t_w, and the temperature of the interface is t_i. The slope of the line QR is given by Eqn [6.32]. This line is sometimes called the *tie-line*. In the simplified analysis (see Example 6.5), the tie-line is a vertical line; note that this implies a slope of infinity, which is given by Eqn [6.32] when R_w is zero.

Figure 6.16 Specific enthalpy vs temperature chart for non-vertical tie line

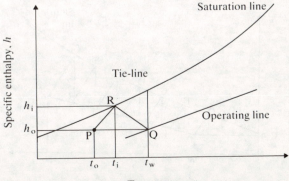
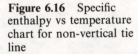

Saturation line

Specific enthalpy, h

Tie-line

R

Operating line

h_i

h_o

P   Q

t_o t_i t_w

Temperature, t

Note that Eqn [6.27] gives the slope of the condition line at any point, as before, i.e.

$$\frac{dh_o}{dt_o} = \frac{(h_i - h_o)}{(t_i - t_o)} = \text{Slope of PR}$$

Problems can therefore be solved as in Example 6.5 but with the tie-line at a slope of $-1/(R_w \beta_\omega)$ instead of as a vertical line.

Example 6.6

Calculate the height of the tower for the data of Example 6.5 but taking the product $(R_w \beta_\omega)$ as 0.01 kg K/kJ.

Solution

The operating line for the specific enthalpy–temperature chart is as before. The tie-line slope from Eqn [6.32] is

$$-1/R_w \beta_\omega = -1/0.01 = -100 \text{ kJ/kg K}$$

i.e.   Slope of 0A = Slope of 1B = Slope of 2C, etc. (see Fig. 6.17)
$$= -100 \text{ kJ/kg K}$$

The lines 0A, 1B, 2C, 3D, 4E, 5F and 6G are then drawn on the chart as shown in Fig. 6.17; the values of $(h_i - h_o)$ at each point are given by the vertical distances $(h_A - h_o)$, $(h_B - h_1)$, etc. Drawing right-angle triangles as shown, the vertical distances can be measured directly from the chart. The table below gives the values of these vertical distances in millimetres, and the corresponding reciprocals.

Point	$S(h_i - h_o)$* (mm)	$1/S(h_i - h_o)$* (mm^{-1})
0	23.0	0.044
1	18.5	0.054
2	16.7	0.060
3	17.5	0.057
4	19.5	0.051
5	25.8	0.039
6	35.6	0.028

* S is the scale of the enthalpy axis; in this case 0.5 mm per kJ/kg

Figure 6.17 Specific enthalpy vs temperature chart for Example 6.6

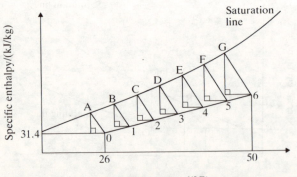

Then applying Simpson's rule using Eqn [6.30], the integral is

$$(27.4 \times 0.5/3)\{0.044 + 0.028 + 4(0.054 + 0.057 + 0.039) \\ + 2(0.060 + 0.051)\}$$

where the interval in h_o is 27.4 kJ/kg K as before, and from the scale of the specific enthalpy axis of the chart used, 1 kJ/kg is equivalent to 0.5 mm.

$$\therefore \quad \frac{\beta_\omega a V}{\dot{m}_a} = 4.08$$

Then, using the data given in Example 6.5,

$$\text{Height} = \frac{4.08 \times 0.1831}{0.288} = 2.59 \text{ m}$$

It can be seen from the table that a value for $(R_w \beta_\omega)$ of 0.01 kg K/kJ given in the Example implies a temperature difference of $23 \times 0.01/0.5 = 0.46$ K between the water temperature and the interface temperature at the air inlet. The height of the tower, 2.59 m, compares with the previous value of 2.38 m found by neglecting the thermal resistance of the water droplet.

Humidifier with Pumped Re-circulation

When water is pumped into a humidifier and continuously re-circulated as shown in Fig. 6.18, the water will reach an equilibrium value given by the wet bulb temperature of the entering air. Neglecting the energy input from the pump, and assuming an adiabatic process, the process will take place at constant wet bulb temperature as shown on a sketch of the psychrometric chart in Fig. 6.19. This type of humidifier is sometimes known as an air washer.

Figure 6.18 Humidifier with pumped re-circulation

Figure 6.19 Condition line on the psychrometric chart for a humidifier with pumped re-circulation

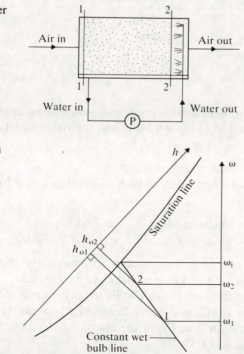

Since the air wet bulb temperature is constant throughout, the water temperature is constant throughout and equal to the wet bulb temperature. It follows that the specific humidity of the air at the air–water interface, ω_i, is also constant throughout. Equation [6.22] can therefore be integrated directly as follows:

$$\int_1^2 \frac{d\omega_o}{\omega_i - \omega_o} = \int_0^V \frac{\beta_\omega a\, dV}{\dot{m}_a}$$

i.e.
$$\ln\left\{\frac{\omega_i - \omega_{o1}}{\omega_i - \omega_{o2}}\right\} = \frac{\beta_\omega a V}{\dot{m}_a} \qquad [6.33]$$

Knowing the end states of the air, the size of the humidifier can then be found from Eqn [6.33].

The size of the humidifier can also be found from Eqn [6.24] in terms of specific enthalpies, but since the specific enthalpy changes are very small the method is not so accurate; the condition line is shown on a specific enthalpy–temperature chart in Fig. 6.20.

Figure 6.20 Condition line on the specific enthalpy vs temperature chart for a pumped re-circulation humidifier

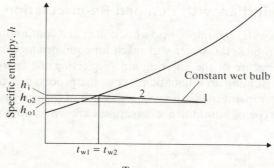

Example 6.7

Air enters a humidifier with pumped re-circulation at the rate of 2.2 m³/s at 32 °C dry bulb and 15 °C wet bulb, and the required saturation at exit is 60%. Calculate:

(i)   the air dry bulb temperature at exit;
(ii) the required length of humidifier assuming a face velocity at inlet no greater than 2 m/s.

Take the value of the product, $\beta_\omega a$ as 2.6 kg/s m³.

Solution

The problem can be solved most simply using the psychrometric chart; the process is shown on a sketch of the chart in Fig. 6.21.

(i)   Point 1 is fixed at the dry bulb temperature of 32 °C and the wet bulb temperature of 15 °C, and point 2 is fixed where the 15 °C wet bulb line cuts the 60% saturation line. From the chart, the dry bulb temperature of the air at exit is found to be 19.7 °C.

Figure 6.21 Process for Example 6.7 on psychrometric chart

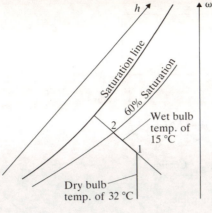

(ii)   From Eqn [6.33],

$$\frac{\beta_\omega aV}{\dot{m}_a} = \ln\left\{\frac{\omega_i - \omega_{o1}}{\omega_i - \omega_{o2}}\right\}$$

From the chart:

$$\omega_{o1} = 0.00345, \ \omega_{o2} = 0.00870, \text{ and } \omega_i = 0.01065$$

Also, $\rho_{ao} = (1/0.869)\,\text{kg/m}^3$

Therefore $\dot{m}_a = \rho_{ao}\dot{V} = 2.2/0.869 = 2.53\,\text{kg/s}$

Therefore,

$$\frac{\beta_\omega aV}{\dot{m}_a} = \ln\left\{\frac{0.010\,65 - 0.003\,45}{0.010\,65 - 0.008\,70}\right\} = 1.306$$

and $$V = \frac{1.306 \times 2.53}{2.6} = 1.27\,\text{m}^3$$

The cross-sectional area of the humidifier is given by $2.2/2 = 1.1\,\text{m}^2$, and therefore the required length is

$$1.27/1.1 = 1.16\,\text{m}$$

De-Humidifier

Air can be de-humidified using a water spray provided the water temperature is low enough to cause condensation from the water vapour in the air onto the water particles; a simple sketch of a direct contact de-humidifier is shown in Fig. 6.22. The theory is exactly the same as for the humidifier supplied with hot water, or the cooling tower, but the water temperatures are now lower than the air temperatures.

The driving force is from the bulk of the air to the air–water interface and hence the operating line is above the saturation line, as shown in Fig. 6.23.

Figure 6.22 Direct
contact de-humidifier

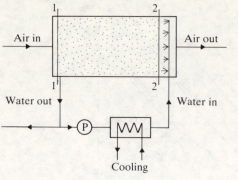

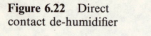

Figure 6.23 Specific
enthalpy vs temperature
chart for a de-humidifier

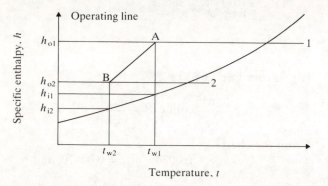

Example 6.8

Air at 40 °C, 20% saturation, flows at a rate of 0.78 m^3/s into a de-humidifier. The water, which flows in counter-current to the air stream, enters at 2 °C and leaves at 8 °C and the water mass flow rate is 1.22 kg/s.

(i) Plot the process line on a psychrometric chart and find the state of the air at exit from the humidifier.

(ii) Assuming a value of $\beta_\omega a$ of 3.6 kg/s m^3, calculate the required length of the humidifier for a cross-sectional area of 0.36 m^2.

Solution

(i) To plot the process on the psychrometric chart, point 1 for the air is first fixed where the 40 °C dry bulb line cuts the 20% saturation line. The slope of the condition line at any point is given by Eqn [6.26], i.e.

$$\frac{dh_o}{d\omega_o} = \frac{(h_i - h_o)}{(\omega_i - \omega_o)}$$

Assuming that the water temperature at any section is equal to the interface temperature at that section, then the interface temperature at the air inlet is 8 °C. The slope of the condition line at point 1 can then be found by connecting point 1 to point i1 at 8 °C on the saturation line as shown on a sketch of the psychrometric chart in Fig. 6.24. The interval $(h_{o2} - h_{o1})$ is given by Eqn [6.29] as,

$$\frac{c_w \dot{m}_w}{\dot{m}_a}(t_{w2} - t_{w1}) = \frac{4.204 \times 1.22 \times (8 - 2)}{(0.78/0.901)}$$

i.e. $h_{o2} - h_{o1} = 35.6$ kJ/kg

Figure 6.24 Condition line for Example 6.8 on the psychrometric chart

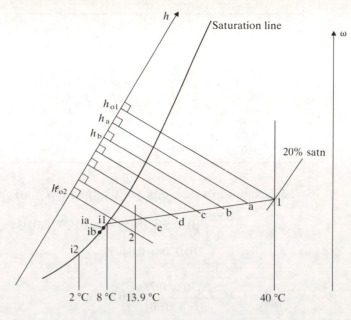

where the mean specific heat of water at $(8 + 2)/2 = 5\,°C$ from tables[6.1] is found to be 4.204 kJ/kg K, and the specific volume of air at point 1 is read from the psychrometric chart as 0.901 m^3/kg.

Divide this interval into six equal parts, say, of $(35.6/6) = 5.93$ kJ/kg, giving seven enthalpy lines on the chart as shown in Fig. 6.24. The saturation line from i1 to i2 is also divided into the same number of equal parts. For example, $h_a = (65.6 - 5.93) = 59.7$ kJ/kg, and point a is located where the line 1–i1 cuts the enthalpy line of 59.7 kJ/kg. Point a is then joined to point ia, and where this line cuts the next enthalpy line, h_b, fixes the point b, and so on until the point 2 is reached at an enthalpy of $(65.6 - 35.6) = 30$ kJ/kg.

From the chart the exit temperature of the air is found to be 13.9 °C with a wet bulb temperature of 10.3 °C.

It is instructive to note that a straight line drawn from point 1 through point 2 cuts the saturation line at a temperature of approximately 4 °C. In Section 1.4 the apparatus dew point for a spray de-humidifier is defined as the average temperature of the spray water; in this case the apparatus dew point is therefore $(2 + 8)/2 = 5\,°C$. It can thus be seen that the approximate method of drawing a straight line between the air inlet state and the apparatus dew point gives an accurate answer. In this example a straight line drawn between state point 1 and the apparatus dew point at 5 °C cuts the 30 kJ/kg enthalpy line at 13.6 °C dry bulb and 10.3 °C wet bulb (compared with the solution above of 13.9 °C dry bulb and 10.3 °C wet bulb; this is excellent agreement).

(ii) Using the same method as before, the size of the de-humidifier can be found by evaluating $\int_1^2 dh_o/(h_o - h_i)$, and then using Eqn [6.25]. Note that for a de-humidifier the air temperature is at all times higher than the water temperature, and hence the sign of Eqn [6.25] changes.

The table overleaf gives the values of h_o, h_i, $(h_o - h_i)$, and $1/(h_o - h_i)$. The values of h_o are as already found from the psychrometric chart; the values of h_i are the saturation values corresponding to the water temperatures from 2 °C to 8 °C.

Point	h_o (kJ/kg)	h_i (kJ/kg)	$h_o - h_i$ (kJ/kg)	$1/(h_o - h_i)$ (kg/kJ)
1	65.60	24.86	40.74	0.025
a	59.67	22.72	36.95	0.027
b	53.74	20.65	33.09	0.030
c	47.81	18.64	29.17	0.034
d	41.88	16.70	25.18	0.040
e	35.95	14.81	21.14	0.047
2	30.02	12.98	17.04	0.059

Using Simpson's rule, Eqn [6.30], the ratio $\beta_\omega aV/\dot{m}_a$ is given by:

$$\frac{35.6}{6 \times 3}\{0.025 + 0.059 + 4(0.027 + 0.034 + 0.047)$$
$$+ 2(0.030 + 0.040)\}$$
$$= 1.30$$

i.e.

$$V = \frac{1.30 \times 0.78}{3.6 \times 0.901} = 0.313 \text{ m}^3$$

Then, for a cross-sectional area of 0.36 m^2, the required length of the humidifier is

$$0.313/0.36 = 0.87 \text{ m}$$

The above example has been solved with references to the psychrometric chart but it could be done just as easily using the specific enthalpy–temperature chart as in the previous examples. Figure 6.25 shows the end states of the air on the specific enthalpy–temperature chart and the operating line AB of slope 5.93.

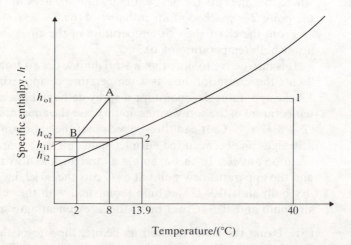

Figure 6.25 Operating line and end states for Example 6.8 on the specific enthalpy vs temperature chart

6.5 COOLING COIL DE-HUMIDIFICATION

De-humidification and cooling of air is more usually accomplished using a finned tube heat exchanger with a refrigerant flowing in the tubes and the air passing over the finned surface. In a direct expansion type, a refrigerant enters the coil and evaporates at constant temperature as it flows through the heat exchanger; in the type using chilled water the temperature of the water rises through the heat exchanger. When the finned tube surface is below the dew point of the air, then condensation of water vapour from the air takes place on to the finned surface; the surfaces are therefore covered with a thin film of water throughout the heat exchanger in a correctly designed de-humidifier.

At any cross-section through the de-humidifier there is a thermal resistance between the refrigerant and the wall of the tube, a thermal resistance of the tube wall itself, and a thermal resistance of the film of water on the outside surface. In most cases the latter two resistances are small compared with the resistance of the refrigerant film on the inside tube surface and can be neglected. That is,

$$d\dot{q} = \frac{(t_i - t_R)}{R_{Rw}} dA_w = \beta_\omega \, dA_o(h_o - h_i) \qquad [6.34]$$

where $R_{Rw} = 1/\alpha_{Rw}$ is the thermal resistance of the refrigerant film; α_{Rw} is the heat transfer coefficient between the refrigerant and the tube inside surface; dA_w is a small element of the inside surface area in the direction of the air flow; dA_o is a small element of the effective outside surface area in the direction of the air flow.

The effective outside surface area is given by

$$A_o = A_b + \eta_F A_F$$

where η_F is a fin efficiency for combined heat and mass transfer, A_b is the area of the tube outside base surface in between the fins, and A_F is the total finned surface area. If we assume that a value of R_{Ro} is defined with reference to the effetive outside area, dA_o, such that $R_{Ro} = R_{Rw} dA_o/dA_w$, then we have

$$\frac{h_i - h_o}{t_i - t_R} = -\frac{1}{\beta_\omega R_{Ro}} \qquad [6.35]$$

Equation [6.35] is similar to Eqn [6.32] for the spray-type humidifier which allows for the temperature differences between the water droplet and the interface.

The same tie-line method can be used (see Example 6.6), using Eqn [6.35] to draw sloping lines on the specific enthalpy–temperature chart as shown in Fig. 6.26. Points 1 and 2 represent the inlet and outlet conditions of the air, and the line AB is the operating line. The chart shown is for a counter-flow

Figure 6.26 Coil de-humidifier on the specific enthalpy vs temperature chart

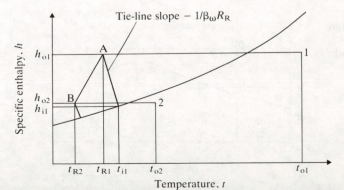

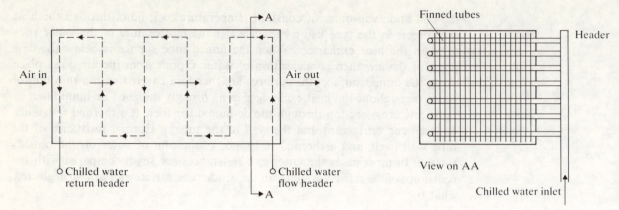

Figure 6.27
Diagrammatic
arrangement of a
six-coil cooling coil

de-humidifier using chilled water, shown diagrammatically in Fig. 6.27. The slope of the operating line, AB, is given by Eqn [6.29] as before, i.e.

$$(h_{o1} - h_{o2})/(t_{R1} - t_{R2}) = c_w \dot{m}_R / \dot{m}_w$$

where $\dot{m}_R$ is the mass flow rate of the chilled water, and c_w is its mean specific heat.

In the case of a de-humidifier using an evaporating refrigerant in the coils, the temperature t_R remains constant throughout. AB then has an infinite slope and is therefore a vertical line on the specific enthalpy–temperature chart as shown in Fig. 6.28; for an evaporating vapour the heat transfer coefficient is very large and hence R_R is very low and the tie-line slope is very high.

It can be seen from Fig. 6.27 that for a chilled water cooling coil an increase in area through the system is in effect an increase in the number of rows of finned tubes. To design a system therefore requires an iterative procedure similar to a dry heat exchanger of the same type. Threlkeld[6.8] has shown that the solution for wet finned surfaces is the same in all respects as the solution for dry finned surfaces but using a modified heat transfer coefficient in place of the outside surface heat transfer coefficient. The modified heat transfer coefficient allows for the mass transfer and also uses a fictitious enthalpy for the inside fluid; this is considered in more detail in the next section.

Figure 6.28 Specific
enthalpy vs temperature
diagram for a direct
expansion cooling coil

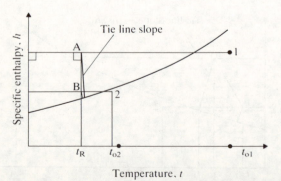

Fictitious Enthalpy Method

With reference to Fig. 6.10 (page 256), it can be seen that over a limited temperature range the saturation line is approximately straight, and we can write as a good approximation:

$$\text{Specific enthalpy of air at saturation} = a + bt_{Sat} \qquad [6.36]$$

where a and b are constants.

Now for a coil de-humidifier, we have from Eqn [6.34]:

$$\frac{(t_i - t_R)\,dA_w}{R_{Rw}} = d\dot{q} \qquad\qquad [1]$$

where dA_w is a small element of the inside surface area.

Assuming fixed values of a and b over the temperature range t_i to t_R, using Eqn [6.36] we can write:

$$t_i - t_R = \frac{(h_i - h_R)}{b_m}$$

where b_m is a mean value of the constant b over the temperature range, and the difference in the values of a_i and a_R is taken to be negligible.

Then, substituting in Eqn [1] above,

$$\frac{d\dot{q}}{dA_w} = \frac{(h_i - h_R)}{R_{Rw}b_m} = \frac{\dot{Q}}{A_w} \qquad\qquad [2]$$

The specific enthalpy h_R is an equivalent specific enthalpy for saturated air at the refrigerant temperature at any section, and is known as the fictitious enthalpy.

Also, we have from Eqn [6.34],

$$h_o - h_i = \frac{d\dot{q}}{\beta_\omega dA_o} = \frac{\dot{Q}}{\beta_\omega A_o} \qquad\qquad [3]$$

where dA_o is a small element of the effective outside area.

Then adding the specific enthalpy difference, $(h_o - h_i)$, from Eqn [3] to the specific enthalpy difference, $(h_i - h_R)$, from Eqn [2], we have:

$$h_o - h_R = \dot{Q}\left(\frac{1}{\beta_\omega A_o} + \frac{b_m R_{Rw}}{A_w}\right)$$

Then an overall coefficient, B, based on the actual outside surface area, A, can be defined as:

$$\dot{Q} = BA(h_o - h_R)$$

where

$$\frac{1}{BA} = \frac{1}{\beta_\omega A_o} + \frac{b_m R_{Rw}}{A_w} \qquad\qquad [6.37]$$

(Note that $A = A_b + A_F$; also the units of B are kg/m² s.)

We now have a system which is exactly analogous to that of a dry heat exchanger but with a specific enthalpy difference in place of a temperature difference. By an exactly similar procedure to that for the derivation of logarithmic mean temperature difference, $\Delta \bar{t}_{ln}$ (see for example Eastop and McConkey[6.9]), an expression for logarithmic mean enthalpy difference, $\Delta \bar{h}_{ln}$ can be derived, i.e.

$$\Delta \bar{h}_{ln} = \frac{(h_{o1} - h_{R1}) - (h_{o2} - h_{R2})}{\ln \left\{ (h_{o1} - h_{R1})/(h_{o2} - h_{R2}) \right\}}$$ [6.38]

and

$$\dot{Q} = \dot{m}_a (h_{o1} - h_{o2}) = BA(\Delta \bar{h}_{ln})$$ [6.39]

Using this approximate method, the required area of tube surface can be calculated without the need to plot the reciprocal of specific enthalpy differences as in the operating line method.

Note: The logarithmic mean enthalpy difference applies only to the counter-flow or parallel flow cases; for cross-flow a correction factor can be used; when a direct expansion coil is used the logarithmic mean enthalpy difference applies regardless of the flow pattern since the refrigerant temperature remains constant throughout.

Example 6.9

A cooling coil consisting of finned tubes in which chilled water flows in counter-flow to the air stream is to be designed for a load of 40 kW. The air enters the coil at 30 °C dry bulb, 21.2 °C wet bulb (sling), at a flow rate of 1.2 m³/s. The chilled water enters at 5 °C and leaves at 10 °C.

Using the data below, calculate:
(i) the required total surface area of the air side using the operating line method;
(ii) the required total surface area on the air side using the fictitious enthalpy method.

Data

Mass transfer coefficient, 0.05 kg/m² s; heat transfer coefficient from tube wall to chilled water, 2.7 kW/m² K; ratio of effective finned tube outside area to tube inside area, 14; ratio of effective finned tube outside area to actual outside finned area, 0.8.

Solution

The cooling coil is as shown in Fig. 6.27; (page 270)
(i) From tables[6.6], at a dry bulb temperature of 30 °C and a wet bulb temperature of 21.2 °C, we have:

$$h_{o1} = 60.89 \text{ kJ/kg}$$

Specific volume, $v = 0.875 \text{ m}^3/\text{kg}$

Then, $\dot{m}_a = 1.2/0.875 = 1.37 \text{ kg/s}$

and $40 = \dot{m}_a (h_{o1} - h_{o2})$

i.e. $h_{o2} = 60.89 - (40/1.37) = 31.7 \text{ kJ/kg}$

Also, $R_{Rw} = 1/\alpha_{Rw} = 1/2.7 = 0.370 \text{ m}^2 \text{ K/kW}$
               $\therefore \quad R_{Ro} = 0.37 \times 14 = 5.18 \text{ m}^2 \text{ K/kW}$

The tie-line slope is then given from Eqn [6.35] as $-1/\beta_\omega R_{Ro}$,

i.e. $-1/0.05 \times 5.18 = -3.86 \text{ kJ/kg K}$

The operating line slope is given by Eqn [6.29] as

$$(h_{o2} - h_{o1})/(t_{w2} - t_{w1})$$

i.e. $$(60.89 - 31.7)/(10 - 5) = 5.84 \text{ kJ/kg K}$$

The system can then be drawn on the specific enthalpy–temperature chart as shown on Fig. 6.29. The line AB can be divided into six equal intervals and the table below gives the values of $(h_o - h_i)$ and of $1/(h_o - h_i)$ at each interval.

Point	h_o (kJ/kg)	h_i (kJ/kg)	$h_o - h_i$ (kJ/kg)	$1/(h_o - h_i)$ (kg/kJ)
A	60.9	43.5	17.4	0.058
P	56.0	40.0	16.0	0.063
Q	51.2	36.5	14.7	0.068
R	46.3	33.0	13.3	0.075
S	41.4	29.5	11.9	0.084
T	36.5	26.0	10.5	0.095
B	31.7	22.5	9.2	0.109

Figure 6.29 Specific enthalpy vs temperature chart for Example 6.9

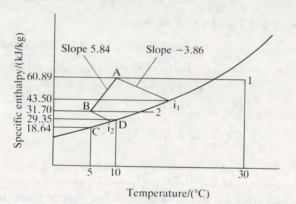

Then, applying Simpson's rule from Eqn [6.30], we can calculate the integral $\int_1^2 dh_o/(h_o - h_i)$ as

$$\frac{(60.9 - 31.7)}{6 \times 3}\{0.058 + 0.109 + 4(0.063 + 0.075 + 0.095)$$
$$+ 2(0.068 + 0.084)\}$$
$$= 2.28$$

Then from Eqn [6.25], with the area A_o in place of the term aV, we have:

$$\frac{\beta_\omega A_o}{\dot{m}_a} = 2.28$$

i.e. $$A_o = \frac{2.28 \times 1.37}{0.05} = 62.5 \text{ m}^2$$

Actual area required $= 62.5/0.8 = 78.1 \text{ m}^2$

(ii)   To use the fictitious enthalpy method we make the approximation that the specific enthalpy of saturated air varies linearly with temperature over the range of chilled water temperatures.

Referring to Fig. 6.29, this implies that the saturation curve between C and D is a straight line, and b_m is given by

$$b_m = (29.35 - 18.64)/(10 - 5) = 2.14$$

where at 10 °C the saturation specific enthalpy is 29.35 kJ/kg, and at 5 °C the saturation specific enthalpy is 18.64 kJ/kg, taken from tables.[6.6]

Then from Eqn [6.37],

$$\frac{1}{BA} = \frac{1}{\beta_\omega A_o} + \frac{b_m R_{Rw}}{A_i}$$

i.e.

$$\frac{1}{B} = \frac{A}{\beta_\omega A_o} + \frac{b_m R_{Rw} A}{A_i} = \frac{1}{0.05 \times 0.8} + \frac{2.14 \times 0.37 \times 14}{0.8}$$

$$\therefore \quad B = 0.0257 \text{ kg/m}^2 \text{ s}$$

The fictitious specific enthalpy values h_{R1} and h_{R2} are the saturation values at temperatures of 10 °C and 5 °C respectively, i.e.

$$h_{R1} = 29.35 \text{ kJ/kg}; \, h_{R2} = 18.64 \text{ kJ/kg}$$

Then using Eqn [6.38], we have

$$\Delta \bar{h}_{\ln} = \frac{(60.89 - 29.35) - (31.70 - 18.64)}{\ln\{(60.89 - 29.35)/(31.7 - 18.64)\}}$$

$$= 21 \text{ kJ/kg}$$

Therefore from Eqn [6.39],

$$\dot{Q} = 40 = BA(\Delta \bar{h}_{\ln})$$

i.e.

$$A = \frac{40}{0.0257 \times 21} = 74.1 \text{ m}^2$$

This compares with the value of 78.1 m² found from the operating line method. Pure counter-flow has been assumed but this would not be strictly true in practice and hence the value of $\Delta \bar{h}_{\ln}$ should be multiplied by a fraction, thus giving a larger area required, and making a smaller difference between the answers from the two methods.

In a well-designed cooling coil, the air at entry should be immediately cooled below its dew point on the first row of finned tubes so that the finned surface is covered with a water film throughout. On a psychrometric chart the dew point is easily seen by drawing a horizontal line of constant specific humidity through the air inlet point until it cuts the saturation line. The coil surface temperature at the air inlet must then be below the dry bulb temperature corresponding to this point. On the specific enthalpy–temperature chart the dew point temperature is not immediately obvious. The specific humidity of the air at inlet is known, and from tables the temperature at saturation with that value of specific humidity can be read by interpolation. For the previous Example, the specific humidity at air inlet at 30 °C dry bulb and 21.2 °C wet bulb is read from tables[6.6] as 0.01202. At a temperature of 17 °C the specific humidity at saturation is 0.012 17, and at 16.5 °C the specific humidity at saturation is 0.011 79; hence, by interpolating, the dew point temperature of the entering air is 16.7 °C. In Example 6.9 the temperature of the coil surface at entry is the temperature at point i1 (see Fig. 6.29), which from the enthalpy–temperature chart is read as 15.5 °C. Therefore in the case of the cooling coil in Example 6.9, the temperature of the coil is at all points below the air dew point.

Effectiveness of a Cooling Coil

Using fictitious enthalpies the coil can be treated exactly as a dry heat exchanger. An effectiveness, E, can be defined as the actual enthalpy decrease of the air divided by the maximum possible decrease; the maximum possible decrease occurs when the air is cooled to the fictitious specific enthalpy of the coolant at air exit. That is,

$$E = \frac{h_{o1} - h_{o2}}{h_{o1}' - h_{R2}} \qquad [6.40]$$

By a similar analysis to that for a dry heat exchanger (see Eastop and McConkey[6.9]), the effectiveness, E, of a counter-flow cooler can be derived as follows.

The ratio of the mass flow rate of cooling water to that of air is given by:

$$\frac{\dot{m}_R}{\dot{m}_a} = \frac{h_{o1} - h_{o2}}{(t_{R1} - t_{R2})c_w} = \frac{(h_{o1} - h_{o2})b_m}{(h_{R1} - h_{R2})c_w}$$

where $(t_{R1} - t_{R2}) = (h_{R1} - h_{R2})/b_m$ from Eqn [6.36], and c_w is the specific heat of the cooling water. That is,

$$\frac{h_{R1} - h_{R2}}{h_{o1} - h_{o2}} = \frac{b_m \dot{m}_a}{\dot{m}_R c_w} = y \qquad [6.41]$$

Also, from the definition on page 253,

$$\text{Number of transfer units, } N_{tu} = \frac{BA}{\dot{m}_a} = \frac{h_{o1} - h_{o2}}{\Delta \bar{h}_{ln}} \qquad [6.42]$$

From Eqn [6.38],

$$\Delta \bar{h}_{ln} = \frac{(h_{o1} - h_{R1}) - (h_{o2} - h_{R2})}{\ln\left\{\dfrac{h_{o1} - h_{R1}}{h_{o2} - h_{R2}}\right\}}$$

Therefore we have:

$$N_{tu}\left\{\frac{(h_{o1} - h_{o2}) - (h_{R1} - h_{R2})}{(h_{o1} - h_{o2})}\right\} = \ln\left\{\frac{h_{o1} - h_{R1}}{h_{o2} - h_{R2}}\right\}$$

Then using Eqn [6.41],

$$N_{tu}(1 - y) = \ln\left\{\frac{(h_{o1} - h_{R2}) - (h_{R1} - h_{R2})}{(h_{o1} - h_{R2}) - (h_{o1} - h_{o2})}\right\}$$

Then introducing the definition of effectiveness given by Eqn [6.40], we have

$$N_{tu}(1 - y) = \ln \left\{ \frac{\dfrac{(h_{o1} - h_{o2})}{E} - (h_{o1} - h_{o2})y}{\dfrac{(h_{o1} - h_{o2})}{E} - (h_{o1} - h_{o2})} \right\}$$

i.e.

$$N_{tu}(1 - y) = \ln \left\{ \frac{1 - Ey}{1 - E} \right\}$$

Re-arranging we then have,

$$E = \frac{\exp\{N_{tu}(1 - y)\} - 1}{\exp\{N_{tu}(1 - y)\} - y} \qquad [6.43]$$

Note that when a direct expansion coil is used with an evaporating refrigerant, then $h_{R1} = h_{R2}$, and from Eqn [6.41], $y = 0$. Hence, for a direct expansion cooling coil,

$$E = 1 - \exp(-N_{tu}) \qquad [6.44]$$

The relationship between effectiveness, E, and N_{tu} can be plotted as shown in Fig. 6.30. It can be seen that for a high effectiveness the value of y, $(= b_m \dot{m}_a / c_w \dot{m}_R)$, must be as low as possible, and the value of N_{tu}, $(= BA/\dot{m}_a)$, as high as possible.

Figure 6.30
Effectiveness vs N_{tu}
for a cross-flow cooling
coil

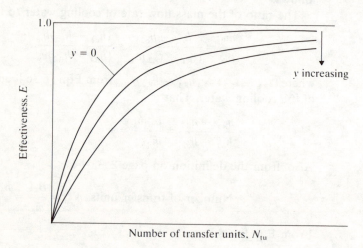

Coil Off the Design Point

When the enthalpy of the outside air is less than the design value, then the cooling coil must be controlled to maintain the required conditions in the conditioned space. This is usually done by reducing the chilled water mass flow rate while maintaining its inlet temperature; the flow is reduced using a by-pass valve so that the chiller providing the water to the cooling coil runs with a constant mass flow rate and outlet temperature. When the cooling load for the conditioned space is reduced, even when the outside air enthalpy is at the design value, a similar control of the coil is required.

It can be seen that for a constant air flow rate the value of y increases when the chilled water flow rate is reduced, thus reducing the effectiveness of the coil.

Using Eqn [6.37],

$$\frac{1}{BA} = \frac{1}{\beta_\omega A_o} + \frac{b_m R_{Rw}}{A_w}$$

the number of transfer units can be written as

$$N_{tu} = \frac{BA}{\dot{m}_a} = \frac{1}{\dfrac{\dot{m}_a}{A_o}\left(\dfrac{1}{\beta_\omega} + \dfrac{A_o b_m R_{Rw}}{A_w}\right)} \qquad [6.45]$$

The mass transfer coefficient, β_ω, is much lower than the heat transfer coefficient on the liquid side, $\alpha_R = 1/R_{Rw}$, but since the outside of the tubes are finned the area ratio A_o/A_w tends to make the two terms in the above expression of the same order. When the flow rate of cooling water is reduced with the air flow rate kept constant, the heat transfer coefficient is reduced and hence the value of R_{Rw} increases; it follows that the N_{tu} is less and hence the effectiveness is reduced. The amount by which the effectiveness is reduced depends on the relative magnitudes of the various terms.

Taking the data of Example 6.9, we have

$$N_{tu} = \frac{1}{\dfrac{\dot{m}_a}{A_o}\left\{\dfrac{1}{0.05} + \dfrac{14 \times 2.14}{2.7}\right\}} = \frac{74.1 \times 0.8}{1.37(20 + 11.1)}$$

$$\therefore \quad N_{tu} = 1.391$$

Assume a reduction in chilled water flow of, say, 30%. From the expression for heat transfer in turbulent flow in a tube we can assume that the heat transfer coefficient on the water side will be reduced by a factor of $0.7^{0.8} = 0.752$.

Therefore the second term in the expression for N_{tu} in the above becomes $11.1/0.752 = 14.76$, and the N_{tu} becomes 1.245; this gives a decrease in N_{tu} of 10.5%.

The initial value of y is $2.14 \times 1.37/8 = 0.367$, where $\dot{m}_R c_w = \dot{Q}/(10-5) = 40/5 = 8\ \text{kW/K}$.

The new value is $0.367/0.7 = 0.524$.

Then using Eqn [6.43], the initial effectiveness is

$$E = \frac{\exp\{(1 - 0.367)1.391\} - 1}{\exp\{(1 - 0.367)1.391\} - 0.367} = 0.691$$

The new effectiveness is

$$E = \frac{\exp\{(1 - 0.524)1.245\} - 1}{\exp\{(1 - 0.524)1.245\} - 0.524} = 0.630$$

It can be seen, therefore, that a 30% decrease in the water flow reduces the effectiveness by only about 9%.

For a direct expansion coil, the resistance of the refrigerant film on the inside surface is negligible and hence in Eqn [6.45] the second term in the denominator is zero, i.e.

$$N_{tu} = \beta_\omega A_o / \dot{m}_a$$

Therefore, when the air mass flow rate is constant, β_ω remains constant and N_{tu} is unchanged. Hence from Eqn [6.44] it can be seen that the effectiveness does not alter when the load changes, provided the air mass flow rate stays the same.

Coil Contact Factor

The coil contact factor is defined in Chapter 1 as

$$\frac{h_{o1} - h_{o2}}{h_{o1} - h_{ADP}}$$

where h_{ADP} is the enthalpy of the air per unit mass of dry air at the apparatus dew point.

For a coil it was also stated in Chapter 1 that the apparatus dew point is usually taken as the mean coil surface temperature, i.e.

$$t_{ADP} = (t_{i1} + t_{i2})/2$$

It follows that coil contact factor and effectiveness are not synonymous; the contact factor is always greater than the effectiveness.

In practice the mean coil surface temperature is not usually known but the mean refrigerant temperature can be found; these temperatures would be approximately the same for a direct expansion coil but not for one using chilled water. This is because the thermal resistance of the fluid film on the inside surface is negligible for an evaporating vapour; the thermal resistances of the metal wall and the film of water on the outside surface are also usually negligible. If the coil contact factor is evaluated taking the enthalpy, h_{ADP}, as the enthalpy of saturated air at the mean refrigerant temperature, this effectively neglects the resistance, R_{Rw}, and hence the coil contact factor is given approximately by Eqn [6.44], i.e.

$$\text{Coil contact factor} = 1 - \exp(-N_{tu})$$

where $N_{tu} = \beta_\omega A_o / \dot{m}_a$.

Example 6.10

For the coil of Example 6.9 compare the coil effectiveness with the coil contact factor (based on an apparatus dew point taken as the mean temperature of the chilled water).

Compare also the condition of the air leaving the coil calculated by the method of this section, and the psychrometric method, which assumes a mean apparatus dew point and a straight line joining the initial and final states of the air on the psychrometric chart.

Solution

The effectiveness calculated above is 0.691.

The mean temperature of the chilled water is $(5 + 10)/2 = 7.5\,°C$; hence,

$$\text{Coil contact factor} = \frac{h_1 - h_2}{h_1 - h_{ADP}} = \frac{60.9 - 31.7}{60.9 - 23.8}$$

$$= 0.787$$

This is substantially different from the effectiveness of 0.691. In Example 6.9 the operating line is divided into six equal intervals and the tie-lines drawn for each point to give seven interface enthalpy values; these values are given in the table for the solution of Example 6.9 (see page 273). The graphical method is then to join point 1 to point i1 and where this line cuts the enthalpy line, h_P, the first point on the condition line is fixed. This point is then joined to point ia and where this cuts the enthalpy line h_Q fixes the next point on the condition line. The condition line of the process can be plotted in this way either on an enthalpy–temperature chart or on a psychrometric chart as shown in Fig. 6.31. Not all the construction lines have been drawn but the method gives the state of the air at exit, point 2, at 11.5 °C at an enthalpy value of 31.7 kJ/kg. The psychrometric method consists of joining point 1 to a point on the saturation line corresponding to the mean temperature of the chilled water. From the chart this line is found to cut the enthalpy line, $h_{o2} = 31.7$ kJ/kg, at 12.3 °C, as shown by point 2′ on Fig. 6.31. The agreement is reasonably good with the enthalpies the same and a 7% difference in the temperatures.

Figure 6.31 Process of Example 6.10 plotted on the psychrometric chart

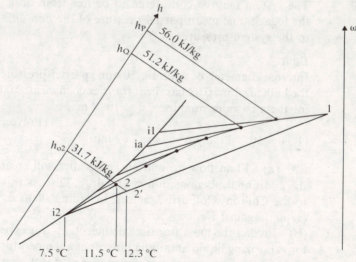

A cooling coil is not a true counter-flow heat exchanger; in fact it consists of a series of cross-flow heat exchangers, as shown in Fig. 6.27. In Examples 6.9 and 6.10 the process was divided into six increments; this is approximately equivalent to the cooling coil of Fig. 6.27 with one increment representing one bank of tubes. For a cooling coil of four banks of tubes, say, the process could be divided into four increments, but the smaller the number of increments the less accurate is the method.

In practice a manufacturer of cooling coils will specify the performance in a way which allows the process to be drawn as a straight line on the psychrometric chart for different coolant temperatures and air and water mass flow rates. The psychrometry of systems with cooling coils operating at part-load is considered again in Chapter 8.

PROBLEMS

6.1 A vertical shaft of 200 mm diameter, length 2 m, leads from a reservoir of water at 10 °C up to atmospheric air at outlet which is at 10 °C, 40% saturation. The system pressure is 1.013 bar. Taking a diffusion coefficient for water vapour diffusing into air of $2.6 \times 10^{-5} \, \text{m}^2/\text{s}$, calculate the mass flow rate of water which is diffused into the atmosphere from the reservoir.

$(0.315 \times 10^{-6} \, \text{kg/s})$

6.2 An experimental wetted wall column has an annular-shaped prism of water flowing vertically downwards under the influence of gravity. Air, in counter-flow to the water, is pumped through the centre of the prism. Use the data below to determine, for a section of the column where the system pressure and temperature are 1.0133 bar and 27 °C respectively, the following:

(i) an expression for the air phase mass transfer coefficient, β_G, assuming the Colburn analogy, $j_M = j_H = St(Pr)^{2/3}$;

(ii) the value of the mass transfer coefficient, β_G, and the thickness of the air boundary layer at the air–water interface.

The system may be considered to be free from drag, but do not assume that the logarithmic mean partial pressure of the non-diffusing constituent is equal to the system pressure.

Data

Internal diameter of column, 50 mm; partial pressure of the water vapour in the bulk air flow, 0.02 bar; heat transfer coefficient, $0.0035 \, \text{kW/m}^2 \, \text{K}$; diffusivity for the air–water mixture, $2.5 \times 10^{-5} \, \text{m}^2/\text{s}$.

(Polytechnic of the South Bank)

$(1.34 \times 10^{-4} \, \text{kmol/m}^2 \, \text{s bar}; \, 7.69 \, \text{mm})$

6.3 (a) Fluid flow around curved bodies will create skin friction and often the additional phenomenon called drag. Discuss this occurrence with respect to the Chilton–Colburn j factors, including sketches of graphs of j_M, j_H, and $(\tau/\rho u^2)$ against Re.

(b) Predict the mass transfer coefficient, β_G, for experimental equipment used for vaporizing liquid ammonia into air at a temperature of 300 K and a pressure of 1 bar. The mean heat transfer coefficient for this equipment at these gas and liquid flow rates is $3 \, \text{kW/m}^2 \, \text{K}$. The following analogy has been used for similar predictions at various other flow rates:

$$j_M = j_H = St(Pr)^{2/3}$$

Assume a dilute ammonia-in-air mixture and take the diffusivity of ammonia through air as $2.6 \times 10^{-5} \, \text{m}^2/\text{s}$. Take all other properties from tables.[6.1]

(Polytechnic of the South Bank)

$(0.113 \, \text{kmol/m}^2 \, \text{s bar})$

6.4 By applying basic mass and heat transfer theory to the case of a wet and dry bulb hygrometer, show that the measured specific humidity of the humid air is given by:

$$\omega_o = \omega_i - \frac{Le \, c_{pma}}{h_{fg}} \left\{ 1 + \frac{\alpha_r}{\alpha_c} \right\} (t_{db} - t_{wb})$$

where ω_i is the specific humidity of saturated air at the wet bulb temperature; Le is the Lewis number; c_{pma} is the mean specific heat of humid air per unit mass of dry air; h_{fg} is the enthalpy of vaporization at the wet bulb temperature; α_r is the heat transfer coefficient for radiation; α_c is the heat transfer coefficient for convection; t_{db} is the dry bulb temperature; t_{wb} is the wet bulb temperature.

Hence show that when radiation is negligible and when $Le = 1$:

$$\frac{\omega_i - \omega_o}{t_{db} - t_{wb}} = \frac{c_{pma}}{h_{fg}}$$

6.5 Readings from a wet and dry bulb hygrometer are 14 °C and 21 °C. Using steam tables[6.1] only, calculate the specific and relative humidities assuming a barometric pressure of 1.013 25 bar, and compare these values with those given in the CIBSE tables.[6.6] Take c_{pma} and h_{fg} at the saturated wet bulb temperature, and state any other assumptions made.
(0.007 06; 45.7%; 0.007 02; 45.3%)

6.6 Air enters a well-insulated parallel flow humidifier at the rate of 1.4 m³/s at 15 °C, 20% saturation. The water enters at 24 °C and leaves at 18 °C and the rate of flow of water is 1.5 kg/s. Calculate the length of the humidifier if the air face velocity is 1.8 m/s, and plot the process on the psychrometric chart, indicating clearly the state of the air at exit. Assume a mean value of composite coefficient of the humidifier, $\beta_\omega a$ of 0.9 kg/m³ s.
(2.29 m; 18 °C DB, 15.2 °C WB)

6.7 250 kW are to be dissipated from the condenser cooling water of an air conditioning plant by means of a vertical induced-draught packed cooling tower. The cooling water enters the top of the tower at a temperature of 35.5 °C and a cooling range of 15.5 K is required. The value of βa is 0.25 s⁻¹ for the tower packing for a water flow rate of 0.35 kg/m² s when the air/water ratio is 1.5 times the minimum.

Air enters the base of the tower at wet and dry bulb temperatures of 8.1 °C (sling), and 16 °C.
(a) Construct the air condition curve on the specific enthalpy–temperature chart and by numerical integration compute the required packed height of the tower and its area of cross-section. The resistance to heat and mass transfer in the liquid phase may be considered negligible compared with that in the gas phase.
(b) Calculate the dry bulb temperature of the air in the tower at a section where the water temperature is 25 °C.

(Polytechnic of the South Bank)

(2 m, 11 m²; 18.2 °C)

6.8 A vertical induced-draught packed cooling tower is the heat sink for dissipating 200 kW from the condenser cooling water of an air conditioning plant.
(a) Use the following data to construct the air condition curve on the specific enthalpy–temperature chart and by numerical integration compute the required packed height of the tower. Do not assume a vertical tie-line.
(b) A vertical tie-line could have been assumed in the solution to part (a). Explain the principle that makes this approximation acceptable.

Data

Temperature of water entering tower, 34 °C; cooling range of tower, 15 K; composite coefficient (βa) of the tower packing, 0.25 s^{-1}; cross-sectional area of packing, 10 m^2; air approach condition to tower, 16 °C and 8.1 °C (sling); slope of tie-line, -18 kJ/kg K; air flow rate, 0.32 kg/m^2 s.

(Polytechnic of the South Bank)

(3 m)

6.9 Water flowing at the rate of 0.75 kg/s is to be cooled in a small cooling tower from 28 °C to 24 °C when the ambient conditions are 24 °C dry bulb and 18.2 °C wet bulb. Taking the ratio of the rate of mass flow of water to air as 0.75, calculate:

(i)   the required tower volume;
(ii) the condition of the air at exit;
(iii) the height of the tower if the air velocity at exit is 1.5 m/s.

Take the mean value of $\beta_\omega a$ for the tower as 0.45 kg/m^3 s, and assume that the temperature difference between the water and the water–air interface at the water inlet is 0.5 K.
(1.31 m^3; 25.3 °C db, 21.8 °C wb; 2.25 m)

6.10 (a) Use an enthalpy–temperature graph, together with the data below, to calculate the composite mass transfer coefficient, $\beta_\omega a$, of a vertical induced-draught packed cooling tower when operating at a cooling duty of 300 kW. Air is drawn into the base of the tower at wet and dry bulb temperatures of 9.5 °C (sling) and 20 °C respectively.
(b) Determine the value of dh_o/dt_o (the enthalpy temperature gradient) and also the value of the moisture content of the air at the point in the packing where the air dry bulb temperature is 21 °C.

Data

Height of packing, 3.42 m; cross-sectional area of packing, 13.28 m^2; water temperature at entry to the tower, 40 °C; water cooling range, 18 K; water flow rate over packing, 0.3 kg/m^2 s; tie-line slope, -18.26 kJ/kg K.

Air/water ratio may be taken as 1.4 times the minimum.

(Engineering Council)

Note: The original examination question gave an extract from the CIBSE tables;[6.6] the only data required from the tables is the enthalpy of humid air at 20 °C dry bulb, 9.5 °C wet bulb (sling), which may be taken as 27.6 kJ/kg K; to solve the last part of the problem, Eqn [1.29] may be used if tables or the psychrometric chart are not available.
(0.278 kg/m^3 s; 6.7 kJ/kg K; 0.0114)

6.11 (a) Discuss the purpose and the principle of operation of a cooling tower used in air conditioning systems.
(b) Compare and contrast the different types of cooling towers available in the UK.
(c) A cooling tower has been installed having the following characteristics:

Area, 4 m^2; height, 1 m; water flow, 11.5 kg/s; air flow, 10 kg/s; value of $\beta_\omega a$ at the design air and water flow rates, 11 kg/m^3 s.

By a trial and error method, using the Stevens chart, estimate the tower water exit temperature if the inlet water temperature is 38 °C, and the ambient air at inlet has a wet bulb temperature of 20 °C.

(Newcastle upon Tyne Polytechnic)

(25.1 °C)

6.12 (a) State Merkel's enthalpy theory as applied to cooling towers.

(b) A cooling tower is required for the condenser cooling water in a chemical process. If 30 kg/s of water at 27 °C is required at the condenser, and the water leaves the condenser at 50 °C, show that for a packing height of 1.5 m, the area required by the cooling tower lies between 9 m^2 and 12 m^2. The plant is situated at a town in Lincolnshire where the average design wet bulb temperature may be taken as 18.5 °C.

The packing specification is:

Air flow rate, 2.2 kg/s m^2; $\beta_\omega a = 3(\dot{m}_w/A)^{0.2}(\dot{m}_a/A)^{0.27}$ kg/m^3 s
where A is the cooling tower area in m^2, and $\dot{m}_w$ and $\dot{m}_a$ are the mass flow rates of water and air in kg/s.

Use the Stevens chart.

(Newcastle upon Tyne Polytechnic)

6.13 Air at 27 °C dry bulb, 13.2 °C wet bulb (sling), enters a washer using pumped re-circulation of the spray water. The air enters the washer at a rate of 1.9 m^3/s with a mean velocity of 3 m/s, and leaves the washer with a percentage saturation of 50%. Calculate the required face area and length of the humidifier, taking the value of $\beta_\omega a$ as 3.2 kg/m^3 s.

(0.633 m^2; 0.925 m)

6.14 Air at 32 °C, percentage saturation 50%, enters a spray-type de-humidifier at a rate of 4.7 m^3/s. The chilled spray water enters at 4 °C and leaves at 11 °C; the water spray is in parallel-flow with the air, and the spray ratio of water mass flow rate to air mass flow rate is 1.2. The face velocity of the air is 3.4 m/s, and the value of the product $\beta_\omega a$ is 2.6 kg/m^3 s. Calculate the required length of the humidifier and the state of the air at exit.

(2.61 m; 13.8 °C, enthalpy, 36.30 kJ/kg)

6.15 Calculate the required length of the humidifier of Problem 6.14 if the air and water are in counter-flow, with the same flow rates, the same inlet air conditions, the same air enthalpy at exit, and the same inlet water temperature.

(1.81 m; 14 °C, enthalpy, 36.30 kJ/kg)

6.16 (a) The specific enthalpy–temperature chart of Fig. 6.32 shows the operating line and air condition curve for a finned-tube, air cooling coil when transferring 70 kW to chilled water which has terminal temperatures of 5 °C and 9 °C. At entry the humid air has dry and wet bulb temperatures of 27 °C and 22.9 °C (sling), and an equivalent dry air mass flow rate of 4.58 kg/s.

Assuming that the air flows in counter-flow to the chilled water, that $Le = 1$, and using Fig. 6.32 and the data below, determine:

(i) the total external tube surface area and the number of tube rows, using a method of numerical integration;

(ii) the temperature difference across the air laminar sub-layer at a particular zone of the chiller where the chilled water temperature is 6 °C.

Figure 6.32 Specific enthalpy vs temperature chart for Problem 6.18

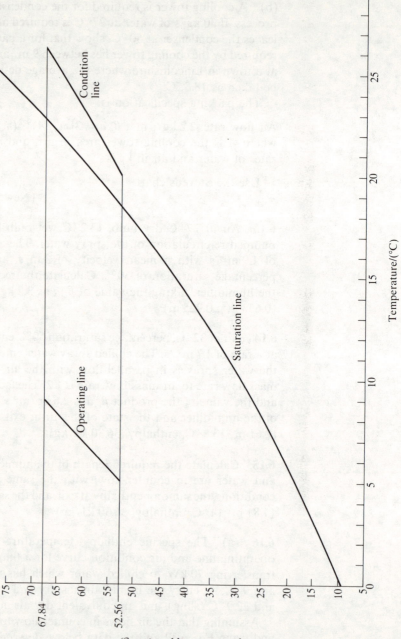

Data

Cooler face area, 0.6 m²; specific heat of humid air, 1.03 kJ/kg K; air-side heat transfer coefficient, 0.05 kW/m² K; refrigerant-side heat transfer coefficient, 2.4 kW/m² K; condensate heat transfer coefficient, 2 kW/m² K; fin area per metre run of tube, 1.3 m²; fin efficiency, 82%; effective outside bare tube area per metre run of tube, 0.055 m²; tube pitch, 60 mm; inside diameter of tube, 25 mm.

Ignore the thermal resistance of the tube wall.

(b) The system operating line of some coolers causes the initial stage of the air-side heat transfer surface to remain dry. The dry surface is considered to be less effective in terms of heat transfer than the wet surface, in spite of the latter offering additional resistance to heat transfer due to the condensate film. Explain why this is so, describing the transport mechanism involved.

(Polytechnic of the South Bank)

(81.6 m²; 6; 9.25 K)

6.17 (a) Outline how you would adapt the principle of heat exchanger effectiveness to the theory of moist air de-humidification using fictitious enthalpy potentials.

(b) An air conditioning system is designed for a cooling load of 80 kW with an entering air state of 30 °C, 88% saturation. In this case the air leaving state is 14 °C, 100% saturation. Determine the part-load heat transfer rate and air leaving state (assumed saturated), if the air and refrigerant mass flow rates and thermal resistances remain largely unchanged. The refrigerant evaporating temperature is 5 °C, and the part-load entering condition is 24 °C, 66% saturation.

(Polytechnic of the South Bank)

(41 kW; 10 °C saturated)

6.18 (a) Explain what you understand by the term 'fictitious enthalpy potential'.

(b) A direct expansion coil is designed to transfer 75 kW of heat when operating with an evaporating temperature of 4 °C and incoming air at 28 °C, 80% saturation. The outlet air condition is then 15 °C, 100% saturation.

Assuming that the heat transfer resistances in the coil, the evaporating temperature, and the air mass flow rate, do not change appreciably at part-loads, determine:

(i) the heat transfer rate at a seasonal part-load with an inlet air state of 22 °C, 60% saturation;

(ii) the temperature of the air leaving the coil to the nearest 0.5 K, assuming it is saturated.

(Polytechnic of the South Bank)

(38.1 kW; 10 °C)

6.19 A direct-expansion, finned-coil cooler is used to de-humidify air initially at 38 °C, 20% percentage saturation. The coil has an effective cooling surface per row of 40 m² and the air mass flow rate is 5.2 kg/s. Refrigerant at 3.5 °C flows in the coil.

Taking the mean thermal resistance from the air–water interface on the outside surface to the bulk of the refrigerant in the tubes as 3.7 m² K/kW, and

the mass transfer coefficient as $0.07 \, \text{kg/m}^2 \, \text{s}$, estimate the coil effectiveness and the de-humidification load, assuming that there are four coil rows.
(0.75; 174 kW)

6.20 Air at 32 °C, 50% saturation, enters a finned-coil de-humidifier at a rate of $4.5 \, \text{m}^3/\text{s}$. The refrigerant in the coils is at 4.5 °C throughout and the effective coil surface area per row is $42 \, \text{m}^2$. The mass transfer coefficient, β_ω, is $0.08 \, \text{kg/m}^2 \, \text{s}$ and the thermal resistance from the air–water interface to the refrigerant in the tubes is $3.6 \, \text{m}^2 \, \text{K/kW}$.

Calculate the de-humidification load and the required number of coil rows assuming an overall effectiveness of 0.8.
(219 kW; 4)

REFERENCES

6.1 Rogers G F C and Mayhew Y R 1988 *Thermodynamic and Transport Properties of Fluids* 4th edn Basil Blackwell

6.2 Welty J R 1984 *Fundamentals of Momentum, Heat and Mass Transfer* 3rd edn John Wiley

6.3 Colburn A P 1933 *Tr Amer Inst Chem Eng* **29**: 174

6.4 Chilton T H and Colburn A P 1934 *Ind Eng Chem* **26**: 1183

6.5 Douglas J F, Gasiorek J M and Swaffield J A 1985 *Fluid Mechanics* 2nd edn Longman

6.6 CIBSE 1986 *Guide to Current Practice* volume C1

6.7 Carey W F and Williamson G J 1950 Gas cooling and humidification: design of packed towers from small-scale tests. *Proc I Mech E* **163**: 49

6.8 Threlkeld J R 1970 *Thermal Environmental Engineering* Prentice Hall

6.9 Eastop T D and McConkey A 1986 *Applied Thermodynamics for Engineering Technologists* 4th edn Longman

7 REFRIGERATION PLANT

In any heat pump or refrigeration cycle heat is input to the working fluid, the refrigerant, at a low temperature and rejected from the working fluid at a higher temperature. The second law of thermodynamics shows that for heat to flow against a temperature gradient there must be an additional energy input. This energy input may be in the form of heat or work or both, depending on the system.

There is no difference in the operation of heat pump and refrigeration cycles. The distinction is the use to which the cycle is put. If it is used to remove heat from a cold fluid or space it is termed a refrigerator. If it is used to supply heat to a warm fluid or space it is termed a heat pump.

Both heating and cooling effects are present in any system and the secondary effect may or may not be useful. For example, a domestic refrigerator or freezer is designed to maintain a confined space at a low temperature. At the same time it must act as a heat pump and contribute to heating the room. The heat rejected in this case is useful in winter but not very effective as a kitchen heating system. If a heat pump were designed to produce hot water by extracting heat from a canal, the side effect would be to cool the water, which would not be useful and could cause problems if too much heat were extracted. Sometimes both heating and cooling are essential—the classic example being a sports complex with a swimming pool and an ice rink. In such cases the term 'heat pump' is normally used but with both effects being equally important the term 'reversed heat engine' would be more appropriate. Situations where heating and cooling are provided by the same plant are increasing as people become more energy-conscious.

At the time of writing, the refrigeration industry is in turmoil. The most commonly used refrigerants are the chlorofluorocarbons (CFCs) which are now known to be damaging the Earth's ozone layer. In 1987 the Montreal Protocol, agreeing to limit the production of CFCs, was signed by 24 countries. The Protocol classified the refrigerants R11, R12, R113, R114 and R115 as Group 1 substances.

Note: CFC refrigerants (R) are given a two or three digit number with the first digit representing one less than the number of carbon atoms, the second digit representing one more than the number of hydrogen atoms, and the third digit representing the number of fluorine atoms; if one of the digits is zero then it is omitted e.g. R12 is (CCl_2F_2), so the first digit is zero and is omitted.

The Protocol laid down a programme to cut the production and consumption of Group 1 refrigerants to 50% of the 1986 level by 1999. The main concern for the refrigeration industry is now to replace R11 and R12. The latter has been the main refrigerant used for domestic units for many years and both are widely used for commercial plants.

One solution is to change to alternative CFCs which are not on the Group 1 list. For example, R22 could replace R12 in new plants (but not in existing ones). For some applications an alternative plant may already be on the market, but for others new designs are required. However, this might only be a temporary solution because alternative CFCs are not completely 'ozone friendly'. R22 is only 5% as destructive as either R11 or R12, but that is far too much for some people and it could be added to the Group 1 list at a future meeting.

Another immediate option is to use ammonia, which does not affect the ozone layer. Ammonia has many excellent properties and is already widely used. Its disadvantages are that it is flammable and toxic, and it is therefore not currently used for applications such as air conditioning. Hydrocarbons such as propane and butane are also well established but their use to date has been restricted to petrochemical plants because they are highly flammable.

One option for air conditioning applications is to use vapour absorption systems. The absorption system most likely to be used has the ultimate in ozone-friendly substances as its refrigerant—water. The choice between vapour compression and vapour absorption systems has largely been economic in the past, and the balance could change in the future.

Of course, any enforced change to current practice will result in thermally less efficient systems, and any improvement in the ozone situation will be partly paid for by making the greenhouse effect worse. In the long term there is the possibility of developing new ozone-friendly refrigerants which will also be economic. Further details of the probable impact of new CFC regulations on the Building Services industry can be found in Ref. 7.1.

Whatever changes may take place, the demand for refrigeration is unlikely to diminish and new plant will continue to be installed. Although the refrigerants may change, the basic systems are unlikely to do so and the principles developed in this chapter will still be applicable.

7.1 VAPOUR-COMPRESSION CYCLES

An ideal vapour-compression cycle is a reversed Carnot cycle although a cycle with no pressure drops in the heat exchangers or pipework with an isentropic compressor is often termed an ideal cycle. The development of simplified practical cycles from the Carnot cycle is covered in many standard textbooks on basic thermodynamics such as Ref. 7.2.

The *p–h* Chart

A *p–h* diagram is the most appropriate means of illustrating refrigeration cycles and a *p–h* chart reduces the calculations required to analyse a system. Pressure is chosen as one axis because the heat exchange processes ideally take place at constant pressure. Specific enthalpy is the property most commonly read from the chart so this is chosen as the other axis.

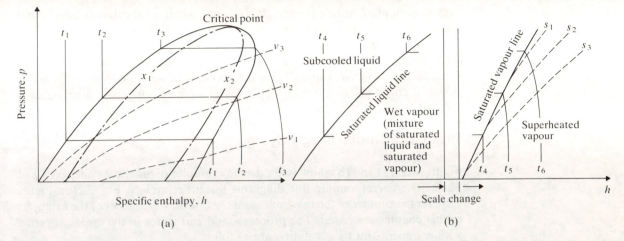

Figure 7.1 The p–h chart

The essential features of a p–h chart are shown on Fig. 7.1. Figure 7.1(a) is a relatively full diagram with a pressure scale ranging from pressures too low for normal refrigeration work to pressures above the critical value. The lowest pressure in a refrigeration cycle should ideally be above ambient, and the highest pressure well below the critical value where the enthalpy of vaporization is still substantial. The top and bottom of Fig. 7.1(a) are therefore not used in practice. Also, all the salient points in a cycle are close to either the saturated liquid or the saturated vapour lines so that only the areas around these lines are required. Figure 7.1(b) shows only the essential areas of the chart, thus substantially increasing the accuracy for a fixed size of paper. A common feature of these restricted charts is a scale change, which makes the slope of the saturated vapour line less steep and further increases the accuracy of enthalpy values to and from the compressor.

In addition to lines of constant pressure (isobars) and lines of constant enthalpy (isenthalps)—the essential lines are those of constant temperature (isotherms), constant entropy (isentropes), and constant specific volume (isochores). These are represented by the letters t, s and v respectively in the diagrams. Some charts also have lines of constant dryness fraction, x, but these are not usually required. In many charts only the ends of the horizontal section of the isotherms are drawn to avoid confusion with isobars. It is also common practice to restrict isochores and isentropes to the superheat region.

All charts and tables with enthalpy and entropy values must have a datum. In the case of water there is general agreement that the saturated liquid should have an enthalpy and entropy value of zero at the triple point (0.01 °C and 0.006 112 bar). Brine, which is below 0 °C, therefore has negative values of both enthalpy and entropy. There is no general agreement on suitable datum points for refrigerants other than water. Some charts use the triple point as a datum of 0 for both enthalpy and entropy. The tables in Ref. 7.3 use −40 °C as a zero datum for both enthalpy and entropy. The charts in the 1986 edition of the CIBSE *Guide* use the critical point as a datum, allocating specific enthalpy a value of 1000 kJ/kg and entropy a value of 1 kJ/kg K. In this book examples are used for many sources and different charts have been used. If the reader tries to follow an example he may find that values at each state point are different from the values on his chart. The change in specific enthalpy across each item of plant must of course be the same, whatever the datum used. When analysing vapour power plants it is usual to take some enthalpy values from tables and others from a chart such as an $h-s$ chart. This can only be done with refrigeration plants if tables and charts are used which are based on the same datum. No chart can ever give the accuracy of tabulated values but a well designed chart will give satisfactory answers to most refrigeration problems.

Practical Single-Stage Cycles

Figures 7.2, (a) and (b) show two extreme examples of practical cycles on a $p-h$ plane. The corresponding line diagrams for either cycle is Fig. 7.2(c). Figure 7.2(a) is the simpler of the two cycles and the one on which most of the examples in this chapter are based. The processes and end states in the cycles, together with comments on their validity, are as follows.

Cycle shown in Fig. 7.2(a)

Process 1–2 (compression):
Starting with a superheated vapour, the temperature and entropy increase as the pressure increases. The process is shown as a broken line because only the end states are defined. The ideal compression is an isentropic process from 1

Figure 7.2 Basic vapour compression cycles

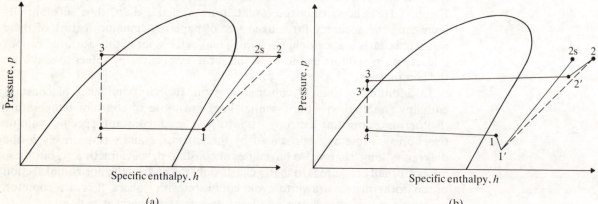

Specific enthalpy, h

(a)

Specific enthalpy, h

(b)

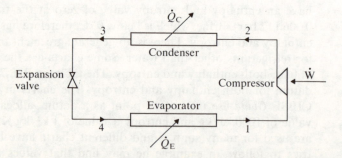

(c)

to 2 s. Throughout refrigeration work, isentropic compression is used as a basis of comparison for practical processes. This is standard practice for rotodynamic machines in other thermodynamic applications; the justification for using it for reciprocating machines is explained in Section 7.3. An isentropic efficiency, η_s, is used to compare the actual process with the ideal isentropic process as follows:

$$\eta_s = \frac{\text{Isentropic work}}{\text{Actual work}} = \frac{\dot{m}(h_{2s} - h_1)}{\dot{m}(h_2 - h_1)} = \frac{h_{2s} - h_1}{h_2 - h_1}$$

Since the specific enthalpy at the end of compression is normally the unknown, it is convenient to transpose the equation as follows:

$$h_2 = h_1 + \frac{h_{2s} - h_1}{\eta_s} \qquad [7.1]$$

State 2:

The superheated vapour at state 2 has the highest temperature in the cycle. High temperatures are undesirable in compressors because of lubrication problems. The temperature at this point in the cycle is important for the design case and for some methods of capacity control which are discussed later in this chapter.

It is assumed that the refrigerant enters the condenser at the same state as it leaves the compressor. Even with the condenser adjacent to the compressor, there must be some pressure drop across the discharge value and the connecting pipework.

Process 2–3 (Heat rejection in the condenser):

The complete process is assumed to take place at constant pressure. The temperature falls during the de-superheating, is constant during the condensation and then falls again during the liquid sub-cooling.

In many refrigerant condensers the de-superheating is not true de-superheating as far as the heat transfer is concerned. A vapour will condense whenever it is in contact with a surface below its dew point. The surface temperature of the tubes in a water-cooled condenser are invariably well below the dew point of the vapour and condensation will take place at entry. When condensation takes place outside tubes, which is the usual arrangement for water-cooled condensers, there is very little change in pressure. Typically the pressure drop might be sufficient to change the condensing temperature by about 0.1 °C between inlet and outlet. Such a small change could not be detected on a p–h chart.

Sub-cooling takes place naturally in any condenser (in a steam surface condenser one of the design problems is how to minimize sub-cooling). There is therefore no extra pressure drop unless a separate sub-cooler is fitted. In refrigeration, sub-cooling is wholly beneficial because it increases the specific refrigeration effect (h_1-h_4) without changing the specific work input required $(h_2 - h_1)$.

State 3:

The sub-cooled liquid is assumed to leave the condenser and enter the expansion device without any loss in pressure or change in temperature. This is very close to reality when the evaporator and condenser are close together.

Process 3–4 (expansion):

The expansion or throttling process in a refrigeration cycle is always illustrated as a broken or dotted line because only the end states are defined. During the process there may be appreciable conversion of enthalpy into kinetic energy which is re-converted back into enthalpy as the velocity decreases. As expansion proceeds there is a substantial fall in temperature. This occurs because some of the liquid vaporizes and extracts its enthalpy of vaporization from the remaining liquid. Increasing the pressure drop across an expansion device increases the proportion of 'flash' vapour produced. This mixture of saturated liquid and saturated (flash) vapour has the same enthalpy as the sub-cooled liquid at state 3. Having reduced the temperature of the remaining liquid, the flash vapour has nothing more to contribute to the cycle and if the proportion of vapour becomes too great multi-stage compression should be considered.

It is assumed that the refrigerant leaves the expansion device and enters the evaporator at the same state. This assumption is realistic because the expansion device is always close to the evaporator. Also, any pipework and discontinuities can be regarded as part of the expansion device.

Process 4–1 (heat input in the evaporator):
The complete process is assumed to take place at constant pressure. Most of the process is also assumed to be isothermal with a small increase in temperature as superheating takes place before the refrigerant leaves.

The term 'evaporator' for this item of plant is unfortunate. Evaporation is a process which takes place at the surface of a liquid such as in a cooling tower. In this case the heat transfer involves the formation of bubbles on the surface of the tubes or passages—it is boiling, not evaporation, and this component should be designated a boiler. Since the term evaporator is generally accepted it will continue to be used in this text: however, the reader should think of evaporators in a refrigeration plant as boilers.

When boiling takes place outside tubes, e.g. in a flooded shell-and-tube unit, there is a negligible change in pressure and a practical process will be very close to the ideal one.

State 1:
The vapour entering a compressor should always be superheated. This ensures that no liquid enters the compressor. Superheat can also be used as a means of controlling the flow of refrigerant.

Cycle shown in Fig. 7.2(b)

Process 1′–2 (compression):
The compression is the same as for cycle (a).
Process 2–2′ (flow in pipework):
State 2 is the refrigerant at the end of compression and state 2′ is the refrigerant entering the condenser. There must always be some pressure drop between these two points and if the condenser is remote from the compressor it will be quite significant. With a remote condenser there will also be a decrease in temperature of the refrigerant because of the substantial difference in temperature between the refrigerant in the pipe and the surroundings.
Process 2′–3 (heat rejection in the condenser):
When condensation takes place inside tubes the pressure drop is far higher than for condensation outside tubes. The difference is typically by a factor of about ten, and a pressure drop of about 0.2 bar would result in a change in saturation temperature of $1-2$ K which is detectable on a $p-h$ chart. Air-cooled condensers always have the refrigerant inside the tubes. With air cooling it is also more likely that de-superheating will take place before condensation starts because

the condenser tube walls will be at a higher temperature than in a water-cooled condenser, and this will lead to a further increase in pressure drop.

Process 3–4 (expansion):

If the evaporator is remote from the condenser there will be some pressure drop in the liquid line between 3 and 3'. The change in temperature will however be negligible because the liquid leaving the condenser will be close to ambient temperature. Any pressure drop in the liquid line is deducted from the pressure drop across the expansion device so that there is no net effect on the cycle.

Process 4–1 (heat input in the evaporator):

The expansion device is always close to, and often an integral part of, the evaporator (boiler). If boiling takes place inside tubes the pressure drop is far higher than when boiling takes place outside tubes. The difference in magnitude is similar to that for condensation; however, a pressure drop of 0.2 bar in an evaporator will cause a change in saturation temperature of about 2–4 K and is very easy to detect on a p–h chart (the pressure scale is logarithmic; the temperature scale is not). Boiling always takes place inside tubes while air passes over the tubes in direct expansion coils (often abbreviated to DX) which are used in air conditioning.

Process 1–1' (flow in pipework):

If the evaporator is remote from the compressor there must be an appreciable pressure drop in the pipework. Since the refrigerant will normally be well below ambient temperature during this process, there will also be an increase in temperature. There may also be a substantial pressure drop across a valve at this point, which may be used as a means of capacity control (see Section 7.3 and Example 7.11).

Summary

The cycle shown in Fig. 7.2(a) is not too far from reality for a compact plant with both heat transfer processes taking place outside tubes.

The cycle shown in Fig. 7.2(b) occurs in practice when each item of plant is some distance from other items of plant and both heat transfer processes take place inside tubes.

Most plants will have practical cycles which are a combination of the two cycles discussed above.

Coefficients of Performance

The performances of different heat engine cycles are compared by the ratio of the required output to the input which must be paid for, as follows:

$$\text{Performance} = \frac{\text{Required output}}{\text{Costed input}} \qquad [7.2]$$

In the case of a forward heat engine cycle such as a vapour power plant, the required output is mechanical work and the input which must be paid for is the heat supplied to the steam in the boiler. The Second Law of Thermodynamics can be used to show that this ratio must always be less than 1 and it is termed the *cycle thermal efficiency*.

The performance of complete plants rather than thermodynamic cycles can be compared using the same equation: however, the output and costed input will be different. The required output from the plant is normally electrical power, which will be less than the shaft power from the turbine. The costed input is the rate at which chemical energy in the fuel is supplied to the boiler. This is always greater than the rate of energy supply to the steam because some energy is always lost to the exhaust. The plant efficiency is therefore always less than the cycle thermal efficiency.

Equation [7.2] is also used to compare reversed heat engines and reversed heat engine cycles. However, in this case the ratio can result in numbers greater than 1 and the required output depends on the use to which the plant is put. The term 'efficiency' is therefore not used, and the ratio is called *coefficient of performance*, which is usually abbreviated to COP.

Any reversed heat engine cycle has two possible coefficients of performance depending on whether it is being used for heating or cooling. When the cycle is used for cooling it is termed a refrigeration cycle, and when it is used for heating it is termed a heat pump cycle. Figure 7.2(c) shows the basic components of a reversed heat engine cycle with the relevant work and heat flow rates. Applying the steady-flow energy equation to each of the energy flow rates gives:

Rate of heat *supply* to evaporator $\quad \dot{Q}_E = \dot{m}(h_1 - h_4)$ [7.3]

Rate of heat *rejection* from condenser $\quad \dot{Q}_C = \dot{m}(h_2 - h_3)$ [7.4]

Rate of work *input* to the compressor $\quad \dot{W} = \dot{m}(h_2 - h_1)$ [7.5]

It should be noted that the direction of the energy flow rates is with respect to the working fluid (the refrigerant). The heat which flows into the evaporator is the heat which has been extracted from the refrigerated air or chilled water. The heat which is rejected from the condenser is the useful output from a heat pump or the heat supplied to the cooling medium from a refrigerator. Equation [7.2] can now be applied to each situation as follows:

For a refrigeration cycle:

$$\text{Performance} = \frac{\text{Required output}}{\text{Costed input}}$$

$$\text{COP}_{\text{ref}} = \frac{\dot{Q}_E}{\dot{W}} = \frac{\dot{m}(h_1 - h_4)}{\dot{m}(h_2 - h_1)} = \frac{h_1 - h_4}{h_2 - h_1} \qquad [7.6]$$

For a heat pump cycle:

$$\text{Performance} = \frac{\text{Required output}}{\text{Costed input}}$$

$$\text{COP}_{\text{HP}} = \frac{\dot{Q}_C}{\dot{W}} = \frac{\dot{m}(h_2 - h_3)}{\dot{m}(h_2 - h_1)} = \frac{h_2 - h_3}{h_2 - h_1} \qquad [7.7]$$

The above analysis applies to heat pump and refrigeration *cycles* and the expressions developed for the COPs are *cycle* or *internal* COPs. The *external* COPs for the *plant* would be lower in both cases. In general the heat flow rates would be very similar but the costed input would be much higher. This is mainly because of the losses incurred between the electrical power supply and the mechanical power input to the cycle, but ancillaries such as fans and pumps should also be taken into account. External (plant) COPs will always be substantially less than internal (cycle) COPs and the difference will increase as the size of plant decreases.

Relationship between COP_{ref} and COP_{HP}

Applying the first law of thermodynamics to the cycle in Fig. 7.2(c) gives:

$$\dot{Q}_C = \dot{Q}_E + \dot{W}$$

Dividing through by $\dot{W}$,

$$\frac{\dot{Q}_C}{\dot{W}} = \frac{\dot{Q}_E}{\dot{W}} + \frac{\dot{W}}{\dot{W}}$$

$$COP_{ref} = COP_{HP} + 1$$

This relationship only applies to the cycle and not to the complete plant and will only be correct if there are no heat losses or gains in the pipework.

Example 7.1

A Refrigerant 12 refrigeration plant has a constant condenser pressure of 9 bar and a constant evaporator pressure of 2 bar. The vapour leaving the evaporator is superheated by 10 K and the liquid leaving the condenser is subcooled by 5 K. If the compressor has an isentropic efficiency of 80%, determine the coefficient of performance of the plant assuming:
(i) no pressure drops in any pipes;
(ii) a pressure drop of 0.2 bar and a temperature rise of 3 K in the pipe between the evaporator and compressor.

The two cycles are shown in Fig. 7.3. Cycle 1234 is the one with no pressure drops in the pipes and 1′2′34 is the cycle with the pressure drop in the compressor suction line. Taking the enthalpy values from a p–h chart gives:

$$h_3 = h_4 = 130.2 \text{ kJ/kg} \qquad h_1 = 252.3 \text{ kJ/kg} \qquad h_{2s} = 281 \text{ kJ/kg}$$
$$h_1' = 254.3 \text{ kJ/kg} \qquad h_{2s}' = 285 \text{ kJ/kg}$$

Figure 7.3   Example 7.1

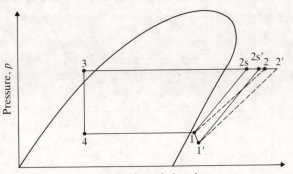

For cycle 1234, Eqns [7.1] and [7.6] give:

$$h_2 = h_1 + \frac{h_{2s} - h_1}{\eta_{IS}} = 252.3 + \frac{281 - 252.3}{0.8} = 288.2 \text{ kJ/kg}$$

$$COP_{ref} = \frac{h_1 - h_4}{h_2 - h_1} = \frac{252.3 - 130.2}{288.2 - 252.3} = 3.4$$

For cycle 1'2'34, Eqns [7.1] and [7.6] give:

$$h_2' = h_1' + \frac{h_{2s}' - h_1'}{\eta_{IS}} = 254.3 + \frac{285 - 254.3}{0.8} = 292.7$$

$$COP_{ref} = \frac{h_1 - h_4}{h_2' - h_1'} = \frac{252.3 - 130.2}{292.7 - 254.3} = 3.2$$

In this example the difference between the two COP values is about 6%. For some cycles, taking into account all pressure drops can easily make over 10% difference to the COP.

Example 7.2

A heat pump using Refrigerant 12 extracts heat from groundwater for space heating. The compressor is driven by a gas engine which has a thermal efficiency of 25%. A large flow rate of water is available and the temperature may be considered to be a constant 10 °C. The temperature difference between the water and the refrigerant in the evaporator is 10 K.

Saturated vapour enters the compressor, which has an isentropic efficiency of 85%. The condenser operates at a constant pressure of 9 bar with sub-cooled condensate leaving at 25 °C.

Calculate the percentage saving in fuel compared with a simple gas-fired heating plant of 80% thermal efficiency.

Solution

The best approach to this type of problem is to begin by analysing the heat pump cycle in isolation. The p–h diagram is shown in Fig. 7.4(a). With a heat

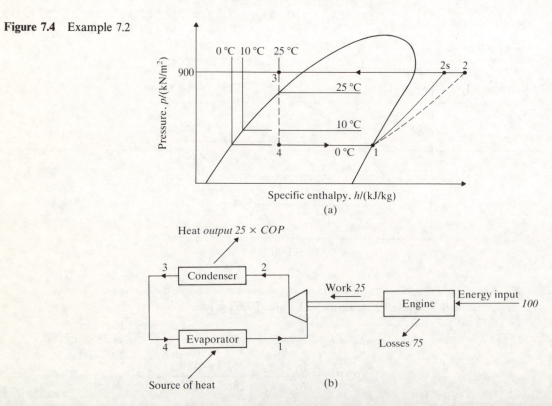

Figure 7.4   Example 7.2

source at 10 °C and a 10 K temperature difference, the evaporator is operating at 0 °C. Taking specific enthalpy values from a chart gives:

$$h_1 = 187 \text{ kJ/kg}, \ h_{2s} = 207.5 \text{ kJ/kg}, \ h_3 = h_4 = 59 \text{ kJ/kg}$$

Using Eqn [7.1],

$$h_2 = h_1 + \frac{h_{2s} - h_1}{\eta_{\text{IS}}} = 187 + \frac{207.5 - 187}{0.85} = 211.1 \text{ kJ/kg}$$

The internal COP_{HP} for the cycle is found using Eqn [7.7] as follows:

$$\text{Internal COP}_{\text{HP}} = \frac{h_2 - h_3}{h_2 - h_1} = \frac{211.1 - 59}{211.1 - 187} = 6.31$$

The high COP in this example is because of the relatively small temperature difference between the heat source and the heat sink. The condenser is probably supplying heat to warm water at about 25 °C, which suggests an underfloor heating system. This is an ideal application for a heat pump.

It is now necessary to combine the heat pump with the gas engine. This is illustrated in Fig. 7.4(b). To analyse the complete system, consider what happens to 100 units of energy supplied to the engine: 25 units of work are delivered to the heat pump compressor with the remaining 75 units lost. Each unit of energy supplied to the compressor results in 6.31 units being extracted from the refrigerant in the condenser. The energy output from the condenser is therefore $25 \times 6.31 = 157.8$ units.

If a gas-fired boiler had been used instead of the heat pump system, 100 units of energy input would have only provided 80 units of heat output. When both systems provide the same output, the input required for the heat pump system is only $80/157.8 = 0.507$ of the input required from the boiler; the saving is therefore $(1 - 0.507) = 0.493$, or 49.3%.

It should be noted that in this example the plant COP is 1.578, which is considerably lower than the internal COP of 6.31. If an electrical drive with an efficiency of 80% had been used, the plant COP would have been 5.05. However, if the efficiency of electrical power generation and distribution had been taken into account the difference between the two COP values might have been even greater.

Whilst a full financial analysis would probably show this scheme to be viable, it could be improved by recovering heat from the engine cooling system and exhaust. Heat recovery is covered in Chapter 9.

7.2 MODIFICATIONS TO BASIC CYCLES

The Addition of a Sub-cooler

Figure 7.5 shows a plant with the addition of a sub-cooler which is positioned between the condenser and the expansion valve. As mentioned previously, some sub-cooling will always take place in a condenser. However, since subcooling ideally improves the specific refrigeration effect without increasing the work input, it should be maximized.

Figure 7.5 System with sub-cooler

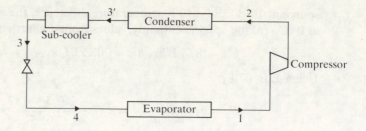

When the same coolant is used for the condenser and the sub-cooler, it should pass through the sub-cooler first. However, one advantage of a separate sub-cooler is that the condenser can be air-cooled and the sub-cooler water-cooled. The p–h diagram is the same shape as for the basic cycle with some pressure drop during the sub-cooling between 3′ and 3.

The Addition of a Heat Exchanger

Figure 7.6(a) shows a plant with a heat exchanger and Fig. 7.6(b) the corresponding p–h diagram. The sub-cooling between 3 and 4 is no longer limited by the temperature of an external cooling medium but it now has to be paid for in terms of either reducing the specific refrigeration effect or increasing the work input. There is also the extra penalty of the pressure drops required for the convective heat transfer on each side of the heat exchanger. The net effect is to decrease the COP. The main purpose of any heat exchanger is to ensure that no liquid enters the compressor.

To analyse the system it is necessary to carry out an energy balance on the heat exchanger. The change in specific enthalpy must be the same on both sides, i.e. $\Delta h = h_1 - h_6 = h_3 - h_4$. A simple heat exchanger can be designed by bringing together the pipes from the condenser and evaporator and covering them in insulation. This is quite common on small plants.

Figure 7.6 System with heat exchanger

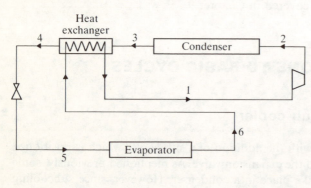

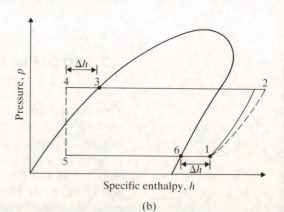

(a)   (b)

Multi-stage Compression

A closer examination of the throttling process is necessary to understand why some plants use multi-stage compression. During throttling the sub-cooled liquid expands slightly until it is saturated and then any further expansion results in some of the saturated liquid changing into saturated (flash) vapour. When liquids evaporate, heat must be extracted from the surroundings to provide the enthalpy of evaporation for the vapour. The surroundings in this case are the remaining liquid and any vapour already present, which is therefore cooled. As expansion proceeds, more and more saturated liquid must be changed into saturated vapour and the temperature will continue to fall.

Once the refrigerant has been changed into flash vapour it has fulfilled its purpose in the cycle since only the remaining liquid will provide cooling in the evaporator. Vapour produced early in the expansion consumes useful liquid as it is cooled to the evaporator temperature, increases the pressure drop as it passes through the evaporator and connecting pipework, and then absorbs power as it is recompressed.

The greater the pressure difference between the condenser and evaporator, the more flash vapour is produced and the more worthwhile it is to separate the vapour from the remaining liquid at an intermediate pressure so that it can be made to by-pass the low-pressure part of the system. Figures 7.7 and 7.8 show two methods of achieving this. Both methods use two compressors and a flash chamber which is a pressure vessel with baffles to minimize entrainment. The system shown in Fig. 7.7 mixes the flash vapour with the discharge from the LP compressor in the HP compressor suction pipe so that the refrigerant entering the HP compressor is superheated. To analyse the system it is necessary to carry out an energy balance on both the flash chamber and the HP compressor suction pipe.

An alternative arrangement is shown in Fig. 7.8. In this system the LP discharge is passed to the flash chamber and the vapour entering the HP compressor is saturated. The advantage of this system is that the maximum cycle temperature is reduced. The disadvantage is that a more sophisticated flash chamber may be required to ensure that no liquid enters the compressor.

Figure 7.7   Two-stage compression system

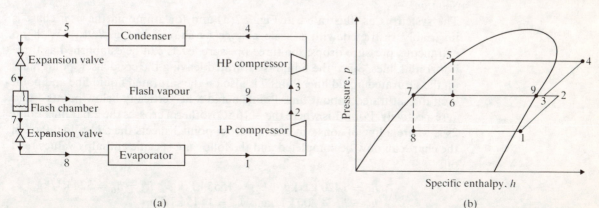

(a)   (b)

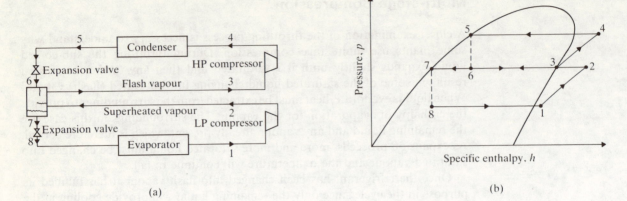

(a)

(b)

Figure 7.8 Alternative two-stage compression system

There is no significant difference between the coefficient of performance of the two basic systems and there are many other variants. One way of improving the COP of a two-stage system is to pass the discharge from the LP compressor through a water-cooled intercooler.

Example 7.3

An ammonia two-stage vapour-compression refrigeration plant operates with a condenser pressure of 12 bar, a flash chamber pressure of 5 bar and an evaporator pressure of 2 bar. Saturated liquid leaves the condenser and after being throttled to 5 bar the saturated liquid and saturated vapour are separated in the flash chamber. The saturated vapour is then mixed with the superheated vapour from the LP compressor discharge before it enters the HP compressor while the saturated liquid is throttled down to the evaporator pressure.

The vapour leaving the evaporator is at $-16\,°C$. If each stage of compression has an isentropic efficiency of 90% calculate:

(i)   the mass fraction of vapour leaving the flash chamber;
(ii) the coefficient of performance of the plant;
(iii) the mass flow rate of refrigerant through the condenser when the refrigeration load is 400 kW.

Use the $p-h$ chart provided and hand the chart in with your solution.

(Wolverhampton Polytechnic)

Solution
The system is as illustrated in Fig. 7.7(a) and for ammonia the $p-h$ chart is invariably restricted with a scale change as shown in Fig. 7.9(a). With no extraneous pressure drops, the three pressure levels can be established as three horizontal lines. Since the liquid is saturated leaving the condenser, point 5 lies on the saturated liquid line. Point 7 is also on the saturated liquid line and point 9 on the saturated vapour line. Points 6 and 8 lie vertically below points 5 and 7, respectively. Point 1 is where the $-16\,°C$ isotherm crosses the 2 bar line. Point 2s is where a line of constant entropy from point 1 meets the 5 bar line. Part of the chart can now be completed and the following specific enthalpy values read off:

$$h_1 = 1420\ \text{kJ/kg}, \quad h_{2s} = 1555\ \text{kJ/kg}, \quad h_5 = h_6 = 324\ \text{kJ/kg},$$
$$h_7 = h_8 = 200\ \text{kJ/kg}, \quad h_9 = 1445\ \text{kJ/kg}.$$

Figure 7.9 Example 7.3: (a) p–h diagram; (b) flash chamber; (c) adiabatic mixing

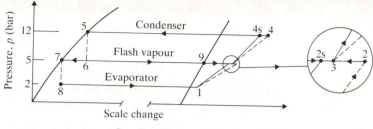

(a) p–h diagram

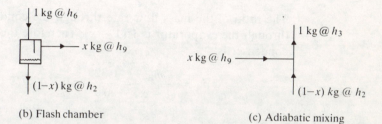

(b) Flash chamber (c) Adiabatic mixing

Equation [7.1] can now be used to calculate h_2 as follows:

$$h_2 = h_1 + \frac{h_{2s} - h_1}{\eta_{IS}} = 1420 + \frac{1555 - 1420}{0.9} = 1570 \text{ kJ/kg}$$

It is now necessary to carry out an energy balance on the flash chamber where adiabatic separation takes place. The flash chamber is shown in detail in Fig. 7.9(b). Considering unit mass flow rate through the condenser with x kg of flash vapour, the energy balance becomes:

$$1 \times h_6 = x \times h_9 + (1 - x)h_7$$
$$324 = x.1445 + (1 - x)200$$
$$\therefore \quad x = 0.1$$

An energy balance is now required on the HP compressor suction pipe where the flash vapour at state 9 and the LP compressor discharge at state 2 are adiabatically mixed. This detail is shown on Fig. 7.9(c) and the energy balance is:

$$1 \times h_3 = x \times h_9 + (1 - x)h_2$$
$$h_3 = (0.1 \times 1445) + (0.9 \times 1570)$$
$$\therefore \quad h_3 = 1558$$

Point 3 can now be established on the chart and h_{4s} found where the isotherm from 3 meet the 12 bar line; h_{4s} is 1690 kJ/kg and h_4 can be found using Eqn [7.1] as follows:

$$h_4 = h_3 + \frac{h_{4s} - h_3}{\eta_{IS}} = 1558 + \frac{1690 - 1558}{0.9} = 1705 \text{ kJ/kg}$$

The COP can now be calculated by modifying Eqn [7.6] to take into account the variations in mass flow rates as follows:

$$
\begin{aligned}
\text{COP}_{\text{ref}} &= \frac{(1-x)(h_1 - h_8)}{(1-x)(h_2 - h_1) + (h_4 - h_3)} \\
&= \frac{0.9(1420 - 200)}{0.9 \times 150 + 147} \\
&= 3.9
\end{aligned}
$$

The mass flow rate through the evaporator can be found using Eqn [7.3] as follows:

$$
\begin{aligned}
\dot{Q}_E &= \dot{m}_E(h_1 - h_8) \\
400 &= \dot{m}_E(1420 - 200) \\
\therefore \qquad \dot{m}_E &= 0.328 \text{ kg/s}
\end{aligned}
$$

The ratio of the mass flow rate through the condenser to the mass flow rate through the evaporator is $1:(1-x)$. Therefore the mass flow rate through the condenser is:

$$
\dot{m}_C = \frac{\dot{m}_E}{(1-x)} = \frac{0.328}{0.9} = 0.364 \text{ kg/s}
$$

Multi-evaporator Systems

A refrigeration plant may have more than one evaporator. Figure 7.10 shows two possible arrangements. Figure 7.10(a) shows two evaporators which will operate at the same temperature and Fig. 7.10(b) a system for two evaporators which will operate at different temperatures. The former arrangement is suitable for air conditioning plants using direct expansion coils where a single refrigeration plant serves more than one air-handling unit. The latter arrangement could be used where there is a requirement for air conditioning (high-temperature evaporator) and food storage (low-temperature evaporator).

Figure 7.10 Multi-evaporator systems

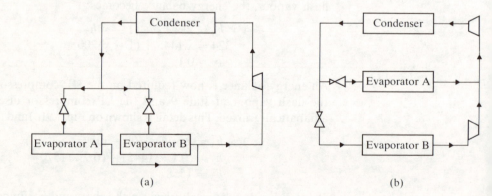

(a)　　　　　　　　　　　　　　　　(b)

Cascade Systems

Multi-stage compression using a single refrigerant is limited by the physical properties of refrigerants. Ideally the operating pressures for a refrigeration plant should be moderate in the condenser and just above ambient in the evaporator.

This minimizes the compression work and prevents moist air leaking into the system. As the temperature in the evaporator falls, so does the pressure, producing a corresponding increase in specific volume. A refrigerant with satisfactory properties at normal condensing temperatures therefore becomes increasingly unsatisfactory as the evaporator temperature falls below some value which depends on the refrigerant used. Conversely a refrigerant which is suitable for use in a low-temperature evaporator may have an unacceptably high pressure at typical heat rejection temperatures. It may also be unable to reject heat by condensation if the critical temperature is below that of the coolant (e.g. Refrigerant 13 has a critical temperature of 29 °C).

The limitations of single refrigerants led to the development of cascade systems. A simple cascade system with two refrigerants is shown in Fig. 7.11. The low-temperature refrigerant receives the heat from the low-temperature space in the evaporator and rejects it to a heat exchanger which is the condenser for the low-temperature refrigerant and the evaporator for the high-temperature refrigerant. The high-temperature refrigerant receives heat in the heat exchanger and rejects it from its condenser to the cooling medium. With this arrangement each refrigerant operates within its own ideal range. The thermodynamic disadvantage of such an arrangement is the temperature difference required in the heat exchanger, which increases the work required for compression.

Figure 7.11 Cascade system

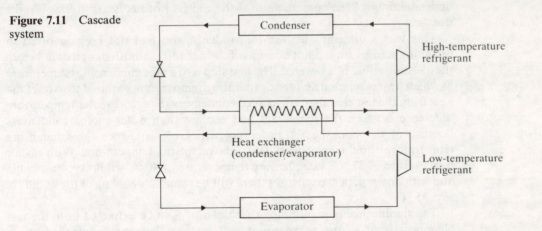

Cascade systems are only used for low-temperature refrigeration work. Two refrigerants which are suitable only for the low-temperature cycle of a cascade system are Refrigerants 13 and 14. Refrigerant 13 has an evaporating temperature of −81 °C at ambient pressure and a critical temperature of 29 °C, and Refrigerant 14 an evaporating temperature of −128 °C at ambient pressure and a critical temperature of −46 °C. Any refrigerant suitable for a single-stage system could be used as the high-temperature refrigerant.

Example 7.4
A test chamber is to be maintained at −35 °C by the extraction of 15 kW of energy. The final sink for heat rejection is outside air at a temperature of 27 °C. The following alternative vapour-compression direct-expansion refrigeration systems have been suggested:

(a) a two-stage compression system using standard R22 single-stage, semi-hermetic compressors; or

(b)   a cascade system employing R12 in the high-temperature cycle and R22 in the low-temperature cycle, both cycles using semi-hermetic compressors.

Sketch line diagrams of feasible circuits and make estimates of the temperature, pressures and efficiencies at each stage of the cycle, and hence calculate the energy absorbed by the compressors in each system. From the results of your calculations, and taking into consideration any other relevant factors, recommend the system that you consider should be adopted.

Note: Neglect pressure drops through pipelines and heat exchangers.

(CIBSE)

Solution

An open-type compressor is one where the drive is in a separate housing to the compressor with the shaft connecting them passing through a gland seal to prevent the leakage of refrigerant to atmosphere. To avoid reliance on a seal, the compressor and drive can be installed in the same housing with the refrigerant in contact with the motor. This is termed a hermetically sealed unit. Hermetically sealed units are used in most small refrigerators such as domestic units. A semi-hermetic unit is one where the compressor and drive are in the same housing but there is some access for maintenance such as removable cylinder heads. An understanding of the construction of the unit is not necessary to answer the question.

This was a difficult question for students who had not been involved in designing refrigeration plant because of the need to establish design data before the question could be answered. The first step is to fix the condensing temperature at which heat is rejected and the evaporating temperature required to extract the heat from the test chamber. Air-cooled condensers require a greater temperature difference between the refrigerant and coolant than water-cooled condensers because of the poor heat transfer properties of air; hence the condensing temperature should be at least 10 K above the air temperature. With an air temperature of 27 °C, a condensing temperature of 40 °C would be reasonable and with any practical condenser there will be some sub-cooling of the liquid to, say, 35 °C.

The significance of mentioning that the heat is to be extracted from the test chamber using a direct expansion coil is that these are controlled using thermostatic expansion valves (TEVs). A TEV is operated by the degree of superheat of the vapour leaving the evaporator (see Section 7.4) and the superheat would therefore be greater than with a flooded evaporator (with the refrigerant outside the tubes the vapour leaving the evaporator could be saturated—see Section 7.7). The evaporating temperature will therefore need to be at least 10 K below the chamber temperature. If the evaporating temperature is −45 °C, the vapour leaving the evaporator could be at −37 °C, giving 8 K of superheat to operate the TEV.

A reasonable isentropic efficiency for each compressor is 80% and the design data established so far will be the same for both systems.

(a)   Two-stage system using R22:

Figure 7.12 shows a line diagram and a *p–h* diagram for the system selected. This is one of the systems described earlier but there are several alternatives which could have been used. When a gas is compressed in two stages from a

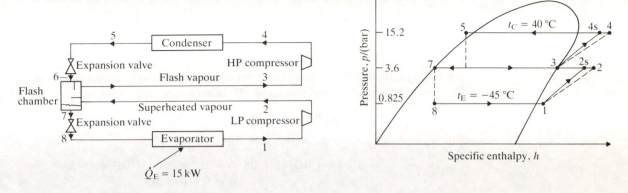

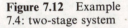

Figure 7.12 Example
7.4: two-stage system

pressure p_1 to a pressure p_3 it can be shown that for minimum work the intermediate pressure p_2 can be found from the following equation:

$$p_2 = \sqrt{p_1 p_3} \qquad [7.8]$$

The same equation would apply to a refrigerant if the mass flow rate were the same in each stage. However, for a two-stage refrigeration system, the mass flow rates are not the same and the ideal intermediate pressure will always be slightly above that predicted by Eqn [7.8]. (The intermediate pressure is higher because the greater mass flow rate passes through the high-pressure stage.) To determine the optimum intermediate pressure would require sufficient calculations to plot a graph of the rate of work input against intermediate pressure. This is clearly not intended and any reasonable value *above* that predicted by Eqn [7.8] would be acceptable.

From the p–h chart the pressure corresponding to the condensing temperature of 40 °C is 15.2 bar and the pressure corresponding to the evaporating temperature of -45 °C is 0.82 bar. Equation [7.8] can now be used as follows:

$$p_2 = \sqrt{15.2 \times 0.82} = 3.5 \text{ bar}$$

The intermediate pressure must be above this value—say 3.6 bar. There is now sufficient information to establish all the points on the p–h diagram except 2 and 4, which must be calculated and need not be included on the diagram. Specific enthalpy values from the chart are:

$$h_1 = 236.0 \text{ kJ/kg}, \quad h_{2s} = 274.0 \text{ kJ/kg}, \quad h_3 = 246.3 \text{ kJ/kg},$$

$$h_5 = h_6 = 87.6 \text{ kJ/kg}, \quad h_7 = h_8 = 33.7 \text{ kJ/kg}, \quad h_{4s} = 286 \text{ kJ/kg}$$

Equation [7.1] can be used to calculate h_2 and h_4 as follows:

$$h_2 = h_1 + \frac{h_{2s} - h_1}{\eta_{IS}} = 236.0 + \frac{274 - 236}{0.8} = 283.5 \text{ kJ/kg}$$

$$h_4 = h_3 + \frac{h_{4s} - h_3}{\eta_{IS}} = 246.3 + \frac{286 - 246.3}{0.8} = 295.9 \text{ kJ/kg}$$

The mass flow rate of R22 through evaporator and LP compressor can now be found from an energy balance on the evaporator using Eqn [7.3] as follows:

$$\dot{Q}_E = \dot{m}_1(h_1 - h_8)$$
$$15 = \dot{m}_1(236 - 33.7)$$
$$\dot{m}_1 = 0.0741 \text{ kg/s}$$

An energy balance on the flash chamber can now be used to calculate the mass flow rate through the condenser and HP compressor as follows:

$$\dot{m}_6 h_6 + \dot{m}_2 h_2 = \dot{m}_3 h_3 + \dot{m}_7 h_7$$

But $\dot{m}_6 = \dot{m}_3$ and $\dot{m}_7 = \dot{m}_2 = \dot{m}_1$; therefore

$$\dot{m}_3 \times 87.6 + 0.0741 \times 283.5 = \dot{m}_3 \times 246.3 + 0.0741 \times 33.7$$
$$\therefore \qquad \dot{m}_3 = 0.1167 \text{ kg/s}$$

Using Eqn [7.5] for the work input to each compressor:

$$\dot{W} = \dot{m}_3(h_4 - h_3) + \dot{m}_1(h_2 - h_1)$$
$$= 0.1167(295.9 - 246.3) + 0.0741(283.5 - 236)$$
$$= 9.3 \text{ kW}$$

(b)   Cascade system using R12 and R22:

Figure 7.13 shows a line diagram and a p–h diagram for this system. The first problem is to fix the condensing and evaporating temperatures in the heat exchanger. With good heat transfer rates at either side of the heat exchanger, a design with a 5 K difference between the condensing and evaporating temperatures should be satisfactory. With a different refrigerant in each compressor Eqn [7.8] is no longer relevant. However, the intermediate temperatures in the heat exchanger will be similar to the saturation temperature in the flash chamber of the previous system ($-9.7\,°C$). Selecting a condensing temperature of $-5\,°C$ for the R22 will give an evaporating temperature of $-10\,°C$ for the R12. These values are only reasonable estimates and many calculations would be necessary to determine optimum values. Figure 7.13(b) shows the two cycles on the same diagram to illustrate the processes in the heat exchanger but it must be remembered that the scales are not the same and the values are taken from different charts.

Figure 7.13   Example 7.4: cascade system

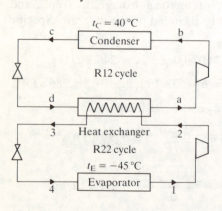

(a)

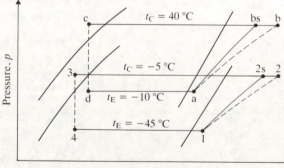

(b)

For the low-temperature cycle with R22, point 1 is the same as for the two-stage cycle and point 3 can be established allowing 2 K sub-cooling. The cycle can now be drawn and the following values read from the chart:

$$h_1 = 236 \text{ kJ/kg}, \quad h_{2s} = 279 \text{ kJ/kg}, \quad h_3 = h_4 = 36.5 \text{ kJ/kg}$$

Using Eqn [7.1],

$$h_2 = h_1 + \frac{h_{2s} - h_1}{\eta_{IS}} = 236.0 + \frac{279 - 236}{0.8} = 289.8 \text{ kJ/kg}$$

The mass flow rate of R22 through the evaporator can now be found from Eqn [7.3] as follows:

$$\dot{Q}_E = \dot{m}_1(h_1 - h_4)$$
$$15 = \dot{m}_1(236 - 36.5)$$
$$\dot{m}_1 = 0.0752 \text{ kg/s}$$

The R12 cycle can now be drawn allowing 2 K superheat for point a and the same sub-cooling as in the two-stage cycle (5 K) to establish point c. (It would have made little difference if no sub-cooling or superheating had been allowed for in the heat exchanger.) From the p–h chart:

$$h_a = 184.6 \text{ kJ/kg}, \quad h_{bs} = 213 \text{ kJ/kg}, \quad h_c = h_d = 69 \text{ kJ/kg}$$

Using Eqn [7.1],

$$h_b = h_a + \frac{h_{bs} - h_a}{\eta_s} = 184.6 + \frac{213 - 184.6}{0.8} = 220.1 \text{ kJ/kg}$$

The mass flow rate of refrigerant R12 can now be found from an energy balance on the heat exchanger as follows:

$$\dot{m}_a(h_a - h_d) = \dot{m}_1(h_2 - h_3)$$
$$\dot{m}_a(186.4 - 69) = 0.0752(220.1 - 36.5)$$
$$\therefore \quad \dot{m}_a = 0.118 \text{ kg/s}$$

The rate of work input to the compressors can now be found by applying Eqn [7.5] to each as follows:

$$\dot{W} = \dot{m}_a(h_b - h_a) + \dot{m}_1(h_2 - h_1)$$
$$= 0.118(220.1 - 184.6) + 0.0752(289.8 - 236)$$
$$= 8.23 \text{ kW}$$

The calculations show that the cascade system requires a rate of work input of 8.23 kW compared with 9.3 kW for the two-stage system, which is an energy saving of about 13%. Whether this would justify selecting the cascade system is not easy to judge without a knowledge of the capital costs, which would be higher for the cascade system.

7.3 RECIPROCATING COMPRESSORS

Reciprocating machines are widely used for refrigeration plants. They can be used in plants ranging from the smallest to those with hundreds of kilowatts of refrigeration capacity.

Figure 7.14
Reciprocating
compressor

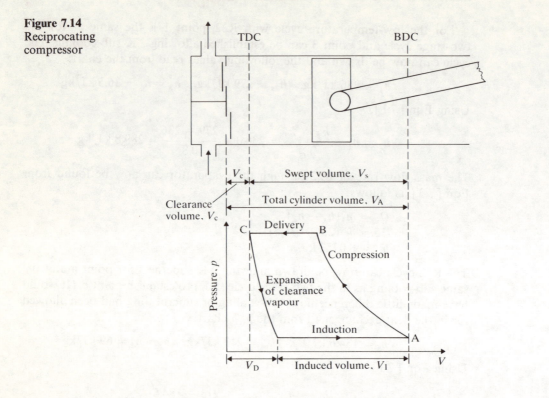

The operation of a reciprocating compressor is illustrated in Fig. 7.14. This shows a cross-section through the cylinder above a p–V diagram. The operation is classed as two-stroke, which means that there is one upward and one downward stroke of the piston for each revolution of the crank and shaft.

Starting at A with the piston at bottom dead centre (BDC), the first part of the upward stroke compresses the vapour to the delivery pressure at point B. The delivery valve then opens and for the remainder of the stroke the temperature and pressure ideally remain constant as the vapour is delivered to the discharge pipe. In practice the pressure in the cylinder must always be slightly above the delivery pressure and the line B–C is wavy due to movement of the spring-loaded valve.

Point C is at top dead centre (TDC), and there is some vapour left in the cylinder clearance volume which is at the delivery pressure. The clearance volume is the minimum volume necessary to allow freedom of valve and piston movement without risking collision. This is typically about 5% of the swept volume.

The downward stroke now starts with both valves closed. The first part of the stroke is expansion of the clearance until the pressure falls to the inlet value at point D. The inlet valve now opens and vapour is drawn into the cylinder during the remainder of this stroke. In practice the cylinder pressure must be slightly below that of the vapour in the suction pipe and line D–A is wavy due to valve movement.

Some of the definitions and equations necessary to investigate reciprocating compressors are:

Induced Volume, V_I

This is the volume of vapour, at the compressor inlet state, which enters each cylinder during each complete cycle.

Induced Volume Flow Rate, $\dot{V}_I$

This is the volume flow rate of vapour, at the compressor inlet state, which passes through the machine.

Swept Volume, V_s

This is the volume through which the piston moves during one stroke. If the piston diameter is d and the stroke L, the swept volume of each cylinder, V_s, becomes:

$$V_s = \frac{\pi d^2 L}{4} \qquad [7.9]$$

Volume Swept per second or Compressor Displacement Rate, $\dot{V}_s$

Assuming a single-acting machine such as the one illustrated in Fig. 7.14, there is one stroke of the piston during which vapour is induced for each revolution of the drive shaft. If the compressor has n cylinders which are driven by a shaft which rotates at N revolutions per second, the volume swept per second is:

$$\dot{V}_s = V_s N n \qquad [7.10]$$

If $\dot{V}_s$ is in m^3 the units of $\dot{V}_s$ are m^3/s. Unfortunately compressor displacement rates are often specified in terms of litres per second and the shaft speeds in terms of revolutions per minute. These should be converted into basic SI units before being used in any equation.

Double-acting Machines

When a compressor is double-acting there are sets of valves on either side of the piston so that the cycles per revolution and hence the volume swept per second are increased by a factor of two if the thickness of the piston rod is ignored.

Volumetric Efficiency, η_v

The actual volumetric efficiency of a reciprocating compressor can be defined using either volumes or volume flow rates as follows:

$$\eta_v = \frac{\text{Inlet volume induced per cycle per cylinder } V_I}{\text{Swept volume of each cylinder } V_s} \qquad [7.11]$$

$$\eta_v = \frac{\text{Volume flow rate of inlet vapour } \dot{V}_I}{\text{Total volume swept per second } \dot{V}_s} \qquad [7.12]$$

The inlet vapour will always be at a slightly higher pressure and lower temperature than the vapour in the cylinder during induction. However it is useful to investigate the volumetric efficiency, $\eta_{v\,CYL}$ based on idealized cylinder conditions—induction and delivery pressures constant—as illustrated in Fig. 7.14.

The expansion of the clearance vapour will be a polytropic process, i.e. it will follow the law $pV^n = $ constant. The volume occupied by the expanded clearance vapour, V_D, can therefore be expressed in terms of the clearance volume and pressure ratio as follows:

$$V_D = V_c \left(\frac{p_2}{p_1} \right)^{1/n}$$

The volume induced per cycle, V_I, is therefore:

$$V_I = V_A - V_D$$
$$= V_s + V_c - V_D$$
$$= V_s + V_c - V_c \left(\frac{p_2}{p_1} \right)^{1/n}$$
$$= V_s - V_c \left\{ \left(\frac{p_2}{p_1} \right)^{1/n} - 1 \right\}$$

Dividing by the swept volume, V_s, gives:

$$\eta_{v\,CYL} = 1 - \frac{V_c}{V_s} \left\{ \left(\frac{p_2}{p_1} \right)^{1/n} - 1 \right\} \qquad [7.13]$$

This expression for the volumetric efficiency based on idealized cylinder conditions will always be higher than the actual volumetric efficiency found using Eqn [7.11] or [7.12]. However, the same factors will affect both and the expression for $\eta_{v\,CYL}$ is useful for investigating compressor performance. For a particular machine the clearance to swept volume ratio, V_c/V_s, is fixed and the index of expansion, n, will be relatively constant, which leaves the volumetric efficiency depending only on the pressure ratio.

Comparison with Air Compressors

Most students have studied air compressors before refrigeration plant and it is useful to compare these two applications of reciprocating machines. In a refrigeration plant the overall performance is specified by an isentropic efficiency, whereas air compressors use an isothermal efficiency.

An isothermal process is used as a basis of comparison in air compressors because it represents a condition for minimum work which can be approached in practice, particularly if the compressor is water-cooled. In a refrigeration plant, isothermal compression is not a practical ideal—firstly because it would result in condensation with the subsequent problems of compressing a two-phase fluid, and secondly because it would require a coolant colder than the refrigerant, and if such a coolant were available there would be no need for the system.

Example 7.5

(a) Discuss briefly the safety and toxicity aspects of modern fluorocarbon refrigerants which are widely used for an air conditioning application.

(b) A heat pump is used to heat the water in a public swimming baths. The refrigerant is R12 operating between the range $320 \, kN/m^2$ and $1000 \, kN/m^2$.

The vapour entering the compressor has 10 K of superheat and the liquid leaving the condenser is undercooled to 35 °C before it enters the expansion valve.

The heat pump obtains its low-grade heat from 0.8 kg/s of water from a nearby river; the water drops in temperature from 15 °C to 10 °C in passing through the evaporator. Assuming that the compressor isentropic efficiency is 0.8 and that there are no losses of heat or pressure drops due to friction in the cycle, sketch the circuit on a pressure/enthalpy diagram and calculate:

(i) the mass flow rate of R12;
(ii) the required power input to the compressor assuming a mechanical efficiency of 0.75;
(iii) the coefficient of performance based on the power input to the compressor;
(iv) the swept volume of the cylinders if the compressor has four single-acting cylinders and rotates at 1500 rev/min.

You may assume the volumetric efficiency is 0.8.

(CIBSE)

Solution

(a) Fluorocarbon refrigerants have typical long-term exposure limits of 1000 ppm. This compares with an alternative such as ammonia which has a limit of only 25 ppm. They are therefore classed as non-toxic and are also non-flammable (Ref. 7.1). These are two important reasons for their widespread use for both domestic and industrial refrigeration units.

Because they are so inert they can reach the Earth's stratosphere without decomposing. Once there, solar radiation releases chlorine which acts as a catalyst which speeds up the breakdown of ozone into oxygen. This has led to a complete reassessment of what was for many years thought to be a harmless group of chemical compounds.

(b) Figure 7.15 is a sketch of the $p–h$ diagram for the system. From the $p–h$ chart:

$$h_1 = 195 \, kJ/kg, \quad h_{2s} = 217 \, kJ/kg, \quad h_3 = h_4 = 69 \, kJ/kg$$
$$v_1 = 0.057 \, m^3/kg$$

Figure 7.15 Example 7.5

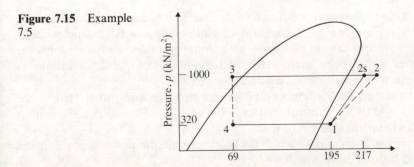

Specific enthalpy, $h/(kJ/kg)$

Using Eqn [7.1]

$$h_2 = h_1 + \frac{(h_{2s} - h_1)}{\eta_s} = 195 + \frac{(217 - 195)}{0.8} = 222.5 \text{ kJ/kg}$$

For the evaporator:
From property tables c_p for water between 10 °C and 15 °C is 4.19 kJ/kg K.

$$\text{Heat flow rate from the water} = \text{Heat supply rate to the R12}$$
$$\dot{m}_w c_{pw}(t_i - t_o) = \dot{m}_R(h_1 - h_4)$$
$$0.8 \times 4.19(15 - 10) = \dot{m}_R(195 - 69)$$
$$\therefore \qquad \dot{m}_R = 0.133 \text{ kg/s}$$

For the compressor:

$$\text{Power input to the refrigerant} = \dot{m}(h_2 - h_1)$$
$$= 0.133(222.5 - 195)$$
$$= 3.66 \text{ kW}$$
$$\text{Power input to the compressor} = \frac{3.66}{0.75} = 4.88 \text{ kW}$$

For the condenser:

$$\dot{Q}_C = \dot{m}_R(h_2 - h_3) = 0.133(222.5 - 69) = 20.4 \text{ kW}.$$

Coefficient of performance:

$$\text{Overall COP} = \frac{\dot{Q}_C}{\text{Power input}} = \frac{20.4}{4.88} = 4.18$$

Swept volume:

Induced volume flow rate $\dot{V}_I = \dot{m}v_1 \qquad = 0.133 \times 0.057$
$$= 0.007\,58 \text{ m}^3/\text{s}$$

Induced volume per cylinder per cycle $V_I = \dfrac{\dot{V}_I}{nN} = \dfrac{0.007\,58 \times 60}{4 \times 1500}$
$$= 7.58 \times 10^{-5} \text{ m}^3$$

Swept volume per cylinder $\qquad = \dfrac{V_I}{\eta_v} = \dfrac{7.58 \times 10^{-5}}{0.8}$
$$= 9.48 \times 10^{-5} \text{ m}^3$$

Example 7.6
The two-stage vapour-compression refrigeration plant shown diagrammatically in Fig. 7.16 is charged with Refrigerant 12. Both compressors are single-acting, single-stage reciprocating machines with four cylinders and an isentropic efficiency of 80%. They each have a stroke/bore ratio of 1.1 and run at 480 rev/min with a volumetric efficiency of 75%.

The condenser, flash chamber and evaporator pressures are 10 bar, 3 bar and 1 bar respectively. Saturation states exist at points 5 and 3 and the vapour at point 1 is superheated by 10 K.

If the refrigeration capacity is 200 kW, determine the power input and the stroke and bore of each cylinder.

(Wolverhampton Polytechnic)

Figure 7.16 Example 7.6

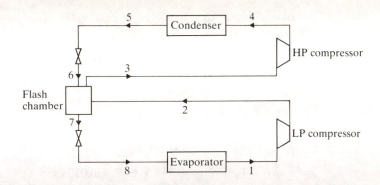

Solution

Figure 7.17 is a sketch of the p–h diagram. From the p–h chart, the specific enthalpy values and the specific volumes are:

$$h_1 = 244 \text{ kJ/kg}, \quad h_3 = 251 \text{ kJ/kg}, \quad h_5 = h_6 = 140 \text{ kJ/kg},$$
$$h_{2s} = 264 \text{ kJ/kg}, \quad h_{4s} = 271 \text{ kJ/kg}, \quad h_7 = h_8 = 100 \text{ kJ/kg},$$
$$v_1 = 0.17 \text{ m}^3/\text{kg}, \quad v_3 = 0.058 \text{ m}^3/\text{kg}$$

Figure 7.17 Example 7.6

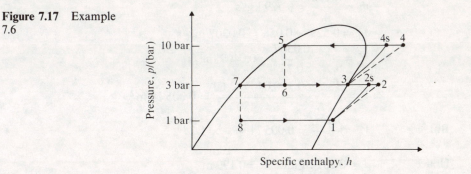

Using Eqn [7.1],

$$h_2 = h_1 + \frac{(h_{2s} - h_1)}{\eta_s} = 244 + \frac{(264 - 244)}{0.8} = 269 \text{ kJ/kg}$$

$$h_4 = h_3 + \frac{(h_{4s} - h_3)}{\eta_s} = 251 + \frac{(271 - 251)}{0.8} = 276 \text{ kJ/kg}$$

Applying Eqn [7.3] for the evaporator gives:

$$\dot{Q}_E = \dot{m}_E(h_1 - h_8)$$
$$200 = \dot{m}_E(244 - 100)$$
$$\therefore \quad \dot{m}_E = 1.39 \text{ kg/s}$$

Carrying out an energy balance on the flash chamber gives:

$$\dot{m}_C h_6 + \dot{m}_E h_2 = \dot{m}_E h_7 + \dot{m}_C h_3$$
$$\dot{m}_C 140 + (1.39 \times 269) = (1.39 \times 100) + \dot{m}_C 251$$
$$\therefore \quad \dot{m}_C = 2.116 \text{ kg/s}$$

The power input can now be calculated by applying Eqn [7.5] to each compressor as follows:

$$\dot{W} = \dot{m}_E(h_2 - h_1) + \dot{m}_C(h_4 - h_3)$$
$$= 1.39(269 - 244) + 2.116(276 - 251)$$
$$= 87.7\ \text{kW}$$

For each LP cylinder,

$$\dot{m} = \frac{1.39}{4} = 0.348\ \text{kg/s}$$
$$\dot{V}_I = \dot{m}v = 0.348 \times 0.17 = 0.0591\ \text{m}^3/\text{s}$$
$$\dot{V}_s = \frac{\dot{V}_I}{\eta_v} = \frac{0.0591}{0.75} = 0.0788\ \text{m}^3/\text{s}$$
$$V_s = \frac{\dot{V}_s}{N} = \frac{0.0788 \times 60}{480} = 0.009\,85\ \text{m}^3$$

But $$V_s = \frac{\pi d^2 L}{4} = 0.009\,85 = \frac{\pi}{4} \times 1.1 d^3$$

Hence $d = 0.225$ m and $L = 0.248$ m

For each HP cylinder,

$$\dot{m} = \frac{2.116}{4} = 0.529\ \text{kg/s}$$
$$\dot{V}_I = 0.529 \times 0.058 = 0.0307\ \text{m}^3/\text{s}$$
$$\dot{V}_s = \frac{\dot{V}_I}{\eta_v} = \frac{0.0307}{0.75} = 0.0409\ \text{m}^3/\text{s}$$
$$V_s = \frac{\dot{V}_s \times 60}{N} = \frac{0.0409 \times 60}{480} = 0.005\,11\ \text{m}^2$$

But $$V_s = \frac{\pi d^2 L}{4} = 0.005\,11 = \frac{\pi}{4} \times 1.1 d^3$$

Hence $d = 0.181$ m and $L = 0.199$ m

Reciprocating Compressor Characteristics

The most important quantities in any refrigeration plant are the refrigeration capacity, $\dot{Q}_E$, and the power input, $\dot{W}$. Since compressors are designed for particular refrigerants it is possible to use these two parameters in specifying a compressor. However, both quantities vary as the conditions in the evaporator and condenser vary, so it is useful to present the characteristics graphically.

The central section of Fig. 7.18(d) is typical of the way compressor characteristics are presented. The vertical scale is the refrigeration capacity and power input in kilowatts. The horizontal scale is normally the evaporator saturation temperature, t_E. Values of $\dot{Q}_E$ and $\dot{W}$ are plotted for different values of condensing temperature, t_C. There is only one value of saturation pressure for each value of saturation temperature for any particular refrigerant; the horizontal scale could therefore have been the evaporator saturation pressure. A compressor in isolation would be thought of as working from a suction to

Figure 7.18
Reciprocating
compressor
characteristics

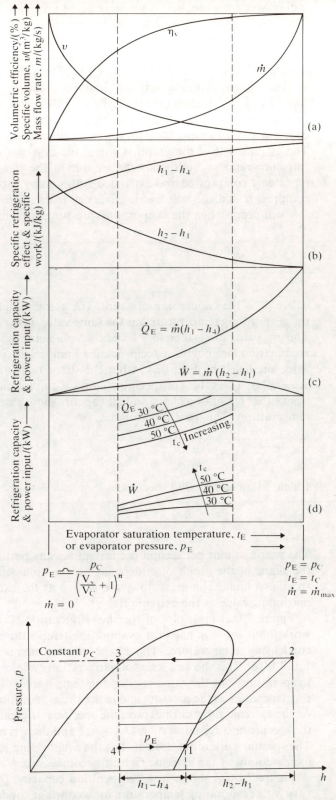

a discharge pressure, but in a refrigeration system it is convenient to think of it as working from an evaporating temperature to a condensing temperature.

To understand the form of the characteristics it is necessary to analyse the basic principles involved and build up the graphs in stages. Figure 7.18(a) has three graphs: volumetric efficiency, η_v; specific volume, v; and mass flow rate, $\dot{m}$. The shape of the graph of volumetric efficiency can be plotted from Eqn [7.13], which gives the ideal cylinder volumetric efficiency for any pressure ratio for a particular compressor. In this case the condensing temperature is constant and the evaporating pressures are plotted as evaporating temperatures. The extreme left of the graph is where the evaporator temperature is so low (pressure ratio so high) that the vapour in the clearance volume would be repeatedly compressed and expanded without any vapour passing through the compressor and the volumetric efficiency is therefore zero. From Eqn [7.13] this will occur when the evaporating pressure, p_E, falls to the following value:

$$p_E = \frac{p_C}{\left(\dfrac{V_s}{V_c} + 1\right)^n}$$

where p_C is the condensing pressure. The extreme right of the graph is where the evaporating temperature has the same value as the condensing temperature. The pressure ratio is therefore 1 and the volumetric efficiency 100%. The actual volumetric efficiency of a reciprocating compressor is always lower than the value predicted by Eqn [7.13] but the shape of the graph and the restraints are similar. The values of specific volume, v, to plot on the graph can be taken from either tables or a chart, and all vapours show a steep rise at low saturation temperatures. The mass flow rate is:

$$\dot{m} = \frac{\dot{V}_1}{v}$$

which, using Eqn [7.12], becomes:

$$\dot{m} = \frac{\dot{V}_s \eta_v}{v}$$

The swept volume per second is constant for any particular machine; therefore the shape of the graph of $\dot{m}$ depends on the volumetric efficiency divided by specific volume. The mass flow rate is zero at the extreme left and reaches a maximum value at the extreme right.

Figure 7.18(b) is a plot of specific refrigeration effect $(h_1 - h_4)$ and specific work input $(h_2 - h_1)$ against evaporating temperature for a constant value of condensing temperature. The shape of the curves can be deduced from Fig. 7.18(e). This shows a series of refrigeration cycles with different evaporating temperatures but the same condensing temperature. The specific refrigeration effect increases as the pressure and hence the evaporating temperature increases. At the extreme right of the graph the specific refrigeration effect is the same as the specific heat rejection effect $(h_2 - h_3)$, and there is no thermodynamic cycle. The specific work input decreases as the evaporating pressure and temperature increase until at the extreme right it becomes zero.

Figure 7.18(c) shows how refrigeration capacity, $\dot{Q}_E$, and power input, $\dot{W}$, vary with evaporating temperature for a constant condensing temperature. The values for these curves are obtained by multiplying values from the graphs in Fig. 7.18(b) by the corresponding mass flow rate from Fig. 7.18(a). The capacity curve increases from zero at the lowest evaporating temperature to a maximum when the evaporating temperature is equal to the condensing temperature. The power input has a value of zero at either extreme of the graph. It is zero at the extreme left because the mass flow rate is zero and at the extreme right because $(h_2 - h_1)$ is zero. The point at which the maximum value occurs depends on the condensing temperature.

In Figs 7.18(a)–(c) the condensing temperature was constant. To get a full set of characteristics it is necessary to repeat this process for a number of different condensing temperatures. Also, in practice compressors operate within a much narrower range than is theoretically possible. The results are characteristics which look like the graph in Fig. 7.18(d). The capacity curves always increase as the evaporating temperature increases. Most of the power curves will also increase with an increase in evaporating temperature, but some at the lowest condensing temperatures may have a maximum value within the working range.

Capacity Control of Reciprocating Compressors

The most common methods of controlling the capacity of reciprocating compressors are the following.

(1) Cycling

The machine is stopped and then re-started as required. It is therefore always operating at full capacity when running. This method works well for small systems such as domestic machines but is unsuitable for larger installations.

(2) Cylinder Unloading

Compressors with more than one cylinder connected in parallel can be controlled in steps by progressive unloading. The suction valve is held open, making the cylinder inoperative, and as the load falls more cylinders can be unloaded until finally the machine is stopped. Reference 7.5 gives details of a typical unloader mechanism. This method has the advantage of saving power at reduced loads, although the saving is not proportional to the number of cylinders inoperative because there are still friction losses.

To investigate the use of cylinder unloading in plants producing chilled water, it is first necessary to examine what will happen in the evaporator at reduced capacity. The equation governing the heat flow rate from the chilled water is:

$$\dot{Q}_E = \dot{m}c_p(t_i - t_o)$$

where t_i is the chilled water inlet temperature to the evaporator (the return from the chilled water system) and t_o is the temperature at outlet from the evaporator (the flow to the chilled water system). The term $\dot{m}c_p$ is normally constant so that the capacity depends only on the change in chilled water temperature. Figure 7.19 is drawn for a system with chilled water flow and return temperatures of 6 °C and 14 °C respectively at the design capacity. The

Figure 7.19

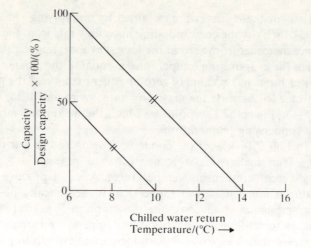

Chilled water return
Temperature/(°C) ⟶

vertical scale is capacity as a percentage of the design value, and the horizontal scale the return temperature. A line is drawn from the design capacity to the design return temperature. If the flow temperature is maintained constant the temperature difference and hence the capacity depend only on the return temperature. Any new capacity and corresponding return temperature can be found from a line drawn parallel to the original one. The line shown is at 50% capacity with a new return temperature of 10 °C. The new temperature difference is now 4 K, compared with 8 K for the design case. The same reasoning can be applied to the supply temperature with the return temperature fixed. In practice both the flow and return temperatures will change as the capacity changes, but new values of each can be found by drawing lines from the new balance point parallel to those for the design case.

It was shown earlier how compressor capacities were plotted against the evaporator saturation temperature. If the temperature scale is continued, the chilled water temperature can be represented on the same diagram. Also, when cylinder unloading is used the new capacity is proportional to the number of working cylinders, which enables a part-load curve to be drawn. Figure 7.20 shows a plot of capacity in kilowatts against temperature in °C. The capacity

Figure 7.20

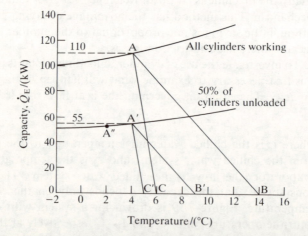

Temperature/(°C)

curve with all cylinders working could have been one of the lines from Fig. 7.18(d). The curve with only half the cylinders working is drawn by multiplying the capacity at any temperature by 0.5. The design capacity in this case is 110 kW and point A is where the compressor characteristic has this value. Point A is known as a system balance point.

A line can be drawn from A to B, the chilled water return temperature of 14 °C, and from A to C, the chiller water flow temperature of 6 °C. The lines together with the temperature axis form a triangle ABC. The horizontal scale also represents the evaporator saturation temperature. This is the temperature at point A, which in this case is 4 °C. When the plant is operating at 50% capacity the load will be 55 kW and only half the cylinders will be working. The new balance point is at A′. The new chilled water return temperature is found by drawing a line A′B′ parallel to line AB and B′ in this case is 9 °C. The new chilled water flow temperature is found by drawing a line A′C′ parallel to line AC and C′ in this case is 5 °C. A′B′C′ and ABC are now similar triangles. The original temperature difference is represented by line CB, which in this case is $(14 - 6) = 8$ K and the new temperature difference by line C′B′, which in this case is $(9 - 5) = 4$ K.

The above two balance points have the same evaporating temperature of 4 °C. If however the capacity should fall below 55 kW to (say) 52 kW, the new balance point will be at A″ and the evaporator saturation temperature will fall to about 2 °C. New flow and return water temperatures can be found by drawing another triangle similar to the first two with its apex at A″. Most chilled water systems operate with untreated water, which must be prevented from freezing. It is therefore necessary to maintain the flow temperature within quite a narrow band of, say, 4–6 °C. The cooling water return temperature will vary over a wider range and is easier to control. When cylinder unloading is used the signal to unload is therefore taken from temperature sensors in the chilled water return line.

Example 7.7

(a)   For a refrigeration system:

(i)   Sketch a graph of specific refrigeration effect against evaporating temperature for a constant condensing temperature.

(ii)   Sketch a graph of specific refrigeration effect against condensing temperature for a constant evaporating temperature. With the aid of further sketches explain the difference between the two graphs.

(b)   Sketch and describe a hot gas by-pass control for a reciprocating compressor suitable for a plant with an evaporator remote from the rest of the plant.

(c)   For a controlled condensing temperature, a reciprocating compressor has the following characteristic:

Evaporator saturation temperature/(°C)	0	5	10
Refrigeration capacity/(kW)	55	66	85

The four-cylinder machine is controlled by a thermostat which senses the chilled water temperature entering the evaporator and cuts out cylinders as the capacity falls.

The water leaving the evaporator is required to have a minimum temperature of 5 °C and a maximum of 7 °C. The mass flow rate of chilled water remains constant at 2 kg/s and the design flow and return temperatures are 7 °C and 15 °C respectively.

Determine the return water temperatures at which the thermostat must be set to cut out the cylinders and the minimum plant capacity.

(Wolverhampton Polytechnic)

Solution

(a) The relevant sketches are shown in Fig. 7.21. With the condensing temperature constant and the evaporating temperature varied, the specific refrigeration effect increases as the evaporating temperature increases. The shape of the curve is explained by Fig. 7.21(b) which shows a number of simple cycles with a common condensing temperature. Since the state of the refrigerant leaving the condenser (point 3) is the same for each cycle, the change in the specific refrigeration effect ($h_1 - h_4$) is controlled by the slope of the saturated vapour line, which is governed by the specific heat at constant pressure of the saturated vapour—the lower the value of specific heat, the steeper the slope of the saturation line.

Figure 7.21   Example 7.7

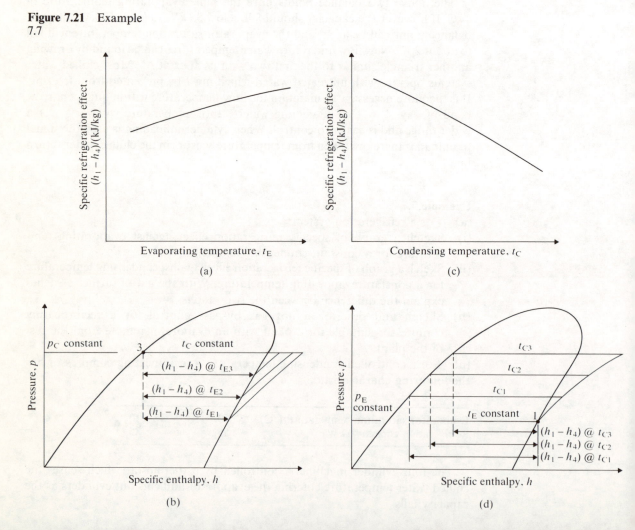

(a)

(c)

(b)

(d)

Figure 7.21(c) shows that when the evaporating temperature is fixed and the condensing temperature varied, the specific refrigeration effect falls with an increase in condensing temperature. Not only is the slope of this graph the opposite to the slope of Fig. 7.21(a); it is also steeper. This is explained by Fig. 7.21(d), in which the evaporating temperature is kept constant and the condensing temperature varied. In this case the state of the refrigerant leaving the evaporator (point 1) is the same for each cycle and the change in specific refrigeration effect depends on the slope of the saturated liquid line, which is governed by the specific heat at constant pressure of the saturated liquid. This graph therefore has a steeper slope because the liquid specific heat is greater than the vapour specific heat.

(b)   This is covered later in the text (see page 322).

(c)   A four-cylinder compressor has capacity curves at 100%, 75%, 50% and 25% of the design curve. The characteristic at each capacity is tabulated below.

	Evaporator saturation temperature		
	0 °C	5 °C	10 °C
100% capacity/(kW)	55	66	10
75% capacity/(kW)	41.3	49.5	63.8
50% capacity/(kW)	27.5	33	42.5
25% capacity/(kW)	13.8	16.5	21.3

These compressor characteristics are plotted on Fig. 7.22. A mean specific heat value for the chilled water can be found from tables such as Ref. 7.3, and the

Figure 7.22   Example 7.7

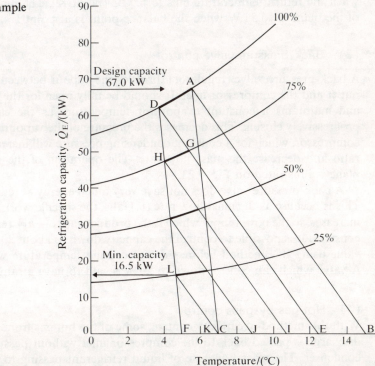

design capacity of the plant may be calculated from an energy balance on the evaporator as follows:

$$\dot{Q}_E = \dot{m}c_p(t_i - t_o) = 2 \times 4.19(15 - 7) = 67\,\text{kW}$$

The design operating point is at A, where the compressor 100% capacity line has a value of 67 kW. This fixes the evaporating temperature, which is 5.5 °C. Lines are then drawn from A to the chilled water return temperature at B and the chilled water flow temperature at C to complete the triangle ABC. Triangles similar to ABC can be drawn for any new balance point.

The minimum flow temperature is 5 °C so a line is drawn from 5 °C (point F) parallel to AC to meet the 100% capacity line at D. Starting with the compressor operating at its design point A, as the load falls the balance point will move down the 100% capacity curve until it reaches point D. A new triangle DEF can now be constructed which is similar to triangle ABC. To prevent the flow temperature falling below 4 °C, one cylinder must now be cut out. This is controlled by the return water temperature, which has now fallen to 12.5 °C at point E. The return temperature will remain constant due to the thermal inertia of the cooling load, but the flow temperature will rapidly rise as a new balance point is established along the 75% capacity curve at point G. For each new balance point there is a new value of evaporator and chilled water flow temperature but only the chilled water return temperatures are required to answer Example 7.7(c).

As the load continues to fall, the operating point will move down the 75% capacity line until the minimum flow temperature is again reached at point H. A second cylinder is then cut out, controlled by the new return temperature of 10.8 °C at point I. This process continues with the third cylinder cut out when the return temperature is 8.8 °C (point J) and the compressor is shut down when the return temperature falls to 6.9 °C (point K). The minimum capacity of the plant is 16.5 kW when the balance point is at point L.

(3) Back-pressure valve control

A back-pressure valve (regulator) throttles the vapour between the evaporator outlet and the compressor inlet. It should be fully open for the design capacity and maintains a constant evaporator temperature as the capacity falls by progressively closing. This decreases the pressure of the vapour delivered to the compressor, which for a constant condensing pressure will increase the pressure ratio and decrease the mass flow rate. The operation of the valve on a p–h plane is illustrated on Fig. 7.23.

A back-pressure valve does not save very much energy at reduced capacities. This is because as the mass flow rate ($\dot{m}$) falls, the specific work input ($h_2 - h_1$) increases, so the power input which is the product, $\dot{m}(h_2 - h_1)$, remains relatively constant. It is possible to control the capacity down to about 20% of the design value using this method before the evaporating temperature will start to fall. A valve which can control down to 20% is said to have a rangeability of 5.

(4) Hot gas by-pass control

With this method of capacity control, some of the hot gas from the compressor discharge is passed back to the compressor inlet without passing through the condenser. The mass flow rate of liquid refrigerant passing to the evaporator

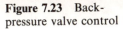

Figure 7.23 Back-pressure valve control

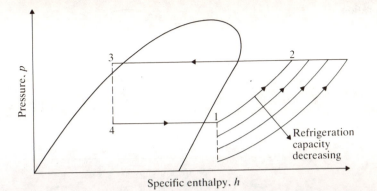

is therefore reduced but the mass flow rate through the compressor is not. This reduces the capacity but does not reduce the rate of work input required at part-load. Figure 7.24 shows three possible methods of hot gas by-pass control.

The arrangement in Fig. 7.24(a) is the most basic one. As the capacity starts to fall the by-pass valve is progressively opened so that some vapour is only passing through the compressor and by-pass. Figure 7.25 shows what is

Figure 7.24 Hot gas by-pass control arrangements

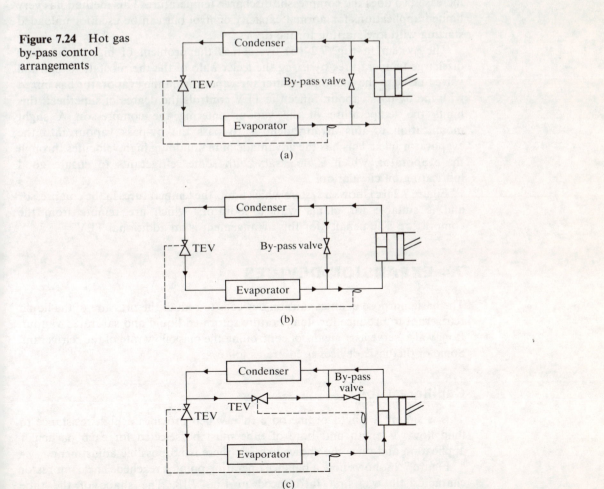

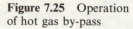

Figure 7.25 Operation of hot gas by-pass

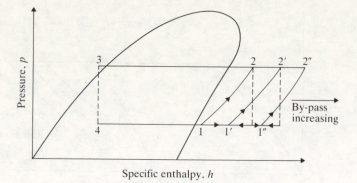

happening on a p–h plane. The compression for the design capacity is 1–2. At a reduced capacity, some of the vapour at state 2 is throttled back to the suction pressure and then mixes with the vapour at state 1 to give a new compressor inlet state at 1′. This results in a new discharge temperature of 2′. A further decrease in capacity will result in a new discharge temperature of 2″. The disadvantage of this system is that as the flow rate of the vapour by-passed increases, so does the compressor discharge temperature. This method has very limited applications for normal capacity control but can be used for unloaded starting with low starting torque drives.

The system in Fig. 7.24(b) overcomes the problem of high compressor discharge temperatures by fixing the feeler bulb of the thermostatic expansion valve (TEV) to the suction line after the vapour from the evaporator has mixed with the by-pass vapour. Since the TEV controls the degree of superheat, this limits the temperature of the vapour entering the compressor. A slight modification to this arrangement is to pass the by-pass vapour into the evaporator inlet. This has the advantage of maintaining high velocities through the evaporator, which is necessary with some refrigerants to ensure good lubricating oil circulation.

Figure 7.24(c) shows a system which limits the temperatures in the compressor and is suitable for plants with evaporators which are remote from the compressor. The penalty for this arrangement is an additional TEV.

7.4 EXPANSION DEVICES

The basic purpose of an expansion device is to reduce the pressure of the liquid refrigerant to produce low-temperature saturated liquid and saturated vapour. It may also serve as a means of controlling the mass flow rate of the refrigerant. Some of the basic devices used are as follows.

Capillary Tube

This is simply a length of fine-bore tube which provides a high resistance to fluid flow. A length and bore of tube must be selected for each particular application and once the choice is made there is no possible adjustment.

Figure 7.26 shows how a balance operating point is reached. The compressor characteristic was first introduced in Fig. 7.18. The shape of the tube

Figure 7.26 Capillary tube expansion

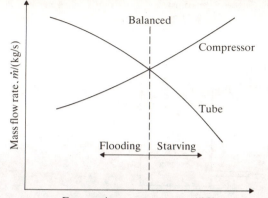

characteristic can be deduced from the knowledge that pressure drop in a tube is approximately proportional to the square of the velocity. As discussed earlier, a change in pressure also represents a change in saturation temperature, so for a fixed condensing temperature both characteristics can be plotted against the evaporator saturation temperature. Where the two characteristics cross represents a balance point where the mass flow rate through each item of plant is constant. The system is now in equilibrium. It should be noted that the balance point fixes the evaporating temperature for a particular value of condensing temperature. If the condensing temperature should change due to a change in coolant temperature, then the evaporating temperature must change. Capillary tubes can therefore only be used where a varying evaporator temperature can be tolerated.

With this system a sudden change in load can cause the evaporator to flood or starve. This is also illustrated in Fig. 7.26. If the evaporator load suddenly increases, the evaporating temperature will rise. The mass flow through the tube will therefore decrease and the mass flow rate through the compressor will increase. There is therefore more fluid leaving the evaporator than entering, and it is said to be starving. If, on the other hand, the evaporator load suddenly falls, the evaporator temperature will also fall and there will be more fluid passing into the evaporator than leaving, which will cause flooding. If any new load remains constant, then the system will adjust itself until a new balance point is reached. The tendency to flood means that the evaporator should be able to hold the full system charge of refrigerant, to avoid the possibility of liquid entering the compressor. This limits the size of plant in which capillary tubes can be used; most domestic systems use them.

Constant-pressure Expansion Valve

This is also called an automatic expansion valve; its function is to maintain a constant pressure in the evaporator. Figure 7.27 shows the basic construction. The valve opening is maintained by the position of a diaphragm or bellows, one side of which is exposed to the evaporator pressure and the other side to the pressure exerted by a spring. The pressure exerted by the spring can be adjusted; once set, it will fix the evaporator pressure and hence the evaporating

Figure 7.27 Constant-pressure expansion valve

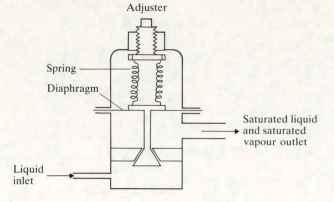

Figure 7.27 Constant-pressure expansion valve

Adjuster

Spring

Diaphragm

Saturated liquid and saturated vapour outlet

Liquid inlet

temperature. If the pressure starts to fall below the set value the valve will open wider and allow more refrigerant to pass into the evaporator, and if the evaporator pressure starts to rise the valve will start to close. Clearly this method of expansion involves the same problems of starving and flooding as the simple capillary coil. Its application is therefore also limited to small sealed plants where the charge of refrigerant is closely controlled. This limitation, together with the constant evaporator temperature, make it ideal for small chilled water plants.

Hand Expansion Valve

These are manually operated needle valves which must be adjusted each time the load changes. They are only suitable for systems with relatively constant loads with operators available to alter the setting.

Float Valve

This maintains a constant liquid level in shell-and-tube evaporators which have the refrigerant outside the tubes. If the liquid level starts to fall, the valve will open and vice versa. The operation is similar to a ball cock valve in a water tank. It is normally located in a separate chamber which is connected to the evaporator. With a constant liquid level in the evaporator, no liquid can enter the compressor and the mass flow rate through the valve and compressor will be the same. Float valves are used mainly on large plants.

Thermostatic Expansion Valve (TEV)

The name is misleading because it is not temperature which is controlled but the degree of superheat of the vapour leaving the evaporator (defined as the number of degrees difference in temperature between the actual temperature and the saturation temperature corresponding to the pressure). A more appropriate term for this widely used expansion valve would be 'superheat control valve'. The basic construction is shown in Fig. 7.28 and the operation on a p–h plane in Fig. 7.29.

Figure 7.28
Thermostatic expansion
valve (TEV)

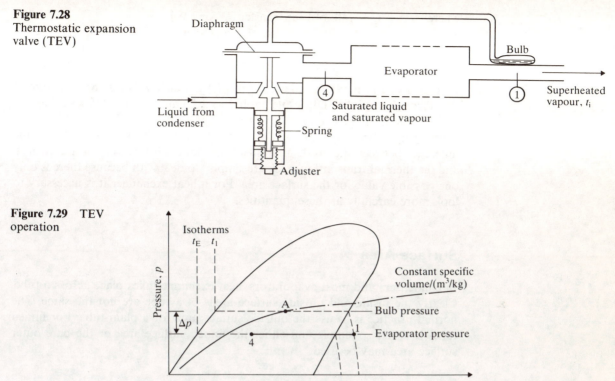

Figure 7.29 TEV
operation

The diaphragm which operates the needle valve is exposed to the evaporator pressure on one side and the pressure of a power fluid on the other side. The power fluid is normally the same fluid as the refrigerant. The side of the diaphragm containing the power fluid is connected to a bulb which is attached to the outlet from the evaporator. The space filled with the power fluid contains sufficient charge to ensure that there is always some liquid present. With the power fluid in the bulb and the superheated vapour at the same temperature, the pressure in the bulb must be higher. This is illustrated on the p–h diagram. The refrigerant enters the evaporator at state 4 and leaves at state 1 without any significant loss of pressure. The power fluid moves along a line of constant specific volume and the state point A will be where the specific volume line crosses the isotherm corresponding to the temperature of state 1. The higher the degree of superheat, the greater will be the difference in pressure Δp. The valve is set by means of the spring and adjuster to have a specific opening for a particular degree of superheat. If the degree of superheat tries to rise, the valve will open and increase the mass flow rate into the evaporator. If the degree of superheat tries to fall the valve will close and restrict the flow into the evaporator. The evaporator temperature is not controlled.

7.5 HEAT EXCHANGERS

The basic equation governing the heat transfer between two fluids is:

$$\dot{Q} = U \times A \times \Delta \bar{t}$$

[7.14]

REFRIGERATION PLANT

where $\dot{Q}$ = rate of heat transfer (W or kW)

U = overall heat transfer coefficient (W/m² K or kW/m² K)

A = surface area perpendicular to the direction of heat flow (m²)

$\Delta \bar{t}$ = the mean temperature difference between the fluids (K)

Equation [7.14] is very similar to Eqn [1.1], which governs the flow rate of heat through a building fabric and has been used many times in different chapters of this book. In fact Eqn [1.1] is a simplification of Eqn [7.14]. When considering fabric heat flows, the mean temperature difference, $\Delta \bar{t}$, is the difference between the inside and outside air temperatures; the surface area, A, and the thermal transmittance, U, are simpler to deal with because there is only one possible value for the surface area. For a heat exchanger it is necessary to look more carefully at these quantities.

Surface Area, A

In condensers and most evaporators, heat exchange takes place across a tube. Clearly the inside and outside surface areas of a tube are not the same. The normal practice is to use the outside surface area for a plain tube. For finned tubes it is not so simple and either the finned surface area or the bare outer surface area may be used.

Overall Heat Transfer Coefficient, U

Considering a plain round tube, U can be found from the following equation:

$$\frac{1}{U} = \frac{1}{\alpha_o} + f_o + \frac{\Delta x}{k} + \frac{1}{\alpha_i} \cdot \frac{r_o}{r_i} + f_i \frac{r_o}{r_i} \qquad [7.15]$$

The fouling factors f_o and f_i are resistances due to the build-up of deposits on the tube surfaces and are particularly important with fluids such as cooling water. They are either fixed by a designer's experience in dealing with particular fluids or selected from a data source such as Ref. 7.6. The two thermal resistances inside the tube are referred to the outer surface. This increases their value, which compensates for the outer surface area being greater than the inside surface area; i.e. the product αA is the same at both surfaces. It should also be noted that the thermal resistance of the tube is so small in comparison with the other resistance that plain boundary theory is sufficiently accurate.

Mean Temperature Difference, Δt̄

For any heat exchanger, $\Delta \bar{t}$ is found from the following equation:

$$\Delta \bar{t} = F \times \Delta \bar{t}_{ln} \qquad [7.16]$$

where $\Delta \bar{t}_{ln}$ is the logarithmic mean temperature difference and F is a correction factor which depends on the flow paths of the two fluids. If the flow paths are co-current or counter-current or if one side of the heat exchanger is isothermal, the correction factor is 1. This is the case for many refrigeration evaporators and condensers. The equation for $\Delta \bar{t}_{\mathrm{ln}}$ is as follows:

$$\Delta \bar{t}_{\mathrm{ln}} = \frac{\Delta t_1 - \Delta t_2}{\log_e(\Delta t_1 / \Delta t_2)} \qquad [7.17]$$

The proof of this equation can be found in most Heat Transfer books and in many books on basic Thermodynamics such as Ref. 7.2. Δt_1 and Δt_2 are the terminal temperature differences. It does not matter which is which, but there is less likelihood of a mistake if Δt_1 is taken to be the greater of the two. The terminal temperature differences are the differences in temperature between the two fluids at each end of the exchanger.

Refrigeration Condensers and Evaporators

A Building Services Engineer is unlikely to have to design a new heat exchanger but he may well want to know what will happen if conditions or use of an existing plant change. With condensers and evaporators there are some assumptions that can be made to simplify calculations without introducing unacceptable inaccuracies.

Figure 7.30 shows the temperature profile for an evaporator, with no superheating, producing chilled water. This is realistic for an evaporator with the refrigerant boiling outside the tubes. Working out $\Delta \bar{t}_{\mathrm{ln}}$ is straightforward and the correction factor is 1. For some purposes it may even be acceptable to use an arithmetic mean temperature difference instead of $\Delta \bar{t}_{\mathrm{ln}}$.

Figure 7.30 Evaporator temperature profile

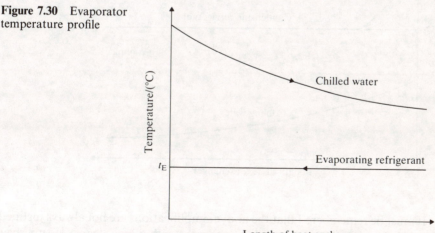

Example 7.8
A flooded evaporator operates with an evaporating temperature of 2 °C when producing chilled water at 6 °C from return water at 10 °C. Determine the error involved in using an arithmetic mean temperature difference instead of a log mean temperature difference.

Solution

The terminal temperature differences are calculated as follows:

$$10\,°C \quad \rightarrow \quad 6\,°C \text{ (Chilled water)}$$
$$2\,°C \quad \leftrightarrow \quad 2\,°C \text{ (Refrigerant)}$$
$$\Delta t_1 = \quad 8\,K \qquad \Delta t_2 = 4\,K$$

Equation [7.14] is used to calculate Δt_{ln}:

$$\Delta \bar{t}_{\text{ln}} = \frac{\Delta t_1 - \Delta t_2}{\log_e(\Delta t_1 / \Delta t_2)} = \frac{8 - 4}{\log_e 8/4} = 5.77$$

The arithmetic mean temperature, $\Delta \bar{t} = (8 + 4)/2 = 6$

The error in this case is 4%.

A typical temperature profile in a condenser is shown in Fig. 7.31. This appears to be much more complicated than the diagram for the evaporator. However, for a water-cooled condenser the tubes will be below the dew point of the refrigerant and condensation will start at inlet. It is therefore reasonable to assume a condensing coefficient but not to take advantage of the extra temperature difference. If the sub-cooling is relatively small and may be neglected, the refrigerant temperature may be considered to be isothermal and the calculation of the mean temperature difference is straightforward.

Figure 7.31 Condensor temperature profile

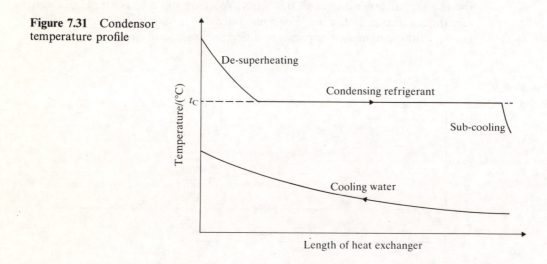

It must be remembered that the above simplifications are not always justified. Consider for example an air-cooled condenser with a high degree of superheat at inlet. The tube wall temperature might be well above the dew point and the de-superheating section would have to be treated separately from the main condensing section. Each section would then have its own mean temperature difference and heat transfer coefficient.

Example 7.9

(a) Calculate the surface area required for an air-cooled condenser in an R12 refrigeration plant, given the following design conditions and data:

Compressor displacement	$0.1 \ \mathrm{m^3/s}$
Evaporating temperature	5 °C
Condensing temperature	45 °C
Suction vapour superheat	5 K
Liquid subcooling at expansion valve	3 K
Compressor volumetric efficiency	70%
Compressor isentropic efficiency	70%
Air ON to condenser	28 °C
Air OFF condenser	35 °C
Condenser heat transfer coefficient	$20 \ \mathrm{W/m^2 \ K}$

Use the R12 pressure/enthalpy chart provided. Ignore the effects of de-superheating and sub-cooling on the heat transfer coefficient for the condenser.

(b) 'Packaged refrigeration equipment for air conditioning systems generally contains condensers and evaporators which are sized as small as possible.'

Comment on the economic, energy-consumption and practical aspects of this statement with respect to the refrigeration plant.

(CIBSE)

Solution

(a) Taking values from a *p*–*h* chart as shown in Fig. 7.32:

$$v_1 = 0.051 \ \mathrm{m^3/kg}, \quad h_1 = 193 \ \mathrm{kJ/kg}, \quad h_{2s} = 214 \ \mathrm{kJ/kg},$$
$$h_3 = h_4 = 76 \ \mathrm{kJ/kg}$$

Figure 7.32 Example 7.9

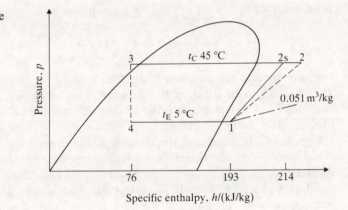

From Eqn [7.1],

$$h_2 = h_1 + \frac{h_{2s} - h_1}{\eta_{1s}} = 193 + \frac{214 - 193}{0.7} = 223 \ \mathrm{kJ/kg}$$

At inlet to the compressor, using Eqn [7.12],

$$\dot{m} = \frac{\dot{V}_s \eta_v}{v} = \frac{0.1 \times 0.7}{0.051} = 1.37 \ \mathrm{kg/s}$$

Using Eqn [7.4] for the condenser,

$$\dot{Q}_C = \dot{m}(h_2 - h_3) = 1.37(223 - 76) = 201.4 \ \mathrm{kW}$$

When one fluid undergoes an isothermal process $\Delta \bar{t}$ is the same as $\Delta \bar{t}_{ln}$ which can be found as follows:

$$
\begin{array}{lcl}
45\,°C & \leftrightarrow & 45\,°C\ (\text{Refrigerant}) \\
35\,°C & \leftarrow & 28\,°C\ (\text{Cooling water}) \\
\Delta t_2 = 10\ K & & \Delta t_1 = 17\ K
\end{array}
$$

Using Eqn [7.17]:

$$
\Delta \bar{t}_{ln} = \frac{\Delta t_1 - \Delta t_2}{\log_e(\Delta t_1 / \Delta t_2)} = \frac{17 - 10}{\log_e(17/10)} = 13.2\ K
$$

It should be noted that in this case $\Delta \bar{t}_{ln}$ is only 2% different from the arithmetic mean temperature difference of $(10 + 17)/2 = 13.5$ K.

Using Eqn [7.14],

$$
A = \frac{\dot{Q}_C}{U \times \Delta \bar{t}} = \frac{201.4 \times 10^3}{20 \times 13.2} = 763\ m^2
$$

The reader may consider the surface area to be rather high but the low value of heat transfer coefficient makes it clear that the surface area referred to is a finned surface area. (Air-cooled condensers are always finned because of the poor heat transfer characteristics of air.) The coefficient referred to the bare tube surface would be about 200 W/m² K and the corresponding bare tube surface 76.3 m².

(b) The size of any heat exchanger is a function of the surface area required. Equation [7.14] can be applied to a condenser and evaporator as follows:

$$
A_C = \frac{\dot{Q}_C}{U_C \times \Delta \bar{t}_C} \quad \text{and} \quad A_E = \frac{\dot{Q}_E}{U_E \times \Delta \bar{t}}
$$

For any specific refrigeration plant the capacity $\dot{Q}_E$ is fixed. The other terms are variables to be fixed by the designer. The condenser overall heat transfer coefficient, U_C, depends on the refrigerant used and the cooling medium (air, cooling water or evaporating water). For a particular application the refrigerant to be used and the cooling medium are fixed at an early stage of the design. The size therefore basically depends on the mean temperature difference.

The evaporator overall heat transfer coefficient, U_E, depends on the refrigerant used, the heating medium (air or chilled water) and the mean temperature difference. In this case the mean temperature difference is important because the refrigerant boiling coefficient increases as the temperature difference increases, which increases the value of U_E, Thus increasing the mean temperature difference in the evaporator has an even greater effect on the surface area and hence the size of heat exchanger required.

The coefficient of performance of any refrigeration plant depends on the refrigerant used and the condensing and evaporating temperatures, t_C and t_E. The smaller the difference between t_C and t_E, the greater becomes the coefficient of performance. The cooling medium temperature is always less than t_C and the chilled water temperature is always greater than t_E. Therefore the larger the mean temperature difference becomes in the condenser and evaporator, the lower is the coefficient of performance. Large mean temperature differences in the condenser and evaporator thus result in a cheap capital cost plant with high running costs, and vice versa. In a competitive market it is in the interests of vendors to offer the cheapest plant possible which satisfies the customers' requirements, as the customers are all too often restricted on capital expenditure. The cheapest plant is unlikely to be the most economic, and it is up to the client to balance his capital costs carefully against the energy costs over the life of the plant.

Example 7.10

(a)   Comment briefly on the viability of using a heat pump to recover waste heat in a building as compared with other heat-recovery systems.

(b)   An R12 packaged water chiller with a water-cooled condenser has the following rated performance data.

Refrigeration capacity	250 kW
Evaporation temperature	2 °C
Mean chilled-water temperature	9 °C
Condensing temperature	40 °C
Mean condenser cooling-water temperature	30 °C
Suction vapour superheat	8 K
Liquid sub-cooling	2 K

(i)   For the design conditions quoted, determine the refrigerant enthalpies, the refrigerant mass flow rate and the compressor displacement.

(ii) In order to make good use of the waste heat in the condenser cooling water, a mean water temperature of 55 °C is required, while the chilled water temperature requirements remain unchanged. By use of successive iterations using estimates of the reduced refrigeration capacity or evaporator heat load, calculate the reduced temperature differences across both evaporator and condenser, together with the consequent evaporating and condensing temperatures at each stage. Continue to iterate for two to three times in total until a convergence of $+4\%$ of refrigeration capacity has been secured.

Use Fig. 7.33(a) to determine the volumetric efficiency. Assume isentropic compression.

Assume that the heat transfer rates in the condenser and evaporator are directly proportional to the temperature difference between refrigerant and mean water temperatures in each case.

(Engineering Council)

Solution to part (b)

(i)   Figure 7.33(b) is a p–h chart which shows the cycles on a p–h plane. For the design condition, from the chart,

$$h_1 = 194 \text{ kJ/kg}, \quad h_2 = 216 \text{ kJ/kg}, \quad h_3 = h_4 = 72 \text{ kJ/kg},$$

$$p_1 = 325 \text{ kN/m}^2, \quad p_2 = 980 \text{ kN/m}^2, \quad v_1 = 0.056 \text{ m}^3/\text{kg}$$

For the evaporator, using Eqn [7.3],

$$\dot{m} = \frac{\dot{Q}_E}{h_1 - h_4} = \frac{250}{194 - 72} = 2.049 \text{ kg/s}$$

Figure 7.33 Example
7.10

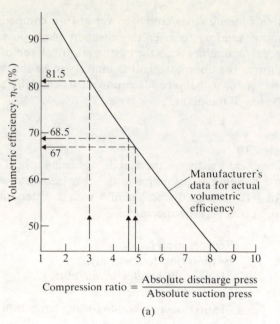

(a)

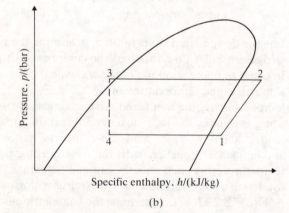

(b)

For the compressor,

$$\dot{V}_1 = v_1 \dot{m} = 0.056 \times 2.049 = 0.115 \text{ m}^3/\text{s}$$
$$\frac{p_2}{p_1} = \frac{980}{325} = 3.015$$

From the chart of volumetric efficiency against compression ratio (Fig. 7.33(a)), $\eta_v = 81.5\%$; therefore, using Eqn [7.12] gives:

$$\text{Displacement, } \dot{V}_s = \frac{\dot{V}_1}{\eta_v} = \frac{0.115}{0.815} = 0.1411 \text{ m}^3/\text{s}$$

(ii)   For the original condenser design the mean cooling water temperature is 30 °C and the condenser heat rejection rate, $\dot{Q}_C$, can be found from Eqn [7.4] as follows:

$$\dot{Q}_C = \dot{m}(h_2 - h_3) = 2.049(216 - 72) = 295 \text{ kW}$$

The new mean cooling water temperature is 55 °C but the new heat rejection rate is unknown. Assuming the same sub-cooling and superheat it is now necesary to estimate new condensing and evaporating temperatures. Since the plant is now operating between wider temperature limits, the loads on both the condenser and the evaporator will be reduced. And since for each of these the load is proportional to the mean temperature difference, the mean temperature differences will also be reduced. It is now necessary to estimate new values for the condensing and evaporating temperatures.

Try $t_C = 63\,°C$ and $t_E = 3\,°C$. For the new cycle, from the chart,

$$h_1 = 195\,kJ/kg, \quad h_2 = 226\,kJ/kg, \quad h_3 = h_4 = 97\,kJ/kg$$
$$p_1 = 330\,kN/m^2, \quad p_2 = 1620\,kN/m^2, \quad v_1 = 0.055\,m^3/kg$$
$$\frac{p_2}{p_1} = \frac{1620}{330} = 4.9$$

Hence $\eta_v = 67\%$

$$\dot{V}_1 = \dot{V}_s \eta_v = 0.1411 \times 0.67 = 0.0945\,m^3/s$$

$$\dot{m} = \frac{\dot{V}_1}{v_1} = \frac{0.0945}{0.052} = 1.72\,kg/s$$

Since the heat rejection rate depends only on the mean temperature difference:

$$\frac{\dot{Q}_{C2}}{\dot{Q}_{C1}} = \frac{\Delta \bar{t}_2}{\Delta \bar{t}_1}$$

Rearranging,

$$\dot{Q}_{C2} = \dot{Q}_{C1} \frac{\Delta \bar{t}_2}{\Delta \bar{t}_1} \qquad\qquad [7.18]$$

Values for the new condenser heat rejection rate can now be evaluated from Eqns [7.4] and [7.18] as follows:

$$\dot{Q}_{C2} = \dot{m}(h_2 - h_3) = 1.72(226 - 97) = 222\,kW$$

$$\dot{Q}_{C2} = \dot{Q}_{C1} \frac{\Delta \bar{t}_2}{\Delta \bar{t}_1} = 295 \frac{(63 - 55)}{(40 - 30)} = 236\,kW$$

A similar equation to [7.18] will apply to the evaporator, i.e.

$$\dot{Q}_{E2} = \dot{Q}_{E1} \frac{\Delta \bar{t}_2}{\Delta \bar{t}_1} \qquad\qquad [7.19]$$

Two values for the evaporator capacity, $\dot{Q}_E$, can now be evaluated from Eqns [7.3] and [7.19]:

$$\dot{Q}_E = \dot{m}(h_1 - h_4) = 1.72(195 - 97) = 169\,kW$$

$$\dot{Q}_{E2} = \dot{Q}_{E1} \frac{\Delta \bar{t}_2}{\Delta \bar{t}_1} = 250 \frac{(9 - 3)}{(9 - 2)} = 214\,kW$$

The two expressions for the condenser heat rejection rate are within about 6% and the evaporator capacities differ by 27%; a new estimate is necessary. The evaporating temperature is too low and any increase will increase the condenser heat rejection rate of 222 kW because the pressure ratio will decrease, the volumetric efficiency will increase and hence the mass flow rate will increase.

The condensing temperature is therefore retained and a new t_E estimated. Try $t_C = 63\,°C$ and $t_E = 4\,°C$. For this new cycle, from the chart,

$$h_1 = 197\,\text{kJ/kg}, \quad h_2 = 227\,\text{kJ/kg}, \quad h_3 = 97\,\text{kJ/kg}$$
$$p_1 = 350\,\text{kN/m}^2, \quad p_2 = 1620\,\text{kN/m}^2, \quad v_1 = 0.053\,\text{m}^3/\text{kg}$$
$$\frac{p_2}{p_1} = \frac{1620}{350} = 4.63$$

Hence, from Fig. 7.33, $\eta_v = 68.5\%$

$$\dot{V}_1 = \dot{V}_s \eta_v = 0.1411 \times 0.685 = 0.0967\,\text{m}^3/\text{s}$$
$$\dot{m} = \frac{\dot{V}_1}{v_1} = \frac{0.09767}{0.053} = 1.83\,\text{kg/s}$$

$$\dot{Q}_{C2} = \dot{m}(h_2 - h_3) = 1.83(227 - 97) = 238\,\text{kW}$$
$$\dot{Q}_{C2} = \dot{Q}_{C1}\frac{\Delta \bar{t}_2}{\Delta \bar{t}_1} = 295\frac{(63 - 55)}{(40 - 30)} = 236\,\text{kW}$$
$$\dot{Q}_E = \dot{m}(h_1 - h_4) = 1.83(197 - 97) = 183\,\text{kW}$$
$$\dot{Q}_{E2} = \dot{Q}_{E1}\frac{\Delta \bar{t}_2}{\Delta \bar{t}_1} = 250\frac{(9 - 4)}{(9 - 2)} = 179\,\text{kW}$$

The agreement is now within 1% for the condenser heat rejection rate and within 3% for the evaporator capacity, which is satisfactory.

7.6 CONDENSER CONSTRUCTION AND CHARACTERISTICS

The basic construction of a condenser is dictated by the cooling medium. If the coolant is water, a shell-and-tube construction would be used for a reasonably sized unit. Figure 7.34 shows a typical example. Tubes are fixed into tubesheets which are then enclosed in a cylindrical shell. This prevents the water inside the tubes (tube side) from coming into contact with the refrigerant outside the tubes (shell side). The unit shown is a two-pass unit because a partition in the channel on the right divides the tube bank into two sections. The designer will select a number of passes to optimize heat transfer and available pressure drop. The tubes would be supported on the shell side to prevent mechanical failure but unless a de-superheating section was fitted there would be no other baffles

Figure 7.34 Water-cooled condenser

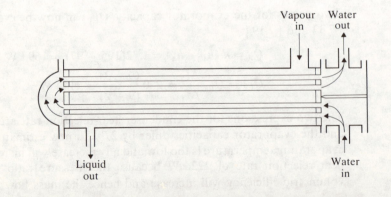

to minimize the pressure drop. One advantage of water cooling is the compactness of the units, which are generally capable of operating at lower condensing temperatures than air-cooled units. The disadvantages depend on the cost and source of the water.

A simple air-cooled condenser is shown in Fig. 7.35. Finned tubes are expanded into headers which may have pass partitions to increase the velocity of the refrigerant inside the tubes. The condenser shown has a single pass, so the vapour would enter one header and liquid would leave the other one. The disadvantages of air cooling are the size of the unit, the higher condensing temperatures and the running cost of the fan.

Figure 7.35   Air-cooled condenser

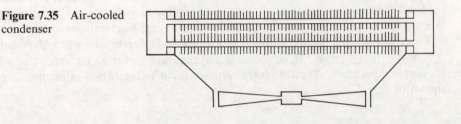

Another possible way to reject heat is by evaporative cooling. This is a compromise between air cooling and water cooling. The refrigerant is passed through tubes while water is sprayed over them so that some of it evaporates. The evaporative condenser is therefore a combined condenser and cooling tower.

Previous examples have shown that for refrigeration work an arithmetic mean temperature difference is often sufficiently accurate. For a condensing temperature t_C and coolant temperatures of t_i and t_o, $\Delta\bar{t}$ becomes:

$$\Delta\bar{t} = t_C - \frac{(t_o + t_i)}{2}$$

$$= t_C - t_i - \frac{(t_o - t_i)}{2} \qquad [1]$$

Also,

$$\dot{Q}_C = \dot{m}c_p(t_o - t_i) \qquad [2]$$

Combining Eqns [1] and [2] gives:

$$\Delta\bar{t} = t_C - t_i - \frac{\dot{Q}_C}{2\dot{m}c_p} \qquad [3]$$

From Eqn [7.14]:

$$\dot{Q}_C = UA \times \Delta\bar{t} \qquad [4]$$

Combining Eqns [3] and [4] gives

$$\dot{Q}_C = UAt_C - UAt_i - \frac{UA\dot{Q}_C}{2\dot{m}c_p}$$

Rearranging,

$$\dot{Q}_C = \frac{UA}{(1 + UA/2\dot{m}c_p)}t_C - \frac{UA}{(1 + UA/2\dot{m}c_p)}t_i \qquad [7.20]$$

For a particular condenser the surface area is fixed and the mass flow rate is normally constant. The variation in the specific heat is negligible over the temperature range considered. The overall heat transfer coefficient is found from Eqn [7.15], i.e.

$$\frac{1}{U} = \frac{1}{\alpha_o} + f_o + \frac{\Delta x}{k} + \frac{1}{\alpha_i} \cdot \frac{r_o}{r_i} + f_i \frac{r_o}{r_i}$$

The fouling factors and tube wall resistance are constant and if the coolant mass flow rate is constant its heat transfer coefficient will also be constant (α_i for water cooling or α_o for air cooling). The remaining thermal resistance is due to the condensing coefficient. Since this is normally a small proportion of the total thermal resistance, any minor changes due to variations in condensate loading will not have any significant effect on the overall heat transfer coefficient.

Equation [7.20] therefore predicts a linear relationship between the heat rejection rate, $\dot{Q}_C$, and the condensing temperature t_C for any given value of coolant temperature. Typical characteristics for a refrigeration condenser are shown in Fig. 7.36.

Figure 7.36 Condensor characteristics

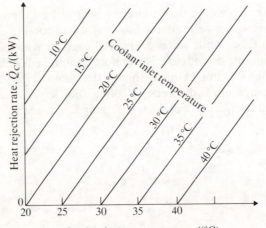

7.7 EVAPORATOR CONSTRUCTION AND CHARACTERISTICS

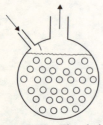

Figure 7.37 Flooded shell-and-tube evaporator

The construction of an evaporator is also dictated by the heating medium. Figure 7.37 shows an evaporator which would be used to produce chilled water. The basic construction is similar to that of the water-cooled condenser. The differences lie in a space without tubes that has been left above the refrigerant liquid level to provide disengaging space, and the nozzle arrangements. With evaporators, the vapour outlet nozzle would be at the centre of the shell and a liquid inlet with flash vapour would be above the liquid level behind a baffle as shown. The liquid level is controlled by a float valve. The diagram shows only one tube pass for the chilled water but there could be any number which would optimize heat transfer and pressure drop.

Figure 7.38 Plate evaporator

Figure 7.38 shows a section through a plate evaporator. The refrigerant goes through passages formed by welding or braising two plates together. The reader should be familiar with this construction, which is used in domestic refrigerators. It is also used in large cold stores and refrigerated vehicles. The direct expansion coil or dry expansion coil which is illustrated in Fig. 7.39 is often abbreviated to 'DX coil'. Only one tube is shown but any number may be fixed between the two headers. They are used in air conditioning units and are always finned to compensate for the low air-side heat transfer coefficient. They are normally controlled by a thermostatic expansion valve.

Figure 7.40 shows typical characteristics for an evaporator. It is interesting to compare condenser and evaporator characteristics. For the evaporator the chilled water inlet temperature (the return from the system) replaces the coolant inlet temperature. The lines have opposite slopes because in an evaporator capacity increases as the evaporating temperature decreases, whereas in a condenser the heat rejection rate increases as the condensing temperature increases. As long as there is a substantial difference between the inlet temperature and evaporating temperature, the evaporator characteristics are linear. This can be demonstrated using a similar analysis to that used for a condenser. However, at low-temperature differences the characteristics are no longer linear. This is because boiling coefficients are strongly dependent on temperature difference (Ref. 7.7 recommends a correlation with $\alpha \propto \Delta t^3$). With

Figure 7.39 Direct expansion coil

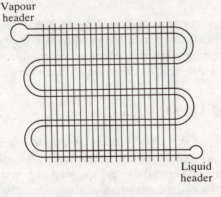

Figure 7.40 Evaporator characteristics

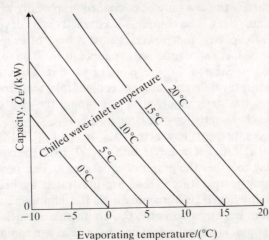

large temperature differences the boiling coefficient is high enough to ensure that the thermal resistance is a small proportion of the total. As the temperature difference falls, the thermal resistance of the boiling coefficient increases and the change in the overall heat transfer coefficient, U, is sufficient to make the characteristic non-linear.

7.8 SYSTEM ANALYSIS

In previous sections the characteristics of reciprocating compressors, condensers and evaporators were investigated. For a large plant the evaporator and condenser might be specially designed around a selected compressor. At the other end of the market, for very small plants such as domestic units, one specific design would be carried out (and possibly tested) before the units were mass-produced. Between these extremes are a great number of individual specifications which do not justify detailed design work and do not occur often enough to warrant a standard plant. This gap is filled by packaged plants. The packaging of a refrigeration plant involves selecting a suitable compressor, condenser and evaporator which will together make up a suitable combination to carry out the specified duty. Each item of plant will be a standard unit with its own design performance and characteristics. The characteristics of each item must then be combined with the characteristics of the other items selected to predict the performance of the packaged plant. This process is termed 'system analysis'.

Another use of system analysis is when it is proposed to change the service of an existing plant or to investigate what will happen at part-load.

When two components of a plant are in equilibrium, there will be a system balance point which will be common to the characteristics of each item.

Condensing Units

A condensing unit is a combination of a condenser and a compressor. These are often sold as combined units, particularly when the condenser is air-cooled. If they are specified separately the first step in a system analysis is to combine the characteristics of the compressor and condenser to form a condensing unit.

Characteristics can only be combined graphically when they are both plotted on the same axes. The compressor characteristics developed in Section 7.3 are graphs of capacity and power against a base of evaporating temperature for different values of condensing temperature. The condenser characteristics explained in Section 7.6 are plots of heat rejection rate against condensing temperature for different values of coolant inlet temperature. Figure 7.41(a)–(d) illustrates the steps involved in combining the two sets of characteristics to form the characteristics of a condensing unit.

The compressor characteristics from Fig. 7.18(d) are reproduced in Fig. 7.41(a). Lines of heat rejection rate, $\dot{Q}_C$ from the compressor are then drawn by adding together the graph of capacity, $\dot{Q}_E$, and power, $\dot{W}$, for each value of condensing temperature. These are illustrated by broken lines. It should be noted that when $\dot{W}$ is a maximum $\dot{Q}_E$ is a minimum, and vice versa. The lines of constant heat rejection rate, $\dot{Q}_C$, are therefore much closer together than the capacity lines. The compressor now has one axis in common with the condenser.

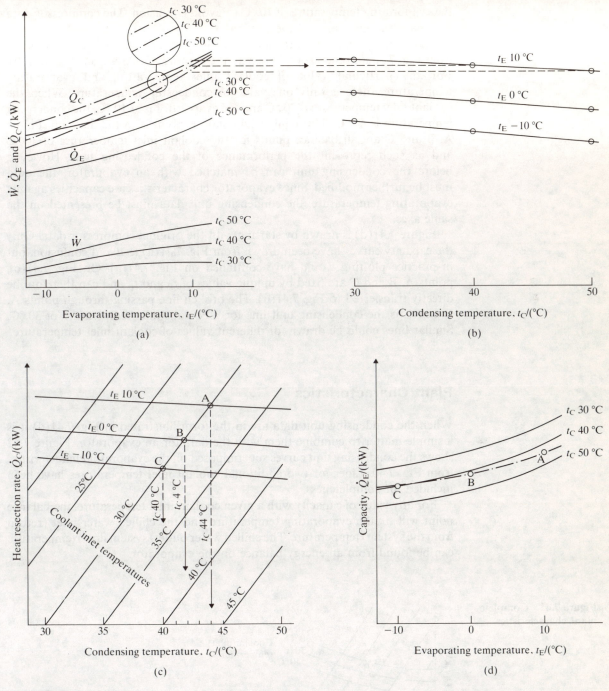

Figure 7.41 Condensing
unit characteristics

A new graph, Fig. 7.41(b), must now be drawn to rearrange the compressor
data so that the horizontal axis is condensing temperature. This is done by
selecting values of evaporating temperature, t_E, in this case $-10\,°C$, $0\,°C$ and
$10\,°C$, from Fig. 7.41(a) and transferring values of condensing temperature to
the new graph for fixed values of t_E. The construction has only been shown for

the evaporating temperature of 10 °C to avoid confusion. The compressor data now have the same axes as the condenser.

Figure 7.41(c) has been drawn by starting with the condenser characteristics from Fig. 7.36 and then superimposing the compressor data from Fig. 7.41(b). For any particular value of coolant inlet temperature and evaporating temperature, there is only one value of condensing temperature. When the coolant inlet temperature is 30 °C and the evaporator is at 10 °C, the condensing temperature is 44 °C. This equilibrium state is point A in the diagram. Points A, B and C are all balance points for the coolant inlet temperature of 30 °C and as such represent the performance of the condensing unit. However, before the condensing unit can be matched with an evaporator, the data must be further modified. Since evaporator characteristics are capacities against evaporating temperature, the condensing unit data must be presented on the same axes.

Figure 7.41(d) is drawn by starting with the original compressor data. Only the capacity curves have been drawn from Fig. 7.41(a) to avoid confusion, but in practice plotting would have continued on Fig. 7.41(a). On Fig. 7.41(c), points A, B and C are fixed by unique values of t_E and t_C and can therefore be directly transferred to Fig. 7.41(d). The broken line passing through points A, B and C is the condensing unit line for a coolant inlet temperature of 30 °C. Similar lines could be drawn for different values of coolant inlet temperature.

Plant Characteristics

When the condensing unit data are in the form illustrated in Fig. 7.41(d) it is a simple matter to combine them with the data for an evaporator. Figure 7.42 shows the condensing unit curves superimposed on the evaporator characteristics from Fig. 7.40. Lines for two additional coolant inlet temperatures have been included for completeness.

For any value of capacity with a given coolant inlet temperature, the balance point will fix the evaporating temperature and the chilled water inlet (return from the system) temperature. The chilled water outlet (system flow) temperature can be found from an energy balance on the evaporator.

Figure 7.42 Complete plant characteristics

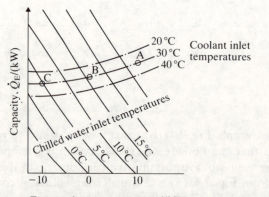

Example 7.11

(a) Figure 7.43(a) shows the performance curves for the compressor and condenser of a Refrigerant 22, vapour compression, water chilling plant.

When operating at its maximum duty, the system will cool water from 12 to 9 °C at a water flow rate of 6.0 litres/s. A back-pressure valve having a rangeability of 3.5:1 is the only means of capacity control.

Use Fig. 7.43(a) and the data tabulated below to determine:
(i) the maximum allowable pressure drop in the suction pipeline;
(ii) the minimum stable capacity of the plant when it is attempting to maintain a constant evaporating temperature;
(iii) the chilled water outlet temperature at the minimum stable capacity of the plant.

(b) Figure 7.43(a) shows that the compressor performance improves with decreasing condensing temperatures at any refrigerant evaporating temperature, t_E. Conversely, the condenser performance decreases with decreasing condensing temperatures. Explain why these phenomena occur.

Evaporator Data
Rated capacity 100 kW when operating at an evaporating temperature of 6.5 °C and a chilled water outlet temperature of 8 °C.

Back Pressure Valve Data
Rangeability, 3.5:1.

Pressure drop across BPV/(bar)	0.05	0.1	0.2	0.3	0.75
Capacity/(kW)	40	66	108	136	240

Note: It may be assumed that the chilled water flow rate and the pressure drop in the suction pipeline are constant throughout the operating range of the plant.

(South Bank Polytechnic)

Figure 7.43 Example 7.11

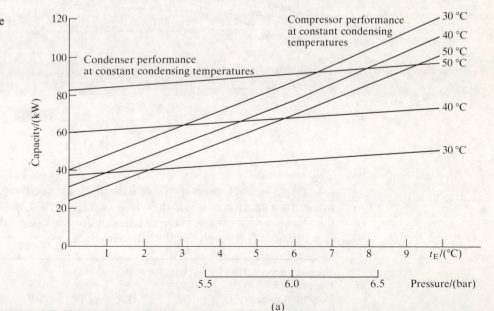

(a)

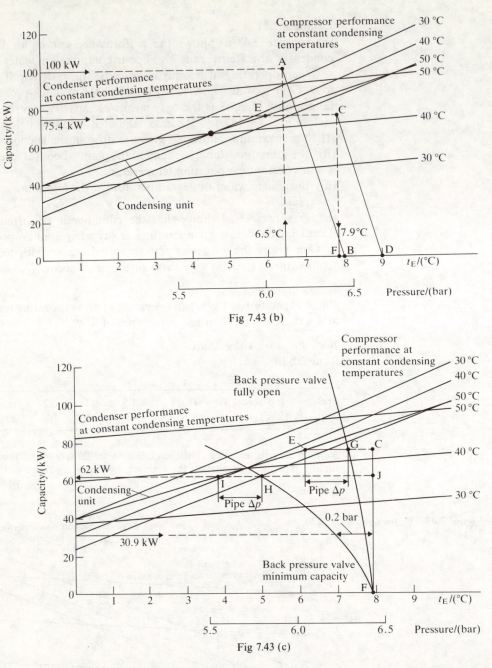

Fig 7.43 (b)

Fig 7.43 (c)

Solution

(a) (i) The back-pressure valve data given are for the fully open valve. The minimum capacity is when the valve is fully turned down. To establish the minimum capacity, each value of capacity for the open valve is divided by the rangeability. This results in the following table.

Pressure drop across BPV/(bar)	0.5	0.1	0.2	0.3	0.75
Capacity/(kW)	40	66	108	136	240
Minimum capacity/(kW)	11.4	18.9	30.9	38.9	68.6

Figure 7.43(a) incorporates both the compressor and condenser characteristics. Combining the compressor and condenser characteristics to form a condensing unit is simply a matter of identifying the three points where the characteristics have the same value, i.e. when the 50 °C condenser line crosses the 50 °C compressor line, etc., and drawing a curve through these points. This has been done on Fig. 7.43(b).

The evaporator design point A is then established on the same figure from $\dot{Q}_E = 100$ kW and $t_E = 6.5$ °C. A line from A to cut the chilled water outlet temperature of 8 °C at point B completes the evaporator design data.

In the actual plant the evaporator capacity can be found from the heat flow rate from the chilled water as follows:

$$\dot{Q}_E = \dot{m}c_p(t_i - t_o) = 6 \times 4.19(12 - 9) = 75.4 \text{ kW}$$

Also in the packaged plant the chilled water outlet temperature is 9 °C and the evaporator must work at this value and not at its original design value of 8 °C. With the same mass flow rate of chilled water through the evaporator, the new evaporator operating point must be where the new chilled water outlet temperature line, C–D, crosses the plant capacity line of 75.4 kW. The line C–D is parallel to the line A–B and crosses the temperature axis at 9 °C. Point C is the evaporator operating point in the actual plant.

In a plant with no pressure drop between the evaporator and compressor point, C would lie on the condensing unit line. In this case the condensing unit operating point is at E where the actual plant capacity line crosses the condensing unit line. Between C and E there is a pressure drop, of which part takes place across the back pressure valve and part takes place in the pipe between the evaporator and condenser.

Reading vertically down from C gives the evaporator operating temperature of 7.9 °C at point F. Having established points C, E and F, the solution is continued on Fig. 7.43(c) to avoid the confusion of too many lines.

The horizontal axis of Fig. 7.43 has scales of both temperature and pressure. This simplifies the solution because the pressure drops across the back pressure valve do not need to be converted into corresponding changes in saturation temperature.

The characteristics of the back pressure valve can now be drawn on Fig. 7.43(c), starting from point F. Each pressure drop is taken from the 7.9 °C line for the corresponding value of capacity. A sample point for the back pressure valve minimum capacity of 30.9 kW with a pressure drop of 0.2 bar is illustrated.

When the plant is operating at design capacity, the back-pressure valve is fully open, the evaporator is operating at point C and the condensing unit at point E. The pressure drop from C to G is across the valve and the remaining pressure drop from G to E must be across the pipework. The maximum allowable pressure drop across the pipework is therefore 0.25 bar.

(ii) If the pressure drop across the pipe, E–G, is constant, a line can be drawn through G parallel to the condensing unit line which cuts the back pressure valve minimum stable capacity line at point H. The condensing unit is now operating at point I and the evaporator at point J. The minimum plant capacity of 62 kW can now be read off the diagram. If the capacity falls below this value the evaporating temperature will start to fall.

(iii)   The chilled water temperature at minimum capacity is now found by drawing a line from J to cut the temperature axis at point K such that line J–K is parallel to lines A–B and C–D, (Fig. 7.43(b)). The outlet temperature at point K is 8.8 °C. This line has not been drawn on the diagram, to avoid confusion.

(b)   For any fixed evaporating temperature the compressor performance will increase as the condensing temperature decreases, because of the reduced pressure ratio and consequent increase in volumetric efficiency. This will cause the mass flow rate of refrigerant, and hence the evaporator capacity, to increase. However, the condenser capacity will decrease as the condensing temperature decreases because of the reduction in mean temperature difference between the refrigerant and cooling medium.

Example 7.12

(a)   A packaged water chilling plant is required to produce 4 kg/s of chilled water at 8 °C with cooling water available at 25 °C. The following components are to be used for the plant:

Evaporator: Designed to cool 4 kg/s of water from 14 °C to 6 °C with an evaporating temperature of 0 °C.

Condenser: Designed to produce a condensing temperature of 45 °C when supplied with 5 kg/s of cooling water which enters at 30 °C and leaves at 40 °C.

Compressor: This is a reciprocating machine which has the characteristics shown in Fig. 7.44(a). Its heat rejection rate may be assumed to vary linearly with condensing temperature.

Establish the system balance point and hence determine the refrigerating capacity, the evaporating temperature and the chilled water return temperature.

It may be assumed that the pressure drops between the evaporator and compressor and between the compressor and condenser are negligible.

(b)   If the minimum chilled water temperature for the plant in part (a) is to be 4 °C, determine the minimum plant capacity and the setting and position of a thermostat to stop the compressor. Discuss briefly the positioning of the thermostat.

(Wolverhampton Polytechnic)

Solution

(a)   From the compressor characteristics in Fig. 7.44(a), values of evaporator capacity $\dot{Q}_E$ and work input $\dot{W}$ can be read off for two condensing temperatures, t_C, for each of three different evaporating temperatures, t_E. These values are then used to determine the heat rejection rates in the condenser, $\dot{Q}_E$, using the first law, i.e. $\dot{Q}_C = \dot{Q}_E + \dot{W}$. This is shown in the table below.

t_e (°C)	t_c (°C)	$\dot{Q}_E$ (kW)	$\dot{W}$ (kW)	$\dot{Q}_C$ (kW)
10	20	161.0	26.0	187.0
10	50	95.0	39.5	134.5
0	20	89.0	22.5	111.5
0	50	44.0	27.7	71.7
−6	20	60.3	18.7	79.0
−6	50	22.5	22.0	44.5

Figure 7.44 Example
7.12

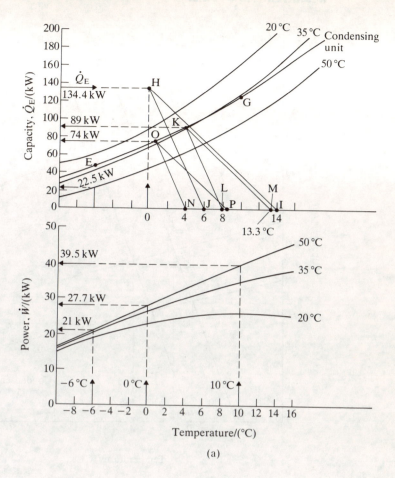

Figure 7.44 Example 7.12

(a)

Only sample values have been shown on Fig. 7.44(a), to avoid confusion.

Another graph is now plotted of the heat rejection rate $\dot{Q}_C$ against condensing temperature for each evaporating temperature, t_E. This is shown in Fig. 7.44(b). The horizontal axis of temperature is used for both the condensing temperature and the cooling water temperature.

For the original design of the condenser:

$$\dot{Q}_C = \dot{m}c_p(t_o - t_i) = 5 \times 4.18(40 - 30) = 209 \text{ kW}$$

The condenser design point, A, is now located on Fig. 7.44(b) using $\dot{Q}_C = 209$ kW and $t_C = 45$ °C. A line A–B is then drawn from point A to the cooling water inlet temperature of 30 °C at point B. For the packaged plant the cooling water is available at 25 °C. This will lower the condensing temperature for any fixed value of evaporating temperature. This is represented by drawing a line C–D parallel to line A–B which cuts the temperature axis at 25 °C, which is point D.

New values of $\dot{Q}_C$ and t_C are then read off line C–D for each value of t_E (points E, F and G). These values are then transferred back to Fig. 7.44(a) (only points E and G are shown, to avoid confusion). A line is then drawn through these three points: this is the condensing unit line.

For the original design of the evaporator,

$$\dot{Q}_E = \dot{m}c_p(t_i - t_o) = 4 \times 4.2 \times (14 - 6) = 134.4 \text{ kW}$$

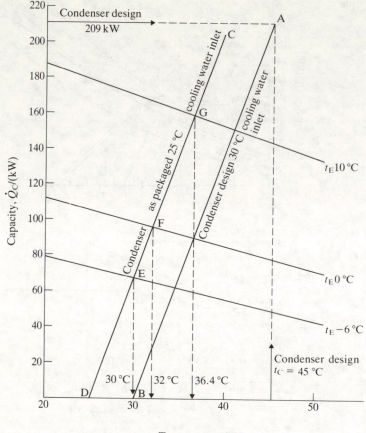

Fig. 7.44 (b)

The evaporator design point, H, can now be superimposed on Fig. 7.44(a) using $\dot{Q}_E = 134.4$ kW and $t_E = 0$ °C. A line from H to the chilled water return temperature of 14 °C at I is then drawn. Another line can be drawn from H to the chilled water outlet temperature of 6 °C at point J. These are the conditions for the evaporator as originally designed.

It is now necessary to combine the evaporator characteristics with those of the condensing unit to establish the behaviour of the combined packaged plant. The capacity of the evaporator must fall until it matches that of the condensing unit. A balance point for the complete plant is found when the evaporator is operating on the condensing unit line. The balance point for the plant is established by drawing a line parallel to H–J which passes through the chilled water flow temperature of 8 °C at point L and cuts the condensing unit line at point K. Point K is the design balance point for the packaged plant. The chilled water return temperature is now found by drawing a line from K parallel to line H–I to cut the temperature axis at point M, which is at 13.3 °C. It should be noted that JHI and LKM are similar triangles and the chilled water temperature drops I–J and M–L are proportional to the capacities at H and K respectively.

At the balance point K:

$$\dot{Q}_E = 89 \text{ kW}$$
$$t_E = 4 \,°C$$
$$t_R = 13.3 \,°C$$

(b)   Draw a line from the minimum supply temperature of 4 °C, point N, parallel to H–J which cuts the condensing unit line at point O. This is the new balance point at the minimum capacity. A line O–P parallel to line H–I cuts the temperature line at P, the minimum return temperature.

At the balance point:

$$\dot{Q}_E = 74 \text{ kW}$$
$$t_R = 8.4 \,°C$$

The variation in return temperature is greater than the variation in supply temperature and is therefore more sensitive to thermostatic control. The compressor is stopped when the return temperature reaches 8.4 °C.

Example 7.13

An R12 refrigeration compressor has a displacement of 0.1 m³/s and is part of an air-to-water heat pump package which includes a condenser rated at 200 kW heating capacity when the condensing temperature is 35 °C and water enters the condenser at 25 °C. The water flow rate is 6 litres/s.

Determine the cost of heating, in pence/kW heating, when the evaporating temperature is:

(a)   10 °C,

(b)   − 10 °C,

and when supplying water from the condenser at 50 °C. Briefly comment on the problems of determining a true seasonal coefficient of performance.

Use the Refrigerant 12 pressure–enthalpy and volumetric efficiency charts (Figs. 7.45(a) and (b)), and assume the following:

(i)   electricity costs 5 p/kWh;

(ii) compressor capacity and power characteristics are linear;

(iii) 10 K suction superheating and 2 K liquid sub-cooling;

(iv) the compression is isentropic.

(CIBSE)

Solution

The data for the condenser in this example show that it was originally rated for use in a refrigeration plant but is now to be used in a heat pump.

The range of cooling water temperatures are between 25 °C and 50 °C. A mean specific heat value of 4.18 kJ/kg can be taken from property tables.

For the condenser design case the cooling water outlet temperature can be found from the heat rejection rate as follows:

$$\dot{Q}_C = \dot{m}c_p(t_o - t_i)$$
$$200 = 6 \times 4.18(t_o - 25)$$
$$\therefore \qquad t_o = 33 \,°C$$

This is only 2 K below the condensing temperature, so when the condenser is used in the heat pump the condensing temperature will be only a few degrees above the water temperature of 50 °C. It is therefore reasonable to select two condensing temperatures between 50 °C and 60 °C to plot the new condenser and compressor characteristics. (With a linear characteristic more diverse values could have been selected.)

It is now necessary to analyse two cycles for each evaporator temperature. The cycles are shown in Fig. 7.45(a) and property values are from a chart. In the following text, subscripts 1, 2 and 3 refer to the compressor inlet, the compressor outlet and the condenser outlet respectively.

Evaporating temperature 10 °C, condensing temperature 60 °C:

$$h_1 = 198 \text{ kJ/kg}, \quad h_2 = 223.3 \text{ kJ/kg}, \quad h_3 = 92 \text{ kJ/kg},$$
$$p_1 = 420 \text{ kN/m}^2, \quad p_2 = 1540 \text{ kN/m}^2, \quad v_1 = 0.045 \text{ m}^3/\text{kg}$$
$$\frac{p_2}{p_1} = \frac{1540}{420} = 3.67$$

Figure 7.45 Example 7.13

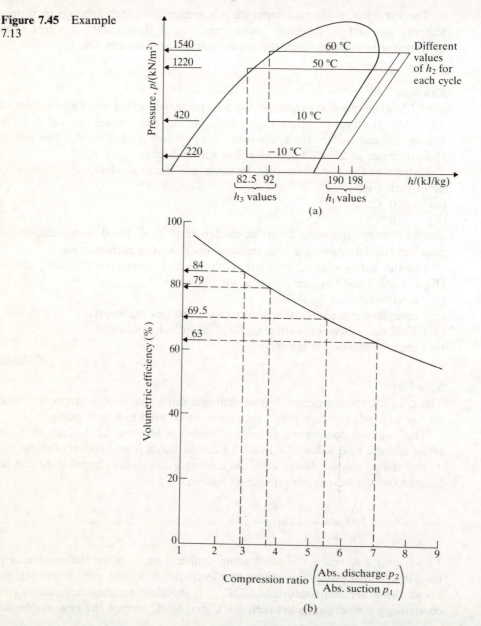

(a)

(b)

Hence, from Fig. 7.45(b) $\eta_v = 0.79$

$$\dot{m} = \frac{\dot{V}_s \eta_v}{v_1} = \frac{0.1 \times 0.79}{0.045} = 1.755 \text{ kg/s}$$

Using Eqns [7.4] and [7.3]:

$$\dot{W} = \dot{m}(h_2 - h_1) = 1.755(223.3 - 198) = 44.4 \text{ kW}$$
$$\dot{Q}_C = \dot{m}(h_2 - h_3) = 1.755(223.3 - 92) = 230.4 \text{ kW}$$

Evaporating temperature 10 °C, condensing temperature 50 °C:

$$h_1 = 197 \text{ kJ/kg}, \quad h_2 = 218 \text{ kJ/kg}, \quad h_3 = 82.5 \text{ kJ/kg},$$
$$p_1 = 420 \text{ kN/m}^2, \quad p_2 = 1220 \text{ kN/m}^2, \quad v_1 = 0.045 \text{ m}^3/\text{kg},$$
$$\frac{p_2}{p_1} = \frac{1220}{420} = 2.9,$$

Hence $\eta_v = 0.84$

$$\dot{m} = \frac{\dot{V}_s \eta_v}{v_1} = \frac{0.1 \times 0.84}{0.045} = 1.867 \text{ kg/s}$$
$$\dot{W} = \dot{m}(h_2 - h_1) = 1.867(218 - 197) = 39.2 \text{ kW}$$
$$\dot{Q}_C = \dot{m}(h_2 - h_3) = 1.867(218 - 82.5) = 253 \text{ kW}$$

Evaporating temperature −10 °C, condensing temperature 60 °C:

$$h_1 = 190 \text{ kJ/kg}, \quad h_2 = 227.4 \text{ kJ/kg}, \quad h_3 = 92 \text{ kJ/kg},$$
$$p_1 = 220 \text{ kN/m}^2, \quad p_2 = 1540 \text{ kN/m}^2, \quad v_1 = 0.082 \text{ m}^3/\text{kg},$$
$$\frac{p_2}{p_1} = \frac{1540}{220} = 7,$$

Hence $\eta_v = 0.63$

$$\dot{m} = \frac{\dot{V}_s \eta_v}{v_1} = \frac{0.1 \times 0.63}{0.082} = 0.768 \text{ kg/s}$$
$$\dot{W} = \dot{m}(h_2 - h_1) = 0.768(227.4 - 190) = 28.7 \text{ kW}$$
$$\dot{Q}_C = \dot{m}(h_2 - h_3) = 0.768(227.4 - 92) = 104 \text{ kW}$$

Evaporating temperature −10 °C, condensing temperature 50 °C:

$$h_1 = 190 \text{ kJ/kg}, \quad h_2 = 223 \text{ kJ/kg}, \quad h_3 = 82.5 \text{ kJ/kg},$$
$$p_1 = 220 \text{ kN/m}^2, \quad p_2 = 1220 \text{ kN/m}^2, \quad v_1 = 0.082 \text{ m}^3/\text{kg},$$
$$\frac{p_2}{p_1} = \frac{1220}{220} = 5.5,$$

Hence $\eta_v = 0.695$

$$\dot{m} = \frac{\dot{V}_s \eta_v}{v_1} = \frac{0.1 \times 0.695}{0.082} = 0.848 \text{ kg/s}$$
$$\dot{W} = \dot{m}(h_2 - h_1) = 0.848(223 - 190) = 28 \text{ kW}$$
$$\dot{Q}_C = \dot{m}(h_2 - h_3) = 0.848(223 - 82.5) = 119.1 \text{ kW}$$

The compressor characteristics of $\dot{Q}_C$ and $\dot{W}$ for evaporating temperatures of 10 °C and −10 °C can now be plotted on a base of temperature which is shown in Fig. 7.45(c).

Figure 7.45
(*continued*)

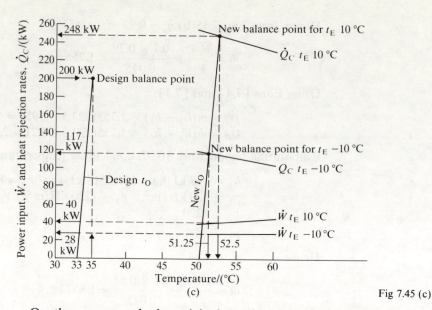

Fig 7.45 (c)

On the same graph the original condenser design balance point can be established at $\dot{Q}_C = 200$ kW and $t_C = 35$ °C. From this point a line can be drawn to cut the temperature axis at 33 °C. This is the original design cooling water outlet temperature. It should be noted that the same temperature scale is used for both the condensing temperature and the cooling water temperatures. A line is then drawn parallel to the design cooling water outlet temperature line which cuts the temperature axis at 50 °C, the new cooling water outlet temperature. This line cuts the new condenser characteristics at the new balance points. The new condenser capacities and the new compressor powers are then read directly off the graph from the points where vertical lines from the new balance points cut the compressor characteristics.

The results and costs are as follows:

For evaporating temperature 10 °C:

$$\dot{Q}_C = 248 \text{ kW}, \quad \dot{W} = 40 \text{ kW}$$
$$\text{Cost} = 40 \times 5\text{p} = 200\text{p for 1 hour}$$
$$\text{Cost/kW heating} = \frac{200}{248} = 0.81 \text{ p/kW}$$

For evaporating temperature −10 °C:

$$Q_C = 117 \text{ kW}, \quad W = 28 \text{ kW}$$
$$\text{Cost} = 28 \times 5\text{p} = 140\text{p for 1 hour}$$
$$\text{Cost/kW heating} = \frac{140}{117} = 1.2 \text{ p/kW}$$

To calculate a seasonal coefficient of performance requires repetition of the above calculations for a series of evaporating temperatures corresponding to the outdoor air temperatures encountered during the year when the heat pump is running. Then weather frequency data are used to predict how often each condition will occur. For example, an air temperature of 3 °C might correspond to an evaporating temperature of −5 °C; this value could be selected to cover air temperatures from 2 °C to 4 °C, which might have a frequency of occurrence, f, of 0.05 for the year. The seasonal coefficient of performance can then be found from the following equation:

$$\text{Seasonal COP} = \frac{\Sigma(\text{COP} \times f)}{\Sigma f}$$

Clearly such a calculation would be very tedious without the use of a computer program.

The same approach can then be used to calculate the average loads on cooling coils and is covered in Section 8.11 of Chapter 8.

7.9 CENTRIFUGAL COMPRESSORS

Centrifugal compressors are suitable for large volume flow rates and the larger the plant the more likely it is that a centrifugal machine will be used. Because of their suitability for high volume flow rates they can handle the low specific volumes used in low-temperature work but their characteristics also make them suitable for large chilled water plants.

Operation

Figure 7.46 shows two simplified sections through the main components of a centrifugal machine. The impeller is rotated at high speed inside a casing. Typical speeds range from 3000 to over 20 000 revolutions per minute. The vapour enters the eye of the impeller in an axial direction. The flow is then turned through 90° as the vapour passes through curved passages which are separated by guide vanes. As the vapour flows outward its pressure increases due to centrifugal effects and it leaves the impeller tip at an increased pressure and a very high velocity. From the impeller it is discharged to diffuser passages where the velocity is reduced, resulting in a further increase in pressure before it enters the discharge pipe.

Figure 7.46 Centrifugal compressor

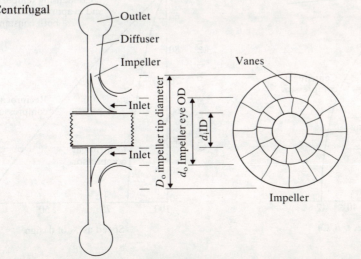

Centrifugal compressors are compared by an isentropic efficiency and in general the larger the machine the higher will be its isentropic efficiency. This is because the losses due to friction between the fluid and the impeller and casing are much greater through narrow passages than through large ones (this is comparable with what happens in pipes and ducts). Typical values of isentropic efficiency range between 70 and 85%.

Characteristics

The general characteristics for any centrifugal compressor are shown in Fig. 7.47(a). For any particular speed there are two limiting values of mass flow rate. The high value occurs when the velocity at some point in the system becomes sonic and the system is said to be choked.

Figure 7.47 Centrifugal compressor characteristics

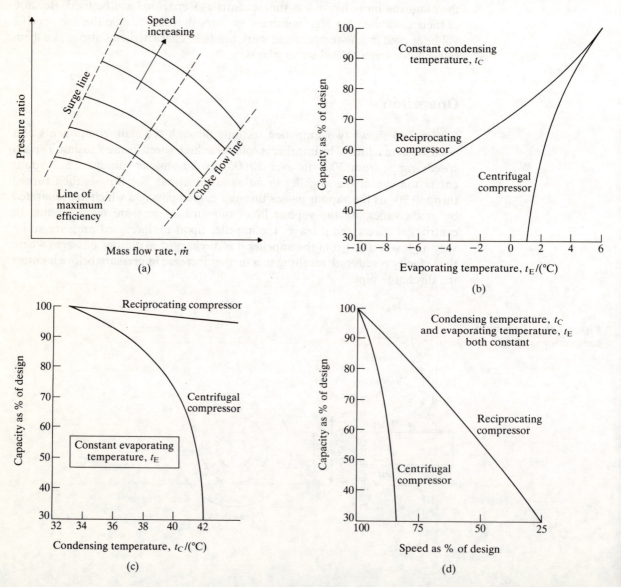

(a)

(b)

(c)

(d)

The low value occurs when the vapour starts to flow back from the region of high pressure to that of low pressure. This only occurs momentarily before the fluid is again flowing in the right direction. This oscillation in the direction of flow is called surging. To the left of the point where surging starts is an increasingly unstable region where oscillations can quickly build up to the point where they can cause mechanical failure. One reason for surging is that as the mass flow rate falls the direction of flow into and out of the compressor increasingly diverges from the direction of impeller blade angles.

Within the limits of stable operation there is a point of maximum isentropic efficiency for each value of speed and the machine should ideally operate close to this point. When a series of characteristics are drawn for different speeds a surge line, a choke line and lines of constant efficiency can be drawn.

It is useful to compare some of the characteristics of a centrifugal compressor with those of a reciprocating compressor. Characteristics for both types of machine are drawn on Fig. 7.47(b), (c) and (d)). In each case the vertical axis is refrigeration capacity as a percentage of the design value and the characteristics for both types of machine meet at the design point.

In Fig. 7.47(b) the horizontal axis is evaporating temperature and the condensing temperature is kept constant. The curve for the reciprocating compressor was first introduced in Section 7.3 under the heading 'Reciprocating compressor characteristics'. A modest change in capacity produces a substantial change in evaporating temperature. The evaporating temperature with a centrifugal compressor, on the other hand, is relatively constant over a wide range of capacities. This is a very desirable characteristic for the compressor of any chilled water plant.

Figure 7.47(c) is plot of capacity against condensing temperature for a constant value of evaporating temperature. Once again the reciprocating compressor curve was introduced earlier, in this case in Section 7.8 on 'system analysis'. A considerable change in condensing temperature produces very little change in capacity and the line is almost linear. The centrifugal compressor is quite different, and modest changes in condensing temperatures produce considerable changes in capacity.

Finally, Fig. 7.47(d) is a plot of capacity against rotational speed with both the condensing and evaporating temperatures constant. For the reciprocating compressor the capacity is almost directly proportional to speed. For the centrifugal compressor a relatively small change in speed is sufficient to modulate through the capacity range.

Capacity Control

The most usual method of capacity control is to vary the speed, particularly if the drive is a steam or gas turbine. When constant-speed drives are used, the compressor speed can be varied with a magnetic or hydraulic clutch. With speed control the power required is approximately proportional to the capacity.

Another popular method is to use pre-rotation vanes. These are movable vanes situated at the inlet to the compressor. At full load the vapour enters the eye of the impeller in an axial direction. To reduce the load the vanes are moved so as to induce progressively a tangential component to the velocity at inlet. A hot gas by-pass may be incorporated into the control system to prevent surging.

Since changing the condensing temperature produces an appreciable change in capacity, this can also be used as a means of control. To change the condensing temperature the coolant must be throttled. This is relatively simple with a water-cooled condenser. With an air-cooled condenser the choice is between variable-speed fans, variable-pitch fans or a variable-speed drive. This method relies on reducing mass flow rate by increasing the pressure ratio (which is comparable with the use of a back-pressure regulating valve with a reciprocating compressor). It is clearly less desirable than the other methods because it does not produce any saving in power.

Velocity Triangles

Assuming radial vanes and axial entry, the velocity triangles are as illustrated in Fig. 7.48. The inlet triangle (a) is the simpler of the two. A whirl velocity is a velocity tangential to the impeller. With axial entry there is no whirl velocity at inlet; this is termed 'no pre-whirl'. The flow velocity, u_{f1}, is therefore the same as the inlet velocity, u_1. The velocity u_{b1} is the blade (impeller) velocity at inlet. This is taken at the mean eye radius. Therefore,

$$u_{b1} = \frac{\pi(d_o + d_i)}{2} N \qquad [7.21]$$

Figure 7.48 Velocity triangles

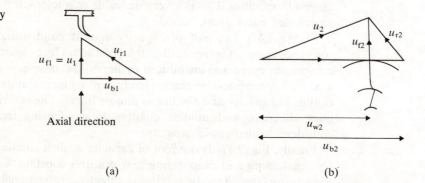

(a)   (b)

where d_o and d_i are the outer and inner eye radii respectively and N the speed in revolutions per second. The velocity u_{r1} is the velocity at inlet relative to the moving impeller. A stationary observer would see the vapour entering the eye in an axial direction but the impeller 'sees' the vapour moving along the path of u_{r1}. This is why the vanes in the eye must be curved so that the vapour has a smooth entry. If the inlet velocity is reduced with the speed constant, the relative velocity will change direction and the entry will no longer be smooth. At some point this will cause surging.

Applying the mass continuity equation to the eye gives:

$$\dot{V}_1 = \dot{m}v_1 = \frac{\pi}{4}(d_o^2 - d_i^2)u_1 \qquad [7.22]$$

where v_1 is the vapour specific volume at inlet.

At the impeller tip the outlet velocity triangle is more complex because of a phenomena called slip. Without slip the tangential component of the vapour velocity, u_{w2}, would have the same value as the impeller tip velocity, u_{b2}. A slip factor, σ, is therefore defined as follows:

$$\sigma = \frac{u_{w2}}{u_{b2}} \qquad\qquad [7.23]$$

where $\qquad u_{b2} = \pi D_o N \qquad\qquad [7.24]$

As a consequence of slip, the absolute velocity at outlet is u_2 and the relative velocity at outlet is u_{r2}. The flow velocity at outlet is u_{f2} and since this is the component of velocity in the radial direction, it is the one which must be used when continuity is applied to the outlet vapour.

Assuming that the vanes occupy a negligible proportion of the outlet flow area (which is not unreasonable since they are intentionally tapered towards the tip), and the distance across the impeller at the tip is y, continuity is applied as follows:

$$\dot{V}_2 = \dot{m}v_2 = \pi D_o y u_{f2} \qquad\qquad [7.25]$$

Power Input $\dot{W}$

Since the impeller is constrained to rotate in one direction, there is only one direction in which work can be transferred from the impeller to the vapour—tangentially. The components of velocity in this direction are the whirl velocities. An expression for the power input can be developed as follows.

$$\text{Momentum} = \text{Mass} \times \text{velocity} = mu_w$$
$$\text{Force} \qquad = \text{Rate of change of momentum} = \dot{m}u_{w1} - \dot{m}u_{w2}$$

But with no pre-whirl u_{w1} is 0. Therefore,

$$\text{Force} \qquad = \dot{m}u_{w2} \ (\text{kg/s} \times \text{m/s} = \text{N})$$
$$\text{Power} \qquad = \text{Force} \times \text{distance moved per second}$$
$$\qquad\qquad = \dot{m}u_{w2}u_{b2} \ (\text{N} \times \text{m/s} = \text{W})$$

It is not practice to specify a compressor as having a negative power. If the term 'power input' is used the direction of work flow is implied; therefore,

$$\text{Power input, } \dot{W} = \dot{m}u_{w2}u_{b2}$$

Introducing the slip factor from Eqn [7.23] gives:

$$\text{Power input, } \dot{W} = \dot{m}\sigma u_{b2}^2 \qquad\qquad [7.26]$$

Equation [7.5] is also still applicable. Therefore,

$$\text{Power input, } \dot{W} = \dot{m}(h_2 - h_1)$$

Care must be taken with the units of these two expressions for $\dot{W}$ since the natural units for Eqn [7.26] are watts but with specific enthalpy values in kJ/kg the natural units of Eqn [7.5] are kilowatts. Equations [7.5] and [7.26] can be combined to give

$$\dot{m}\sigma u_{b2}^2 = \dot{m}(h_2 - h_1)$$

Rearranging,

$$u_{b2} = \sqrt{\frac{(h_2 - h_1)}{\sigma}}$$

[7.27]

Stages of Compression

Equation [7.27] shows that if the slip factor remains constant, the impeller tip velocity depends only on the change in specific enthalpy across the compressor. However, the impeller tip velocity is limited by the local sonic velocity of the refrigerant (shock waves develop above the speed of sound). This provides a limit to the specific enthalpy increase and hence to the pressure rise which can be achieved in a single stage. It is therefore frequently necessary to use a number of stages in series to achieve a desired pressure ratio. If no other information is available, it is reasonable to assume that when more than one stage of compression is used the change in specific enthalpy will be the same for each stage.

When more than one stage is used, the width of the impeller tips will be reduced as the pressure increases to accommodate the changes in the specific volume of the vapour. The impeller tip diameters may be the same for large machines, but if the width of the impeller tip becomes too small at the higher pressure stages it may be necessary to reduce the diameters so that the width can be maintained at a reasonable value.

Because it is necessary to use more than one stage for all but the most modest pressure ratios, centrifugal machines are particularly suited to multi-stage compression systems.

Example 7.14
(a)   For a centrifugal compressor with the absolute inlet flow velocity in an axial direction and the exit flow velocity relative to the impeller in a radial direction, show that, neglecting slip:

$$\text{Power input, } \dot{W} = \dot{m}(\pi ND)^2$$

where $\dot{m}$ = mass flow rate;
N = rotational speed;
D = impeller tip diameter.

(b)   A refrigerating plant uses Refrigerant R11 which enters a centrifugal compressor dry saturated at 0.4 bar. The compressor has an isentropic efficiency of 0.75 with a pressure ratio of 2.5 at a rotational speed of 3600 rev/min. It may be assumed that the refrigerant enters the impeller in an axial direction and that its velocity relative to the impeller at exit is in the radial direction without slip.

Sketch the layout of the plant and, for a cooling load of 400 kW and no under-cooling in the condenser, calculate:
(i)   the coefficient of performance;
(ii)   the mass flow rate of refrigerant;
(iii) the diameter of the impeller of the compressor;
(iv)   the outside diameter of the eye of the impeller inlet when the inside diameter is 60 mm and the axial velocity of the refrigerant at impeller inlet is 80 m/s.

Use the p–h chart for Refrigerant R11. State any assumptions made.

(CIBSE)

Solution

(a) This involves a substitution of Eqn [7.24] into Eqn [7.26] with a value of 1 for the slip factor.

(b) The following values can be found from a p–h chart as shown in Fig. 7.49:

$$h_1 = 960 \text{ kJ/kg}, \quad h_{2s} = 977 \text{ kJ/kg}, \quad h_3 = h_4 = 790 \text{ kJ/kg},$$
$$v_1 = 0.4 \text{ m}^3/\text{kg}$$

Figure 7.49 Example 7.14

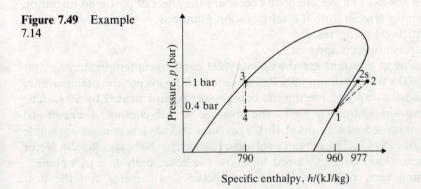

Using Eqn [7.1]:

$$h_2 = h_1 + \frac{h_{2s} - h_1}{\eta_s} = 960 + \frac{977 - 960}{0.75} = 982.7 \text{ kJ/kg}$$

Using Eqn [7.6]:

$$\text{COP} = \frac{h_1 - h_4}{h_2 - h_1} = \frac{960 - 790}{982.7 - 960} = 7.5$$

Using Eqn [7.3]:

$$\dot{m} = \frac{\dot{Q}_E}{h_1 - h_4} = \frac{400}{960 - 790} = 2.353 \text{ kg/s}$$

Using Eqn [7.27]:

$$u_{b2} = \sqrt{\frac{(h_2 - h_1)}{\sigma}} = \sqrt{(982.7 - 960)10^3} = 151 \text{ m/s}$$

Using a rearranged form of Eqn [7.24]:

$$D_o = \frac{u_{b2}}{\pi N} = \frac{151}{\pi \times 60} = 0.8 \text{ m}$$

The volume flow rate at inlet can now be found as follows:

$$\dot{V}_1 = \dot{m}v_1 = 2.353 \times 0.4 = 0.941 \text{ m}^3/\text{s}$$

Using Eqn [7.22]:

$$\dot{V}_1 = \dot{m}v_1 = \frac{\pi}{4}(d_o^2 - d_i^2)u_1$$

$$0.941 = \frac{\pi}{4}(d_o^2 - 0.06^2)80$$

$$\therefore \qquad d_o = 0.136 \text{ m}$$

Example 7.15

(a)   With the aid of a sketch, briefly compare the effect of part-load operation on evaporator temperature if a refrigeration plant has:

(i)   a centrifugal compressor;

(ii) a reciprocating compressor;

both running at constant speed with constant condensing temperature.

(b)   An 800 kW heat pump uses Refrigerant 11. The evaporating temperature is 0 °C and the vapour entering the compressor is superheated by 10 K. The condensing temperature is 80 °C and there is no sub-cooling. A centrifugal compressor is used which runs at 10 080 rev/min and has a maximum allowable tip speed of 175 m/s. The overall isentropic efficiency is 80% and the slip factor 0.9. Compression may be assumed to follow the linear path on a p–h plane.

Assuming each stage has the same impeller tip diameter and the flow component of velocity at the outlet is 20 m/s, calculate:

(i)   the number of stages required;

(ii) the impeller tip diameter;

(iii) the width of the impeller passages of each impeller at outlet.

(Wolverhampton Polytechnic)

Solution

(a)   Figure 7.47(b), (page 354), shows a sketch of the two characteristics.

With a centrifugal machine there is little change in evaporator temperature over a wide range of capacity.

With a reciprocating machine the capacity is reduced by increasing the pressure ratio, which decreases the volumetric efficiency. A change in capacity therefore produces a substantial change in evaporating pressure and hence in temperature.

(b)   From the p–h diagram shown in Fig. 7.50, the relevant enthalpy values are:

$$h_1 = 966 \text{ kJ/kg}, \quad h_{2s} = 1016 \text{ kJ/kg}, \quad h_3 = h_4 = 846 \text{ kJ/kg}$$

Using Eqn [7.1]:

$$h_2 = h_1 + \frac{h_{2s} - h_1}{\eta_s} = 966 + \frac{1016 - 966}{0.8} = 1028.5 \text{ kJ/kg}$$

Using Eqn [7.4]:

$$\dot{m} = \frac{\dot{Q}_C}{h_2 - h_3} = \frac{800}{1028.5 - 846} = 4.384 \text{ kg/s}$$

Figure 7.50 Example 7.15

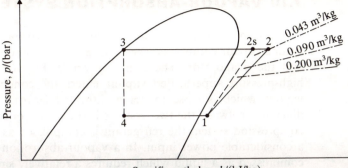

Equation [7.27], giving the blade speed for a single-stage compressor, is:

$$u_{b2} = \sqrt{\frac{(h_2 - h_1)}{\sigma}}$$

With equal enthalpy changes across each of n stages,

$$u_{b2} = \sqrt{\frac{(h_2 - h_1)}{\sigma n}} = \sqrt{\frac{(1028.5 - 966) \times 10^3}{0.9n}}$$

With u_{b2} at the maximum value of 175 m/s, n is 2.27 so that three stages must be used. With $n = 3$ the value of u_{b2} is 152 m/s. The speed in revolutions per second is $10\,080/60 = 168$; therefore, using a rearranged form of Eqn [7.24],

$$D_o = \frac{u_{b2}}{\pi N} = \frac{152}{\pi \times 168} = 0.288 \text{ m}$$

The compression line 1–2 on the p–h chart illustrated in Fig. 7.50 can now be divided into three equal sections representing each stage of compression. Specific volume values at outlet, v_o, can now be used with Eqn [7.25] to determine the width of the impeller tips. From the chart:

Stage 1 $v_o = 0.2$ m³/kg
Stage 2 $v_o = 0.09$ m³/kg
Stage 3 $v_o = 0.043$ m³/kg

For Stage 1,

$$\dot{V}_o = \dot{m}v_o = 4.384 \times 0.2 = 0.8768 \text{ m}^3/\text{s}$$

$$A_o = \frac{V_o}{u_f} = \frac{0.8768}{20} = 0.043\,84 \text{ m}^2$$

$$\text{Width } y = \frac{A_o}{\pi D} = \frac{0.04384}{\pi \times 0.288} = 0.0485 \text{ m}$$

For Stages 2 and 3, the only change is in the specific volume; therefore

$$\text{Width of Stage 2} = 0.0485 \times \frac{0.09}{0.2} = 0.0218 \text{ m}$$

$$\text{Width of Stage 3} = 0.0485 \times \frac{0.043}{0.2} = 0.0104 \text{ m}$$

7.10 VAPOUR-ABSORPTION SYSTEMS

Three of the basic components of any vapour-absorption system are the same as for a vapour-compression system, namely the condenser, the expansion device and the evaporator. These are shown in Fig. 7.51. With both systems a high-pressure superheated vapour enters the condenser and a low-pressure vapour which is close to being saturated leaves the evaporator. The only difference between the two systems is the means of compression. In a vapour-compression system the refrigerant is compressed as a vapour which requires a considerable power input. In a vapour-absorption system the refrigerant is compressed as a liquid which requires a relatively small power input.

Figure 7.51

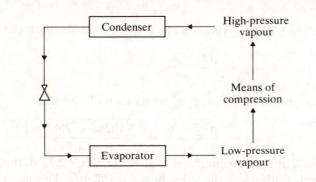

The only way to carry out a compression with the refrigerant as a liquid is to absorb it first into a suitable absorbent and after compression to release the vapour at high pressure. This absorption and subsequent release of vapour are made possible by the amount of vapour which can be absorbed by a liquid being dependent on the liquid temperature. At low temperatures large amounts of vapour can be absorbed and at higher temperatures much less. In vapour-absorption plants, when the solution contains its maximum mass of refrigerant it is termed a 'rich' solution and when it contains its minimum mass of refrigerant it is known as a 'lean' solution.

There are two established vapour-absorption systems:

	Refrigerant	Absorbent
Lithium bromide/water	Water	Lithium bromide
Aqua/ammonia	Ammonia	Water

Lithium Bromide/Water

This is much the simpler of the two systems because the lithium bromide (LiBr) does not evaporate in the temperature ranges used. It was developed after the aqua/ammonia system but is now preferred where either system could be used. The most important limitation is that with water as the refrigerant it cannot be used with evaporating temperatures below about 3 °C. Its main application is therefore for water chilling plants.

Figure 7.52 Simple vapour absorption system

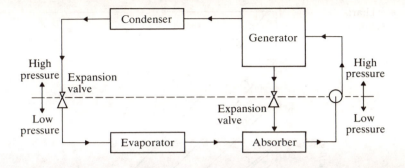

In the absence of water, pure lithium bromide is solid at room temperatures but its strong affinity for water will quickly turn it into a liquid. The most simple possible system is illustrated in Fig. 7.52. The compressor in a vapour-compression system is replaced by a generator, an expansion valve, an absorber and a pump. Only the water passes through the condenser and evaporator. The water vapour leaving the evaporator is mixed with cool, lean (low water content) liquid in the absorber. The vapour is absorbed producing a rich (high water content) mixture which then passes to the pump. The absorber is maintained at a low temperature by a coolant, which is normally cooling water. The absorber should be maintained at as low a temperature as possible to maximize the absorption of vapour.

The high-pressure rich solution from the pump is passed into the generator, where heat is supplied to raise the temperature. As the temperature increases, water vapour is driven off from the solution. The water vapour then leaves at the top of the generator and a lean solution leaves at the bottom. The lean solution is throttled before passing back to the absorber.

The complete system is divided into two pressure levels. The condenser and generator are at high pressure and the evaporator and absorber are at low pressure. The terms 'high' and 'low' are strictly relative since the complete plant operates under a vacuum. The two pressure levels are obvious on actual plants because the condenser and generator are inside one pressure vessel and the evaporator and absorber in another. As with any vacuum system, it is important to be able to expel any air which might leak into the system.

Properties of LiBr/water solutions

To analyse any refrigeration plant, it is necessary to establish specific enthalpy values at the salient points around the system. The easiest way to do this for LiBr/water solutions is to use an enthalpy–concentration chart. The basic construction of such a chart is shown in Fig. 7.53. The vertical axis is specific enthalpy in kJ/kg and the horizontal axis is the mass fraction or concentration of LiBr in the LiBr/water solution. A value of x of 1 would be pure LiBr, which would be a solid. The chart only shows values up to 0.8: before this value is reached crystals are starting to form from the solution. The broken line is the crystallization line, which represents a limit to the solutions which can be used. A value of x of 0 is pure water, on the left-hand side of the diagram. Figure 7.54 is a restricted chart which increases the scales within the normal working areas.

Figure 7.53 *h–x* chart
for LiBr/water solutions

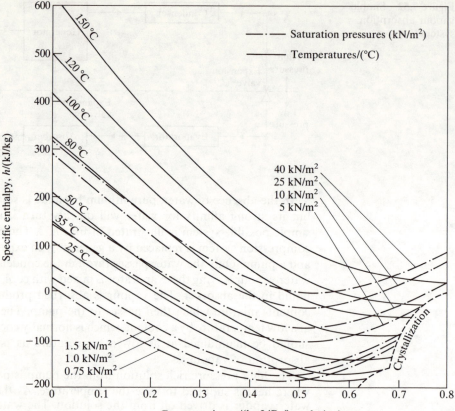

There are two sets of curved lines on the charts—lines of constant *saturation* pressure, and isotherms. It is important to understand the difference between the pressures which are only applicable to saturation states and the temperatures which can be used for saturated and unsaturated (sub-cooled/pressurized) states.

Suppose a solution is maintained at a temperature of 25 °C while exposed to water vapour at a pressure of 1 kN/m². The solution will become saturated when it has a concentration of 0.48 (when it contains 0.48 kg of LiBr and 0.52 kg of water in each kilogram of solution). This is point A on Fig. 7.55.

If the temperature is kept constant at 25 °C and the vapour pressure is increased, the solution will pass through a series of saturation states in which the vapour and solution are in equilibrium. As the vapour pressure increases, each saturation state will have a lower concentration as more vapour is absorbed. Point B is a saturated state where water vapour at 1.5 kN/m² is in equilibrium with a solution at 25 °C with a concentration of 0.45.

If the vapour pressure is maintained constant and the temperature is increased, water vapour will have to be released from solution to maintain equilibrium. The concentration increases between A and C. At point C the saturated equilibrium state is with vapour at 1.0 kN/m², the temperature at 35 °C and the solution concentration at 0.53.

If the solution at state A is isolated from water vapour, the proportions of LiBr and water must remain constant. Suppose it is now pressurized to

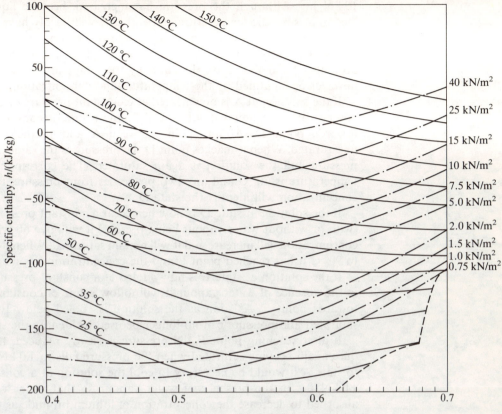

Figure 7.54 Restricted
h–x chart for LiBr/water

Figure 7.55 Solutions
of LiBr/water

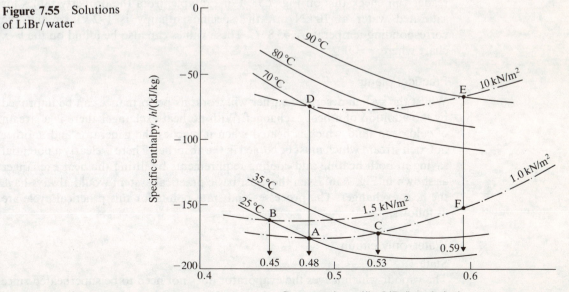

10 kN/m^2. Although the pressure has increased, because liquids are almost incompressible the temperature and enthalpy will hardly have changed. It has been unable to absorb any more water and will still be at the same point A as far as its location on the chart is concerned. This new unsaturated state can no longer be identified on the chart by the temperature and pressure. It can however be identified by the temperature and concentration.

If the solution at A is now heated at a constant pressure of 10 kN/m^2 in the absence of any water vapour, it will follow a line of constant concentration towards point D. At any point between A and D the solution remains unsaturated. When it reaches point D the pressure is the same as the saturation pressure and the solution is again saturated. The corresponding saturation temperature is 70 °C and any two properties from pressure, temperature and concentration will identify the state D.

If the solution at state D is now heated at a constant pressure of 10 kN/m^2 (with or without water vapour being present) it will give off water vapour, the concentration will increase and it will remain saturated. When it has been heated to 90 °C it has reached point E and the concentration has increased to 0.59.

If the solution at state E is now cooled at a constant pressure of 10 kN/m^2 in the absence of water vapour it will follow a line of constant concentration towards point F. As soon as the solution is cooled below E it is no longer saturated and pressure can no longer be used to identify the state.

If after reaching point F the pressure is slowly reduced, the solution will once again become saturated when the pressure falls to 1.0 kN/m^2. To return to state A it would be necessary to cool the solution at a constant pressure of 1.0 kN/m^2 in the presence of water vapour so that water vapour could be absorbed to decrease the concentration of lithium bromide as the temperature fell.

Once state points have been established, it is a simple matter to read off enthalpy values. However, care must be taken when using the values because many will be negative numbers. The saturation pressure, temperature and enthalpy values for pure water are compatible with property table values. The reader can check this on Fig. 7.53. For example, from the property tables for saturated water at 10 kN/m^2, the specific enthalpy is 192 kJ/kg and the corresponding temperature 45.8 °C. These values can also be found on the $h-x$ chart where $x = 0$.

Practical plants

Whilst the system described earlier will work, its performance can be improved by the addition of a heat exchanger. With no heat exchanger there is a stream of cold, lean fluid which is heated when it reaches the generator and another hot, rich stream which must be cooled in the absorber. There is clearly a potential saving in both heating and cooling requirements by fitting the heat exchanger as shown in Fig. 7.56. Even the most basic practical system would always have the heat exchanger. The processes and state points for this practical cycle are as follows.

Water-only circuit

State 1:

The vapour which leaves the evaporator does not need to be superheated since liquid carry-over into the absorber would not cause problems. The vapour at

Figure 7.56 Practical vapour-absorption system

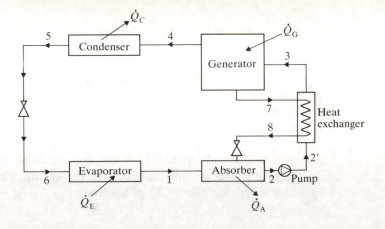

this point is therefore assumed to be saturated and the specific enthalpy is the appropriate h_g value from property tables such as Ref. 7.3. The temperature will be no lower than 3 °C.

Process 4–5 in the condenser:

This is the same as for a simple vapour-compression system—de-superheating followed by isothermal condensation and liquid sub-cooling, all at constant pressure. The heat rejection rate, $\dot{Q}_C$, can be found by modifying Eqn [7.4] as follows:

$$\dot{Q}_C = \dot{m}(h_4 - h_5)$$

The condensing temperature is fixed by the cooling medium, which is invariably water, and it determines the condensing pressure which becomes the pressure for all the high-pressure part of the plant. The condensing temperature must always be above that of the cooling water so that typical values are between 30 °C and 50 °C, with corresponding saturation pressures between 12.33 and 4.24 kN/m².

State 4:

The vapour at this point is highly superheated. The pressure is fixed by the condensing temperature as above, and the temperature is fixed by the temperature in the generator. It is therefore necessary to interpolate in the superheat section of the property tables to establish a specific enthalpy value or to assume that the enthalpy of a superheated vapour is the same as that of a saturated vapour at the same temperature.

State 5:

As with any condenser, some sub-cooling will take place and the temperature will be below the condensing temperature. It is sufficiently accurate to take the specific enthalpy value to be the same as the specific enthalpy of saturated waer, h_f, at the same temperature. If no other information is available, state 5 can be taken to be saturated.

Process 5–6 (expansion):

A typical throttling or expansion process is described in detail in the section on vapour-compression systems. In many designs there may be sufficient resistance to flow in the pipes and fittings so that an expansion valve may not be required.

State 6:

This is a mixture of saturated liquid and saturated vapour which has the same specific enthalpy value as the liquid at state 5. The proportion of vapour is much less than in vapour-compression systems because the high enthalpy of vaporization of water means that less needs to evaporate to produce a given fall in temperature.

Process 6–1 in the evaporator:

The evaporating temperature must not fall below about 3 °C to prevent freezing. Typical values are therefore between 3 °C and 8 °C. This fixes evaporating pressures between 0.76 and 1.07 kN/m², and the evaporating pressure becomes the pressure for the low-pressure part of the system. The capacity $\dot{Q}_C$ can be found by modifying Eqn [7.3] as follows:

$$\dot{Q}_E = \dot{m}(h_1 - h_6)$$

LiBr/water circuit

Processes in the absorber:

The absorber receives both the vapour from the evaporator at State 1 and the lean solution from the generator: this has been cooled down in the heat exchanger to state 8. After the vapour has been absorbed in the liquid it leaves as a rich solution at State 2. The absorber should be kept at as low a temperature as possible and the cooling water is passed through the absorber first and then the condenser.

An energy balance on the absorber gives:

$$\dot{Q}_A + \dot{m}_2 h_2 = \dot{m}_1 h_1 + \dot{m}_8 h_8 \qquad [7.28]$$

There are three mass balances which can be carried out on the absorber. If $\dot{m}$ denotes a total mass flow rate and x is the concentration of LiBr in solution the equations are:

For the total flows:
$$\dot{m}_1 + \dot{m}_8 = \dot{m}_2 \qquad [7.29]$$

For LiBr only:
$$0 + \dot{m}_8 x_8 = \dot{m}_2 x_2 \qquad [7.30]$$

For water only:
$$\dot{m}_1 + \dot{m}_8(1 - x_8) = \dot{m}_2(1 - x_2)$$

Since the last equation can be formed by subtracting Eqn [7.30] from Eqn [7.29] it is not an independent equation and is therefore not required.

State 2:

The rich solution leaving the absorber is normally considered to be saturated. Its specific enthalpy value can therefore be found on an enthalpy–concentration, $h–x$ diagram by fixing the state point from the temperature and saturation pressure. The concentration, x_2, can also be read from the chart.

State 2′:

After passing through the pump the rich mixture is at the maximum pressure but because the work input in the pump is so small the enthalpy and temperature are almost the same as at state 2. However, the mixture is no longer saturated.

Processes in the heat exchanger:

The pressure drops in the heat exchanger do not represent a significant thermodynamic penalty which offsets some of the gain in plant performance. The pressure drop for the rich stream can be allowed for by an increase in the pump work which is negligible compared with the other energy flows. The pressure drop across the lean stream makes the pressure drop across the expansion valve smaller and may well allow the valve to be omitted.

The energy balance for the heat exchanger is as follows:

$$\dot{m}_3(h_3 - h_2) = \dot{m}_7(h_7 - h_8) \qquad\qquad [7.31]$$

State 3:

Ideally the heat exchanger should heat the rich stream until it is just saturated again at the system high pressure. During this process the concentration x remains constant. If the mixture is assumed to be saturated, the specific enthalpy can be found from the h–x chart by identifying point 3 from the saturation pressure line corresponding to the system high pressure and the concentration, x_3, which must be the same as x_2. In practice the temperature would be a few degrees below saturation to prevent flashing, in which case the temperature must be known to identify the state.

State 7:

The lean mixture leaving the generator must be saturated. Its state point can therefore be fixed on the h–x chart from the generator temperature and the saturation pressure, which is the system high pressure. The concentration as well as the specific enthalpy can be read from the chart.

State 8:

The lean stream at point 8 is sub-cooled. The concentration is the same as at point 7 but without the temperature the state point can not be fixed on the h–x chart. The only way to establish point 8 is to calculate the specific enthalpy from Eqn [7.31].

Processes in the generator:

The generator drives off vapour from the rich stream, which is passed to the condenser, and returns the lean stream to the absorber via the heat exchanger. The operating temperature depends on the heating medium, which is usually hot water or low-pressure steam. This gives operating temperatures in the range 80–120 °C. An energy balance gives the following equation:

$$\dot{Q}_G + \dot{m}_3 h_3 = \dot{m}_4 h_4 + \dot{m}_7 h_7 \qquad\qquad [7.32]$$

Mass balances could be carried out on the generator instead of the absorber: both result in the same equations.

First Law analysis

If the First Law is applied to the system, neglecting the small power input from the pump, the following equation results:

$$\dot{Q}_G + \dot{Q}_E = \dot{Q}_C + \dot{Q}_A \qquad\qquad [7.33]$$

Coefficient of performance

Equation [7.2] can still be applied:

$$\text{Performance} = \frac{\text{Required output}}{\text{Costed input}}$$

As a refrigerator:

$$COP_{ref} = \frac{\dot{Q}_E}{\dot{Q}_G}$$

and as a heat pump:

$$COP_{HP} = \frac{\dot{Q}_C}{\dot{Q}_G}$$

For a basic plant the values are low—normally less than 1. This is much lower than typical values for vapour-compression systems. However, COPs of vapour-absorption systems should not be compared with the COPs of vapour-compression systems without taking into account the fact that the input is heat, not work. If the COPs of vapour-compression systems are multiplied by the power station thermal efficiencies or the thermal efficiency of the engines which provide the mechanical drive, the figures are much closer to those of vapour-absorption systems.

Example 7.16
A water chilling plant uses a lithium bromide absorption system in which the generator is maintained at 75 °C using a hot-water coil. The evaporator and condenser saturation temperatures are 8 °C and 36 °C respectively, and the refrigeration capacity is 50 kW.

It may be assumed that there is no undercooling in the condenser and that the water vapour leaving the evaporator is dry saturated. Assuming that there is a heat exchanger between the absorber and generator, sketch the plant, and calculate:
(i)   the heat supplied to the generator;
(ii) the heat rejected in the condenser;
(iii) the heat rejected in the absorber.
For lithium bromide/water solutions:

	Mass concentration, LiBr/mixture	Enthalpy (kJ/kg)
Weak solution leaving generator	0.60	−90.7
Strong solution leaving absorber	0.51	−172.1
Strong solution entering generator	0.51	−140.7

Take the enthalpy of superheated steam as approximately equal to the saturated value at the same temperature.

Neglect pressure drops and all heat losses.

(CIBSE)

Solution
A sketch of the plant is shown in Fig. 7.56, (page 367). Only water passes through the condenser and evaporator, so that enthalpy values can be found from property tables. Enthalpy values elsewhere for solutions of water in lithium bromide are given in the question.

Condenser:

For a condensing temperature of 36 °C the corresponding saturation pressure is 0.0594 bar. This is the pressure at points 3, 4, 5 and 7.

$$h_5 = h_f \text{ at } 36\,°C = 150.7 \text{ kJ/kg} = h_6$$
$$h_4 \simeq h_g \text{ at } 75\,°C = 2634.7 \text{ kJ/kg}$$

Evaporator:

For an evaporating temperature of 8 °C the saturation pressure is 0.010 72 bar. This is the pressure at points 6, 1 and 2.

$$h_1 = h_g \text{ at } 8\,°C = 2515.5 \text{ kJ/kg}$$

Using Eqn [7.3]:

$$\dot{Q}_E = \dot{m}_1(h_1 - h_6)$$
$$50 = \dot{m}_1(2515.5 - 150.7)$$
$$\therefore \quad \dot{m}_1 = 0.021\,14 \text{ kg/s}$$

Absorber:

Using Eqns [7.29] and [7.30]:

Total mass balance: $\dot{m}_1 + \dot{m}_8 = \dot{m}_2$
$$0.021\,14 + \dot{m}_8 = \dot{m}_2 \qquad [1]$$
Li-Br mass balance: $x_1\dot{m}_1 + x_8\dot{m}_8 = x_2\dot{m}_2$
x_1 is 0 and $x_8 = x_7$ therefore: $0.6\dot{m}_8 = 0.51\dot{m}_2 \qquad [2]$

Solving Eqns [1] and [2] gives $\dot{m}_8 = 0.1198 \text{ kg/s}$ and $\dot{m}_2 = 0.1409 \text{ kg/s}$.

Generator:

Using Eqn [7.32] and knowing that $\dot{m}_4 = \dot{m}_1$ and $\dot{m}_7 = \dot{m}_8$ gives:

$$\dot{Q}_G + \dot{m}_3 h_3 = \dot{m}_4 h_4 + \dot{m}_7 h_7$$
$$\dot{Q}_G + (0.1409 \times -140.7) = (0.021\,14 \times 2634.7)$$
$$+ (0.1198 \times -90.7)$$
$$\dot{Q}_G - 19.82 = 55.7 - 10.87$$
$$\therefore \quad \dot{Q}_G = 64.66 \text{ kW}$$

Condenser:

Now that the mass flow rates are known, Eqn [7.4] can be applied to the condenser as follows:

$$\dot{Q}_C = \dot{m}_4(h_4 - h_5) = 0.021\,14(2634.7 - 150.7) = 52.5 \text{ kW}$$

For the complete system:

Applying the first law in the form of equation 7.33 gives:

$$\dot{Q}_A + \dot{Q}_C = \dot{Q}_G + \dot{Q}_E$$
$$\dot{Q}_A + 52.5 = 64.66 + 50$$
$$\therefore \quad \dot{Q}_A = 62.16 \text{ kW}$$

An alternative method of calculating $\dot{Q}_A$ would have been to carry out an energy balance on the heat exchanger to establish h_8 and then another energy balance on the absorber. Using the First Law reduces the calculations required.

Example 7.17

A 500 KW lithium bromide/water refrigeration plant operates between pressure limits of 10 kN/m² and 1.0 kN/m². The generator is supplied with saturated steam at 2 bar which leaves sub-cooled by 10 K, producing a generator temperature of 100 °C.

The H_2O is saturated leaving the condenser and evaporator. The rich mixture is saturated leaving the absorber and entering the generator and the weak mixture is saturated leaving the generator. If the absorber operates at 40 °C, calculate:

(i)   the total rate of heat rejection;
(ii) the mass flow rate of steam to the generator;
(iii) the specific enthalpy of the lean mixture entering the absorber.

Use the $h–x$ chart for $LiBr/H_2O$ provided.

(Wolverhampton Polytechnic)

Solution
A sketch of the system is shown in Fig. 7.56, (page 367).
    From tables:

$$h_1 = 2514 \text{ kJ/kg (saturated vapour at 0.01 bar)}$$
$$h_5 = h_6 = 192 \text{ kJ/kg (saturated liquid at 0.1 bar)}$$
$$h_4 = 2688 \text{ kJ/kg (superheated vapour at 100 °C and 0.1 bar)}$$

From the enthalpy–concentration chart:

$$h_2 = -170 \text{ kJ/kg (saturated solution at 1 kN/m}^2 \text{ and 40 °C)}$$
$$x_2 = x_3 = 0.56$$
$$h_3 = -75 \text{ kJ/kg (saturated solution at 10 kN/m}^2 \text{ and } x = 0.56)$$
$$h_7 = -50 \text{ kJ/kg (saturated solution at 10 kN/m}^2 \text{ and 100 °C)}$$
$$x_7 = x_8 = 0.645$$

Evaporator:
The mass flow rate through the evaporator can be found using Eqn [7.3]:

$$\dot{Q}_E = \dot{m}_1(h_1 - h_6)$$
$$500 = \dot{m}_1(2514 - 192)$$
$$\therefore \qquad \dot{m}_1 = 0.2153 \text{ kg/s}$$

Absorber:
Using Eqns [7.29] and [7.30]:

Total mass balance
$$\dot{m}_1 + \dot{m}_8 = \dot{m}_2$$
$$0.2153 + \dot{m}_8 = \dot{m}_2 \qquad\qquad [1]$$

LiBr mass balance
$$x_1\dot{m}_1 + x_8\dot{m}_8 = x_2\dot{m}_2$$
$$0 + 0.645\dot{m}_8 = 0.56\dot{m}_2 \qquad\qquad [2]$$

Solving equations [1] and [2] gives $\dot{m}_2 = 1.634 \text{ kg/s}$

and $\dot{m}_8 = 1.419 \text{ kg/s}$

Generator:
Using Eqn [7.32]:

$$\dot{Q}_G + \dot{m}_3 h_3 = \dot{m}_4 h_4 + \dot{m}_7 h_7$$
$$\dot{Q}_G + (1.634 \times -75) = (0.2153 \times 2688) + (1.419 \times -50)$$
$$\therefore \qquad \dot{Q}_G = 630.3 \text{ kW}$$

The heat rejection rate can be found by applying the First Law (Eqn [7.33]) as follows:

$$\dot{Q}_A + \dot{Q}_C = \dot{Q}_G + \dot{Q}_E = 630.3 + 500 = 1130.3 \text{ kW}$$

From property tables the steam has an enthalpy of 2707 kJ/kg at inlet and the condensate an enthalpy of 461.9 kJ/kg at outlet.

$$\dot{m}_s = \frac{\dot{Q}_G}{(h_i - h_o)} = \frac{630.3}{(2707 - 461.9)} = 0.281 \text{ kg/s}$$

Heat exchanger:
Using Eqn [7.31]:

$$\dot{m}_2(h_3 - h_2) = \dot{m}_8(h_7 - h_8)$$
$$1.634(-75 + 170) = 1.419(-50 - h_8)$$
$$h_8 = -159.4 \text{ kJ/kg}$$

The processes on the relevant property diagrams are shown on Fig. 7.57(a) and (b). The p–h diagram in Fig. 7.57(a) shows the pure-water part of the cycle.

Figure 7.57 Processes for Example 7.17

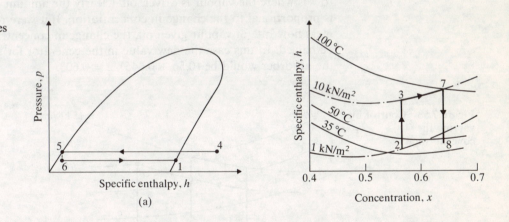

(a)

The h–x diagram shows the processes in the heat exchanger and the generator. In the heat exchanger the rich solution is heated from 2 to 3 while the lean solution is cooled from 7 to 8. The temperature of the rich stream leaving the exchanger (point 3) is about 85 °C. This temperature is increased during the constant-pressure heating in the generator to 100 °C at point 7. It should be noted that the lean solution leaving the heat exchanger (point 8) is below the 1 kN/m² saturation pressure line. However, this is not a saturation state and the actual pressure will be between 10 kN/m² and 1 kN/m², depending on the pressure drops through the heat exchanger and pipes.

The point (8) where the lean solution leaves the heat exchanger is a critical state in the design of a vapour-absorption system because it is the point of minimum temperature from the stream with the highest concentration. It is therefore always the closest point to the crystallization line, where solidification of the solution could cause blockages.

Capacity Control of Absorption Systems

To vary the capacity it is necessary to vary the mass flow rate of water vapour supplied to the condenser and evaporator. There are two fundamental methods of achieving this.

(a) Vary the mass flow rate of lithium bromide

If the mass flow rate of rich solution delivered to the generator by the pump is varied, this will change the mass flow rate around the whole system. Providing the generator temperature is constant, its load will drop in proportion to the change in mass flow rate, and so will the mass flow rate of vapour driven off.

(b) Vary the concentration of lithium bromide

If the mass flow rate through the pump is constant, the mass flow rate of vapour driven off in the generator can still be varied by changing the concentration. Figure 7.58 shows how this is achieved on an $h-x$ diagram. Points 2, 3, 7 and 8 have been taken from Example 7.17. The constant-pressure process from 3 to 7 is where the vapour is driven off. Clearly the amount of vapour driven off is proportional to the change in concentration. If it were required to halve the mass flow rate of vapour given off, the change in concentration would need to be halved. In this case the new value in the generator for the same value from the absorber would be $(0.56 + 0.645)/2 = 0.603$.

Figure 7.58 Control by varying the concentration

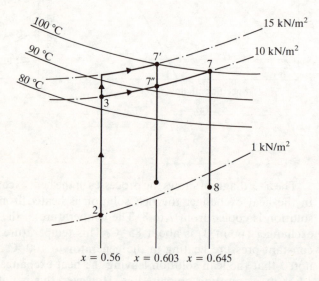

$x = 0.56 \quad x = 0.603 \quad x = 0.645$

There are two ways of achieving this new concentration in the generator. One method is to reduce temperature in the generator by throttling the heating medium. In this case, if the generator temperature is reduced to about 93 °C, the lean solution will leave the generator at 7″ with the required concentration. The other method is to maintain a constant temperature and increase the pressure in the generator. This can only be done by increasing the pressure in the condenser, which means throttling the cooling water. In this case the pressure would need to rise to about 15 kN/m^2 and the lean solution would leave the generator at 7′ with the required concentration. The process from 3 to 7′ would be in the generator at constant pressure. The first part of the process at constant concentration is sensible heating until the new saturation temperature of about 92 °C is reached. Some plants may use a combination of both methods of control.

Aqua/ammonia Systems

In an aqua/ammonia system, the refrigerant is ammonia and the absorbent or carrier is water. With ammonia as the refrigerant it is possible to operate at temperatures below 0 °C. The main problem with this system is that both the ammonia and the water are volatile. It is therefore never possible to separate them completely. To minimize the amount of water vapour which leaves the generator with the refrigerant, a rectifying column and de-phlegmator are fitted which add to the cost and complexity of the plant.

The system analysis is also more complex than for the lithium bromide/water system because of the need to deal with two vapours in the condenser and evaporator. A detailed analysis of binary mixtures and aqua/ammonia plants can be found in Ref. 7.8.

Domestic Gas Refrigerator

This system was first designed by Platen-Munters and has for many years been manufactured under the trade name 'Electrolux'. The term 'gas refrigerator' is now generally used because propane or butane gas is the source of heat for most units. However, like all absorption systems, any source of heat could be used. Unlike other systems the gas refrigerator needs no power input—hence its popularity in situations where there is no guarantee of mains electricity being available (e.g. in caravans).

The essential features of a gas refrigerator are shown in Fig. 7.59. The system is a development of the aqua/ammonia system and ammonia remains the refrigerant and water the absorbent. As far as the refrigerant is concerned, the condenser and generator are operating at high pressure and the evaporator and absorber at low pressure, as in the lithium bromide/water system. However, since there are no physical boundaries preventing fluid flow in any direction, the whole system must be under approximately the same pressure. The difference between the partial pressure of the ammonia in the 'low-pressure' part of the system and system total pressure is made up by introducing a third fluid. The third fluid is hydrogen gas, which acts as an inert medium in the evaporator and absorber. Liquid seals are used to confine the hydrogen to one section of the unit.

The condenser contains almost pure ammonia with a partial pressure equal to the system total pressure. When the liquid ammonia leaves the condenser it is piped directly to the evaporator. On entering the evaporator it undergoes an expansion ratio of about 7:1 because there are six molecules of hydrogen for each molecule of ammonia vapour at this point in the system. This expansion has the same effect as throttling with a valve and produces low-temperature

Figure 7.59 Domestic
gas refrigerator

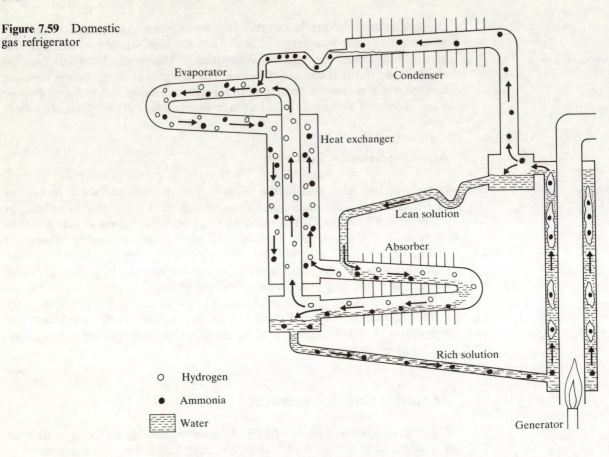

liquid with some flash vapour. The hydrogen and ammonia vapour which leave
the evaporator pass through a heat exchanger where the temperature of the
hydrogen returning to the evaporator is reduced.

From the heat exchanger the ammonia vapour and hydrogen pass into the
absorber where they meet a stream of lean solution. The ammonia is absorbed
in the liquid, which becomes a rich solution which then passes to the generator.
As the ammonia is absorbed, the partial pressure of the hydrogen increases
until it reaches its maximum value as it returns to the evaporator.

The strong solution is heated in the generator which produces the ammonia
vapour, which then passes to the condenser, and a lean liquid which returns to
the absorber.

The only forces inducing fluid flow are those resulting from the slight
difference in hydraulic head around the system. With such small pressure
differences the alignment of these units is critical and they will not work unless
they are reasonably level. The COP of these units is low, even compared with
other absorption systems. This is not too important for domestic units where
the heat input is quite small but it inhibits their development for more substantial
applications.

7.11 STEAM-JET REFRIGERATION

In this system the refrigerant is again water; this limits it to producing chilled
water for applications such as air conditioning. A unique feature of the system

is that the chilled water circulated to the air conditioning loads is also the refrigerant.

The essential items of plant are shown in Fig. 7.60. The pressurized chilled return water at point 6 is expanded down to a very low pressure which produces flash vapour and cooled water. The cooled water becomes the chilled water supply at point 7 and the vapour must be removed from the top of the evaporator at point 3. This expansion and subsequent separation are shown on a p–h plane in Fig. 7.61.

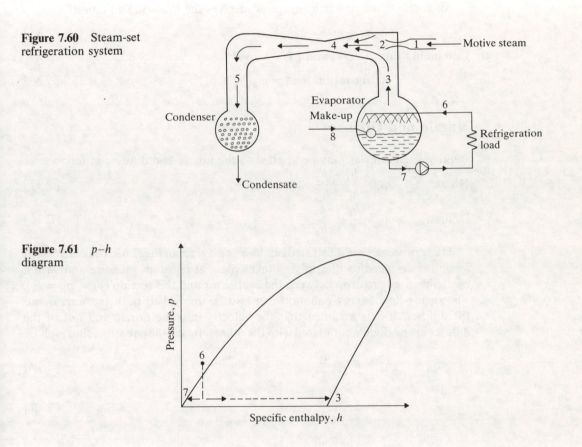

Figure 7.60 Steam-set refrigeration system

Figure 7.61 p–h diagram

The flash vapour produced represents a very small proportion of the chilled water by mass but, because of the very high specific volume (157.3 m³/kg at 4 °C), there is a considerable volume flow rate at point 3. A steam-jet compressor is used to extract the vapour from the evaporator. This consists of a convergent–divergent nozzle which is supplied with medium-pressure steam (termed 'motive steam') and a venturi. The nozzle expands the motive steam to a very high velocity (supersonic) which produces a region of pressure even lower than in the evaporator. The vapour from the evaporator is entrained in the high-velocity, low-pressure jet stream and the mixed stream passes into the diffuser section of the venturi. In the diffuser the velocity falls and the pressure increases before the mixed stream enters a condenser. The condenser operators well below ambient pressure and must have its own steam-jet compressor (air ejector).

Evaporator

The refrigeration load can be found from an energy balance on the chilled water as follows:

$$\text{Refrigeration load} = \dot{m}_7(h_6 - h_7) \qquad [7.34]$$

The mass of vapour lost from the condenser must be continuously replaced; therefore the mass flow rate of make-up, $\dot{m}_8$, must be equal to the mass flow rate of vapour, $\dot{m}_3$.

An energy balance on the evaporator gives the following equation:

$$\dot{m}_3 h_3 + \dot{m}_7 h_7 = \dot{m}_6 h_6 + \dot{m}_8 h_8 \qquad [7.35]$$

Combining Eqns [7.34] and [7.35] gives:

$$\text{Refrigeration load} = \dot{m}_3(h_3 - h_8) \qquad [7.36]$$

Nozzle and Diffuser

Isentropic efficiencies may be applied to the nozzle and diffuser as follows:

Nozzle: $\eta_N = \dfrac{h_1 - h_2}{h_1 - h_{2s}}$ $\qquad [7.37]$

Diffuser: $\eta_D = \dfrac{h_{5s} - h_4}{h_5 - h_4}$ $\qquad [7.38]$

These processes are illustrated on the h–s diagram in Fig. 7.62. The convergent section is designed so that mixing takes place at constant pressure and with a negligible pressure drop between the evapoator and this section ($p_2 = p_3 = p_4$). The steady-flow energy equation can also be applied to both the nozzle and the diffuser. If it is assumed that the velocity into the nozzle and out of the diffuser are negligible compared with the other terms in the equation, the result is:

$$h_1 = h_2 + \frac{u_2^2}{2} \qquad [7.39]$$

$$h_5 = h_4 + \frac{u_4^2}{2} \qquad [7.40]$$

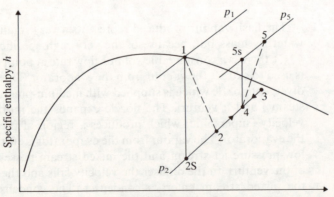

Figure 7.62 h–s diagram

An overall energy balance on the nozzle and diffuser gives the following equation:

$$\dot{m}_1 h_1 + \dot{m}_3 h_3 = (\dot{m}_3 + \dot{m}_1) h_5 \qquad [7.41]$$

Lastly, an entrainment efficiency can be defined as the ratio of the kinetic energy of the mixed stream after entrainment to the kinetic energy of the motive steam before entrainment.

$$\text{Entrainment efficiency} = \frac{\text{KE}_4}{\text{KE}_2} = \frac{(\dot{m}_1 + \dot{m}_3) u_4^2}{\dot{m}_1 u_2^2} \qquad [7.42]$$

The production of dry ice (solid carbon dioxide) utilizes the same principles as steam-jet refrigeration.

Example 7.18
(a) Briefly discuss circumstances which might make it economic to install a steam-jet refrigeration plant for producing chilled water.
(b) A steam-jet refrigeration plant has a capacity of 400 kW when producing chilled water at 7 °C from return water at 12 °C. A supply of saturated steam at 10 bar is available and the condenser operates at 0.1 bar. The make-up water is under-cooled condensate from the condenser at 30 °C.
   If the vapour leaving the evaporator has a dryness fraction of 0.96, calculate:
(i)   the mass flow rate of chilled water;
(ii) the capacity of the steam-jet compressor (the volume flow rate of vapour leaving the evaporator);
(iii) the velocity of the steam leaving the nozzle if the isentropic efficiency is 90%;
(iv) the mass flow rate of motive steam if the entrainment efficiency is 65% and the diffuser isentropic efficiency 85%.

Solution
(a) A steam-jet refrigeration plant is relatively cheap to install, provided the cost of the steam plant is excluded. The running costs depend on the cost of steam and cooling water (for the condenser). They are therefore only viable if a steam plant is available with surplus capacity. This can happen when steam plant is used for heating in winter but must be run with a small base load throughout the year.

(b) The system is illustrated in Fig. 7.60 and Fig. 7.62 represents the processes on an h–s plane. The following data have been taken from property tables.
   With a chilled water supply temperature of 7 °C, the corresponding saturation pressure is 0.01 bar. This is p_3; p_2 and p_4 have the same value.

$$h_6 = 50.4 \text{ kJ/kg}, \quad h_7 = 29 \text{ kJ/kg}, \quad h_8 = 125.7 \text{ kJ/kg},$$
$$h_1 = 2778 \text{ kJ/kg}, \quad s_1 = 6.586 \text{ kJ/kg K}, \quad v_{g3} = 129.2 \text{ m}^3/\text{kg},$$
$$s_{f2} = 0.106 \text{ kJ/kg K}, \quad s_{fg2} = 8.868 \text{ kJ/kg K},$$
$$h_3 = h_f + x h_{fg} = 29 + 0.96 \times 2485 = 2414.6 \text{ kJ/kg}$$

(i) The mass flow rate of chilled water is found from Eqn [7.34] as follows:

$$\text{Refrigeration load} = \dot{m}_7 (h_6 - h_7)$$
$$400 = \dot{m}_7 (50.4 - 29)$$
$$\therefore \qquad \dot{m}_7 = 18.7 \text{ kg/s}$$

(ii)   Using Eqn [7.36]:

$$\text{Refrigeration load} = \dot{m}_3(h_3 - h_8)$$
$$400 = \dot{m}_3(2414.6 - 125.7)$$
$$\therefore \quad \dot{m}_3 = 0.175 \text{ kg/s}$$

Assuming the entrained water occupies a negligible volume:

$$\dot{V}_3 = \dot{m}_3 v_3 = \dot{m}_3 x_3 v_{g3} = 0.175 \times 0.96 \times 129.2 = 21.7 \text{ m}^3/\text{s}$$

The capacity of the steam-jet compressor is therefore 21.7 m³/s.

(iii)   The state of the motive steam leaving the nozzle must be found using the values from property tables because point 2s is outside the range of most h–s charts.

$$s_1 = s_{2s} = 6.586 = 0.106 + x_{2s} \times 8.868$$

Hence $$x_{2s} = 0.731$$

$$h_{2s} = h_f + h_{fg} = 29 + 0.721 \times 2485 = 1844.6 \text{ kJ/kg}$$

Using Eqn [7.37]:

$$\text{Nozzle } \eta_N = \frac{h_1 - h_2}{h_1 - h_{2s}} = 0.9 = \frac{2778 - h_2}{2778 - 1844.6}$$

Hence $$h_2 = 1937.9 \text{ kJ/kg}$$

Using Eqn [7.39]:

$$h_1 = h_2 + \frac{u_2^2}{2} = 2778 = 1937.9 + \frac{u_2^2}{2 \times 10^3}$$

hence $u_2 = 1296$ m/s, which is the velocity of the steam leaving the nozzle.

(iv)   Using Eqn [7.41]:

$$\dot{m}_1 h_1 + \dot{m}_3 h_3 = (\dot{m}_3 + \dot{m}_1)h_5$$
$$\dot{m}_1 \times 2778 + 0.175 \times 2414.6 = (0.175 + \dot{m}_1)h_5$$

Rearranging,

$$\frac{0.175}{\dot{m}_1} = \frac{2778 - h_5}{h_5 - 2414.6} \qquad [1]$$

Using Eqn [7.40]:

$$h_5 = h_4 + \frac{u_4^2}{2}$$

With specific enthalpy values in kJ/kg, rearranging gives:

$$u_4^2 = (h_5 - h_4)2000 \qquad [2]$$

Using Eqn [7.42],

$$\text{Entrainment efficiency} = \frac{KE_4}{KE_2} = \frac{(\dot{m}_1 + \dot{m}_3)u_4^2}{\dot{m}_1 u_2^2}$$

$$0.65 = \left(1 + \frac{0.175}{\dot{m}_1}\right)\frac{u_4^2}{u_2^2} \qquad [3]$$

Substituting Eqns [1] and [2] in Eqn [3] gives:

$$0.65 = \left(1 + \frac{2778 - h_5}{h_5 - 2414.6}\right) \frac{(h_5 - h_4)}{1296^2} 2000$$

This equation simplifies to:

$$h_5 = 7224.6 - \frac{h_4}{0.502} \qquad [4]$$

Equation [7.38] for the diffuser isentropic efficiency also constrains the values of h_4 and h_5, i.e.

$$\eta_D = \frac{h_{5s} - h_4}{h_5 - h_4} = 0.85$$

A trial and error procedure is now necessary using Eqns [4] and [7.38]. Since state 4 is a result of mixing States 2 and 3 it must lie between them, i.e. $2414.6 > h_4 > 1937.9$.
Try $h_4 = 2300$ kJ/kg:
From an h–s chart h_{5s} is 2600 kJ/kg. From Eqn [7.38],

$$0.85 = \frac{2600 - 2300}{h_5 - 2300}$$

Hence $\qquad h_5 = 2653$ kJ/kg

From Eqn [4],

$$h_5 = 7224.6 - \frac{h_4}{0.502} = 7224.6 - \frac{2300}{0.502} = 2642 \text{ kJ/kg}$$

Try $h_4 = 2295$ kJ/kg:
From an h–s chart h_{5s} is 2597 kJ/kg. From Eqn [7.38],

$$0.85 = \frac{2597 - 2295}{h_5 - 2295}$$

Hence $\qquad h_5 = 2650.3$ kJ/kg

From Eqn [4],

$$h_5 = 7224.6 - \frac{h_4}{0.502} = 7224.6 - \frac{2295}{0.502} = 2652.9 \text{ kJ/kg}$$

This is a sufficiently accurate agreement when enthalpy values are taken from a chart.
The mass flow rate of motive steam can now be calculated from Eqn [1]:

$$\frac{0.175}{\dot{m}_1} = \frac{2778 - h_5}{h_5 - 2414.6} = \frac{2778 - 2650.3}{2650.3 - 2414.6}$$

from which the mass flow rate of motive steam, $\dot{m}_1 = 0.323$ kg/s.

7.12 GAS REFRIGERATION CYCLES

A gas can be used as a refrigerant in a cycle which is basically similar to a vapour-compression cycle. The differences are that the constant pressure heat exchange processes are not isothermal and the expansion takes place in a turbine.

The refrigerant used is air and an air system known as the Bell–Coleman refrigerator was once used in buildings. The only modern use of air refrigeration is in the air conditioning of aircraft spaces. Air is bled from the engine compressor and cooled before being expanded in a turbine and delivered to the air-conditioned spaces. An analysis of gas refrigeration systems can be found in Ref. 7.2.

PROBLEMS

7.1   A basic single-stage compression refrigeration system using R12 is known to be operating under the following conditions. The compressor discharge temperature is 85 °C and the compression is isentropic; the evaporator pressure is 2.0 bar; the thermostatic expansion valve is operating with 12.5 K superheat and the expansion is adiabatic; the refrigerant at entry to the thermostatic expansion valve is at saturation; pressure changes in the system are negligible everywhere except at the compressor and expansion valve.

(a)   For the conditions and assumptions stated, plot the cycle on the pressure–enthalpy diagram, label the cycle main processes and calculate the following:
(i)   the specific refrigeration effect;
(ii)   the specific heat of compression;
(iii)   the specific heat of rejection;
(iv)   the coefficient of performance.

(b)   Explain what the effect would be on the cycle if there were a sudden increase in the load on the evaporator. You may assume that the condensing temperature remains constant through external control of its cooling medium.

(University of Liverpool)

(92 kJ/kg; 40 kJ/kg; 132 kJ/kg; 2.3)

7.2   Discuss the effects of condensing and evaporating temperatures on refrigeration system capacity. As part of your discussion, mention the effects of sub-cooling and superheating. (Not more than 300 words.)

A refrigeration plant was designed to cope with a load of 25 kW, with an evaporating temperature of −5 °C and a condensing temperature of 30 °C. It was necessary to change the working conditions to −20 °C, with the same load and condensing temperature. Determine the required increase in volume flow rate if the system uses R12. For simplicity, you may assume saturation conditions at the condenser outlet and the compressor inlet.

(Dublin Institute of Technology)

(0.0104 m^3/s at the compressor inlet)

7.3   A flash freezer unit uses a Refrigerant 22 single-stage vapour-compression plant. Saturated vapour leaves the evaporator and the liquid leaving the condenser is sub-cooled by 5 K. The compressor has an isentropic efficiency of 85% and is supplied with 50 kW of power.

The condenser is water-cooled with the water entering at 19 °C and leaving at 29 °C. The minimum temperature difference for heat exchange in the condenser is 3 K.

The air for the flash freezing is blown across the evaporator coils, entering at 250 K and leaving at 240 K. The minimum temperature difference for heat exchange in the evaporator is 8 K. Calculate:

(i)   the plant coefficient of performance;
(ii) the mass flow rate of cooling water;
(iii) the volume flow rate of air supplied for flash freezing.

(Harper Adams Agricultural College)

(1.8; 3.3 kg/s; 6 m³/s)

7.4 A single-stage Refrigerant 12 vapour-compression refrigeration plant has a condensing temperature of 40 °C and an evaporating temperature of −20 °C. The vapour leaving the evaporator is superheated by 15 K and the compressor has an isentropic efficiency of 85%. Saturated liquid leaves the condenser and then enters a sub-cooler where it is cooled to 15 °C before being throttled to the evaporator pressure. Sketch the plant layout and use the *p–h* diagram to calculate:

(i)   the coefficient of performance;
(ii) the mass flow rate of refrigerant if the capacity is 500 kW;
(iii) the mass flow rate of cooling water to the sub-cooler when the temperature rise of the cooling water is 2 °C;
(iv) the coefficient of performance without the sub-cooler.

(Wolverhampton Polytechnic)

(3.3; 3.6 kg/s; 10.6 kg/s; 2.7)

7.5 A milk processing unit must cool 1000 litres of milk from 37 °C to 5 °C in two hours twice a day.

A Refrigerant 12 plant is used with an evaporating temperature of 265 K and a condensing temperature of 315 K. The liquid leaving the condenser is sub-cooled by 5 K and the vapour leaving the evaporator is superheated by 5 K. The compressor has an isentropic efficiency of 80% and the electric motor an efficiency of 85%.
Calculate:

(i)   the plant coefficient of performance;
(ii) the average electrical power input if no ice bank is used;
(iii) the average electrical power input and the mass of ice formed if an ice bank is built up for six hours before the milk is added.

It may be assumed that milk has the same physical properties as water and that the ice is formed from water at 0.01 °C.

(Harper Adams Agricultural College)

(3.1; 6.1 kW; 1.52 kW, 301 kg)

7.6 (a) Compare the COP of a refrigeration cycle which uses wet compression with one that uses dry compression. In both cases R22 is the refrigerant, the condensing temperature is 45 °C and the evaporating temperature is 0 °C. Assume that the compression is isentropic, that the liquid leaving the condenser is saturated and that for wet compression the refrigerant leaves the compressor as a saturated vapour. Assume also that saturated vapour enters the compressor for dry compression and that throttling occurs between compressor and evaporator. You may use the saturated and superheated property tables given below.

(a)   Properties of saturated vapour and liquid:

t/(°C)	p/(kPa)	Enthalpy (kJ/kg)		Entropy (kJ/kg K)		Specific volume (l/kg)	
		h_f	h_g	s_f	s_g	v_f	v_g
−1	481.57	198.828	404.994	0.99575	1.75326	0.77629	48.6517
0	497.59	200.000	405.361	1.00000	1.75279	0.77834	47.1354
1	514.01	201.174	405.724	1.00424	1.75034	0.78041	45.6757
44	1688.5	255.042	417.174	1.18315	1.69435	0.89828	13.6341
45	1729.0	256.396	417.308	1.18730	1.69305	0.90203	13.2841
46	1770.2	257.756	417.432	1.19145	1.69174	0.90586	12.9436

(b)   Properties of superheated vapour:

t/(°C)	v/(l/kg)	h/(kJ/kg)	s(kJ/kg K)
	Saturation temperature, 45 °C		
45	13.2841	417.308	1.6931
50	13.8136	422.241	1.7084
55	14.3154	427.025	1.7231
60	14.7946	431.693	1.7372
65	15.2550	436.268	1.7509
70	15.6995	440.769	1.7641
75	16.1303	445.209	1.7769

(b)   In the vapour compression cycle, a throttling device is used almost universally to reduce the pressure of the liquid refrigerant. Determine the percentage saving in net work of the cycle per kilogram of refrigerant if an expansion engine could usefully be used to expand saturated R22 isentropically from 45 °C to the evaporation temperature of 0 °C. Assume that compression is isentropic from saturated vapour at 0 °C to a condensing pressure equivalent to a condensing temperature of 45 °C.

(CIBSE)

(Wet 4.7, dry 4.73; 17%)

7.7   A two-stage vapour-compression refrigeration plant using Refrigerant 12 as the working fluid has operating pressures in the condenser, flash chamber and evaporator of 10 bar, 3.2 bar and 1 bar respectively. The liquid leaving the condenser is at 35 °C and in the flash chamber there is complete separation of the saturated vapour and saturated liquid. The vapour from the flash chamber is mixed with the discharge from the LP compressor before it enters the HP compressor. If the vapour leaving the evaporator is at −20 °C and the isentropic efficiency of each stage of compression is 80%, calculate:
(i)   the coefficient of performance;
(ii)   the mass flow rate of refrigerant through the condenser when the rate of heat rejection in the condenser is 50 kW;
(iii) the rate of heat extraction from the cold store.

(Wolverhampton Polytechnic)

(2.6; 0.31 kg/s; 35 kW)

7.8   Figure 7.63 shows a three-stage vapour-compression refrigeration system.

Figure 7.63   Problem
7.8

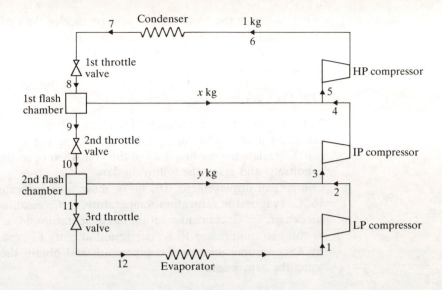

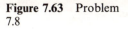
Figure 7.63   Problem 7.8

The working fluid is Refrigerant 12 and the saturation temperatures in the condenser and evaporator are 50 °C and −30 °C respectively. The vapour leaving the evaporator is superheated by 10 K and there is no under-cooling in the condenser. The compression ratios for each stage may be assumed to be equal. If compression is isentropic, calculate the coefficient of performance.

(Wolverhampton Polytechnic)

(2.6)

7.9   (a)   High compression ratios can lead to a serious drop in the volumetric efficiency and other problems with reciprocating compressors. Sketch and describe a multi-stage system that overcomes these problems.
(b)   An air-cooled condensing unit condenses Refrigerant 12 at 45 °C and then sub-cools the liquid by 10 K. The evaporating temperature is 5 °C and the vapour is superheated by a further 5 K before entering the compressor. If the compressor displacement is 575 m³/h, calculate the required flow rate of cooling air over the condensing coil.

Data
Air on to condenser, 20 °C; air off condenser, 30 °C; compressor volumetric efficiency, 82%; compressor isentropic efficiency, 70%; density of air at 20 °C, 1.2 kg/m³; specific heat capacity of air, 1.02 kJ/kg K.

(Dublin Institute of Technology)

(33.6 m³/s)

7.10   (a)   A vapour-compression refrigerating plant employing Refrigerant 12 operates at an evaporating temperature of −15 °C and a condensing temperature of 35 °C. The vapour entering the compressor is superheated by 20 K and the liquid leaving the condenser is saturated.
   The compressor, which is double-acting, has a bore of 220 mm and a stroke of 350 mm and runs at a rotational frequency of 300 rev/min. The volumetric efficiency is 0.8. Assuming the compression process to be isentropic,
(i)   Sketch the cycle on the *p–h* diagram provided.

(ii) Determine the coefficient of performance of the cycle.

(iii) Determine the refrigerating capacity in kW.

(b)   Explain why it is desirable that the vapour entering the compressor should be superheated.

(Dublin Institute of Technology)

(4.2; 134 kW)

7.11   A vapour-compression refrigeration unit operating with refrigerant R12 can be used alternately in a cooling or heating mode, dependent on demand.

(a)   Calculate the cooling duty of the unit when operating under the following conditions and given the following data:

Compressor displacement, 0.01 m³/s; mean air temperature across evaporator, 16 °C; evaporator saturation temperature, 8 °C; mean air temperature across condenser, 35 °C; condenser saturation temperature, 44 °C; refrigerant superheat at entry to compressor 10 K; refrigerant at entry to expansion valve subcooled by 4 K. Assume isentropic compression and obtain the volumetric efficiency using the expression,

$$\eta_v = 1 - 0.15\left[\frac{v_s}{v_d} - 1\right]$$

where v_s and v_d are the specific volumes at suction and discharge respectively. Use the R12 p–h chart attached.

(b)   Determine the heating capacity of the unit when operating under the following conditions.

Mean air temperature across evaporator   0 °C
Mean air temperature across condenser   28 °C

Note that the mass flow capacity of the compressor will be less than that calculated in (a). Use an iterative procedure to establish the evaporator and condenser saturation temperatures, assuming that the heat-transfer rate in each of these components is linearly proportional to the temperature difference between air and saturation values. Provide only one step of iteration and assume initially that the evaporator and condenser duties are reduced to 5/8 of their summer values. State how you would proceed to a closer approximation of equilibrium running.

(Engineering Council)

(21.3 kW; 15.4 kW; new $\dot{m}$, 0.104 kg/s)

7.12   (a)   With the aid of appropriate sketches, explain how the evaporating temperature is determined for a simple plant with a capillary tube and a reciprocating compressor. Also explain how such systems respond to sudden changes in load and why they are only suitable for small-capacity plants.

(b)   Figure 7.64 shows the capacity curves for a reciprocating compressor with four stages of unloading. Additional control is obtained from a back-pressure regulating valve which has a rangeability of 5:1 and is designed to provide a minimum evaporating temperature of 2 °C.

The design capacity of the plant is 130 kW when cooling 4.42 kg/s of water to 7 °C. The minimum chilled water outlet temperature is 4 °C.

Neglecting pressure drops in the suction line and across the fully-open valve, determine:

(i)   the capacity ranges when the back-pressure valve will be modulating;

Figure 7.64 Capacity curves for Problem 7.12

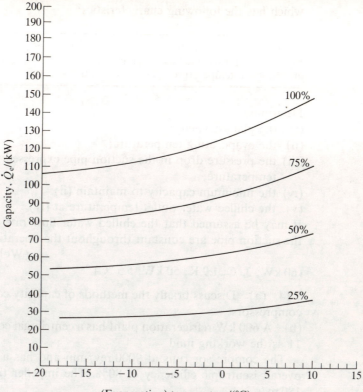

(ii) the return water temperatures to the plant at which cylinders should be cut out;

(iii) the maximum evaporating temperature.

Back-pressure regulator valve data:

Pressure drop expressed as a difference in saturation temperature/(K)	4	8	12	16	20
Capacity fully open/(kW)	240	358	432	490	530

(Wolverhampton Polytechnic)
(Modulate: 130–107 kW—four cylinders, 97.5–85 kW—three cylinders, 60–65 kW—two cylinders, no modulation with one cylinder; cut out cylinders at 12 °C, 9.9 °C, 7.8 °C, 5.9 °C; 4.4 °C)

7.13 (a) Describe the operation and application of float valves to control the expansion of a refrigerant.

(b)   A chilling plant is required to cool 3.57 litres/s of water from 10 °C to 6 °C.

The evaporator to be used in the plant is rated at 100 kW when operating at 0 °C producing chilled water at 5 °C. The condensing unit has a linear characteristic with a capacity of 100 kW at an evaporating temperature of 10 °C and 40 kW at an evaporating temperature of −4 °C.

The only means of control is a back-pressure valve with a rangeability of 4 which has the following characteristics:

Capacity fully open/(kW)	40	66	108	136	240
Pressure drop expressed as a change in saturation temperature/(K)	0.25	0.5	1.0	1.5	3.75

Determine:

(i)   the plant capacity;

(ii) the evaporating temperature;

(iii) the pressure drop in the suction pipe expressed as a change in saturation temperature;

(iv) the minimum capacity to maintain (ii);

(v)   the chilled water outlet temperature at (iv).

It may be assumed that the chilled water flow rate and the pressure drop in the suction pipe are constant throughout the operating range of the plant.

(Wolverhampton Polytechnic)

(60 kW; 3 °C; 1.9 K; 50 kW; 5.5 °C)

7.14 (a)   Discuss briefly the methods of capacity control used for centrifugal compressors.

(b)   A 600 kW refrigeration plant has a centrifugal compressor and Refrigerant 11 as the working fluid.

The compressor runs at 9000 rev/min and has a slip factor of 0.9 and an overall isentropic efficiency of 80%. The impeller tip speed is not to exceed 180 m/s.

The evaporating and condensing temperatures are 270 K and 320 K respectively. The vapour entering the compressor is superheated by 9 K and the liquid leaving the condenser is sub-cooled by 5 K. Calculate:

(i)   the number of stages and the impeller tip diameter;

(ii) the outer diameter of the impeller eye at inlet if the inner diameter is 50 mm and the vapour inlet velocity is 60 m/s in an axial direction;

(iii) the angle at which the blades at inlet are offset from an axial direction.

(Wolverhampton Polytechnic)

(2, 0.326 m; 0.2 m; 44.6°)

7.15 A Refrigerant 11 heat pump operates between pressure limits of 2 bar and 0.5 bar. The liquid leaving the evaporator is sub-cooled by 5 K and the vapour entering the compressor is superheated by 10 K.

The single-stage centrifugal compressor has an isentropic efficiency of 80%, slip factor of 0.9 and an impeller tip diameter of 400 mm. The vapour enters the compressor axially with a velocity of 50 m/s and the mean diameter of the impeller eye is 125 mm.

If the design heat load is 700 kW and the electrical drive design speed 1800 rev/min, calculate:

(i)   the gear box ratio;

(ii) the dimensions of the impeller eye.

(Wolverhampton Polytechnic)

(5.04; 202 mm, 98 mm)

7.16 An 800 kW heat pump uses Refrigerant 11. Saturated vapour at 5 °C enters the compressor and there is no sub-cooling of the condensate.

A centrifugal compressor is used which has a compression ratio of 10/1 and an overall isentropic efficiency of 80%. The speed is 12000 rev/min and the slip factor is 0.9. If the maximum speed of the impeller tip is 180 m/s, calculate:

(i) the number of stages assuming equal changes in enthalpy per stage;
(ii) the impeller tip diameter if all stages are the same;
(iii) the outside diameter of the impeller eye at inlet if the inlet velocity of 60 m/s is axial and the eye inner diameter is 100 mm;
(iv) the width of the impeller passages at outlet if the outlet velocity is 10° from being tangential at each stage.

(Wolverhampton Polytechnic)

(2; 275 mm; 204 mm; 23 mm, 8 mm)

7.17 A lithium bromide/water refrigerator operates between pressure limits of 10 kN/m² and 1.5 kN/m². The rich liquid leaving the absorber is saturated and contains equal proportions of lithium bromide and water by mass. After passing through the heat exchanger it enters the generator sub-cooled by 10 K. The weak liquid leaving the generator, the water leaving the condenser and the water vapour leaving the evaporator are all saturated.

If the generator operates at 100 °C, calculate:

(i) the absorber operating temperature;
(ii) the cycle coefficient of performance.

(Wolverhampton Polytechnic)

(35 °C; 0.79)

7.18 A 100 kW vapour-absorption refrigeration plant operates with a condenser pressure of 0.1 bar and an evaporator pressure of 0.01 bar. The absorber and condenser are both water-cooled in a series circuit. Saturation states exist at the condenser outlet, the evaporator outlet, the rich liquid entering and leaving the heat exchanger and the lean liquid leaving the generator. If the generator operates at 90 °C and the absorber at 37.5 °C, calculate:

(i) the mass flow rate of cooling water required if the total temperature rise is 10 K;
(ii) the coefficient of performance.

(Wolverhampton Polytechnic)

(5.3 kg/s; 0.83)

7.19 A heat recovery plant is to be designed for a simultaneous cooling duty of 500 kW and a heating duty of 1500 kW.

Refrigeration systems to perform this function will evaporate at 7 °C and condense at 60 °C with the vapour leaving the evaporator dry saturated and the liquid leaving the condenser sub-cooled by 5 K. Any heating not performed by the refrigeration plant will be made up from direct air heating using gas-fired heaters with a 90% thermal efficiency.

Compare the hourly energy costs for the two alternatives:

(a) An R12 vapour-compression plant in which the power absorbed in isentropic compression is 60% of the electrical power supplied to the compressor motor.
(b) A single-stage lithium bromide/water absorption plant in which a direct gas-fired generator operates with a thermal efficiency of 90% and produces a generator temperature of 125 °C; solution leaves the absorber at 45 °C and enters the generator saturated at 100 °C.

Data
Electricity cost   3 p/kW h
Gas cost   12 p/m³
Gas calorific value   41 500 kJ/m³

(CIBSE)

(£16.5; £17.4)

7.20   A steam-jet refrigerator is required to supply a cooling coil load of 50 kW. The evaporator pressure is 0.01 bar and the vapour extracted has a dryness fraction of 0.95. The make-up water is at 20 °C. If the water return temperature is 11 °C, calculate the flow rate of chilled water and the capacity of the steam jet compressor in m³/s.
(2.91 kg/s, 2.66 m³/s)

7.21   (a)   A diagrammatic sketch of a steam-jet refrigeration plant is shown in Fig. 7.65. Give a brief description of the operation of such a plant.

(b)   In a plant such as that shown in Fig. 7.65 the kinetic energies at sections 1, 3 and 5, are negligible. Using the data below and the $h-s$ chart supplied and neglecting heat losses, calculate:
(i)   the mass flow rate of entrained vapour;
(ii)   the mass flow rate, $\dot{m}$, of motive steam required;
(iii) the temperature of the steam entering the condenser.

Figure 7.65   Problem 7.21

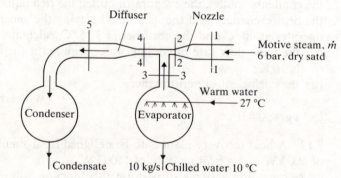

Data
Mass flow rate of chilled water, 10 kg/s; temperature of chilled water, 10 °C; temperature of warm water, 27 °C; condition of steam at entry to the nozzle, 6 bar saturated; isentropic efficiency of nozzle, 0.9; velocity of steam entering the diffuser, 700 m/s; isentropic efficiency of diffuser, 0.8; condition of entrained vapour leaving the evaporator is 0.03 bar and dryness fraction 0.96; entrainment efficiency = (kinetic energy after entrainment)/(kinetic energy before entrainment) 0.65; mean specific heat of water, 4.18 kJ/kg K.

(CIBSE)

(0.304 kg/s; 0.384 kg/s; 67 °C)

REFERENCES

7.1 Johansson M 1988 *Chloroflorocarbons (CFCs) and the Building Services Industry* BSRIA TN 2/88

7.2 Eastop T D and McConkey A 1986 *Applied Thermodynamics for Engineers and Technologists* 4th edn Longman

7.3 Rogers G F C and Mayhew Y R 1988 *Thermodynamic and Transport Properties of Fluids* 4th edn Basil Blackwell

7.4 Stoecker W F and Jones J W 1983 *Refrigeration and Air Conditioning* 2nd edn McGraw-Hill

7.5 Dossat R J 1990 *Principles of Refrigeration* 2nd edn John Wiley

7.6 *Standards of Tubular Exchanger Manufacturers Association* 6th edn TEMA

7.7 Janna W S 1988 *Engineering Heat Transfer* SI edn Van Nostrand–Reinhold

7.8 Threlkeld J L 1970 *Thermal Environmental Engineering* Prentice-Hall

8 AIR CONDITIONING

This chapter deals with the psychrometrics involved in air conditioning. The construction of the CIBSE psychrometric chart, together with an analysis of the basic processes of heating, cooling, dehumidifying and humidifying, is covered in Chapter 1.

The basic processes can be combined and controlled in many different ways, leading to a number of basic systems that can be combined to form more complex systems. This can become confusing to a student and a thorough understanding of the basic systems should be gained before anything more complex is attempted.

One of the challenges of designing air conditioning systems is that there is seldom a clear-cut correct answer to a problem. For a given application there will always be some systems which are without doubt unsuitable, but this frequently leaves a number of possibilities with little to choose between them. Some guidance on selection is offered at the end of this chapter in Section 8.13 but it is *only* guidance; no matter how long a problem is studied, experienced engineers may still come to different solutions.

The inexperienced engineer may find this uncertainty rather daunting and the easy way out is to 'pass the buck backwards' and 'do it the way it was done last time'. There are two problems with this approach: firstly, it assumes that the system selected for the last job was the optimum one, which may well be so but equally it may have been chosen because it was the way it had been done before, an attitude which can result in many years passing without anyone making a decision; secondly, in many cases the last job is not quite the same as the present one and the differences might be enough to justify the selection of a different system.

All air conditioning systems have internal and external design states for both the winter and summer operation. These design conditions will determine the size of the plant.

However, the design states are only encountered or exceeded for a very small part of the year. To determine the annual energy consumption it is necessary to analyse the plant operation throughout the year. It is therefore necessary to investigate part-load cycles as well as design cycles. Section 8.12 covers annual energy consumption.

Before analysing full cycles it is useful to examine in detail some basic parameters which are used in all systems.

8.1 SENSIBLE AND LATENT HEAT LOADS

The room sensible heat load consists of the algebraic sum of all sources of heat gain or heat loss which have no effect on the moisture in the room. Room sensible heat loads are covered in Chapter 3. Room latent heat loads are due to moisture inputs into the room and are always a heat gain. The main moisture inputs into a room are from the occupants and air infiltration; there may also be an input from equipment in rooms such as kitchens. Table A7.1 in the CIBSE *Guide*[8.1] (reproduced here as Table 8.1 by permission of the CIBSE) gives figures for the sensible and latent heat outputs from adult males. CIBSE recommends that figures for adult females and children are 0.85 and 0.75 respectively of adult male values. CIBSE does not state how the values in Table 8.1 were established but since many of them are the same as those in Table 16 of ASHRAE[8.2] it is reasonable to assume that the same method was used. (The values could not all be the same because the ASHRAE table has 11 different categories.)

Table 8.1   Heat emissions from people

| Application | | | Sensible ($\dot{S}$) and latent ($\dot{L}$) heat emissions/W at the stated dry-bulb temperatures/°C | | | | | | | | | |
| | | | 15 | | 20 | | 22 | | 24 | | 26 | |
Degree of activity	Typical	Total	$\dot{S}$	$\dot{L}$	$\dot{S}$	$\dot{L}$	$\dot{S}$	$\dot{L}$	$\dot{S}$	$\dot{L}$	$\dot{S}$	$\dot{L}$
Seated at rest	Theatre, hotel lounge	115	100	15	90	25	80	35	75	40	65	50
Light work	Office, restaurant*	140	110	30	100	40	90	50	80	60	70	70
Walking slowly	Store, bank	160	120	40	110	50	100	60	85	75	75	85
Light bench work	Factory	235	150	85	130	105	115	120	100	135	80	155
Medium work	Factory, dance hall	265	160	105	140	125	125	140	105	160	90	175
Heavy work	Factory	440	200	200	190	250	165	275	135	305	105	335

* For restaurants serving hot meals, add 10 W sensible and 10 W latent for food

The experimental method used to establish the latent heat gains in ASHRAE[8.2] involved an accurate weighing of subjects to determine the rate of moisture loss, which was then multiplied by the enthalpy of vaporization of water (h_{fg}) at body temperature. It is the application of h_{fg} to establishing figures for *latent heat losses from people* that has led to the misuse of h_{fg} values for the *latent heat gain to the air in a room*. During experimental work the system boundary was around the subject and the water started as a liquid which was then vaporized. For air conditioning work, the system boundary is around the mass of air in the room and the water which enters the system enters as a vapour with an enthalpy value h_g. The error in using h_{fg} instead of h_g is about 3.7%. The values in Table 8.1 should be multiplied by h_g/h_f, to establish the latent heat inputs to the room.

In Chapter 1 it was shown how the sensible heat gains and the latent heat gains can be related by means of a room ratio line (*RRL*). Referring to Fig. 8.1, the room ratio line may be found from the following equation:

$$RRL = \frac{h_X - h_S}{h_R - h_S} = \frac{\dot{S}}{\dot{S} + \dot{L}}$$

[8.1]

Figure 8.1

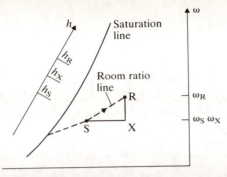

This equation gives the slope of the line between the supply air, S and the room air, R. The point X is where the room dry bulb temperature line crosses the supply air specific humidity line.

The value of h_g used to establish Eqn [8.1] was 2555.7 kJ/kg, which corresponds to 30 °C. This is because 30 °C was chosen as the vertical dry bulb isotherm on the psychrometric chart. The use of the *RRL* is strictly only accurate if the room is also at 30 °C, which is unlikely. The error in using it for normal room temperatures is however very small in terms of the general accuracy of chart work.

Moisture which enters a room from people will be at a mean body temperature. It then mixes with the room air and is cooled to the room temperature. As it cools it will make a small contribution to the room sensible heat gain. It can therefore be argued that a small percentage of the latent gain from people (about 0.7%) should be added to the sensible heat gain. However, since the values in Table 8.1 only have an accuracy of 5 W, such an exercise would be pointless.

It makes little difference which temperature is used to determine an h_g value. However, there are two good reasons for using the room temperature. Firstly it must be remembered that not all moisture gains are at body temperature; for example, infiltration gains are at the outside air temperature, but all moisture gains end up at the room temperature. Secondly, the use of an h_g value at room temperature gives an exact agreement between the equations for room latent heat gains in terms of enthalpy differences and specific humidity differences (Eqn [1.38]), i.e.

$$\dot{L} = \dot{m}_a(h_R - h_X) = \dot{m}_a h_g(\omega_R - \omega_X)$$

Typical room temperatures vary within the range of 19 °C (with a corresponding h_g value of 2535.7 kJ/kg) to 23 °C (with a corresponding h_g value of 2543.0 kJ/kg). An average room temperature value for h_g of 2540 kJ/kg has been used throughout this chapter. In some of the examples and problems the original value may have been different, and these have been changed.

8.2 COOLING COIL CONTACT FACTOR

This was defined and briefly described in Chapter 1. Chapter 6 included an analysis of the heat and mass transfers which take place. In Fig. 8.2(a), if the

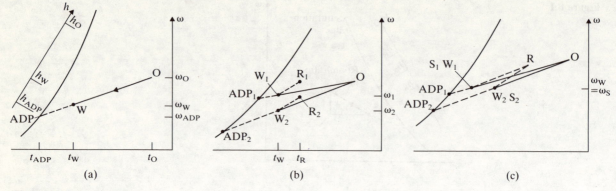

Figure 8.2 Cooling coil contact factors

air enters a cooling coil at a state O and leaves at a state W, the point where the continuation of the line OW would meet the saturation line is known as the apparatus dew point (ADP) and the contact factor can be found using Eqn [1.42] in any of the following forms.

$$\text{Contact factor} = \frac{h_O - h_W}{h_O - h_{ADP}} = \frac{\omega_O - \omega_W}{\omega_O - \omega_{ADP}} = \frac{\text{Line OW}}{\text{Line OADP}}$$

It is also a good approximation to use dry bulb temperatures to find the contact factor:

$$\text{Contact factor} = \frac{t_O - t_W}{t_O - t_{ADP}}$$

The apparatus dew point is the mean surface temperature of the cooling coil, which is a function of the air and coolant temperatures and the thermal resistances between the two fluids. When there is no other information available it is a reasonable approximation to assume that the thermal resistance between the cooling medium and the coil surface is small compared with the thermal resistance between the surface and the air, so that the apparatus dew point has the same value as the mean coolant temperature.

It is useful to consider what determines the value of the contact factor and how this value affects psychrometric cycles. The contact between an air stream and the surface of a coil depends on the number of tube rows crossed in the direction of flow, the physical dimensions of each coil, and the thermal resistances between the air and coolant. The thermal resistances of the coolant and air change as their respective flow rates change. However, assuming the thermal resistance of the air is very large compared with the other thermal resistances, it was shown in Chapter 6 that the contact factor may be expressed as follows:

$$\text{Contact factor} = 1 - e^{-\beta_\omega A_0/\dot{m}_a}$$

where $\dot{m}_a$ is the mass flow rate of air through the coil, A_o is the outside surface area of the coil and β_ω is a mass transfer coefficient. An approximation for a particular coil is:

$$\text{Contact factor} = 1 - e^{-k/\dot{m}_a} \qquad [8.2]$$

where k is a constant for a particular coil. Since there is a very small change in the density of the air passing through a coil, $\dot{V}$ is often substituted for $\dot{m}$ in

Eqn [8.2]. This equation can be used to predict how the contact factor will change if the flow rate of air changes.

One of the most important variables in the design of a cooling coil is the number of tube rows crossed. A single row of tubes would allow considerable by-passing and give a low contact factor; the greater the number of tube rows crossed, the higher the contact factor becomes.

Consider a cooling process with a fixed on-coil state (the state of the air entering the coil) and a fixed off-coil temperature (the temperature of the air leaving the coil). Two possible cooling coils are available, one with a very high contact factor, (process OW_1) and the other with a very low contact factor (process OW_2). The two processes are illustrated in Fig. 8.2(b) and can be compared as follows.

High contact factor (say 0.9):

More tube rows	Increased coil cost, high pressure drop, increased fan power
High ω	Results in high room percentage saturation
High ADP	High chilled water temperature, low load ($h_O - h_{W1}$), low refrigeration plant cost

Low contact power (say 0.6):

Less tube rows	Reduced coil cost, low pressure drop, reduced fan power
Low ω	Lower room percentage saturation
Low ADP	Low chilled water temperature, high load ($h_O - h_{W2}$), high refrigeration plant cost

Most coils have a contact factor which is a compromise between the above extremes.

Another consideration when selecting a cooling coil contact factor is the room latent heat gains compared with the room sensible heat gains.

In Fig. 8.2(c) the specific humidity of the off-coil air is fixed and the state of the room air is the same for both cycles. The two room ratio lines have quite different slopes, the room being supplied with air at state S_1 from the coil with the high contact factor has much lower latent gains (as a proportion of the total gains) than the room supplied with the air at S_2 from the coil with the low contact factor. As room latent gains increase, coils with lower contact factors may be used to avoid using excessive reheat.

It is important to appreciate the consequences of using a particular value of contact factor and essential to check that the coil installed has the same contact factor as that used in the design calculations.

8.3 SUPPLY TEMPERATURE DIFFERENTIAL

It is also useful to consider the consequences of varying the supply air temperature differential, defined as the difference between the room temperature and the supply air temperature, $t_R - t_S$.

This can be investigated by fixing the on-coil state, the contact factor, the room temperature and the room ratio line (shown as broken lines with the same slope) for two different supply temperatures. This is shown in Fig. 8.3. In this diagram the off-coil state (W) is not the same as the supply state (S), the

Figure 8.3 Temperature differentials

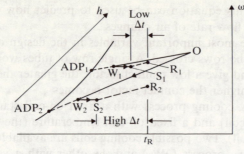

difference being due to fan and duct gains. The consequences of using extreme values of temperature differential in summer are listed below.

Low differential:
High $\dot{m}$ High fan power and/or large ductwork
High ω High room percentage saturation
High ADP   High chilled water temperature, low refrigeration plant cost

High differential:
Low $\dot{m}$ Low fan power and/or smaller ductwork
Low ω Lower room percentage saturation
Low ADP   Low chilled water temperature, high refrigeration plant cost

The maximum supply differential in summer is limited by the mixing of the supply and room air and is normally about 11 K. In winter it is possible to use larger values of differential without any mixing problems, but in practice the summer design normally controls the mass flow rate of air, resulting in lower differentials in winter than in summer.

8.4 VOLUME FLOW RATES AND MASS FLOW RATES

The total air flow rate supplied to the room is calculated on a mass basis using Eqn [1.37]

$$\dot{S} = \dot{m}_a(h_x - h_1) = \dot{m}_a c_{pma}(t_R - t_s)$$

normally for the summer design case. If leakage from ducts is neglected, for the design case, the mass flow rate of dry air remains constant whatever the changes in temperature or humidity through the system. However, the volume flow rate through the system will change as predicted by Eqn [1.33] ($\dot{m} = \dot{V}/v$). An inspection of a psychrometric chart shows that the specific volume varies with both temperature and humidity.

A constant-speed fan will deliver a constant volume and the mass flow rate through the fan will change if the specific volume changes. If a fan handles outside air this can result in considerable changes in mass flow rate. For example, at a typical winter design state of $-5\,°C$ saturated the specific volume is about $0.76\ m^3/kg$ whilst at a typical summer design state of $29\,°C$ dry bulb and $22\,°C$ wet bulb it is $0.875\ m^3/kg$. This difference of over 15% in specific volume could result in the mass flow rate in winter being 15% higher than in summer.

Large differences in mass flow rate between summer and winter will take place if supply fans are used to ventilate a building mechanically in such a way that the fan handles outside air. In these cases the change in mass flow rates must be taken into account.

The situation is quite different in most air conditioning systems where the air passes over a cooler in summer or a pre-heater in winter before it reaches the fan. The result is that in most systems the air passes through the fan between temperatures of 10 °C and 15 °C, and specific humidities between 0.006 and 0.009 throughout the year. Within this area of the chart the maximum change in specific volume is only 2% and for most air conditioning systems the change is much smaller. For these systems it is not unreasonable to assume that the mass flow rate of air through the system is constant throughout the year.

When some of the room air is re-circulated, the total mass flow rate is still predicted by Eqn [1.37] but the fresh air component is frequently expressed as a volume flow rate. For example a table in the CIBSE *Guide*[8.3] recommends

Table 8.2 Recommended outdoor air supply rates for air-conditioned spaces. Reproduced from CIBSE *Guide*.[8.3]

| | | Outdoor air supply (litre/s) | | |
| | | Recommended | Minimum (Take greater of two) | |
Type of space	Smoking	Per person	Per person	Per m² floor area
Factories*†	None			0.8
Offices (open plan)	Some			1.3
Shops, department stores and supermarkets	Some	8	5	3.0
Theatres*	Some			—
Dancehalls*	Some			—
Hotel bedrooms†	Heavy			1.7
Laboratories†	Some			—
Offices (private)	Heavy	12	8	1.3
Residences (average)	Heavy			—
Restaurants (cafeteria)†‡	Some			—
Cocktail bars	Heavy			—
Conference rooms (average)	Some			—
Residences (luxury)	Heavy	18	12	—
Restaurants (dining rooms)†	Heavy			—
Board rooms, executive offices and conference rooms	Very heavy	25	18	6.0
Corridors	A *per capita* basis is not appropriate to these spaces			1.3
Kitchens (domestic)†				10.0
Kitchens (restaurant)†				20.0
Toilets*				10.0

* See statutory requirements and local bye-laws.
† Rate of extract may be overriding factor.
‡ Where queueing occurs in the space, the seating capacity may not be the appropriate total occupancy.

Notes:

1. For hospital wards, operating theatres see Department of Health and Social Security Building Notes.
2. The outdoor air supply rates given take account of the likely density of occupation and the type and amount of smoking.

outdoor air supply rates for air-conditioned spaces in terms of litres/s per person or litres/s per m² of floor area.

The state of the air in the CIBSE table is not given but can be assumed to be at standard atmospheric conditions of 1 atmosphere and 15 °C. Since the normal fan conditions are in the range 10–15 °C (see above), in both winter and summer, then the volume flow rates of fresh air from the table can be taken to apply to the state of the air leaving the fan, which is the summer supply state for most systems.

8.5 SINGLE-ZONE SYSTEMS WITH 100% FRESH AIR

Summer Cycles

In summer, air must be cooled and de-humidified. It must then be supplied to the room at a state S which will result in a room state R. For any cycle the room sensible heat gain, $\dot{S}$, and latent heat gain, $\dot{L}$, must be satisfied by Eqns [1.37] and [1.38]:

$$\dot{S} = \dot{m}_a c_{pma}(t_R - t_S) \qquad [1.37]$$
$$\dot{L} = \dot{m}_a(\omega_R - \omega_S)h_g \qquad [1.38]$$

Also $\qquad \dot{S} + \dot{L} = \dot{m}_a(h_R - h_S)$

The ratio of the sensible heat load $\dot{S}$ to the total heat load $(\dot{L} + \dot{S})$ is the room ratio line which determines the slope of the line which joins point S and R on the psychrometric chart for any room loading. It is not unusual for the sensible heat gains to vary widely whilst the latent heat gains remain relatively constant (for example, when a room occupancy remains constant during a period of changing solar heat gains). It is clear from Eqn [1.38] that if the latent heat gain and the air mass flow rate remain constant the increase in specific humidity between the room and supply states must remain constant. However, from Eqn [1.37], if the sensible heat load falls the supply temperature must rise to maintain the room temperature constant.

Any of the following systems will give the desired room state for the design cycle and the temperature for any part-load cycle; the difference between them is how the room humidity changes at part load.

Off-coil (dew point) system with 100% fresh air

Figure 8.4(a) shows the active items of plant and Fig. 8.4(b) the psychrometric cycles at design and part-loads. The design cycle is OWSR, whilst $O_1W_1S_1R_1$ and $O_2W_2S_2R_2$ are two part-load cycles. The cooling coil controller receives an input from a dry bulb temperature detector, T1, in the ductwork between the cooling coil and fan (point W), and as the load falls a three-port control valve progressively reduces the mass flow rate of chilled water to the cooling coil, thus maintaining a constant off-coil temperature. Figure 8.4(a) shows a mixing valve being used to divert the flow. The use of three-port diverting valves is not recommended by CIBSE.[8.4] There is no theoretical basis for this; it is just a matter of suitable mixing valves being readily available whereas most manufacturers only supply diverting valves which should be used in the open or closed positions only. In addition to the control valve there would also be a regulating valve in the by-pass line, which would be set to the same pressure

Figure 8.4 Off-coil (dew point) system with 100% fresh air

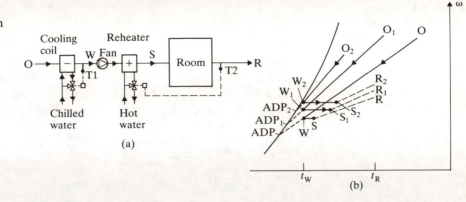

(a)

(b)

drop as the coil, and isolating valves. These have been omitted to simplify the diagram.

The aim of a diverting method of control is to vary the mass flow rate while maintaining the input and outlet temperatures constant. If the energy balance on the chilled water alone were considered, the temperature of the water leaving the coil could be constant and hence the ADP would remain constant. However, the cooling coil contact factor must also be taken into account and, since the air flow rate does not change, Eqn [8.2] (Contact factor = $1 - e^{-k/\dot{m}}$) predicts that the contact factor also should remain constant. An inspection of the cooling processes in Fig. 8.4(b) shows that it is not possible for both the ADP and the coil contact factor to remain constant. The contact factor will fall slightly because of the increase in thermal resistance of the coolant (which is not allowed for in Eqn [8.2]) and the chilled water outlet temperature must rise, which increases the ADP. The change in ADP is not as great as might be assumed from Fig. 8.4(b) which, like many of the psychrometric sketches in this chapter, has been drawn with the vertical axis (specific humidity) increased relative to the horizontal axis (dry bulb temperature) so that cycles may be clearly distinguished. Therefore the change in the specific humidity (and hence the dew point) of the air leaving the coil is quite small—which is how the name 'dew point' for this method of control originated. The term *dew point control* is widely used in industry but *off-coil dry bulb temperature control* is a more accurate description of what takes place.

If a mixing method of control (constant flow rate through the coil with a variable inlet temperature) had been used, the change in the specific humidity of the air leaving the coil would have been more substantial.

For the design case the air is cooled from the outside state O to state W in the cooling coil. It then undergoes a small temperature rise across the fan before being supplied to the room at state S. The reheater is off.

If there was no reheater for part-load cycles, the room temperature would fall as the sensible heat load decreased. A room temperature detector (usually fitted in the return ductwork) therefore provides an output to control the reheater to give the required supply temperature. The reheater may be controlled by either diverting (as shown) or mixing. The two part-load cycles illustrated for outside air states at O_1 and O_2 show a very small increase in room humidity. The penalties for excellent control of humidity are high cooling coil loads as well as the cost of the reheat.

Figure 8.5 Sequence
heating and cooling
system with 100% fresh
air

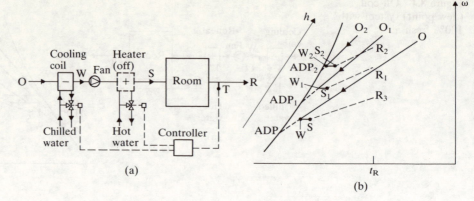

Figure 8.5 Sequence heating and cooling system with 100% fresh air

(a)

(b)

Sequence heating and cooling with 100% fresh air

Figure 8.5(a) shows the active items of plant and Fig. 8.5(b) the psychrometric cycles. The design cycle is OWSR; $O_1W_1S_1R_1$ and $O_2W_2S_2R_2$ are two part-load cycles. The term sequence is used when only the cooling coil *or* heater (drawn with broken lines to show it is not used during summer cycles), but never both, is active at any one time.

A control unit receives an input from a temperature detector in the exhaust duct and sends an output to either the heater *or* the cooling coil. Either diverting (as shown) or mixing systems may be used to control the chilled water at part-load.

With only a cooling coil to control the room temperature, the off-coil temperature must rise as the sensible heat gains fall. To achieve this, the mean coil temperature (ADP) must be increased. This is accomplished by either increasing the chilled water input temperature (mixing) or reducing the mass flow rate and allowing the outlet temperature to rise (diverting). The result is a series of ascending cycles as the room sensible heat gains fall with substantial increases in room humidity.

If the high humidities are tolerable, sequence cooling and heating provides a very economic system. (The cooling coil load is proportional to $h_o - h_w$.)

When only very high humidities are unacceptable, a humidity detector is installed in the exhaust duct and a temperature detector between the coil and fan (point W). The output from both is also fed into the control unit where the room humidity is pre-set to a high-level value. When the pre-set value of humidity is reached the controller switches the plant to a dew point system of control until the humidity has fallen sufficiently to permit a return to sequence control.

Face and by-pass dampers with 100% fresh air

Figure 8.6(a) shows the active items of plant and Fig. 8.6(b) the psychrometric cycles. The design cycle is OWSR; $O_1W_1S_1R_1$ and $O_2W_2S_2R_2$ are two part-load cycles. Face and by-pass dampers are a means of diverting varying proportions of the total flow rate around the cooling coil. This is achieved by dampers across the face of the cooling coil and across the adjacent duct, operating in opposition (i.e. as one opens the other closes). An alternative is to have a separate by-pass duct but this is more expensive and takes up more space in

Figure 8.6 Face and by-pass damper system with 100% fresh air

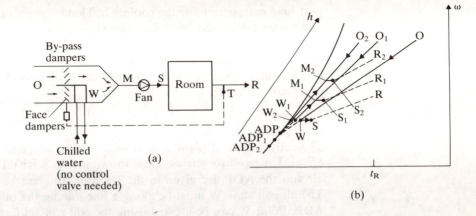

the plant room. The damper control is governed by an input from a temperature detector in the exhaust duct.

The design case is the same as for the previous systems and the by-pass is closed. As the room sensible heat load falls the by-pass is progressively opened and mixing takes place between the off-coil air at state W and the outside air at state O. The result is an increase in room humidity which is higher than for the dew point system but less than for the sequence heating and cooling system.

The cooling coil load is no longer governed by the length of line OW because only part of the total mass flow rate passes through the coil. The overall effect of the by-pass system is to cool the complete air stream from point O to point M: therefore the new cooling coil load is now proportional to line OM. The running costs are much less than for the dew point system but slightly more than for sequence control. The humidity control is slightly better than for sequence control.

The cooling coil could be controlled by a dew point thermostat, in which case point W would be at the same temperature for each cycle. However, there is no need to control the coil at all. If the load on the coil falls owing to some of the air being by-passed, the chilled water return temperature will fall and so will the coil ADP, as shown in Fig. 8.6(b). This is the opposite of what happens with an off-coil or sequence control system when the ADP rises as the load falls.

The cooling coil contact factor will also increase as the flow rate of air through it decreases as predicted by Eqn [8.2] (Contact factor $= 1 - e^{-k/\dot{m}_a}$). For any particular part-load, the off-coil state W will be fixed by the new contact factor and the new ADP. However, for any value of W there is a mixing ratio which will provide supply air at the correct differential to maintain the room temperature. When there is no control on a coil it is said to be *wild*.

Example 8.1

(a)   An air-conditioned room using 100% outside air has design loads of 20 kW sensible and 5 kW latent when the outside air is 28 °C dry bulb and 22 °C wet bulb. The room is maintained at 22 °C and there is a 1 K temperature rise across the fan. The cooling coil has a contact factor of 0.8 and an ADP of 9 °C. Determine the cooling coil load and the room percentage saturation for the design case.

(b)   When the outside air is at 23 °C dry bulb and 21 °C wet bulb the room loads on the plant in part (a) are 10 kW sensible and 5 kW latent. For this part-load case, determine the cooling coil load, the room percentage saturation and heater load if the control system is:

(i)   off-coil (dew point);
(ii) sequence cooling and heating;
(iii) face and by-pass dampers (assume the ADP falls to 7.5 °C and the contact factor increases to 0.9).

Solution

(a)   The design case:

The active items of plant consist of a cooling coil and fan, as shown in Fig. 8.7(a). The psychrometric cycle is shown in Fig. 8.7(b). The state of the outside air and the ADP are given in the question and can be located on the chart. The off-coil state W must lie along a line joining the outside air state and the ADP. Point W can be located using the coil contact factor and Eqn [1.42] in a number of ways as follows:

$$0.8 = \frac{h_O - h_W}{h_O - h_{ADP}} = \frac{64.2 - h_W}{64.2 - 27.5}; \quad \therefore \quad h_W = 34.84 \text{ kJ/kg}$$

$$0.8 = \frac{\omega_O - \omega_W}{\omega_O - \omega_{ADP}} = \frac{0.0141 - \omega_W}{0.0141 - 0.007\,15}, \quad \therefore \quad \omega_W = 0.008\,54$$

$$0.8 = \frac{\text{line O–W}}{\text{line O–ADP}} = \frac{\text{line O–W}}{130 \text{ units}}; \quad \therefore \quad \text{line O–W is 104 units}$$

$$0.8 = \frac{t_O - t_W}{t_O - t_{ADP}} = \frac{28 - t_W}{28 - 9}; \quad \therefore \quad t_W = 12.8 \text{ °C}$$

It was shown in Chapter 1 that the first three methods of establishing point W are rigorous but the fourth method is not because the dry bulb isotherms are not quite straight vertical lines (except the 30 °C isotherm). However the first two methods require three values to be read from the chart and the third method two readings. The last method, using temperatures, requires only one. The accumulation of inaccuracies by the first three methods offsets the inherent inaccuracy of using temperatures when the original data are given in the form of temperatures—which is the usual case. All four forms of Eqn [1.42] are used in this chapter but the most convenient is normally the one using temperatures.

Having established point W with a temperature of 12.8 °C, a horizontal line is drawn to the supply state, S, which is the other side of the fan and at a temperature of 13.8 °C. The mass flow rate of air can now be found using Eqn [1.37] as follows:

$$\dot{m}_a = \frac{\dot{S}}{c_{pma}(t_R - t_S)} = \frac{20}{1.02(22 - 13.8)} = 2.39 \text{ kg/s}$$

and hence the cooling coil load:

$$\dot{Q}_{cc} = \dot{m}_a(h_O - h_W) = 2.39(62.4 - 34.84) = 65.9 \text{ kW}$$

There are three methods which can be used to determine the room state:

(1) The room specific humidity can be found from Eqn [1.38] as follows:

$$\dot{L} = \dot{m}_a(\omega_R - \omega_S)h_g$$
$$5 = 2.39(\omega_R - 0.008\,54)2540$$
$$\therefore \quad \omega_R = 0.009\,36$$

Figure 8.7 Example 8.1

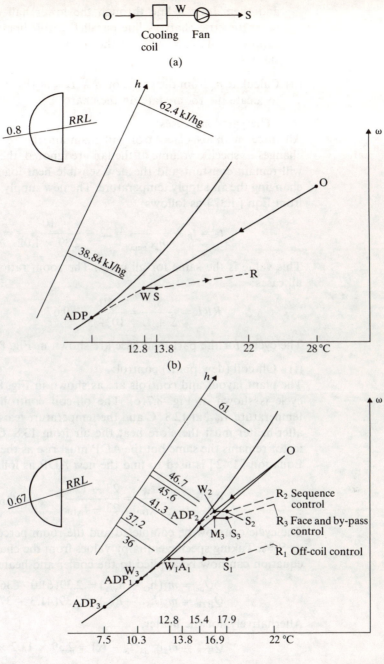

The room state can now be established on the chart and the percentage saturation of 56% read off.

(2) Calculate the room ratio line as follows:

$$RRL = \frac{\dot{S}}{\dot{S} + \dot{L}} = \frac{20}{20 + 5} = 0.8$$

and then draw a line through the lower half of the protractor on the psychrometric chart. A line parallel to this line which passes through the supply air state, S, will cut the room dry bulb temperature at the room state R.

(3) Calculate h_R from the equation $\dot{S} + \dot{L} = \dot{m}_a(h_R - h_S)$ and then use h_R and t_R to locate the room state on the chart.

(b)   The part-load cases:

All three methods of control will maintain the room at 22 °C. If the small changes in specific volume at the fan are ignored, the mass flow rate of dry air will remain constant and the new sensible heat load must be allowed for by changing the air supply temperature. The new supply temperature can be found from Eqn [1.37] as follows:

$$t_S = t_R + \frac{\dot{S}}{\dot{m}_a c_{pma}} = 22 - \frac{10}{2.39 \times 1.02} = 17.9 \,°C$$

This value is the same for all cases. The room ratio line is also the same for all cases:

$$RRL = \frac{\dot{S}}{\dot{S} + \dot{L}} = \frac{10}{10 + 5} = 0.67$$

The cycles for the part-load case are shown in Fig. 8.7(c).

(i)   Off-coil (dew point) control:

The plant layout and controls are as shown in Fig. 8.4(a). The psychrometric cycle is shown in Fig. 8.7(c). The off-coil controller maintains the off-coil temperature, t_{W1}, at 12.8 °C and the temperature leaving the fan is 13.8 °C. The after-heater must therefore heat the air from 13.8 °C to 17.9 °C. The contact factor remains the same but the ADP must rise as the chilled water is throttled. Equation [1.42] is used to find the new ADP as follows.

$$0.8 = \frac{t_O - t_W}{t_O - t_{ADP}} = \frac{23 - 12.8}{23 - t_{ADP}}; \quad \therefore \qquad t_{ADP,1} = 10.3 \,°C$$

The cycle can now be completed and the room percentage saturation read off as 59%. Taking specific enthalpy values from the chart, the steady-flow energy equation can now be applied to the cooler and heater:

$$\dot{Q}_{cc} = \dot{m}_a(h_O - h_{W1}) = 2.39(61.0 - 36.0) = 59.8 \,kW$$
$$\dot{Q}_{HTR} = \dot{m}_a(h_{S1} - h_{A1}) = 2.39(41.3 - 37.2) = 9.8 \,kW$$

Alternatively, for the heater,

$$\dot{Q}_{HTR} = \dot{m}_a c_{pma}(t_S - t_A) = 2.39 \times 1.02 \times (17.9 - 13.8) = 10 \,kW$$

Only enthalpy values can be used for a cooling coil because of the change in specific humidity. It is more rigorous to use enthalpies for the heater also because 1.02 kJ/kg K is only an average value for the specific heat of moist air. However, misreading a specific enthalpy by 0.1 kJ/kg in an enthalpy difference of about 4 kJ/kg introduces an error of 2.5% and in this case the actual value of c_{pma} (calculated from Eqn [1.41], i.e. $c_{pma} = 1.005 + 1.88\omega_S$) is 1.022 kJ/kg K, which is an error of only 0.2%. One of the ironies of using the psychrometric chart is that approximate equations may give more accurate results than rigorous ones.

(ii)   Sequence heating and cooling control:

The plant layout and controls are as shown in Fig. 8.5(a). The psychrometric cycle is shown in Fig. 8.7(c). With sequence control the supply temperature is still 17.9 °C and with no heating the off-coil temperature, t_{W2}, must be 16.9 °C. The contact factor is the same and the new ADP can be found using Eqn [1.42]:

$$0.8 = \frac{t_O - t_W}{t_O - t_{ADP}} = \frac{23 - 16.9}{23 - t_{ADP}}; \quad \therefore \quad t_{ADP2} = 15.4\,°C$$

Points O, W_2, ADP_2 and S_2 can now be located on the chart and drawing the room ratio line from S_2 to R_2 completes the cycle and the room percentage saturation of 76% can be read off the chart. The cooling coil load is:

$$\dot{Q}_{cc} = \dot{m}_a(h_O - h_{W2}) = 2.39(61.0 - 46.7) = 34.2\,kW$$

(iii)   Face and by-pass dampers:

The plant layout and controls are as shown in Fig. 8.6(a). The psychrometric cycle is shown in Fig. 8.7(c). With face and by-pass damper control the supply temperature is still 17.9 °C and with no heating the mixed temperature before the fan, t_{M3}, must be 16.9 °C. The contact factor has increased to 0.9 and the new ADP is 7.5 °C. The new off-coil temperature, t_{W3}, is found using Eqn [1.42]:

$$0.8 = \frac{t_O - t_W}{t_O - t_{ADP}} = \frac{23 - t_{W3}}{23 - 7.5}; \quad \therefore \quad t_{W3} = 10.6\,°C$$

A straight line between the outside air state and the ADP just crosses the saturation line. This is not possible and the off-coil air at point W_3 must be assumed to be saturated. The line between O and W_3 also follows almost the same path as the line between O and ADP_1. Where this line cuts the 16.9 °C dry bulb line locates point M_3. The cycle can now be completed and the percentage saturation read off as 73%. The cooling coil load is:

$$\dot{Q}_{cc} = \dot{m}_a{}'(h_O - h_{M3}) = 2.39(61.0 - 45.6) = 36.8\,kW$$

Summary:

	Room % satn	Q_{cc} (kW)	Q_{HTR} (kW)
Design case	56	65.9	—
Part-load cases			
Dew point control	59	59.8	10
Sequence control	76	34.2	—
Face and by-pass control	73	36.8	—

The dew point system gives a good control of humidity but at a very heavy cost in terms of energy consumption. Not only is there little saving in cooling load, but when the reheater is taken into account the energy consumption at part-load is greater than for the design load.

The face and by-pass system results in a slightly lower humidity than the sequence system at the cost of a slightly higher cooling load. Each has a cooling load which is almost half that of the dew point system. Clearly, if high humidities are not a problem, either of these systems provides a considerable saving. Dew point and sequence control can be combined in the same system. The normal operation is sequence control but a high-limit humidistat switches to dew point control when the humidity tends to rise above a pre-set level.

Winter Cycles

In winter the air should be heated and humidified. Without humidification, systems using 100% fresh air would give very low room humidities when the external temperatures were low. Winter cycles can maintain a room humidity at, or close to, its design value at any part-load.

The following systems can be combined with any of the previously considered summer systems to form a complete plant.

Heating with spray humidification

Figure 8.8(a) shows the active items of plant and Fig. 8.8(b) and Fig. 8.8(c) the psychrometric cycles at design and part-loads for two different methods of control. For both methods of control the design cycle is shown as OABWS and the two part-load cycles have been given the subscripts 1 and 2. The humidifier is shown as a sprayed cooling coil but humidification will follow a line of constant wet bulb temperature for any humidifier supplied with water at or near the wet bulb temperature (see pumped re-circulation systems in Chapter 1, page 33). The two methods of control are:

(i)   Room humidity control:
The first item of plant met by the outside air is a frost coil, which protects the equipment downstream (particularly the filter). If steam is available the frost coil consists of a bank of plane tubes; if not, an electrical heater is used. The frost coil heats the air up to about 4 °C and is switched off when the outside air is above this temperature. It is controlled by the temperature detector, T1, in the air intake.

After the frost coil the air passes over a pre-heater, which is normally a bank of finned tubes supplied with hot water. The pre-heater load is varied by a three-port valve which diverts (illustrated) or mixes the water supplied to the pre-heater. The control system is illustrated in Fig. 8.8(a) and the cycles in Fig. 8.8(b). A humidity detector, H, in the exhaust duct controls the heat input to the pre-heater. The air is then humidified at constant wet bulb temperature until it reaches state W. If the room latent gains are constant, the specific humidity of the air leaving the humidifier will remain constant as shown. If the room latent heat gains fall, the specific humidity of the air leaving the coil will rise to maintain a constant room humidity. With this method of control it is possible to supply air at the correct humidity whatever the room latent load.

Figure 8.8 Winter cycles with spray humidifier

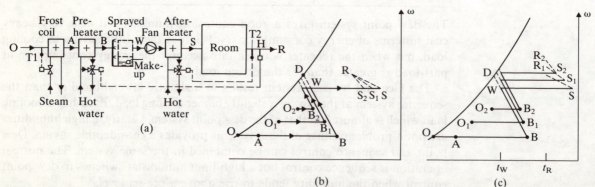

(a)   (b)   (c)

The air leaving the humidifier is then heated again in the after-heater until it reaches the supply temperature, t_S. The after-heater is controlled by a room temperature detector. A system with a room humidity detector can maintain a constant temperature and constant humidity in the room.

(ii)   Off-coil (dew point) temperature control:

The equipment is the same as that shown in Fig. 8.8(a) (but not the controls), and the cycles are shown in Fig. 8.8(c). The operation of the frost coil is the same as for method (i). The only change to the control system compared with the first method is the use of an off-coil temperature detector (not shown) instead of the room humidity detector, to change the heat input to the after-heater. The cycles show that for any room load the temperature of the air leaving the coil has a constant value, t_w. With this method of control the supply specific humidity must vary and it is not possible to maintain a precise room humidity. The variation in room humidity shown is exaggerated by the use of an extended vertical scale to identify each cycle clearly. The small variation in room humidity is not normally a problem and this method has the advantage that temperature detectors are normally more reliable than humidity detectors.

The after-heater is controlled as before and maintains a constant room temperature.

An off-coil (dew point) system may control the temperature at point W to the same value throughout the year. An alternative is to change the set point with an outside temperature detector.

Heating with steam humidification

Figure 8.9(a) shows the active items of plant and Fig. 8.9(b) the psychrometric cycles at design and part-loads. The plant consists of a frost coil, heater and humidifier. The heater is controlled by a temperature detector and the humidifier by a humidity detector, both normally located in the extract duct. The cycles show humidification taking place along a line of constant dry bulb temperature, which was demonstrated to be a reasonable assumption in Chapter 1. However, the system does not rely on constant dry bulb humidification since any change in temperature during humidification will be compensated for by a corresponding change in the temperature of the air leaving the heater. With steam humidification it is possible to maintain a constant room humidity whatever the room latent load.

Throughout the above analysis, fan and duct gains have been omitted to simplify the cycles. Any such gains will produce a corresponding decrease in the load on one of the heaters.

Figure 8.9 Winter cycles with steam humidifier

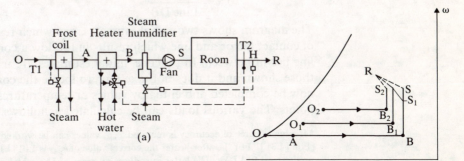

(a)

(b)

Example 8.2

In winter a room is maintained at 21 °C and 50% saturation. The external design temperature is -3.5 °C saturated when the room loads are 15 kW sensible heat loss and 5 kW latent heat gain. The mass flow rate of air of 3 kg/s has been established from the summer cycle. Two systems are proposed for the winter cycle:

(i) a frost coil, a preheater controlled by a room humidity detector, a spray humidifier with an efficiency of 80% and an afterheater;

(ii) a frost coil, heater and an electric steam humidifier which produces saturated steam at 1.1 bar from water at 10 °C.

In each case the frost protection coil is electrically heated and heats the air up to 4 °C. Other heaters use hot water from gas-fired boilers which operate at a thermal efficiency of 81%. Fan and duct gains can be neglected.

Compare the operating costs of the two systems for the design case and a part-load case when the ambient air is at 7 °C and 80% saturation and the sensible heat loss has fallen to 5 kW. Assume that each unit of electrical energy costs three times that for a unit of energy from gas.

Solution

System (i):

The active items of plant are shown in Fig. 8.8(a) and the psychrometric cycle in Fig. 8.10(a). The room state and the two outside states can be established on the chart. The room specific humidity of 0.007 86 is read from the chart and Eqns [1.37] and [1.38] used to determine the state of the supply air as follows:

$$\dot{S} = \dot{m}_a c_{pma}(t_S - t_R)$$
$$15 = 3 \times 1.02(t_S - 21)$$

Hence
$$t_S = 25.9 \text{ °C}$$
$$\dot{L} = \dot{m}_a(\omega_R - \omega_S)h_g$$
$$5 = 3(0.007\,86 - \omega_S)2540$$
$$\therefore \quad \omega_S = 0.007\,20$$

The supply state, S, can now be established on the chart and W, the state of the air leaving the spray, must lie along the same line of constant specific humidity at S. Points A and B must have the same specific humidity as O. Points B and W are found using the spray efficiency of 0.8 by trial and error. This is shown in Fig. 8.10(b). Humidification takes place along a line of constant wet bulb temperature between the two fixed values of specific humidity. There is only one line which will satisfy the relationship:

$$0.8 = \frac{\text{Line WB}}{\text{Line DB}}$$

The diagram shows two initial estimates, one which results in too high a value of contact factor and one which results in too low a contact factor. The correct line lies in between. In practice, initial estimates are usually much closer than those shown and it does not take long to locate the correct line. The cycle can now be completed and enthalpy values or temperatures read off for the salient points. The various loads can be calculated as follows.*

---

* If a high degree of accuracy is required, c_{pma} values can be worked out using $c_{pma} = c_{pma} + \omega c_{ps}$ (Eqn [1.31]). For the after-heater the correct value of c_{pma} is 1.018 kJ/kg and for the frost coil and pre-heater 1.01 kJ/kg. The latter introduces an error of 1%, which is as good as can be expected when reading enthalpy values from a chart.

Figure 8.10 Example
8.2

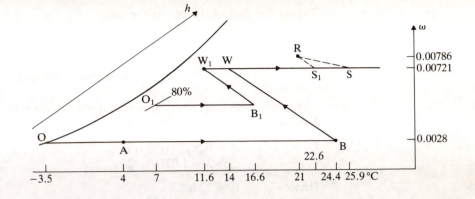

(a)

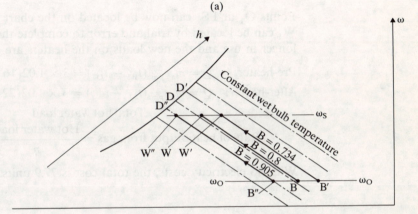

(b)

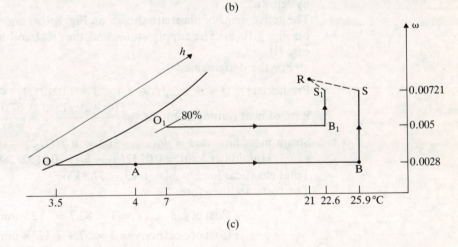

(c)

Pre-heater: $\dot{Q} = \dot{m}_a c_{pma}(t_B - t_A) = 3 \times 1.02(24.4 - 4) = 62.4\,\text{kW}$

After-heater: $\dot{Q} = \dot{m}_a c_{pma}(t_s - t_w) = 3 \times 1.02(25.9 - 14) = 36.4\,\text{kW}$

$\qquad\qquad\qquad$ Total hot water load $\qquad\qquad = 98.8\,\text{kW}$

Rate of heat output from gas $= \dfrac{\text{Hot water load}}{\eta} = \dfrac{98.8}{0.81} = 122\,\text{kW}$

Frost coil: $\dot{Q} = \dot{m}_a c_{pma}(t_A - t_O) = 3 \times 1.02 \times (4 + 3.5) = 23\,\text{kW}$

Taking the cost of the gas as the base:

$$\text{Cost of gas} \qquad = 1 \times 122 = 122 \text{ units}$$
$$\text{Cost of electricity} = 3 \times 23 = \underline{69 \text{ units}}$$
$$\text{Total cost} = 191 \text{ units}$$

For the part-load case the new supply temperature must be calculated as follows:

$$\dot{S} = \dot{m}_a c_{pma}(t_S - t_R)$$
$$5 = 3 \times 1.02(t_s - 21)$$

Hence $t_S = 22.6\,°C$

Points O_1 and S_1 can now be located on the chart (Fig. 8.10(a)), and B_1 and W_1 can be located by trial and error to complete the cycle. The frost coil is no longer in use and the new loads on the heaters are as follows.

Pre-heater: $\dot{Q} = \dot{m}_a c_{pma}(t_{B1} - t_{O1}) = 3 \times 1.02(16.6 - 7) \qquad = 29.4\,kW$

After-heater: $\dot{Q} = \dot{m}_a c_{pma}(t_{S1} - t_{W1}) = 3 \times 1.02(22.6 - 11.6) = \underline{33.7\,kW}$

$$\text{Total hot water load} = 63.1\,kW$$

$$\text{Rate of heat output from gas} = \frac{\text{Hot water load}}{\eta} = \frac{63.1}{0.81} = 77.9\,kW$$

With no electricity costs, the total cost is 77.9 units.

System (ii):

The active items of plant are shown on Fig. 8.9(a) and the psychrometric cycles on Fig. 8.10(c). The supply states and the frost coil load are the same as for case (i).

For the design case,

Pre-heater: $\dot{Q} = \dot{m}_a c_{pma}(t_B - t_A) = 3 \times 1.02(25.9 - 4) = 67\,kW$

$$\text{Rate of heat output from gas} = \frac{\text{Hot water load}}{\eta} = \frac{67.0}{0.81} = 82.7\,kW$$

Steam mass flow rate $= \dot{m}_a(\omega_S - \omega_O) = 3(0.007\,21 - 0.0028) = 0.0132\,kg/s$

$\therefore$ Humidifier load $= 0.0132(h_g - h_f) = 0.0132(2680 - 42) = 34.8\,kW$

Total electrical load $= 34.8 + 23 = 57.8\,kW$

The costs are therefore:

$$\text{Cost of gas} \qquad = 1 \times 82.7 = 82.7 \text{ units}$$
$$\text{Cost of electricity} = 3 \times 57.8 = 173.4 \text{ units}$$
$$\text{Total cost} = 256.1 \text{ units}$$

For the part-load case,

Pre-heater: $\dot{Q} = \dot{m}_a c_{pma}(t_{B1} - t_{O1}) = 3 \times 1.02(22.6 - 7) = 47.7\,kW$

$$\text{Rate of heat output from gas} = \frac{\text{Hot water load}}{\eta} = \frac{47.7}{0.81} = 58.9\,kW$$

Steam flow: $\dot{m}_s = \dot{m}_a(\omega_S - \omega_O) = 3(0.007\,20 - 0.005) = 0.006\,60\,kg/s$

Humidifier load $= \dot{m}_s(h_g - h_f) = 0.006\,60(2680 - 42) = 17.4\,kW$

The costs are:

$$\text{Cost of gas} = 1 \times 58.9 = 58.9 \text{ units}$$
$$\text{Cost of electricity} = 3 \times 17.4 = 52.2 \text{ units}$$
$$\text{Total cost} = 111.1 \text{ units}$$

Cost summary:

	Design load (kW)	Part load (kW)
Spray humidification	191.0	77.9
Steam humidification	256.1	111.1

With a simple choice between the two options in the question, the spray humidification is clearly superior in terms of energy consumption. Steam humidification has one big advantage which is sometimes more important than running costs—that is, its inherent sterility. Water in the temperature range of 20–45 °C is known to favour the growth of the bacterium called *Legionella pneumophilia* which can cause the respiratory illness known as Legionnaires' Disease. *Legionella* is transmitted by inhalation of contaminated water droplets. Most outbreaks of the disease in the UK have been traced to badly maintained cooling towers which operate within the above temperature range for long periods of time. The normal operating temperature for a spray humidifier would be below 20 °C but there may be times when the temperature could rise above this level. There is also another respiratory illness called 'Humidifier Fever' which, as the name suggests, is suspected of having originated in spray humidifiers. Whilst there is little hard evidence against spray humidifiers, there is an understandable reluctance to install them in buildings such as hospitals. Careful maintenance should ensure that spray humidifiers do not become a health hazard. Reference 8.5 is a detailed paper on the problem of Legionnaires' Disease.

Example 8.3

How would the economics of Example 8.2 have changed if steam-raising plant using gas-fired boilers with an efficiency of 77% had been available for the frost coil and steam humidifier? (The efficiency of a steam boiler would be less than for a hot water boiler because of blowdown losses—see Chapter 2.)

Solution

The new rates of energy consumption from the gaseous fuel are as follows.

Spray humidifier:

Design case	Steam boiler	$\dfrac{23}{0.77}$	$= 29.9 \text{ kW}$
	Hot water boiler as before		$= \dfrac{122.0 \text{ kW}}{151.9 \text{ kW}}$
		Total	
Part load	No change		$= 77.9 \text{ kW}$

Steam humidifier:

Design case	Steam boiler	$\dfrac{23 + 34.8}{0.77}$	$= 75.1 \text{ kW}$

		Hot water boiler as before		$= \dfrac{67\,\text{kW}}{142.1\,\text{kW}}$
			Total	
Part load	Steam boiler		$\dfrac{17.4}{0.77}$	$= 22.6\,\text{kW}$
		Hot water boiler as before		$= \dfrac{58.9\,\text{kW}}{81.5\,\text{kW}}$
			Total	

There is now very little difference between the running costs, and one of the most important considerations would be whether the steam boiler plant would be required by other users.

Example 8.4

A room is to be air-conditioned using a dew point, central plant system with 100% fresh air and having a sprayed-coil humidifier of 90% saturation efficiency. The following conditions apply

Room condition	21 °C	55% satn
Summer supply air temperature	14 °C	
Summer design heat gains	10 kW sensible	2 kW latent
Winter design heat loss/gain	3 kW sensible	2 kW latent
Summer design outside conditions	25 °C	75% satn
Winter design outside conditions	−5 °C	100% satn
Plant apparatus dew point	8 °C	

Determine the required air supply flow rate and conditions, the required rating of the pre-heater, reheater and cooler coils, and the required moisture evaporation rate of the humidifier. What is the contact factor of the cooling coil?

If an air-to-air plate heat exchanger of 70% efficiency is fitted between the exhaust and the fresh air ducts, what are the new plant component ratings, assuming that the exhaust flow is equal to the supply flow?

(University of Glasgow)

Solution
The plant is shown in Fig. 8.11(a) and the psychrometric cycles in Fig. 8.11(b) (no heat exchanger) and Fig. 8.11(c) (with heat exchanger). The subscripts used are s for summer and w for winter.

Summer cycle with no heat exchanger:
Points O (also point A), R and the ADP can be established on the chart from the given data. It is also known that the supply state, S_s, lies along the 14 °C dry bulb line. Using the room ratio line,

$$RRL = \frac{\dot{S}}{\dot{S} + \dot{L}} = \frac{10}{10 + 2} = 0.833$$

establishes a second line to locate point S_s. The *RRL* is established on the chart protractor and a parallel line drawn from R to cut the 14 °C line at S_s. A horizontal line from S_s to W_s completes the cycle.

The process between W_s and S_s is after-heating. In practice some of this temperature rise would be due to fan and duct gains and point C_s would be

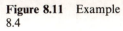

Figure 8.11 Example 8.4

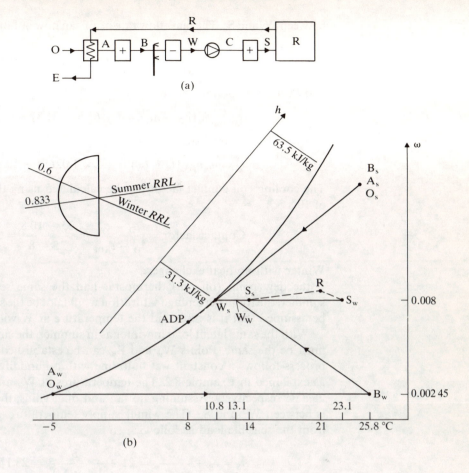

(a)

(b)

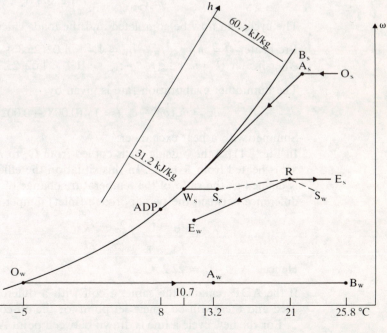

(c)

between W_s and S_s. The mass flow rate of air can now be found from Eqn [1.37]:

$$\dot{m}_a = \frac{\dot{S}}{c_{pma}(t_R - t_S)} = \frac{10}{1.02(21 - 14)} = 1.4 \, \text{kg/s}$$

and hence the cooling coil load:

$$\dot{Q}_{cc} = \dot{m}_a(h_O - h_W) = 1.4(63.5 - 31.3) = 45.1 \, \text{kW}$$

and the after-heater load is:

$$\dot{Q} = \dot{m}_a c_{pma}(t_S - t_W) = 1.4 \times 1.02(14 - 10.8) = 4.6 \, \text{kW}$$

The cooling coil contact factor can be calculated using the temperature at W_s of $10.8 \, °C$:

$$\text{Contact factor} = \frac{t_O - t_W}{t_O - t_{ADP}} = \frac{25 - 10.8}{25 - 8} = 0.835$$

Winter with no heat exchanger:
If the dew point (off-coil) thermostat had the same setting in winter as in summer, the room percentage saturation would not be the same. It must therefore be assumed that it is reset and the temperature at W will change.

With the same latent load in winter as in summer, the supply specific humidity must be the same. Points W_w and B_w can be established by knowing that the process follows a constant wet bulb line with a humidifying efficiency of 90% as explained in Example 8.2. The temperatures at W_w and B_w are $13.1 \, °C$ and $25.8 \, °C$ respectively. Assuming no fan and duct gains, the after-heater process is between W_w and S_w. The winter supply temperature, S_w, can be calculated from the sensible load as follows:

$$t_S = t_R + \frac{\dot{S}}{\dot{m}_a c_{pma}} = 21 + \frac{3}{1.4 \times 1.02} = 23.1 \, °C$$

The cycle can now be completed and the loads calculated as follows:

Pre-heater $\dot{Q} = \dot{m}_a c_{pma}(t_B - t_O(= 1.4 \times 1.02(25.8 + 5) = 44.0 \, \text{kW}$
After-heater $\dot{Q} = \dot{m}_a c_{pma}(t_{S_w} - t_{W_w}) = 1.4 \times 1.02(23.1 - 13.1) = 14.3 \, \text{kW}$

The humidifier evaporation rate is given by

$$\dot{m}_s = \dot{m}_a(\omega_S - \omega_O) = 1.4(0.008 - 0.002\,45) = 0.007\,77 \, \text{kg/s}$$

Summer with a heat exchanger:
In Fig. 8.11(c) the outside air is cooled from O_s to A_s while the room exhaust air is heated from R to E_s. In this situation the efficiency of a heat exchanger is defined as the ratio of the temperature change of the fresh air stream to the difference between the two approach (inlet) temperatures, i.e.

$$0.7 = \frac{t_O - t_A}{t_O - t_R} = \frac{25 - t_A}{25 - 21}$$

Hence $t_A = 22.2 \, °C$

If the ADP remains the same, a coil with a different contact factor must be used and there will be a new set point for the off-coil temperature.

For the new cycle a line is drawn between point A_s and the ADP which cuts the same supply specific humidity as before at a new off-coil point W_s. The line between A_s and the ADP just cuts the saturation line, which is impossible (the situation is in fact possible because condensation is not a straight-line process—see Chapter 6). Taking point W_s as being just saturated gives it a temperature of 10.7 °C, which is almost the same as before, and it is not worthwhile calculating a new after-heater load. The new cooling load and contact factor can now be calculated:

$$\dot{Q}_{cc} = \dot{m}(h_A - h_W) = 1.4(60.7 - 31.2) = 41.3 \, \text{kW}$$

The new contact factor is:

$$\text{Contact factor} = \frac{t_A - t_W}{t_A - t_{ADP}} = \frac{22.8 - 10.7}{22.8 - 8} = 0.818$$

Winter with a heat exchanger:
With a heat exchanger the supply air will be heated from point O_w to point A_w while the room exhaust is cooled from R to E_w. The temperature of point A is found from the exchanger efficiency:

$$0.7 = \frac{t_A - t_O}{t_R - t_O} = \frac{t_A + 5}{21 + 5}$$

Hence $\qquad t_A = 13.2 \, °C$

The new pre-heater load is therefore:

$$\text{Pre-heater } \dot{Q} = \dot{m}_a c_{pma}(t_B - t_A) = 1.4 \times 1.02(25.8 - 13.2) = 18 \, \text{kW}$$

From point B_w the winter cycle is the same as before, i.e. W_w is 13.1 °C and S_w is 23.1 °C. The after-heater load and the mass flow rate of water evaporated will therefore remain unchanged.

It should be noted that there must be some condensation from the exhaust air as it passes through the heat exchanger. The process is shown as a straight line but it would in practice follow a curved path like the process in a cooling coil. If the enthalpy of the exhaust air leaving the heat exchanger at point E_w had been required, it could have been found from an enthalpy balance on the exchanger.

8.6 SINGLE-ZONE SYSTEMS WITH RE-CIRCULATION

Re-circulation of room air is not allowed in some specialized buildings such as hospitals where re-circulation could spread infections. For most buildings, re-circulation is acceptable and results in substantial energy savings. The benefits of using recirculated air are as follows:

(1) reduced cooling loads in summer;
(2) the higher apparatus dew points that can be used, producing a further saving in refrigeration costs;
(3) reduced heating loads in winter;
(4) less humidification required in winter—it may be eliminated altogether in some cases.

Re-circulation ratio

This is defined as the ratio of the mass flow rate of re-circulated air to the mass flow rate of fresh outside air.

$$\text{Re-circulation ratio} = \frac{\dot{m}_R}{\dot{m}_O} \qquad [8.3]$$

Total mass flow rate supplied to the room, $\dot{m}_T = \dot{m}_R + \dot{m}_O$

The minimum mass flow rate of fresh air (FA) is determined by the number of occupants and the room usage (see Section 8.4). For maximum economy the re-circulation ratio should be as high as possible.

Basic Re-circulation

The plant required for a basic re-circulation system is shown in Fig. 8.12(a). A steam humidifier is used in this case; an alternative would be to use a spray humidifier.

Figure 8.12 Basic re-circulation: summer design case

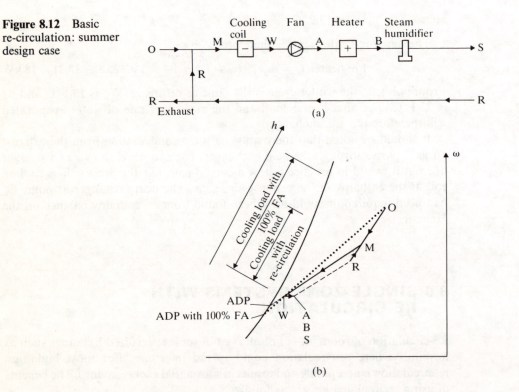

Summer design cycle

The summer design cycle is shown in Fig. 8.12(b). Fresh air at state O is mixed with return room air at state R which results in the mixed state M. The re-circulation ratio is the length of line OM divided by the length of line RM. The closer M is to R, the higher the recirculation ratio. The mixed air then passes over the cooling coil to leave at state W. There is then a small temperature gain across the fan to the supply state S. For comparison a cooling coil process using all outside air to achieve the same off-coil state is shown as a dotted line, demonstrating the reduction in cooling load and increase in ADP.

Winter design cycle

Figure 8.13 shows two winter cycles with re-circulation, (b) and (c), together with a cycle using all outside air for comparison, (a). With all fresh air (a) and no humidification, the specific humidity of the supply air is the same as that of the outside air. With re-circulation and no humidification (b), the supply specific humidity is determined by the mixing ratio OM/MR. Prior to occupancy, the supply specific humidity would be the same as using full fresh air. After occupancy it would steadily increase to reach a balance when the net loss of moisture was equal to the room gains. In cycle (c) steam humidification (process B–S), has been used to increase further the supply specific humidity. The amount of humidification required is quite small compared with what would have been necessary to achieve the same supply state with full fresh air. The higher the recirculation ratio, the less likely it is that humidification will be necessary.

Figure 8.13 Basic re-circulation: winter cycles

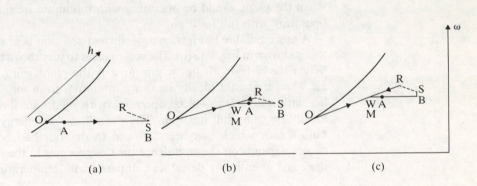

Summer cycles—enthalpy control

For the summer cycle, re-circulation is economic because the enthalpy of the room air is below that of the outside air and the mixed air enters the cooling coil with a lower enthalpy than that of the outside air. If the enthalpy of the outside air is below that required for the room air, it becomes more economic to use full fresh air. For example, if the room conditions were 22 °C and 50% saturation, the specific enthalpy would be 43.4 kJ/kg and the dampers would be controlled to change from minimum to full fresh air at this value of outside air specific enthalpy. This is illustrated in Fig. 8.14(a). The dotted line represents the weather envelope which encloses the normal outside air states.

Enthalpy sensors are now available which measure temperature and humidity and use these values to calculate the enthalpy. A wet bulb temperature detector is a cheaper though less reliable alternative. Because lines of constant enthalpy and lines of constant wet bulb temperature are almost parallel on the psychrometric chart for any value of specific enthalpy, there is an approximate corresponding value of wet bulb temperature. In theory this is a perfectly satisfactory substitute. However, in practice wet bulb sensors can dry out and

Figure 8.14 Summer damper control

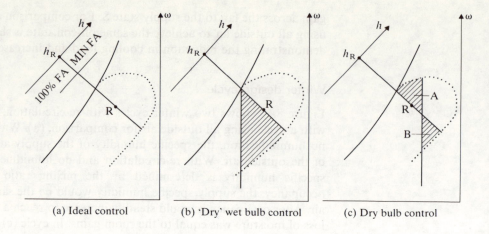

(a) Ideal control   (b) 'Dry' wet bulb control   (c) Dry bulb control

this may not be noticed unless maintenance staff fully appreciate the consequences, or energy consumptions are carefully monitored. Figure 8.14(b) shows the effect of a wet bulb sensor drying out. The area shown hatched in (b) is when the plant would be operating with minimum fresh air when it should be operating with full fresh air.

A more reliable but less energy-efficient solution is to use a dry bulb sensor; this is shown in Fig. 8.13(c). The sensor is set to just above the room temperature. When the outside air lies within area A the plant will operate with full fresh air when it should ideally be using minimum fresh air. When the outside air lies within area B it will be operating with minimum fresh air when it should ideally be using full fresh air. The overall loss in energy is less than if a wet bulb sensor which had been allowed to dry out had been used. Clearly the choice depends on the confidence the designer has in the maintenance staff and the costs of enthalpy detectors compared with temperature detectors.

Mid-season cycles

Figure 8.15(a) shows a cycle using full fresh air. The cooling load is relatively small and will decrease as the enthalpy of the outside air falls. A point will be reached when the temperature of the outside air is sufficiently low to be supplied to the room without being cooled. If the temperature of the outside air falls below this value a supply temperature can be achieved by mixing outside and re-circulated air. There is therefore a weather band during which the dampers will modulate to provide a variable re-circulation ratio. Provided close control of humidity is not required, there is no need either to heat or to cool the supply air for these conditions. A typical mid-season cycle when the room gains are being satisfied solely by ventilation is shown in Fig. 8.15(b).

As the temperature of the outside air continues to fall, the re-circulation ratio will increase until the maximum value is reached (minimum fresh air). The plant will then start to operate on a winter cycle. The operation of the dampers is summarized in Fig. 8.15(c).

In both the illustrated cycles, the outside air temperature is below the room temperature and the room still has a cooling load requirement. This is not unusual. The temperature difference between the inside and outside air governs

Figure 8.15 (a), (b) Basic re-circulation mid-season cycles; (c) basic re-circulation damper control

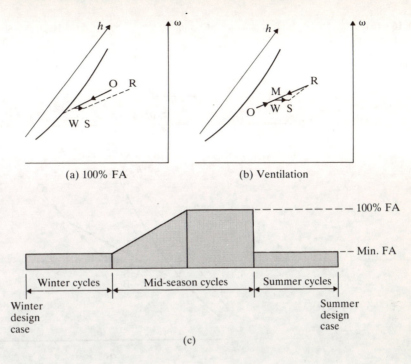

(a) 100% FA

(b) Ventilation

(c)

only the infiltration and fabric heat flow rates, and the latter may have a time lag of up to 9 h (see Chapter 4). Solar gains may be substantial when the outside air temperature is quite low and gains from occupants and machinery do not depend on the weather.

Control is by a signal from a temperature detector in the exhaust duct providing an input to a controller which operates the cooling coil, dampers and heater in sequence (only one is on at any time). As with the 100% fresh air system, a high-level humidity detector can be fitted which will activate an off-coil temperature detector and the plant will over-cool and use reheat until the room humidity has been restored to a satisfactory level.

Re-circulation with Room Air By-pass

An alternative to the basic re-circulation system is to introduce a second mixing of room air with the air which has passed over the cooling coil. The system is illustrated in Fig. 8.16(a) and the summer design cycle in Fig. 8.16(b). The control of humidity is better than for a basic re-circulation system because the secondary mixing has an effect similar to reheating without the extra energy cost.

For the winter design case, the cooler is inoperative so the cycle is exactly the same as for the basic system. The extra complication of this system becomes more worthwhile when the outside air enthalpy is above the room enthalpy for a substantial part of the year; it is unlikely to be economic in the UK.

Example 8.5
(a) Outline the advantages and disadvantages of re-circulation in air conditioning systems.
(b) Two airstreams mix, A from outside, and B re-circulated from an

Figure 8.16 (a) Re-circulation with room air by-pass system; (b) summer design

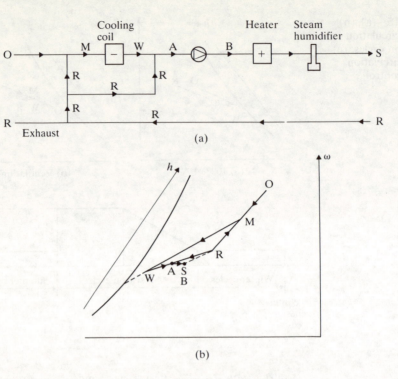

air-conditioned room. The mixed air is cooled and de-humidified by sprayed coil and then reheated and fed to the conditioned room. Determine, using the psychrometric chart and the data below,

(i)   the apparatus dew point of the coil;
(ii)   the cooling load on the coil (kJ/kg of dry mixed air);
(iii) by how many K the cooled air should be reheated to satisfy supply condition;
(iv)   the load (kJ/kg of dry mixed air) on the reheating coil, ignoring any fan heat;
(v)   with the given room latent load, the mass of air that must be supplied per second, and the sensible load that is being satisfied.

Data
Airstream A: 30 °C dry bulb, 27.2 °C wet bulb. Airstream B: 20 °C dry bulb, 50% saturated. Mix ratio: 8 parts (B) to 1 part (A). Supply air condition: 15 °C dry bulb, 0.0046 kg/kg. Spray coil contact factor = 0.9. Room latent load = 10 kW.
(University of Liverpool)

Solution to part (b)
(i)   The system is shown on Fig. 8.17(a) and the cycle on Fig. 8.17(b). Points S, O and R can be fixed on the chart and the mixed point M established either by proportioning lines OM/RM in the ratio of 8:1 or by calculation using an approximate mixing equation as follows:

$$t_M = \left(20 \times \frac{8}{9}\right) + \left(30 \times \frac{1}{9}\right) = 21.1 \,°C$$

Figure 8.17 Example 8.5

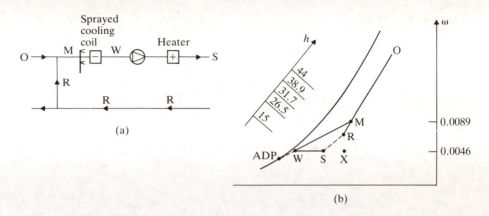

(a)

(b)

The specific humidity is constant across the heater W–S so the coil contact factor can be used as follows:

$$\frac{\omega_M - \omega_S}{\omega_M - \omega_{ADP}} = 0.9 = \frac{0.0089 - 0.0046}{0.0089 - \omega_{ADP}}$$

Hence $\omega_{ADP} = 0.0041$, which corresponds to a temperature of 1.2 °C. The cycle can now be completed on the chart and point X located at the same dry bulb temperature as R and the same specific humidity as S.

(ii)   Specific CC load $= h_M - h_W = 44 - 15 = 29 \text{ kJ/kg}$
(iii)   Temperature rise W–S $= 15 - 3.4 = 11.6 \text{ K}$
(iv)   Reheater load $= h_S - h_W = 26.5 - 15 = 11.5 \text{ kJ/kg}$
(v)   Mass flow rate, $\dot{m}_a = \dfrac{\dot{L}}{(h_R - h_X)} = \dfrac{10}{(38.9 - 31.7)} = 1.39 \text{ kg/s}$

Room sensible heat gain $= \dot{m}_a(h_X - h_S) = 1.39(31.7 - 26.5) = 7.2 \text{ kW}$

Note: The reheater load could have been calculated from $c_{pma}(t_S - t_W) = 1.02(15 - 3.4) = 11.4 \text{ kW}$ and the mass flow rate from $\dot{L} = h_g \dot{m}_a(\omega_R - \omega_S)$ or $\dot{S} + \dot{L} = \dot{m}_a(h_R - h_S)$, i.e. $\dot{m}_a = (10 + 7.2)/(38.9 - 26.5) = 1.39 \text{ kg/s}$.

Example 8.6

(a)   Why do we by-pass the cooling coil with re-circulated air in an air conditioning system and what are the advantages of doing so?

(b)   Discuss a method of controlling the amount of by-pass air by reference to typical psychrometric cycles.

(c)   A room is to be maintained at 22 °C dry bulb, 50% saturation by the air conditioning system shown in Fig. 8.18(a). The room requires 2.2 m³/s of outdoor air, specified at room supply air conditions. Determine the cooling coil apparatus dew point temperature and contact factor under the following conditions.

Room sensible heat gain	90 kW
Room latent heat gain	15 kW
Fan and duct gain	1.5 W
Outdoor air condition	26 °C db, 20 °C wb
Maximum temperature difference between supply and room air	6 K
Dry bulb temperature of the air leaving the cooling coil	11.5 °C

(Newcastle upon Tyne Polytechnic)

Figure 8.18 Example
8.6

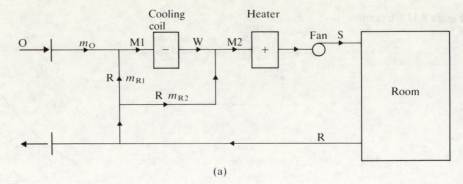

(a)

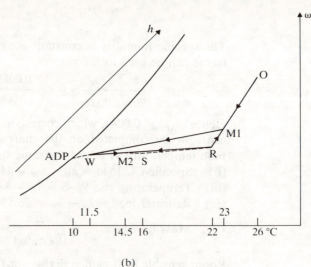

(b)

Solution to part (c)

The psychrometric cycle is shown on Fig. 8.18(b). With a 6 K differential the supply air must be at $(22 - 6) = 16\,°C$. The heater is not in use and the temperature at point M2 must be the supply temperature less the fan and duct gains, i.e. $(16 - 1.5) = 14.5\,°C$. Points O and R can be established on the chart and the temperatures of points W, M2 and S are known.

The total mass flow rate can be calculated from the room sensible heat load using Eqn [1.37]:

$$\dot{m}_a = \frac{\dot{S}}{c_{pma}(t_R - t_S)} = \frac{90}{1.02(22 - 16)} = 14.7\,\text{kg/s}$$

Equation [1.38] can now be used to determine the supply specific humidity using the room specific humidity of 0.0084 from the chart (or the room ratio line could be used):

$$\omega_S = \omega_R - \frac{\dot{L}}{\dot{m}_a h_g} = 0.0084 - \frac{15}{14.7 \times 2540} = 0.008$$

This establishes the supply state S and also point M2 which has the same specific humidity.

The total mass flow rate of recirculated air is therefore $(14.7 - 2.7) = 12\,\text{kg/s}$.

The mass flow rate of outside air is found using Eqn [1.33]:

$$\dot{m}_O = \frac{\dot{V}}{v} = \frac{2.2}{0.83} = 2.7 \text{ kg/s}$$

where the specific volume, v_s, of the supply air is read from the chart as $0.83 \text{ m}^3/\text{kg}$.

The second mixing process where streams at W and R mix to produce stream M2 can now be established on the chart by drawing a line from R through M2 to meet the 11.5 °C db line at W.

The relative masses being mixed are in proportion to the lines W–M2 and M2–R; also, $\dot{m}_{R2} = (12 - \dot{m}_{R1})$, and therefore

$$\frac{\dot{m}_{R2}}{\dot{m}_O + \dot{m}_{R1}} = \frac{12 - \dot{m}_{R1}}{2.7 + \dot{m}_{R1}} = \frac{\text{Line W–M2}}{\text{Line M2–R}} = \frac{13.8 \text{ mm}}{34.7 \text{ mm}}$$

Hence $\dot{m}_{R1} = 7.9 \text{ kg/s}$ and $\dot{m}_{R2} = 12 - 7.9 = 4.1 \text{ kg/s}$

We can now return to the first mixing of the outside air with $\dot{m}_{R1}$ to find the mixed temperature:

$$\dot{m}_O t_O + \dot{m}_{R1} t_R = (\dot{m}_O + \dot{m}_{R1}) t_{M1}$$
$$2.7 \times 26 + 7.9 \times 22 = (2.7 + 7.9) t_{M1}$$
$$\therefore \qquad t_{M1} = 23 \text{ °C}$$

M1, the state after the first mixing, can now be established on the chart on the line between O and R. A line from M1 to W when extended meets the saturation line at the ADP of 10 °C to complete the cycle and the contact factor can be calculated as follows:

$$\text{Contact factor} = \frac{t_M - t_W}{t_M - t_{ADP}} = \frac{23 - 11.5}{23 - 10} = 0.88$$

Iterative Solutions to Problems

For part-load cases the room humidity is normally unknown and the problem is to determine its value. This is particularly difficult with systems using re-circulation of room air, since when the room humidity is not known it is impossible to determine the state of the mixed air entering the cooling coil. One way to overcome this problem is to construct a series of cycles which progressively approach the required solution.

A reasonable estimate is made of the room humidity which enables the state of the mixed air entering the cooling coil to be established. The contact factor can then be used to determine the off-coil state and the supply state can be found by adding any temperature gain across the fan. The room ratio line or the equation $\dot{L} = \dot{m} h_g (\omega_R - \omega_S)$ can then be used to complete the cycle and establish a room humidity. Unless the estimate happened to be the correct answer, the room humidity at the end of the cycle will be different from the value estimated. The room humidity at the end of the first cycle then becomes the starting point for the next cycle. This is repeated until the room humidity is the same at the beginning and end of the cycle. With a close initial estimate two cycles are normally sufficient, and even with a wild initial estimate only three.

Using Simultaneous Equations to Solve Problems

An alternative to using an iterative technique is to establish sufficient equations to calculate the unknown quantities. This takes longer but is more accurate. The two techniques are best demonstrated by means of an example.

Example 8.7
(a)   Discuss the factors which affect the choice of supply-room temperature differential for an all-air air conditioning system.
(b)   An air conditioning system has a recirculation ratio of 3:1 (3 room:1 fresh) by mass. Use an iterative procedure to determine the room percentage saturation for the following part-load case. Fan and duct gains may be neglected.

Room temperature, 22 °C; external dry bulb temperature, 24 °C; external wet bulb temperature, 22 °C; off-coil temperature, 12 °C; coil contact factor, 0.8; sensible heat gain, 70 kW; latent heat gain, 30 kW.
(c)   Write down three equations which would have enabled a solution to part (b) to be found without iteration. A solution is not required.

(Wolverhampton Polytechnic)

Iterative solution
Figure 8.19(a) shows the items of plant in use; with fan gains being neglected, point W is the same as point S. The required cycle will have the shape shown in Fig. 8.19(b). The first step is to establish values for quantities which will not change.

$$t_M = \left(22 \times \frac{3}{4}\right) + \left(24 \times \frac{1}{4}\right) = 22.5\,°C$$

$$\text{Contact factor} = \frac{t_M - t_S}{t_M - t_{ADP}} = \frac{22.5 - 12}{22.5 - t_{ADP}} = 0.8$$

Hence $t_{ADP} = 9.4\,°C$

$$\text{Room ratio line} = \frac{70}{70 + 30} = 0.7$$

Two extreme points of the cycle geometry are now fixed—the outside air state and the ADP. Between these, only the temperatures t_S, t_R and t_M are fixed.

Figure 8.19   Example 8.7

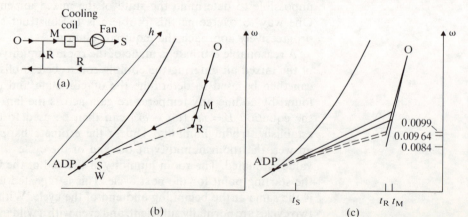

A reasonable value for the room specific humidity is 0.0084, which corresponds to a percentage saturation of 50%; this is the starting point for the first cycle. The cycle ends with a room specific humidity of 0.009 64. Starting from this point the second cycle ends with a room specific humidity of 0.0099. A third cycle starts and ends at this point giving a room percentage saturation of 59%. The first two cycles only are shown in Fig. 8.19(c).

The equation $\dot{L} = \dot{m}_a h_g(\omega_R - \omega_S)$ could have been used to establish ω_R instead of using the room ratio line.

Solution using equations

$$\dot{m}_a = \frac{\dot{S}}{c_{pma}(t_R - t_S)} = \frac{70}{1.02(22 - 12)} = 6.86 \text{ kg/s}$$

From the room latent heat gains:

$$\frac{\dot{L}}{h_g \dot{m}_a} = \frac{30}{2540 \times 6.86} = (\omega_R - \omega_S) \tag{1}$$

From the cooling coil contact factor:

$$\frac{\omega_M - \omega_S}{\omega_M - \omega_{ADP}} = \frac{\omega_M - \omega_S}{\omega_M - 0.007\,35} = 0.8 \tag{2}$$

From the adiabatic mixing:

$$\frac{\omega_O - \omega_M}{\omega_M - \omega_R} = \frac{0.0158 - \omega_M}{\omega_M - \omega_R} = \frac{3}{1} \tag{3}$$

There are now three equations and three unknown specific humidities.

Example 8.8
A room is air-conditioned by an all-air system, comprising a heater, chilled water cooling coil and re-circulating dampers which are controlled in sequence to maintain a room temperature of 22 °C. An enthalpy controller overrides the operation of the fresh-air dampers and closes them to the maximum re-circulation position whenever the outside air enthalpy exceeds the room enthalpy. The summer design conditions occur when the outside air is 27 °C db and 19 °C wb and the room loads are 100 kW sensible and 15 kW latent.
(a) At the summer outside conditions given above, construct the psychrometric cycle and determine the room percentage saturation given the following data:

Minimum fresh air   = 15%
Δt_s max   = 9 K
c_{pma} = 1.02 kJ/kg K
h_g = 2540 kJ/kg
Contact factor (assume constant) = 0.9

(Neglect offsets due to proportional band at controllers.)
(b) When the outside air is at 17 °C db and 16 °C wb the room sensible gain is reduced to 75 kW while the latent gain remains constant. Determine the new percentage saturation of the room air and the coil duty, and plot the psychrometric cycle.

(CIBSE)

Figure 8.20   Example
8.8

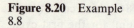

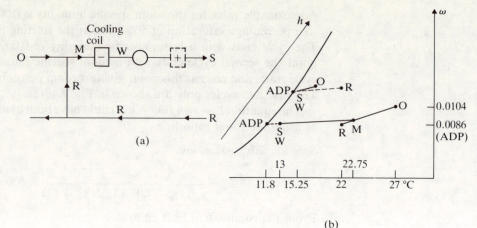

(a)

(b)

Solution

The plant is illustrated in Fig. 8.20(a) and the two psychrometric cycles in Fig. 8.20(b). With sequence operation the heater is off for both cases and the fan and duct gain must be assumed to be negligible.

Part (a) can be solved quite quickly using an iterative method; with an initial estimate of 50% saturation the second cycle gives the required answer and the third cycle confirms this. However, there is no indication in the question that this approach would gain full marks so it is safer to establish and then solve equations.

For part (a) the enthalpy of the outside air is above that of the room air so the dampers will be set to maximum re-circulation.

$$t_M = (22 \times 0.85) + (27 \times 0.15) = 22.75\,°C$$

$$\frac{t_M - t_S}{t_M - t_{ADP}} = \frac{22.75 - 13}{22.75 - t_{ADP}} = 0.9$$

Hence $t_{ADP} = 11.9\,°C$

The ADP and the outside air state can now be established on the chart. Only the temperatures of the room, the mixed and the supply air are known; therefore three equations are required to determine the three unknown specific humidities. From the cooling coil contact factor,

$$\frac{\omega_M - \omega_S}{\omega_M - \omega_{ADP}} = \frac{\omega_M - \omega_S}{\omega_M - 0.0086} = 0.9 \qquad [1]$$

From the adiabatic mixing,

$$\frac{\omega_O - \omega_M}{\omega_O - \omega_R} = \frac{0.0104 - \omega_M}{0.0104 - \omega_R} = 0.85 \qquad [2]$$

The total mass flow rate can be found from the room sensible heat gains,

$$\dot{m}_a = \frac{\dot{S}}{c_{pma}(t_R - t_S)} = \frac{100}{1.02 \times 9} = 10.9\,kg/s$$

From the room latent heat gains,

$$\frac{\dot{L}}{\dot{m}_a h_g} = \frac{15}{10.9 \times 2540} = 0.000\,542 = \omega_R - \omega_S \qquad [3]$$

Solving these equations gives $\omega_R = 0.009\,222$ and $\omega_S = 0.008\,68$, which enables the psychrometric cycle to be completed and gives a room percentage saturation of 56%.

For part (b) the enthalpy of the room air is above that of the outside air so the dampers will be positioned for full fresh air.

$$t_S = t_R - \frac{\dot{S}}{\dot{m}_a c_{pma}} = 22 - \frac{75}{10.9 \times 1.02} = 15.25\,°C$$

Using the coil contact factor

$$\frac{t_O - t_S}{t_O - t_{ADP}} = \frac{17 - 15.25}{17 - t_{ADP}} = 0.9$$

Hence $t_{ADP} = 15.06\,°C$

The line joining point O and the ADP on the chart cuts the 15.25 °C dry bulb line at the supply state S which corresponds to a specific humidity value of 0.0107. With constant latent gains the room, $\Delta\omega$ is the same as before, i.e. 0.000 542. Therefore,

$$\omega_R = \omega_S + \Delta\omega = 0.0107 + 0.000\,55 = 0.011\,24$$

The cycle can now be completed and the new room percentage saturation is found to be 68%.

$$\text{Coil load} = \dot{m}_a(h_O - h_S) = 10.9(45 - 42.4) = 28.3\,kW$$

In this case the cooling coil load is much smaller than the room load, showing that most of the heat gains are being satisfied by the outside air.

8.7 MULTIPLE-ZONE SYSTEMS

A multiple-zone system is defined as one where a single central plant serves more than one zone of a building.

Note: The term *multiple zone* should not be confused with *multi-zone* which some manufacturers use to describe one particular multiple-zone system—the hot-deck/cold-deck system (see page 438). The term 'multi-zone' is therefore unnecessary and is not used in this text.

Multiple-zone systems can be categorized in a number of ways, one of the commonest being according to the terminal cooling fluid. This group's systems are as follows.

All-Air

All the air is treated in a central plant and there is no water piped to individual zones. All the single-zone systems dealt with so far would be classified under this heading, as would the following systems: variable-volume (the supply temperature remains constant and the volume flow rate supplied is reduced as the room sensible heat load falls—see page 469): dual-duct (a constant volume system where the required supply temperature is achieved by mixing air from two ducts adjacent to the room which carry air at different temperatures—see page 438): hot-deck/cold-deck (similar to the dual-duct, but the mixing takes place in the plant room—see page 438) and zonal reheat (separate reheaters serve each zone—see page 432).

Air/Water

Sufficient air is supplied from a central plant to satisfy the fresh air and latent load requirements of each zone. The major part of the zone sensible heat load is satisfied by dry cooling coils (no condensation) and heating coils within the zone. It is therefore necessary to pipe water to each zone. Induction and some fan–coil units would be classified in this way: either an air jet (induction—see page 451) or a fan (fan–coil—see page 461) causes the room air to flow over the secondary room coils.

All-Water

Fresh air is supplied either by natural ventilation or through wall apertures. All the cooling takes place within each zone, which must be supplied with chilled water. Some fan–coil units would be classified in this way.

Unitary

These units are self-contained and require no central plant. The refrigeration plant and air handling unit are combined in a single package. Some of these systems should not strictly be classified as multiple-zone since each zone would have its own units and be independent of any other zone (which is one of their advantages). One unitary system which does depend on other zones is the water-loop heat pump system. It is an ideal system when a building has a near-balance between heating and cooling loads for a substantial part of the year. Water is circulated around the building in a ring main with reversible heat pumps in individual zones receiving heat from or rejecting heat to the water.

Multiple-Zone System Groupings

The above system of classification is the most useful in describing the impact of the services on the building design, for example the ductwork and pipework requirements, the space required for plant rooms, and aesthetics. However it can be confusing when studying the psychrometrics of the various systems. Fan–coils have appeared in two categories and the water-loop heat pump system has much in common with the all-water systems. Reheaters may be in plant rooms or within zones which could be supplied with water: the psychrometric cycle would be the same in both cases. If recool were used instead of reheat, it would clearly be reclassified as an air/water system (a zonal reheat system can use electrical heaters). The following four groupings have therefore been used to analyse the psychrometrics of multiple-zone systems:

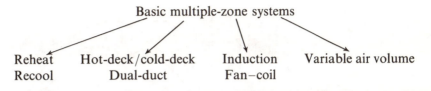

each of the above groups has distinctive psychrometrics and calculation procedures.

Combined systems

Features from more than one basic system may be used in an air conditioning plant. Two common combinations are:

    variable air volume and dual-duct;

    variable air volume and reheat.

It is necessary to understand the basic systems before attempting to design combined plants.

Zone Loads

The load in each zone of a building will vary throughout the day, due to changes in internal loading and external climate. The daily loading will also vary throughout the year. An air conditioning plant must be able to handle the maximum cooling and maximum heating load in each zone. For different zones the maximum cooling load may occur at different times of the day and at different times of the year.

Sometimes internal gains will determine the time of day when the maximum cooling load will occur. For example, a factory canteen would have its maximum gain at lunchtime whereas a theatre would have its maximum gain in the evening.

When internal gains are reasonably constant, solar radiation will determine when the maximum cooling load is required. A zone with substantial glazing which faces east will have its maximum gains in the morning during midsummer. The same zone facing west would have its maximum gains at the same time of year but in the late afternoon. For other orientations the maximum is less predictable since the maximum solar gains do not occur at the same time of the year as maximum air wet and dry bulb temperatures. For example, maximum solar gains for a south-facing zone occur at the spring and autumn equinox when air temperatures are well below their maximum values, so the maximum cooling load is likely to be in late summer.

For a multiple-zone building it is necessary to know the maximum heat gains for the building as well as the maximum heat gains for each zone. The maximum heat gains for the building will always be less than the sum of the maximum gains for each zone. This is illustrated in Fig. 8.21. The building has two zones (E and W), one facing east and the other west, both with appreciable glazing and constant internal gains. This is a very simple situation where the design heat gains for each zone would occur on the same day. In this case the building

Figure 8.21

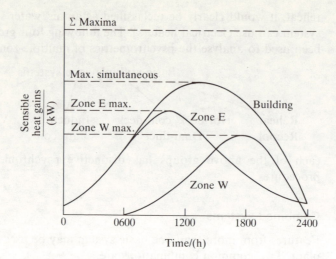

maximum heat gain, which is normally termed the maximum simultaneous heat gain, is substantially lower than the sum of the maximum heat gains for each zone. The ratio of the maximum simultaneous load to the sum of the individual maxima is one of the most important factors to consider when selecting an air conditioning system. For example, if the two ratios are about equal, a zonal reheat system could be economic; on the other hand, the potential of a variable air volume system would not be realized and the extra capital cost could not be justified. The more the two values diverge, the higher will be the running costs of the reheat system, and the variable air volume system would then become increasingly attractive.

Most multiple-zone buildings will be far more complex than the above example. Each zone will have its maximum gain at a set time on a particular day of the year, and these values will be different for each zone. The maximum simultaneous gain will frequently occur when the zone with the greatest load has its maximum gain, but this need not be the case and may occur when none of the zones is at a maximum, as in Fig. 8.21. To analyse the loads throughout the summer for a large number of zones is only practical with the aid of a computer program.

For the winter design loads the absence of solar influence means that all zones are equally affected by the weather. Any difference between the maximum simultaneous load and the sum of the maximum loads is therefore entirely due to varying internal loads. In some buildings this might still be important, but for most buildings the effect is negligible.

8.8 TERMINAL REHEAT SYSTEMS

A typical terminal reheat system is shown in Fig. 8.22. The air is treated in a central plant to provide a common supply state for a number of zones. Thermostats in each zone then adjust the supply temperature by turning on reheaters to match the individual sensible loads. Figure 8.23(a) shows the summer design cycle for a complete plant with an average supply state S and a mixed room/return state R. The distribution ductwork is low-velocity and

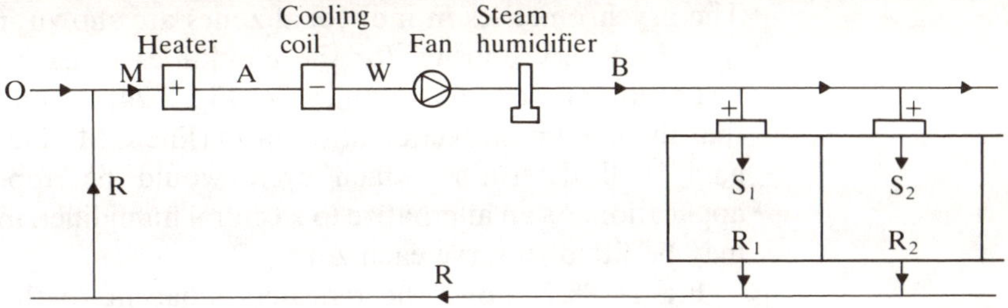

Figure 8.22 Terminal reheat system

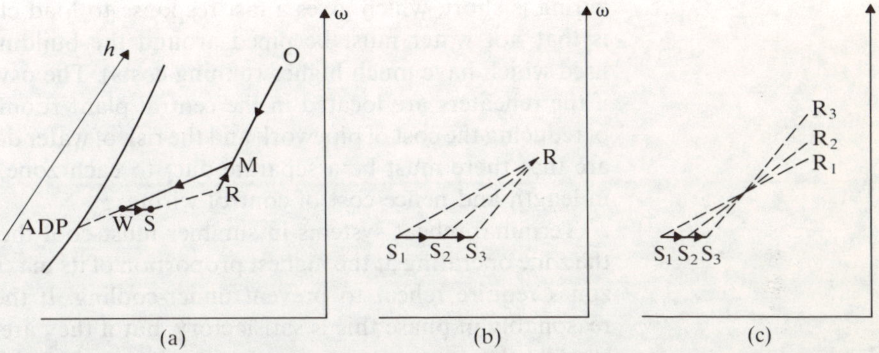

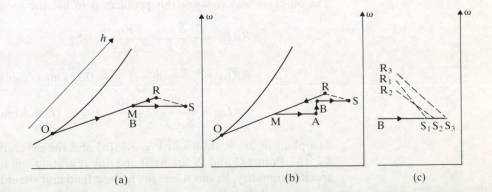

Figure 8.23 Terminal reheat: summer cycles

there is only a small pressure drop across the room diffusers. The temperature rise across the fan is therefore normally no more than 1 K and has been neglected in this section.

Terminal reheat systems give good control of temperature and humidity. For the summer design case it is possible to design for the same humidity in each zone, as shown in Fig. 8.23(b); however this is costly both in reheat and ductwork (the supply/room differentials are low and this results in high flow rates). It is therefore more normal to allow some variation in humidity, as shown in Fig. 8.23(c). For the summer design case there will always be at least one zone which will not require reheat. This also applies at part-load, although the zone not requiring reheat may change. Correct conditions can be maintained by using a load analyser which monitors the control signals to the reheaters and resets the supply temperature until one of the reheaters is off.

Winter cycles are shown in Fig. 8.24. Figure 8.24(a) shows the simplest possible cycle for an overall plant with no pre-heat and no humidification. Figure 8.24(b) shows an overall cycle with pre-heat and a steam humidifier.

Figure 8.24 Terminal reheat: winter cycles

The psychrometrics in individual zones are shown in Fig. 8.24(c). In winter it would be very unusual for the humidities in each zone to be the same. An inspection of the two overall cycles, Fig. 8.24(a) and Fig. 8.24(b), demonstrates that the maximum recirculation ratio (line OM/line RM) and the room latent gains will determine which cycle would be appropriate for a particular application. As an alternative to a central humidifier, individual steam humidifiers may be fitted to serve each zone.

Figure 8.22 shows the reheaters adjacent to the conditioned spaces. The advantages of this are that branching ductwork can be used and the control wiring is short, which gives a fast response to load changes. The disadvantage is that hot water must be piped around the building (or electrical reheaters used which have much higher running costs). The psychrometrics are the same if the reheaters are located in the central plant room. This has the advantage of reducing the cost of pipework and the risk of water damage. The disadvantages are that there must be a separate duct to each zone, and there is an increase in length and hence cost of control wiring.

Terminal reheat systems in summer must cool the air sufficiently to satisfy the zone operating at the highest proportion of its maximum load; the remaining zones require reheat to prevent under-cooling. If the loads in each zone are reasonably in phase this is satisfactory, but if they are not then terminal reheat systems become very uneconomic. Compared with alternative multi-zone systems they are relatively simple to design, operate and maintain.

Example 8.9

Two zones of a building are to be maintained at 22 °C and a maximum of 60% saturation by a terminal reheat system which uses 100% fresh air. The outside air enters the cooling coil at 28 °C db and 22 °C wb and leaves at 14 °C db with a specific humidity of 0.009. The summer design loads are as follows:

Zone	Max. $\dot{S}$	Max. simultaneous $\dot{S}$	$\dot{L}$
A	30 kW gain	30 kW gain	5 kW gain
B	15 kW gain	10 kW gain	5 kW gain

If the temperature rise across the fan is negligible, calculate the percentage saturation in each zone and the heating and cooling loads for maximum simultaneous gains.

Solution

The quickest way to solve this problem is to use the room ratio lines.

$$RRL_A = \frac{\dot{S}}{\dot{S} + \dot{L}} = \frac{30}{30 + 5} = 0.86$$

$$RRL_B = \frac{\dot{S}}{\dot{S} + \dot{L}} = \frac{15}{15 + 5} = 0.75 \text{ (max. gain)}$$

$$RRL_{B1} = \frac{\dot{S}}{\dot{S} + \dot{L}} = \frac{10}{10 + 5} = 0.67 \text{ (max. simultaneous gain)}$$

The plant in use is shown in Fig. 8.25(a) and the psychrometric cycles in Fig. 8.25(b). Points O and W are fixed and any reheating will take place at constant specific humidity. Room A has the highest load and should be investigated first.

Figure 8.25 Example 8.9

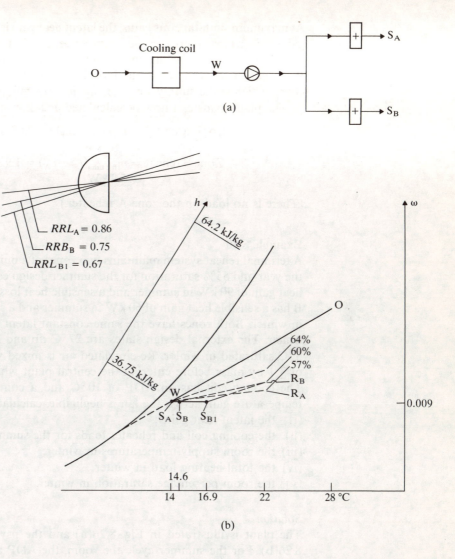

(a)

(b)

If the *RRL* for zone A is drawn from point W to meet the 22 °C db line, the room percentage saturation is 57% (which is satisfactory) and room A will be supplied with air at state W. (Points W and S_A are at the same state.)

$$\dot{m}_A = \frac{\dot{S}}{c_{pma}(t_R - t_S)} = \frac{30}{1.02(22 - 14)} = 3.68 \text{ kg/s}$$

Each zone must be supplied with sufficient air to satisfy its maximum gains. If the *RRL* for room B at maximum gains is drawn from W, the resulting room percentage saturation is 64%, which is too high. Room B must therefore be at 60% saturation. This is state R_B on the chart. The same *RRL* is therefore drawn from R_B to cut the 0.009 specific humidity line at S_B which is at 14.6 °C db. The mass flow rate for room B can now be calculated:

$$\dot{m}_B = \frac{\dot{S}}{c_{pma}(t_R - t_S)} = \frac{15}{1.02(22 - 14.6)} = 1.99 \text{ kg/s}$$

At maximum simultaneous gains, the latent heat gain is the same as for maximum gains (given) and therefore the room percentage saturation must be the same. The room ratio line for maximum simultaneous gains is therefore drawn from R_B to cut the same specific humidity line at 16.9 °C. This is point S_{B1} on the chart, which is the supply state for maximum simultaneous gains.

The plant loads can now be calculated as follows:

$$\text{Cooling coil } \dot{Q} = \dot{m}_T(h_O - h_W) = (3.68 + 1.99)(64.2 - 36.75)$$
$$= 156 \text{ kW}$$
$$\text{Zone B reheat} = \dot{m}_B c_{pma}(t_S - t_W) = 1.99 \times 1.02(16.9 - 14)$$
$$= 5.9 \text{ kW}$$

(There is no load on the zone A reheater.)

Example 8.10

A terminal reheat system maintains two zones of a building at 21 °C throughout the year and 55% saturation for the summer design case. Zone A has a sensible heat gain of 90 kW in summer and a sensible heat loss of 40 kW in winter. Zone B has a sensible heat gain of 60 kW in summer and a sensible heat loss of 30 kW in winter. Both zones have the same constant latent heat gain in summer and winter. The external design states are 27 °C db and 21 °C wb in summer and 5 °C saturated in winter. Re-circulated air is mixed with fresh air in the ratio of 3:1 by mass before entering the central plant which has a cooler and fan. The cooling coil has an ADP of 10 °C and a contact factor of 0.85. If the temperature gain across the fan is negligible, calculate:

(i) the latent heat gains;
(ii) the cooling coil and reheater loads for the summer design case;
(iii) the room supply temperatures in winter;
(iv) the total heating load in winter;
(v) the room percentage saturation in winter.

Solution

The plant is illustrated in Fig. 8.26(a) and the psychrometric cycles in Fig. 8.26(b). For the summer cycle the room, the ADP and the outside air states are fixed. The mixed point M can be found by proportioning line OM and MR in the ratio of 3:1, and the off-coil state W by proportioning line MW to line M–ADP in the ratio of 0.85:1, or by calculation. With no fan gains, state W becomes the supply state from the central plant and has a temperature of 12 °C, a specific enthalpy of 32 kJ/kg and a specific humidity of 0.007 95. The room specific humidity is 0.008 65 and the specific enthalpy of the mixed air entering the coil is 47.5 kJ/kg. The maximum temperature differential (i.e. room − supply) should be used for the zone with the largest sensible heat gain, which in this case is zone A.

$$\dot{m}_A = \frac{\dot{S}}{c_{pma}(t_R - t_S)} = \frac{90}{1.02(21 - 12)} = 9.8 \text{ kg/s}$$

The room latent heat gain can now be found using Eqn [1.38] as follows:

$$\dot{L} = \dot{m}_A h_g(\omega_R - \omega_S) = 9.8 \times 2540(0.008\,65 - 0.007\,95) = 17.4 \text{ kW}$$

or

$$\dot{L} = \dot{m}_A(h_R - h_S) - \dot{S} = 9.8(43 - 32) - 90 = 17.8 \text{ kW}$$

Figure 8.26 Example 8.10

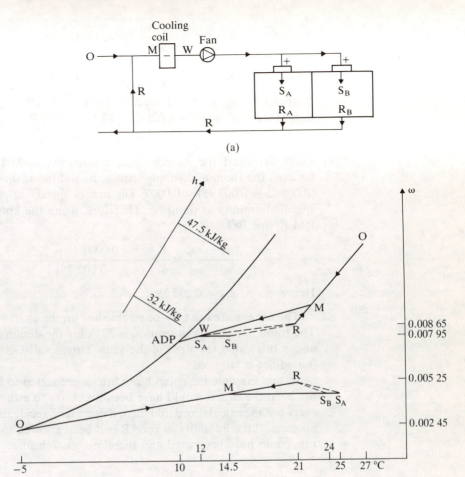

(a)

(b)

With both zones at the same percentage saturation, the increase in specific humidity must be the same and if the latent heat gains are the same the mass flow rates must be the same. The supply temperature to zone B can now be calculated:

$$t_{SB} = t_R - \frac{\dot{S}}{\dot{m}_B c_{pma}} = 21 - \frac{60}{9.8 \times 1.02} = 15\,^\circ\text{C}$$

and hence the cooling coil and reheater loads:

Cooling coil $\dot{Q}_{cc} = \dot{m}_T(h_M - h_W) = 2 \times 9.8(47.5 - 32) = 304\,\text{kW}$
Zone B reheater $\dot{Q} = \dot{m}_B c_{pma}(t_S - t_W) = 9.8 \times 1.02(15 - 12) = 30\,\text{kW}$

In winter the temperature of the air leaving the central plant at W can be found from the mixing process, and the room supply temperatures from the room sensible heat losses.

$$t_M \times 4 = (21 \times 3) + (-5 \times 1)$$
$$\therefore \quad t_M = 14.5\,^\circ\text{C}$$

Zone A: $t_S = 21 + \dfrac{40}{9.8 \times 1.02} = 25\,°\text{C}$

Zone B: $t_S = 21 + \dfrac{30}{9.8 \times 1.02} = 24\,°\text{C}$

Reheater A: $\dot{Q} = 9.8 \times 1.02(24 - 14.5) = 95\,\text{kW}$

Reheater B: $\dot{Q} = 9.8 \times 1.02(25 - 14.5) = 105\,\text{kW}$

$$\text{Total load} = \overline{200\,\text{kW}}$$

From the chart the outside specific humidity is 0.002 45 and the difference between the room and supply specific humidities is the same as in summer, i.e. $(0.008\,65 - 0.007\,95) = 0.0007$. The supply specific humidity is the same as the specific humidity at point M. Therefore, using the known proportions of line RM to line RO:

$$\frac{\omega_R - \omega_S}{\omega_R - \omega_O} = \frac{1}{4} = \frac{0.0007}{(\omega_R - 0.002\,45)}$$

Hence $\omega_R = 0.005\,25$

The room state can now be established on the chart and the cycle completed. The room percentage saturation is 33% for the design case and would be well above this value for most of the year. This is satisfactory for comfort and no humidifier is required.

In this example the room humidities were stated to be the same. If this had not been the case it would have been necessary to estimate the return air state, carry out the calculations and then, if necessary, recalculate using the new return air state. If the humidity in zone B had been allowed to rise, a lower mass flow rate could have been used and therefore less reheat.

8.9 DUAL-DUCT AND HOT-DECK/COLD-DECK SYSTEMS

Dual-duct and hot-deck/cold-deck systems both deliver a constant mass of air to each zone and achieve their supply states by mixing streams of hot and cold air. Although they have quite different physical arrangements, the psychrometrics are the same for the hot-deck/cold-deck system and the basic dual-duct system.

A hot-deck/cold-deck arrangement is shown in Fig. 8.27. Motorized mixing dampers control the proportions of hot and cold air supplied to each zone. The mixing takes place in the plant room so that only one duct is connected to each zone. With this arrangement the maximum number of zones which can be served from one plant is about twelve.

The layout of a basic dual-duct system is shown in Fig. 8.28. Only two ducts leave the plant room and can then be branched so that each zone is connected to both a hot and a cold duct. There is no limit to the number of zones which can be supplied in this way. Each zone is then supplied via its own mixing box. A typical mixing box is shown in Fig. 8.29.

Dual-duct systems are very responsive to changes in load and can simultaneously accommodate heating in some zones and cooling in others. Control of temperature is excellent but there is no control of humidity in individual zones.

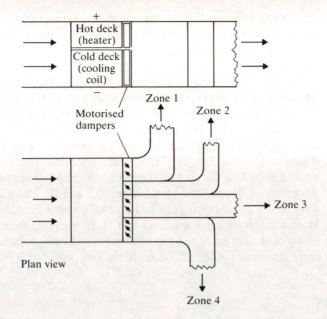

Figure 8.27 Hot-deck/cold-deck system

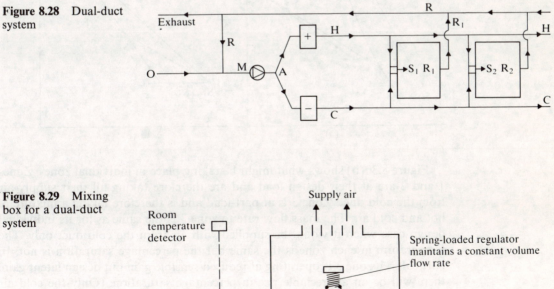

Figure 8.28 Dual-duct system

Figure 8.29 Mixing box for a dual-duct system

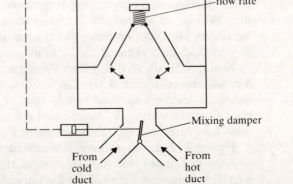

When analysing the psychrometrics it is important to distinguish between the behaviour of individual zones and the behaviour of the system as a whole.

Figure 8.30(a) shows a summer design cycle for a complete system. After mixing there will be a fan gain of about 2–3 K (M to A) and some of the air at this state will be delivered to the hot duct. In summer the heater is off and hence states A and H coincide. With only a small temperature difference between the hot duct and rooms, state A will be the state of the hot air delivered to the terminal boxes. The remaining air will pass through the cooling coil in the cold duct (A to C). A typical off-coil temperature would be about 13 °C and, depending on the length of duct and insulation, there could be a duct gain of up to 3 K before the air is delivered to the most distant mixing boxes. The cold duct state will therefore vary between C and C′. The mean supply state S will then lie on a line joining the hot and mean cold duct states (i.e. on a line from H to a point along the line C to C′). In general, state S will be close to the mean cold duct state but the two states will only coincide when all zones have their maximum gains at the same time.

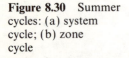

Figure 8.30 Summer cycles: (a) system cycle; (b) zone cycle

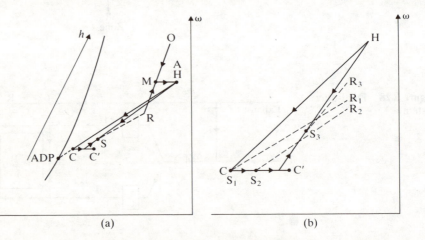

(a)   (b)

Figure 8.30(b) shows what might be taking place in individual zones. Zones 1 and 2 are at their design load and are therefore taking all their supply air from the cold duct. Zone 3 is at part-load and is therefore taking a mixture of hot and cold air. The mass flow rate to zone 3 is the same as for its individual design case when it would be supplied with air from the cold duct only. The temperature in each zone is the same but the percentage saturation is not. In particular, if zone 3 is operating at reduced sensible gain but design latent gain, there will be an appreciable rise in percentage saturation. (Only the cold air which has been de-humidified can take moisture away from the zone, so if the flow of cold air is reduced the humidity must rise.)

Figure 8.31 shows the winter design cycles for a simple system with no humidification. Figure 8.31(a) is for the overall system and 8.31(b) for the individual zones. The mean supply state, S, depends on the temperatures and mass flow rates at S_1, S_2 and S_3 and the mean room state R depends on the mass flow rates and the temperatures of R_1, R_2 and R_3.

Figure 8.32 shows a more sophisticated system with a steam humidifier situated in the hot duct downstream of the heater. Figure 8.32(a) is for the overall system and Fig. 8.32(b) for the individual zones. In both cases there

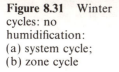

Figure 8.31 Winter cycles: no humidification: (a) system cycle; (b) zone cycle

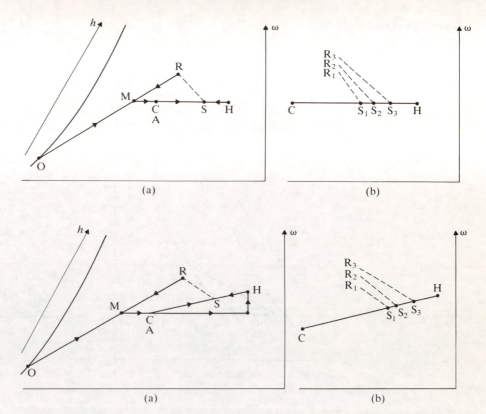

Figure 8.32 Winter cycles with humidification: (a) system cycle; (b) zone cycle

will be little change in temperature along the cold duct but some allowance should be made for heat losses from the hot duct. The humidifier would be controlled by a humidity detector in one of the zones. There would be no control over humidity in the other zones. In winter most zones will be supplied with mixed air. The severity of the external design temperature, the internal latent gains and the minimum tolerable percentage saturation would dictate when humidification was necessary.

The lack of control of humidity in basic dual-duct systems is more of a problem at part-loads than for the summer design case. The problem is most severe when external humidities are high with reduced sensible gains but design latent gains. Such conditions can easily occur in summer when cloud cover eliminates direct solar radiation. One way to overcome this problem is to fit high-limit humidity detectors to critical zones which switch on the heater in the hot duct. This results in zones being supplied with a higher proportion of de-humidified cold air. This is the equivalent of cooling and reheating and must be paid for in increased refrigeration costs as well as the cost of heating.

It is possible to achieve close control over humidity by cooling all the mixed air and then reheating some of the air in the hot duct. Such an arrangement is called a dew point plant, and is illustrated in Fig. 8.33(a). The summer design cycles for the system and typical individual zones are illustrated in Fig. 8.33(b) and (c).

The dew point plant has a much higher refrigeration load than the basic plant. The two plants represented by Figs 8.28 and 8.33(a) represent the extremes of dual-duct systems. Between these extremes there are a number of arrangements (such as a pre-cooler on the fresh air intake) which give varying degrees of

Figure 8.33 Dual-duct: dew point system

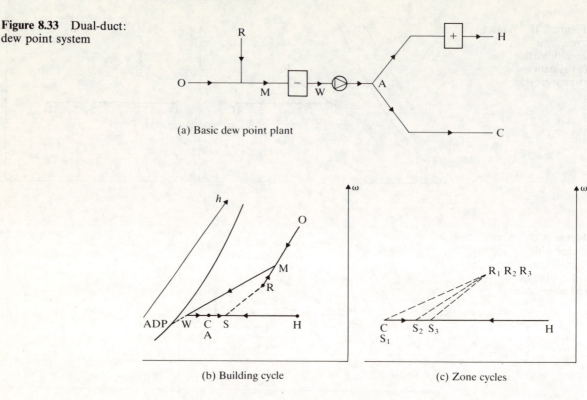

(a) Basic dew point plant

(b) Building cycle

(c) Zone cycles

humidity control. A detailed description of these plants can be found in the ASHRAE *Systems Handbook*.[8.6]

The flow rate for any one zone is the greater of the flows necessary to satisfy the winter and the summer design loads. Because summer differentials (the difference between room and supply temperatures) are much lower than for winter, it is invariably the summer case which will control. The flow through the cold duct is then determined by taking account of the maximum simultaneous gains. Unless all zones have their maximum gains at the same time, there will still be some flow through the hot duct but this will be relatively small. In winter there will be few zones (if any at all) which will require 100% hot air for their design case. In general, the necessary supply temperature will be achieved by mixing, so that there will still be a substantial flow through the cold duct. Hence the hot duct can be sized for a much smaller flow rate than the cold duct.

In the cycles illustrated so far it has been assumed that the temperature in each zone is the same. This need not be the case, since an individual zone thermostat will control its own mixing box.

It should be remembered that although a relatively small number of zones has been used in the above description and in the following examples, in practice a dual-duct system is likely to be used in buildings with a large number of zones. If a building has only two or three zones a hot-deck/cold-deck system would be more appropriate. The CIBSE *Guide*[8.7] states that hot-deck/cold-deck systems are available to serve up to 12 zones. However, since the thermodynamics and psychrometrics of the two systems are the same, the following calculations apply to both systems.

Example 8.11

Two zones of a building are maintained at 21 °C throughout the year by a dual-duct system which has a re-circulation ratio of 4:1 by mass (recirculated air/fresh air). The cooler off-coil temperature is 11 °C and the maximum hot duct temperature is 43 °C. There is a temperature rise across the fan of 2 K, a temperature rise in the cold duct of 2 K in summer and a temperature fall of 3 K in the hot duct in winter. The design cases for both zones occur at the same time and are as follows:

Zone	Summer sensible gain at 28 °C/(kW)	Winter sensible gain at −2 °C/(kW)
A	50	− 30
B	40	− 20

Calculate:
(i)   the mass flow rate through the fan;
(ii) the supply temperature to each zone for the winter design case;
(iii) the maximum mass flow rate through the hot duct;
(iv) the heater design load.

Solution

The plant is shown in Fig. 8.34. In this example there is no reference to humidities and the problem can be solved without using the psychrometric chart. The first step is to determine the design flow rates to each zone.

Figure 8.34 Example 8.11

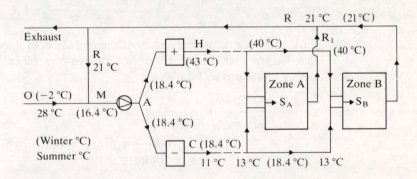

(i)   Summer:
The cold duct temperature will be the supply temperature to both zones and is 2 K above the off-coil temperature because of the duct gain. Therefore $t_c = 11 + 2 = 13 °C$ and Eqn [1.37] can be used to calculate the mass flow rates required:

$$\dot{m}_A = \frac{\dot{S}}{c_{pma}(t_R - t_S)} = \frac{50}{1.02(21 - 13)} = 6.13 \text{ kg/s}$$

$$\dot{m}_B = \frac{\dot{S}}{c_{pma}(t_R - t_S)} = \frac{40}{1.02(21 - 13)} = 4.9 \text{ kg/s}$$

Winter:
The air leaves the heater at 43 °C and with a 3 K temperature drop in the duct the supply temperature will be 40 °C. It should be noted that the higher the

hot duct temperature, the smaller the hot duct becomes, but this saving is offset by higher duct heat losses. To minimize heat losses the hot duct temperature would be scheduled to fall as the room loads decrease.

With all the flow passing through the hot duct, we would have:

$$\dot{m}_A = \frac{\dot{S}}{c_{pma}(t_S - t_R)} = \frac{30}{1.02(40 - 21)} = 1.55 \, \text{kg/s}$$

$$\dot{m}_B = \frac{\dot{S}}{c_{pma}(t_S - t_R)} = \frac{20}{1.02(40 - 21)} = 1.03 \, \text{kg/s}$$

Clearly both zones are controlled by the mass flow rate required for their summer design case. The cold duct must therefore be sized for $6.13 + 4.9 = 11.03 \, \text{kg/s}$, which is also the flow through the fan in this example. (This is usually obvious by inspection without the need for a calculation.)

(ii)   Assuming a negligible change in specific volume at the fan, the mass flow rates supplied to each zone in winter will be the same as in summer. The winter supply temperatures can therefore be calculated as follows:

$$t_{SA} = t_R + \frac{\dot{S}}{c_{pma}\dot{m}_A} = 21 + \frac{30}{1.02 \times 6.13} = 25.8 \, ^\circ\text{C}$$

$$t_{SB} = t_R + \frac{\dot{S}}{c_{pma}\dot{m}_B} = 21 + \frac{20}{1.02 \times 4.9} = 25 \, ^\circ\text{C}$$

(iii)   Mixing of fresh and return air:

$$t_M = \frac{4t_R}{5} + \frac{1t_O}{5} = \frac{4 \times 21}{5} + \frac{-2}{5} = 16.4 \, ^\circ\text{C}$$

With a 2 K temperature rise across the fan, the cold duct temperature becomes 18.4 °C. The mass flow rates of hot air can now be calculated from adiabatic mixing equations as follows.

Zone A mixing box:

$$\dot{m}_H t_H \quad + \dot{m}_C t_C \quad = \dot{m}_T t_S$$
$$\dot{m}_H \times 40 + (6.13 - \dot{m}_H)18.4 = 6.13 \times 25.8$$
$$\therefore \quad \dot{m}_H = 2.1 \, \text{kg/s}$$

Zone B mixing box:

$$\dot{m}_H \times 40 + (4.9 - \dot{m}_H)18.4 = 4.9 \times 25$$
$$\therefore \quad \dot{m}_H = 1.5 \, \text{kg/s}$$

The total mass flow rate in the hot duct is therefore $(2.1 + 1.5) = 3.6 \, \text{kg/s}$. (The mass flow rate for the cold duct for the winter design case is $11.03 - 3.6 = 7.43 \, \text{kg/s}$.)

For the heater,

$$\dot{Q} = \dot{m}_a c_{pma}(t_H - t_C) = 3.6 \times 1.02 \times (43 - 18.4) = 90.3 \, \text{kW}$$

Example 8.12

A hot-deck/cold-deck air conditioning system maintains three zones of a building at 21 °C in winter. In summer, zones A and B are at 21 °C but zone D is at 24 °C. For the summer design case the cold deck is at 15 °C and the hot deck at 23 °C. For the winter design case the cold deck is at 16 °C and the hot deck at 40 °C. The design loads are as follows:

Zone	Max. sensible gain/(kW)	Max. simultaneous sensible gain/(kW)	Minimum sensible gain/(kW)
A	60	60	−20
B	50	25	−20
D	20	20	−50

For both winter and summer cases calculate the supply temperatures to each zone and the mass flow rates through the hot and cold decks.

Solution
The plant is illustrated in Fig. 8.35. The first step in the solution is to establish the mass flow rates to each zone.

Figure 8.35 Example 8.12

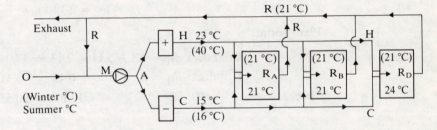

Summer:

$$\dot{m}_A = \frac{60}{1.02(21-15)} = 9.8 \text{ kg/s*}$$

$$\dot{m}_B = \frac{50}{1.02(21-15)} = 8.17 \text{ kg/s*}$$

$$\dot{m}_D = \frac{20}{1.02(24-15)} = 2.18 \text{ kg/s}$$

Winter:

$$\dot{m}_A = \frac{20}{1.02(40-21)} = 1.03 \text{ kg/s}$$

$$\dot{m}_B = \frac{20}{1.02(40-21)} = 1.03 \text{ kg/s}$$

$$\dot{m}_D = \frac{50}{1.02(40-21)} = 2.58 \text{ kg/s*}$$

Note: (*) The asterisk indicates the controlling mass flow rates. The total mass flow rate at the fan is therefore 9.8 + 8.17 + 2.58 = 20.55 kg/s. With experience the winter flow rates to zones A and B could have been dismissed without calculation.

Summer simultaneous gains:

Zone A: Because the design and simultaneous gains are the same,

$$t_S = 15\,°C, \dot{m}_C = 9.8\,\text{kg/s} \quad \text{and} \quad \dot{m}_H = 0\,\text{kg/s}$$

Zone B:

$$t_S = t_R - \frac{\dot{S}}{c_{pma}\dot{m}_B} = 21 - \frac{25}{1.02 \times 8.17} = 18.0\,°C$$
$$\dot{m}_H t_H + \dot{m}_C t_C = \dot{m}_T t_S$$
$$(\dot{m}_H \times 23) + (8.17 - \dot{m}_H)15 = 8.17 \times 18.0$$
$$\therefore \qquad \dot{m}_H = 3.06\,\text{kg/s}$$
$$\therefore \qquad \dot{m}_C = 5.11\,\text{kg/s}$$

Zone C: $\qquad t_S = 24 - \dfrac{20}{1.02 \times 2.58} = 16.4\,°C$
$$(\dot{m}_H \times 23) + (2.58 - \dot{m}_H)15 = 2.58 \times 16.4$$
$$\therefore \qquad \dot{m}_H = 0.45\,\text{kg/s}$$
$$\therefore \qquad \dot{m}_C = 2.13\,\text{kg/s}$$

Plant room:

Cold deck $\dot{m}_T = 9.8 + 5.11 + 2.13 = 17.04\,\text{kg/s}$

Hot deck $\dot{m}_T = 0 \ + 3.06 + 0.45 \ = \ 3.51\,\text{kg/s}$

Check: $\qquad\qquad\qquad\qquad\qquad\qquad$ Fan $\dot{m} = \overline{20.55\,\text{kg/s}}$ (as before)

Winter:

Zone A: $\qquad\qquad t_S = t_R + \dfrac{\dot{S}}{\dot{m}_a c_{pma}} = 21 + \dfrac{20}{1.02 \times 9.8} = 23\,°C$
$$(\dot{m}_H \times 40) + (9.8 - \dot{m}_H)16 = 9.8 \times 23$$
$$\therefore \qquad \dot{m}_H = 2.86\,\text{kg/s}$$
$$\therefore \qquad \dot{m}_C = 6.94\,\text{kg/s}$$

Zone B: $\qquad\qquad\qquad\qquad t_S = 21 + \dfrac{20}{1.02 \times 8.17} = 23.4\,°C$
$$(\dot{m}_H \times 40) + (8.17 - \dot{m}_C)16 = 8.17 \times 23.4$$
$$\therefore \qquad \dot{m}_H = 2.52\,\text{kg/s}$$
$$\therefore \qquad \dot{m}_C = 5.65\,\text{kg/s}$$

Zone C: The winter case controlled the mass flow rate to zone C; it will therefore be supplied at 40 °C with all the flow through the hot duct.

Plant room:

Cold deck $\dot{m}_T = 6.94 + 5.65 + 0 \ = 12.59\,\text{kg/s}$

Hot deck $\dot{m}_T = 2.86 + 2.52 + 2.58 = \ 7.96\,\text{kg/s}$

Check: $\qquad\qquad\qquad\qquad\qquad\qquad$ Fan $\dot{m} = \overline{20.55\,\text{kg/s}}$ (as before)

Example 8.13

Two zones of a building are maintained at 21 °C throughout the year by a dual-duct system which uses re-circulated air. The design criteria are as follows.

Zone	Sensible gain at 27 °C db and 21 °C wb	Sensible gain at −2 °C satn	Min. fresh air
A	60 kW gain	30 kW loss	1.8 kg/s
B	30 kW gain	18 kW loss	1.05 kg/s

If the maximum hot duct temperature is 40 °C and the minimum cold duct temperature is 14 °C, assuming negligible fan and duct gains and a constant mass flow rate through the fan, calculate:

(i)   the recirculation ratio;
(ii) the maximum mass flow rate in each duct;
(iii) the heater maximum load;
(iv) the cooling coil maximum load if the re-circulated air in summer has a percentage saturation of 60% and the cooling coil a contact factor of 0.8;
(v)   the building latent heat load in summer.

(Wolverhampton Polytechnic)

Solution
(i)   The system is shown diagrammatically in Fig. 8.36(a). The mass flow rates must first be established for each zone.

Figure 8.36 Example 8.13: (a) system; (b) summer cycle

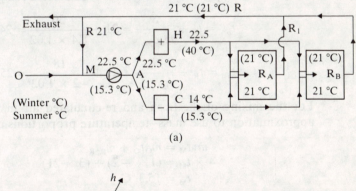

(a)

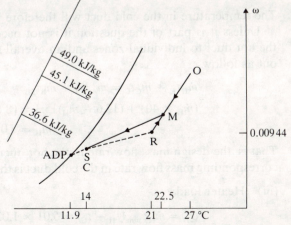

(b)

Summer:

$$\dot{m}_A = \frac{60}{1.02(21-14)} = 8.4 \text{ kg/s}$$

$$\dot{m}_B = \frac{30}{1.02(21-14)} = 4.2 \text{ kg/s}$$

Clearly with lower loads and potentially higher differentials in winter, the mass flow rates will be lower. The fan and cold duct must therefore be sized for $8.4 + 4.2 = 12.6$ kg/s.

Fresh air requirements:

$$\text{Maximum re-circulation ratio for A} = \frac{8.4-1.8}{1.8} = 3.67$$

$$\text{Maximum re-circulation ratio for B} = \frac{4.2-1.05}{1.05} = 3.0$$

The minimum re-circulation ratio for any zone must be used for the system so that zone B controls and the ratio of re-circulated to fresh air is 3.

(ii) Winter:
In winter the supply temperatures can be calculated using Eqn [1.37] as follows:

$$t_{SA} = t_R + \frac{\dot{S}}{\dot{m}_a c_{pma}} = 21 + \frac{30}{8.4 \times 1.02} = 24.5 \,°C$$

$$t_{SB} = t_R + \frac{\dot{S}}{\dot{m}_a c_{pma}} = 21 + \frac{18}{4.2 \times 1.02} = 25.2 \,°C$$

For the mixing of the fresh and re-circulated air in winter it is a sufficient approximation to use mass–temperature proportions as follows:

$$\dot{m}_T t_M = \dot{m}_O t_O + \dot{m}_R t_R$$
$$4 \times t_M = (1 \times -2) + (3 \times 21)$$
$$\therefore \qquad t_M = 15.3 \,°C$$

The temperature in the cold duct will therefore be 15.3 °C.

Unless it is part of the question it is not necessary to calculate flows from the hot duct to individual zones and an overall mixing balance can be carried out as follows:

$$\dot{m}_H t_H + \dot{m}_C t_C = \dot{m}_A t_{SA} + \dot{m}_B t_{SB}$$
$$(\dot{m}_H \times 40) + (12.6 - \dot{m}_H)15.3 = (8.6 \times 24.5) + (4.2 \times 25.2)$$
$$\therefore \qquad \dot{m}_H = 5.01 \text{ kg/s}$$

That is, the design mass flow rate in the hot duct is 5.01 kg/s in winter and the corresponding mass flow rate in the cold duct is therefore $12.6 - 5.01 = 7.59$ kg/s.

(iii) Heater load:

$$\dot{Q}_H = \dot{m}_H c_{pma}(t_H - t_C) = 5.01 \times 1.02 \times (40 - 15.3) = 126.2 \text{ kW}$$

(iv) The psychrometric cycle for the summer case is shown in Fig. 8.36(b).

The temperature of the mixed air can be calculated as follows:

$$4 \times t_M = (1 \times 27) + (3 \times 21)$$
$$\therefore \quad t_M = 22.5 \,°C$$

or the mixed state can be found from the chart by proportioning the lines joining the mixed point. The ADP can now be found by using the cooling coil contact factor:

$$\text{Contact factor} = \frac{t_M - t_C}{t_M - t_{ADP}} = \frac{22.5 - 14}{22.5 - t_{ADP}} = 0.8$$
$$\therefore \quad t_{ADP} = 11.9 \,°C$$

The summer cycle can now be completed on the chart and the cooling coil load calculated as follows:

$$\dot{Q}_{cc} = \dot{m}_T(h_M - h_C) = 12.6(49 - 36.6) = 156 \,\text{kW}$$

(v) The building latent load can now be found from Eqn [1.38]:

$$\dot{L} = \dot{m}(h_R - h_S) - \dot{S}$$
$$= 12.6(45.1 - 36.6) - (60 + 30)$$
$$= 17.1 \,\text{kW}$$

or by, (less accurately), using $\dot{L} = \dot{m}_T h_g(\omega_R - \omega_S)$

Example 8.14

For the system in Example 8.13 the sensible heat gains in zones A and B are 35 kW and 25 kW respectively when the outside air is at 22 °C db and 19 °C wb. The constant building latent heat load is divided equally between each zone and the cooling coil contact factor increases to 0.9. Estimate the percentage saturation in each zone using a trial and error method.

Solution
The new supply temperatures can be calculated as follows:

For zone A: $RRL_A = \dfrac{35}{35 + 6.4} = 0.85$

For zone B: $RRL_B = \dfrac{25}{25 + 6.4} = 0.8$

The room ratio line for each zone is as follows:

For zone A: $RRL_A = \dfrac{35}{35 + 6.4} = 0.85$

For zone B: $RRL_B = \dfrac{25}{25 + 6.4} = 0.8$

Using the re-circulation ratio to calculate the mixed air state at M gives:

$$4 \times t_M = 1 \times 22 + 3 \times 21$$
$$\therefore \quad t_M = 21.25 \,°C$$

The cooling coil contact factor has increased because the flow rate through the coil has decreased. The new coil contact factor can now be used to calculate the new ADP:

$$0.9 = \frac{21.25 - 14}{21.25 - t_{ADP}}$$

$$\therefore \quad t_{ADP} = 13.2\,°C$$

All the relevant temperatures are now known but only the ADP and the outside air state are fixed points. This can be represented on the psychrometric chart as shown in Fig. 8.37(a). Estimating the state of the return air, R, enables a cycle to be completed. An estimate of 70% saturation results in the cycle shown in Fig. 8.37(b) with room percentage saturations of about 68% (the value for room A is slightly higher than for room B). A second trial starting with a return air percentage saturation of 68% gives room values of about 66%. The third trial gives the cycle shown in Fig. 8.37(c) which starts and finishes with percentage saturation values of about 66%, and it is difficult to distinguish between the two zone states and the return air state. The percentage saturation in each zone is therefore about 66%, which is still within comfort limits, so there is no need to heat the hot duct.

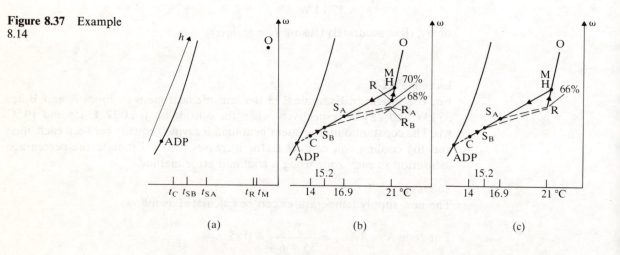

Figure 8.37   Example 8.14

This problem could have been solved without resorting to an iterative procedure. However, this would have involved establishing six equations relating the six unknown specific humidities. It is a useful exercise to establish these equations, but rather tedious to solve them.

8.10 INDUCTION AND FAN–COIL SYSTEMS

These systems involve the mixing of conditioned air and room air within the conditioned spaces. Heating and cooling can take place within the rooms as well as in a central unit. This has the advantage of reducing the size of the central air-handling plant and the supply distribution ductwork. Return air ductwork is also reduced and in some cases dispensed with altogether. The disadvantages are the space occupied by the terminals and potential problems from piping water around the building. When induction and fan–coil systems use conditioned air the overall thermodynamic effect is the same but the psychrometrics are slightly different because mixing does not take place at the same point in the cycle.

Induction Systems

A diagram of a typical room unit is shown in Fig. 8.38. The primary air is expanded from a pressure of about 1.05 bar to produce a high velocity. This causes a reduction in pressure which draws room air over the unit coils. Only the secondary air passes over the secondary (room) coils. The primary and secondary air then mix before being supplied to the room.

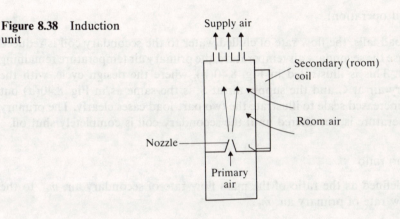

Figure 8.38 Induction unit

Summer cycle

The only items of central plant in operation are a cooling coil and fan. The pressure drops through the ductwork and nozzle are substantial, which results in a temperature rise across the fan of about 2–3 K. The system is illustrated in Fig. 8.39.

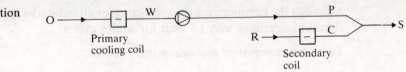

Figure 8.39 Induction systems: summer

The psychrometric cycle in summer is the same for all induction systems and is illustrated in Fig. 8.40(a). The outside air (primary air) is cooled and de-humidified and then increases in temperature through the fan and ductwork until it reaches a state P. The room air is drawn across the secondary (room) coil, which it leaves at a state C. The secondary coil is designed for sensible cooling only and the air leaves at about 3 K above the room dew point. To achieve this, the temperature of the chilled water in the secondary coil is always higher than that supplied to the primary coil. The primary and secondary air are then mixed to give the supply state S.

Figure 8.40 Induction system: summer cycles

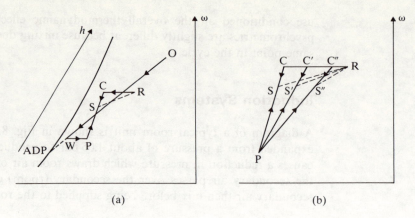

(a) (b)

Part-load operation:

As the load falls, the flow rate of chilled water to the secondary coil is reduced to increase the secondary air temperature, the primary air temperature remaining constant. This is illustrated in Fig. 8.40(b), where the design cycle, with the secondary air at C and the supply air at S, is the same as in Fig. 8.40(a) but with an increased scale to illustrate the two part-load cases clearly. The primary air temperature is not altered until the secondary coil is completely shut off.

Induction ratio

This is defined as the ratio of the mass flow rate of secondary air, $\dot{m}_C$, to the mass flow rate of primary air, $\dot{m}_P$.

$$\text{Induction ratio} = \frac{\dot{m}_C}{\dot{m}_P} \qquad [8.4]$$

Typical values of induction ratio vary between 3 and 8. There is normally a negligible change in induction ratio between summer and winter operation. In Fig. 8.40(a) the induction ratio is the ratio of line P–S to line C–S.

Winter cycles

Although the summer cycles are the same for all induction systems, the following alternative systems may be used for winter cycles.

Two-pipe changeover systems: winter cycles

In a two-pipe changeover system a single room coil is supplied with chilled water in summer and hot water in winter—hence the name. For winter operation the primary air must be heated and humidified. Figure 8.41(a) illustrates a plant with a spray humidifier and Fig. 8.41(b) a plant with a steam humidifier. When steam humidification is used, the humidifier may be either a single unit in the central plant or individual humidifiers serving each zone. A frost coil is shown in both systems but this load is often included in the pre-heater load to simplify examples.

The state of the primary air is close to that used for the summer operation. It is therefore providing cooling rather than heating so that the secondary coil must be capable of heating the primary air up to the room temperature as well

Figure 8.41 Induction changeover systems: winter

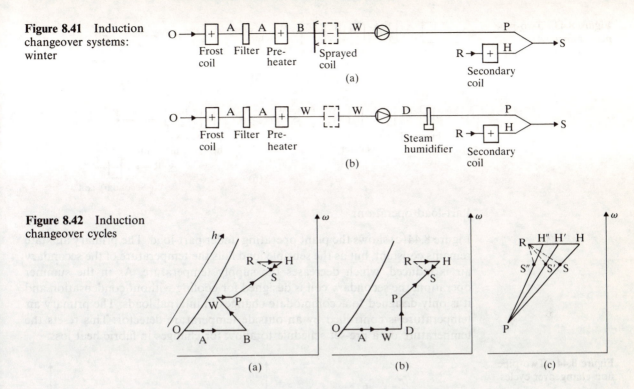

Figure 8.42 Induction changeover cycles

as satisfying the room sensible heat load. The winter design cycles for the plants shown in Fig. 8.41(a) and Fig. 8.41(b) are illustrated in Fig. 8.42(a) and Fig. 8.42(b) respectively.

Part-load operation:

This is illustrated in Fig. 8.42(c). The state of the primary air remains constant while the temperature of the secondary air is progressively reduced as the room sensible load falls. Provided the latent gains remain constant, the room and supply specific humidities will remain constant.

The two-pipe changeover system has either hot or cold water in the secondary coils. It is therefore suitable only for climates where there is a distinct difference between winter and summer. In a country such as the UK a two-pipe changeover induction system would be a problem in mid-season, when some zones may require heating while others require cooling. It would also need frequent changes from one mode to the other.

Two-pipe non-changeover systems: winter cycles

A non-changeover system is so-called because the room coil is supplied with 'cool' water throughout the year. It is not normally necessary to use chilled water from a refrigeration plant for the secondary coils in winter; water from a cooling tower is sufficiently cool. The central plants for typical non-changeover systems are shown in Fig. 8.43(a) and 8.43(b). The after-heaters and the steam humidifier shown could be a number of items serving individual zones. The corresponding winter design cycles for these plants are shown in Figs 8.44(a) and 8.44(b). The primary air satisfies both the sensible and latent load for the winter design case, the secondary coil being switched off.

Figure 8.43 Two-pipe non-changeover systems: winter plant

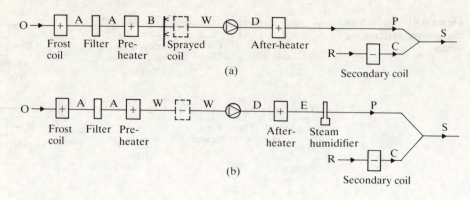

Part-load operation:

Figure 8.44(c) shows the plant operating under part-load. The primary air state remains constant, but as the sensible load falls the temperature of the secondary air is reduced, which decreases the supply temperature. As in the summer operation, the secondary coil is designed for cooling without condensation and it is only designed to accommodate changes in internal loads. The primary air temperature is controlled by an outside temperature detector. This resets the temperature on a pre-set schedule to allow for changes in fabric heat loss.

Figure 8.44 Two-pipe non-changeover cycles

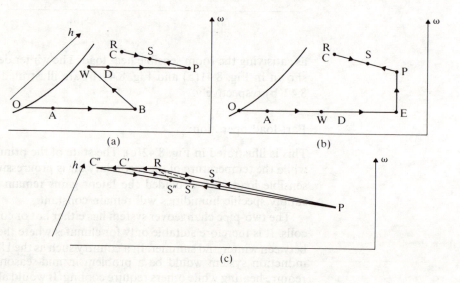

Four-pipe system

Each induction unit can be supplied with either hot or cold water as necessary, usually to separate heating and cooling coils. This allows a single central plant to service a building where some zones are operating in a summer mode and others in a winter mode. This is the induction system which gives the best control, but it is also the most expensive in capital cost.

Three-pipe system

A single coil in each induction unit can be fed with either hot or cold water but there is only one return pipe. The control is therefore the same as for the four-pipe system but there is a substantial saving in capital cost of pipework and heat transfer surface. In summer the return pipe would carry all return chilled water and in winter all return hot water.

The disadvantage of this system is that, in mid-season, the mixed return water is cooler than a normal boiler feed water and warmer than a normal chilled water return. This loss of energy can be recovered by installing a heat pump, as shown in Fig. 8.45. The return water is divided so that the boiler feed water passes over the condenser, where it is heated, and the chilled water return passes over the evaporator, where it is cooled. This of course increases the capital cost, and a three-pipe system with a heat pump would only be economic in buildings with very long pipe runs.

Figure 8.45 Heat pump for a three-pipe induction system

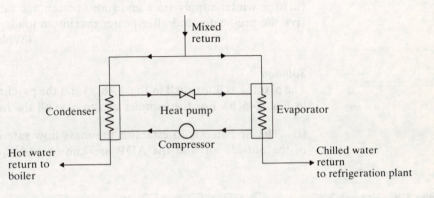

Primary air

To minimize the size of the ductwork, the induction ratio should be as high as is practical. It is therefore necessary to establish the minimum flow rate of primary air. This will be the lowest mass flow rate necessary to satisfy all of the following:
(i)   the fresh air required.
(ii) the latent heat load.
(iii) the heating load for non-changeover systems.
These should all be calculated and the highest value used. If the latent load controls, then the fresh air used will be above the minimum requirement. If the fresh air controls, this will determine the room percentage saturation for a particular cooling coil. The winter design case is rarely the controlling factor.

Secondary air

For the summer design case, as well as satisfying the room latent load, the primary air will also contribute to the room sensible load. The secondary coil is therefore sized by subtracting the contribution of the primary air from the room sensible load. The mass flow rate is then calculated using the maximum permitted enthalpy change across the secondary coil.

Example 8.15

A non-changeover induction system uses 100% outside air through the primary plant and maintains a building at 21 °C throughout the year and at 55%

saturation in summer. The summer design state is 28 °C db and 21 °C wb and the winter design state −2 °C saturated. The latent heat load is 50 kW throughout the year and the sensible heat loads are 500 kW gain in summer and 100 kW loss in winter.

The primary plant consists of a pre-heater, sprayed cooling coil, fan and after-heater. The sprayed cooling coil has a constant contact factor of 0.8 and an apparatus dew point of 6 °C for the summer design case. An off-coil thermostat maintains a constant temperature from the primary cooling coil, and the air leaving the secondary coil must be 3 K above the room dew point. If the fan and duct gain is 3 K and the minimum fresh air required is 10 m³/s, calculate:

(i)   the maximum primary and secondary cooling coil loads;
(ii) the induction ratio and summer supply state;
(iii) the winter supply state and room percentage saturation;
(iv) the pre-heater and after-heater maximum loads.

(Wolverhampton Polytechnic)

Solution
The plant is as illustrated in Fig. 8.46(a) and the psychrometric cycles are shown in Fig. 8.46(b). For the summer design case all the heaters are off.

(i)   The first step is to establish the mass flow rate of primary air. The state of the outside air and the ADP are known and can be fixed on the chart.

Figure 8.46   Example 8.15

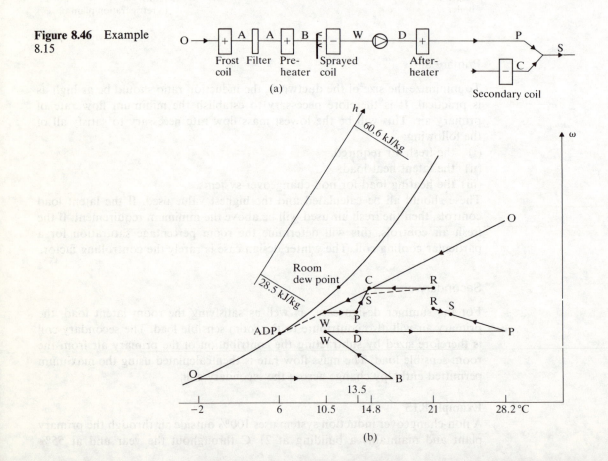

Knowing the cooling coil contact factor, the off-coil specific humidity can be calculated as follows:

$$\frac{\omega_O - \omega_W}{\omega_O - \omega_{ADP}} = 0.8 = \frac{0.0126 - \omega_W}{0.0126 - 0.0058}$$

Hence $\omega_W = 0.007\,16$

Point W can now be fixed on the chart and the temperature read off as 10.5 °C. Point P is 3 K higher (because of the fan gains) at 13.5 °C at the same specific humidity. The minimum mass flow rate of primary air to satisfy the latent load can now be calculated using Eqn [1.38] as follows:

$$\dot{m}_P = \frac{\dot{L}}{h_g(\omega_R - \omega_P)} = \frac{50}{2540(0.0086 - 0.007\,16)} = 13.6\,\text{kg/s}$$

The fresh air requirement must also be satisfied.

$$\dot{m}_P = \frac{\dot{V}}{v_P} = \frac{10}{0.82} = 12.2\,\text{kg/s}$$

The latent load controls the primary air so the room will be over-supplied with fresh air. (With no maximum primary air temperature given, it must be assumed that this does not control. The modest value of the winter primary air temperature calculated later confirms this.) The primary cooling coil load can now be calculated.

$$\text{Primary coil } \dot{Q} = \dot{m}_P(h_O - h_W) = 13.6(60.6 - 28.5) = 437\,\text{kW}$$

The primary air is at 13.5 °C and makes the following contribution towards satisfying the room sensible heat load:

$$\dot{m}_P c_{pma}(t_R - t_P) = 13.6 \times 1.02(21 - 13.5) = 104\,\text{kW}$$

The secondary cooling coil must make up the balance of the 500 kW, i.e. $500 - 104 = 396\,\text{kW}$.

(ii)   The summer room state can be fixed on the chart and the room dew point read off as 11.8 °C. The air leaving the secondary coil at C must be 3 K above this temperature, i.e. 14.8 °C. The mass flow rate of secondary air can now be calculated using Eqn [1.37]:

$$\dot{m}_C = \frac{\dot{Q}_{CC}}{c_{pma}(t_R - t_C)} = \frac{396}{1.02(21 - 14.8)} = 62.6\,\text{kg/s}$$

$$\text{Induction ratio} = \frac{\dot{m}_C}{\dot{m}_P} = \frac{62.6}{13.6} = 4.6$$

The cycle can now be completed by either proportioning line PC or using the room ratio line:

$$RRL = \frac{500}{500 + 50} = 0.909$$

or calculating the mixed temperature:

$$t_S = \left(13.5 \times \frac{1}{5.6}\right) + \left(14.8 \times \frac{4.6}{5.6}\right) = 14.6\,°C$$

The room supply state is 14.6 °C and 79% in summer.

(iii) For the winter cycle, the outside air state is fixed and the off-coil temperature is the same as in the summer, i.e. 10.5 °C. It is now necessary to fix point W by trial and error using a series of wet bulb lines until the contact factor is 0.8. The primary air temperature can now be calculated from the room sensible heat load as follows:

$$t_P = t_R + \frac{\dot{S}}{c_{pma} \dot{m}_P} = 21 + \frac{100}{1.02 \times 13.6} = 28.2 \,°C$$

The cycle can be completed by drawing a horizontal line from W through D to P and then using the winter room ratio line.

$$RRL = \frac{100}{100 + 50} = 0.67$$

The supply state S is established either by proportioning line RP or calculating t_S as follows from the approximate mixing equation:

$$t_S = \left(28.2 \times \frac{1}{5.6} \right) + \left(21 \times \frac{4.6}{5.6} \right) = 22.3 \,°C$$

From the chart the supply state is 22.3 °C and 42% and the room percentage saturation is 47%.

(iv) The pre-heater and after-heater loads can now be calculated.

Pre-heater $\dot{Q} = \dot{m}_P(h_B - h_O) = 13.6(25 - 6) = 258 \,kW$
After-heater $\dot{Q} = \dot{m}_P(h_P - h_D) = 13.6(43.5 - 28.7) = 202 \,kW$

Alternatively,

Pre-heater $\dot{Q} = 13.6 \times 1.02 \times (16.9 + 2) = 262 \,kW$
After-heater $\dot{Q} = 13.6 \times 1.02 \times (28.2 - 13.5) = 204 \,kW$

The difficulty in reading enthalpy values accurately from the chart make the use of specific heat and temperature difference equally acceptable.

Example 8.16

If the plant in Example 8.15 had been fitted with a steam humidifier instead of a sprayed coil, calculate the reduction in the total loading of the pre-heater and after-heater, and the power supplied to the humidifier. Assume the humidifier generates saturated steam at 1.1 bar from feed water at 15 °C.

Solution
The cycle is as illustrated in Fig. 8.44(b). The primary air will be supplied to the room coil at the same state as in Example 8.15. The two heaters must therefore increase the air temperature from $-2\,°C$ to 28.2 °C apart from the 3 K temperature rise across the fan. The new combined load becomes:

$$\dot{Q} = 13.6 \times 1.02 \times (28.2 + 2 - 3) = 377 \,kW$$
$$\therefore \quad \text{Decrease in load} = 262 + 204 - 377 = 89 \,kW$$

From the chart the specific humidity of the treated primary air is 0.006 and the specific humidity of the outside air is 0.0032. The mass flow rate of steam is therefore:

$$13.6(0.006 - 0.0032) = 0.0381 \,kg/s$$

From property tables at 1.1 bar, h_g is $2680 \, \text{kJ/kg K}$ and at 15 °C h_f is $62.9 \, \text{kJ/kg K}$. The power supplied to the humidifier is therefore:

$$\dot{Q} = 0.0381(2680 - 62.9) = 99.7 \, \text{kW}$$

The power required for the humidifier is greater than the saving on the two heaters. If the humidifier had been supplied with electricity the extra running cost would have been substantial, but if the building had been divided into a number of zones each zone could have had its own humidifier. This is not possible with a sprayed coil.

Example 8.17

A building is maintained at 22 °C in summer and 21 °C in winter by a changeover induction system. In summer the percentage saturation for the design case should not be more than 60% and in winter it should not be less than 40%. The external design conditions are 26 °C db and 20 °C wb in summer and −5 °C saturated in winter. In summer the sensible heat gain is 70 kW and in winter the sensible heat loss is 30 kW; the latent heat gain is constant at 10 kW.

The primary plant consists of a frost coil, filter, pre-heater, cooling coil, fan and steam humidifier. The cooling coil has a contact factor of 0.8 and the air leaving the secondary coil is at 17 °C. The primary air is supplied at a constant 15 °C and the temperature rise across the fan is 2 K. If the minimum fresh air required is $2.4 \, \text{m}^3/\text{s}$, calculate:

(i)   the primary and secondary cooling coil loads;
(ii) the induction ratio;
(iii) the summer supply temperature and room percentage saturation;
(iv) the winter supply temperature;
(v)   the mass flow rate of steam from the humidifier;
(vi) the pre-heater load if the frost coil heats the air up to 4 °C and the air leaving the secondary coils is at 25 °C.

(Wolverhampton Polytechnic)

Solution
(i) The plant is illustrated in Fig. 8.47(a) and the two psychrometric cycles in Fig. 8.47(b). For the summer case the fan gain of 2 K is subtracted from the primary air temperature of 15 °C to give the off-coil temperature at W of 13 °C. The cooling coil contact factor can now be used to calculate the ADP temperature.

$$\frac{t_O - t_W}{t_O - t_{ADP}} = 0.8 = \frac{26 - 13}{26 - t_{ADP}}$$

$$\therefore \qquad t_{ADP} = 9.75 \, °\text{C}$$

Points O, W and P can now be located on the chart. With the room at 22 °C and 60% saturation, the specific humidity would be 0.010 07. The mass flow rate of primary air to achieve this is now calculated using Eqn [1.38]:

$$\dot{L} = \dot{m}_P h_g (\omega_R - \omega_P)$$
$$10 = \dot{m}_P \, 2540(0.010\,07 - 0.0084)$$

$$\therefore \qquad \dot{m}_P = 2.36 \, \text{kg/s}$$

The fresh air requirement must now be checked. The primary air leaves the fan

Figure 8.47 Example 8.17

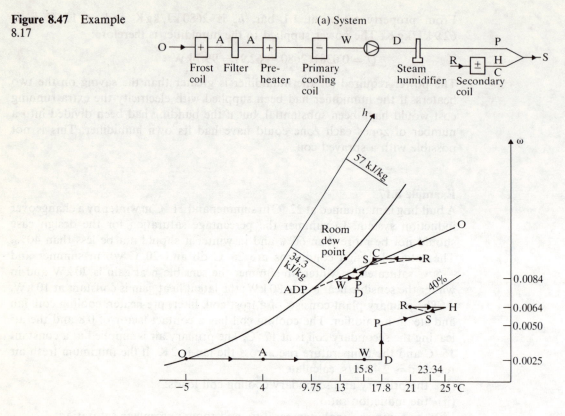

(a) System

Frost coil　Filter　Pre-heater　Primary cooling coil　Steam humidifier　Secondary coil

(b) Cycles

at D with a specific volume of $0.827 \text{ m}^3/\text{kg}$ in summer. (It may be checked later that the value for the winter cycle is almost the same.)

$$\dot{m}_P = \frac{\dot{V}}{v_P} = \frac{2.4}{0.827} = 2.9 \text{ kg/s}$$

The fresh air requirement controls: therefore the room percentage saturation will be less than 60%. The primary and secondary cooling coil loads and the induction ratio can now be calculated as follows:

Primary coil $\dot{Q} = \dot{m}_P(h_O - h_W) = 2.9(57 - 34.3) = 65.8 \text{ kW}$
Primary air room cooling $= 2.9 \times 1.02(22 - 15) = 20.7 \text{ kW}$
Secondary coil load $= 70 - 20.7 = 49.3 \text{ kW}$

(ii)　Secondary coil mass flow rate, $\dot{m}_C = \dfrac{49.3}{1.02(22 - 17)} = 9.67 \text{ kg/s}$

Induction ratio $= \dfrac{\dot{m}_C}{\dot{m}_P} = \dfrac{9.67}{2.9} = 3.33$

(iii)　The room state can now be established from the latent heat gain.

$$\omega_R = \omega_P + \frac{\dot{L}}{\dot{m}_P h_g} = 0.0084 + \frac{10}{2.9 \times 2540} = 0.009\,76$$

The room state is now fixed on the chart and the percentage saturation found to be 58%. The supply state can be found by proportioning line PC, using the room ratio line or using the approximate mixing equation to establish the temperature:

$$t_S = \left(\frac{3.33}{4.33} \times 17\right) + \left(\frac{1}{4.33} \times 15\right) = 16.5\,°C$$

(iv) The winter supply temperature can be found from the building sensible heat loss and the total mass flow rate of $2.9 + 9.67 = 12.57\,kg/s$.

$$t_S = t_R + \frac{\dot{Q}}{\dot{m}_T c_{pma}} = 21 + \frac{30}{12.57 \times 1.02} = 23.34\,°C$$

The winter percentage saturation will be the minimum of 40% which fixes the room state R. Since only the primary air satisfies the room latent heat gains, this can be used to calculate the specific humidity of the primary air.

$$\omega_P = \omega_R - \frac{\dot{L}}{\dot{m}_p h_g} = 0.0064 - \frac{10}{2.9 \times 2540} = 0.0050$$

The outside air has a specific humidity of 0.0025 and the mass flow rate of steam can be found as follows.

$$\begin{aligned}
\text{Mass flow rate of steam} &= \dot{m}_P(\omega_P - \omega_O) \\
&= 2.9(0.0050 - 0.0025) \\
&= 0.007\,25\,kg/s
\end{aligned}$$

The temperature of the primary air can be established by the approximate mixing equation, i.e.

$$(1 \times t_P) + (3.33 \times 25) = 4.33 \times 23.34$$
$$\therefore \quad t_P = 17.8\,°C$$

The psychrometric cycle for the winter design case can now be completed. The pre-heater must heat the air from 4 °C to 17.8 °C, apart from the 2 K rise across the fan, i.e.

$$\text{Pre-heater } \dot{Q} = 2.9 \times 1.02 \times (17.8 - 4 - 2) = 34.9\,kW$$

It should be noted that the question did not specify whether the system was two-pipe, three-pipe or four-pipe. The calculations for the design cases are the same for all three systems; it is only during the mid-season 'changeover period' when they differ. Since the weather data is typical for the UK it may be assumed that a four-pipe or three-pipe system would be appropriate.

Fan–coil Systems

Fan–coil systems are similar to induction systems, the basic difference being that a fan is used to circulate the air instead of a nozzle. The room units consist of a coil (sometimes more than one) mounted below a fan within a cabinet. A drip tray and drain must be fitted below the coil to remove condensate. When used for perimeter zones, they are normally mounted below windows. The four basic arrangements in common use are as follows.

Figure 8.48 Fan–coil
unit with ceiling supply
of primary air

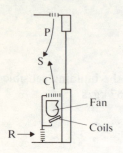

Fan–coil with ceiling supply of primary air

Figure 8.48 illustrates one of the two systems which use primary air from a central plant. Only the room (secondary) air passes over the room coils: mixing then takes place within the room. The sequence of processes is therefore identical to that for induction systems and the psychrometric cycles are the same. In this system, as with induction systems, there should be no condensation in the room coils.

Fan–coil with underfloor supply of primary air

The system shown in Fig. 8.49(a) also uses primary air treated in a central plant and the means of achieving state P is the same as for an induction system. However, in this case the mixing of room and primary air takes place before both pass over the room coil. This results in different psychrometric cycles from the ceiling primary air system. The psychrometric cycles for Fig. 8.49(a) are shown in Fig. 8.49(b) for the summer design case and Fig. 8.49(c) for the winter design case. Although the psychrometrics are different from the ceiling supply system the overall thermodynamic effect is the same. In this system, as with induction systems, there should be no condensation in the room coils.

Figure 8.49 Fan–coil
unit with underfloor
supply of primary air

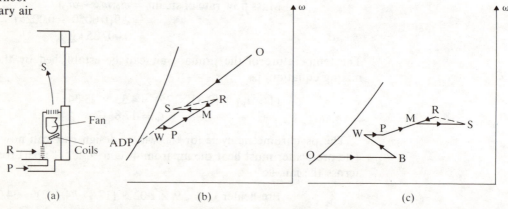

Fan–coil with outside air supply

Figure 8.50(a) shows a system which mixes outside air with room air before passing it over the coil. For this arrangement the summer design cycle is shown in Fig. 8.50(b) and the winter design cycle in Fig. 8.50(c). With no central plant, condensation must take place in the coil to control humidity in summer. Because all the air supplied to the room is de-humidified, the ADP can be higher than is necessary in a central plant. It should also be noted that the fan gain is negligible compared with the fan gain in a system with a central plant. In winter, since the system provides no humidification, separate room humidifiers may need to be installed.

A popular variant of this system is obtained by fixing dampers to the outside air vent so that the units can operate either as in Fig. 8.50, or with 100% room air (see Fig. 8.51), or at any intermediate setting with a reduced flow of outside air.

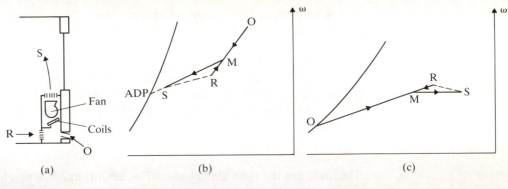

Figure 8.50 Fan–coil unit with outside air

Fan–coil using all room air

Figure 8.51(a) shows a system which uses only natural ventilation. The cycles shown in Figs 8.51(b) and 8.51(c) would only strictly apply to sealed buildings. In practice the summer cycle would be similar to that shown in Fig. 8.50(b) but with points M and R closer together since outside air leaking into the room would mix with the room air. In winter the operation is the same as for a unit heater.

Figure 8.51 Fan–coil unit using all room air

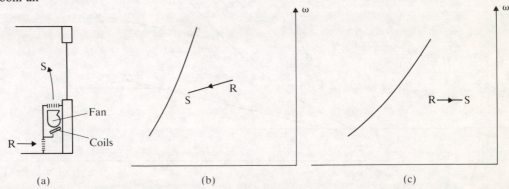

Fan–coil system control

As with induction systems, fan–coil units can be supplied with water in a variety of ways. They can operate on a changeover or non-changeover mode and may have two, three or four pipes. Whilst they may be controlled by varying the temperature of the water supplied to the coils, the most usual method is to run the fans at a number of different speeds (usually three) using a room temperature detector to change the speed.

Example 8.18

A room is maintained at 23 °C in summer and 21 °C in winter by fan–coil units which mix room and outside air in the ratio of 7:1 before cooling or heating takes place. For the summer design case the room percentage saturation is 55% but there is no humidification in winter.

The summer ambient design temperatures are 29 °C db and 23 °C wb when the room gains are 40 kW sensible and 5 kW latent. The winter design ambient temperature is 0 °C saturated when the room sensible heat loss is 20 kW and the room latent gain 5 kW. If the air leaves the coil in summer at a temperature of 15 °C and the temperature rise across the fan is negligible, calculate:
(i) the number of units required if each unit has a 4 kW cooling coil;
(ii) the minimum size of the heating coils and the percentage saturation in winter.

Solution
The units are the type illustrated in Fig. 8.50(a) and the psychrometric cycles are shown in Fig. 8.52.

Figure 8.52 Example 8.18

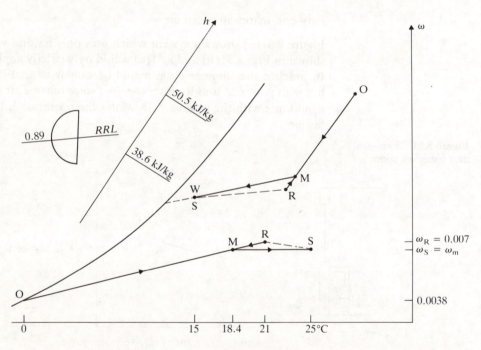

(i) Summer:
Points O and R can be established on the chart. Point M lies along OR and can be located by using the approximate mixing equation for the room and outside air as follows:

$$t_M = \left(29 \times \frac{1}{8}\right) + \left(23 \times \frac{7}{8}\right) = 23.7 \,°C$$

The room ratio line for the summer operation is:

$$RRL = \frac{\dot{S}}{\dot{S} + \dot{L}} = \frac{40}{40 + 5} = 0.89$$

Drawing the room ratio line from the room state, R, locates the supply state, S, at the 15 °C db temperature line, and the summer cycle can be completed.

Equation [1.37] is now used to find the mass flow rate of air as follows:

$$\dot{m}_a = \frac{\dot{S}}{c_{pma}(t_R - t_S)} = \frac{40}{1.02(23 - 15)} = 4.9 \text{ kg/s}$$

The cooling coil load can now be found using enthalpy values from the chart as follows:

$$\text{Cooling coil } \dot{Q} = \dot{m}_a(h_M - h_W) = 4.9(50.5 - 38.6) = 58.3 \text{ kW}$$

Since each unit provides 4 kW of cooling, 58.3/4 = 15 units will be required.

(ii) Winter:
Equation [1.38] can be used to find the difference between the room and supply specific humidity as follows:

$$(\omega_R - \omega_S) = \frac{\dot{L}}{\dot{m}_a h_g} = \frac{5}{4.9 \times 2540} = 0.0004$$

The winter mixing can now be expressed in terms of specific humidities:

$$\frac{\omega_R - \omega_M}{\omega_R - \omega_O} = \frac{0.0004}{\omega_R - 0.0038} = \frac{1}{8}$$

Hence $\qquad \omega_R = 0.007 \quad$ and $\quad \omega_M = 0.0066$

The cycle can now be completed taking t_M from the chart:

$$\text{Heater } \dot{Q} = \dot{m}_a c_{pma}(t_S - t_M) = 4.9 \times 1.02(25 - 18.4) = 33 \text{ kW}$$

Each heating coil therefore needs to be at least 33/15 = 2.2 kW.
From the chart, the room percentage saturation is 44%.

Example 8.19
A zone of a building is air-conditioned by a four-pipe fan–coil system with fresh air supplied from a central plant. The central plant consists of a frost coil, filter, pre-heater, cooling coil, fan and steam humidifier. The primary cooling coil has a contact factor of 0.85, an apparatus dew point of 8 °C and the temperature rise across the primary fan is 2 K. The room units draw primary air, from an underfloor duct, and room air across the secondary coil.

The external design states are 28 °C db and 22 °C wb in summer and −2.5 °C saturated in winter. The internal temperature is to be 21 °C and the design percentage saturations are 60% maximum and 35% minimum for the summer and winter cases. There is a constant latent gain of 10 kW, a summer sensible gain of 100 kW and a winter sensible loss of 60 kW.

If the primary air temperature is maintained constant throughout the year and a minimum of 1.8 m³/s of fresh air is required, draw the summer and winter psychrometric cycles and calculate:
(i) the summer design percentage saturation;
(ii) the primary cooling coil load;
(iii) the induction ratio if the air leaves the room coil at 3 K above the room dew point;
(iv) the secondary cooling coil load;
(v) the secondary heating coil load;
(vi) the maximum mass flow rate of steam from the humidifier.

Solution

(i) The room units are illustrated in Fig. 8.49(a). The central plant and the psychrometric cycles are shown in Fig. 8.53. The outside state and the primary coil ADP can be established on the chart and state W by using the coil contact factor as follows

$$\text{Contact factor} = \frac{t_O - t_W}{t_O - t_{ADP}} = \frac{28 - t_W}{28 - 8} = 0.85$$

$$\therefore \quad t_W = 11\,°\text{C}$$

Figure 8.53 Example 8.19

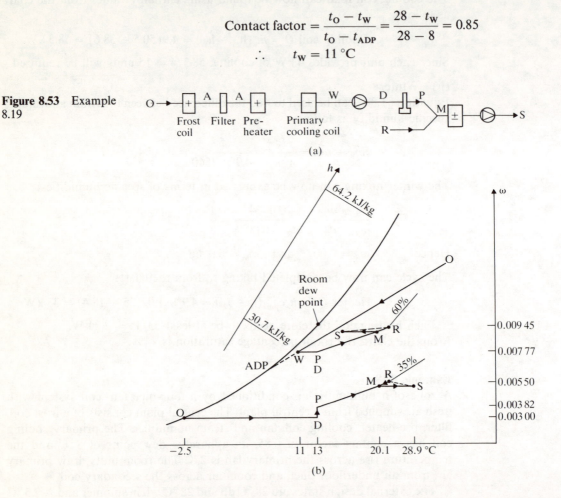

(a)

(b)

Point W can now be established on the chart and its specific humidity of 0.007 77 read off. Allowing for the temperature rise across the fan, the primary air will be at 13 °C and the same specific humidity. At the maximum room percentage saturation of 60% the specific humidity is 0.009 45. The minimum mass flow rate of primary air to satisfy the latent gains is therefore

$$\dot{m}_P = \frac{L}{h_g(\omega_R - \omega_P)} = \frac{10}{2540(0.009\,45 - 0.007\,77)} = 2.34 \text{ kg/s}$$

The specific volume of the air leaving the fan is 0.821 m³/kg (this changes slightly to 0.814 m³/kg in winter) and the minimum mass flow rate of fresh air is therefore

$$\dot{m}_P = \frac{\dot{V}}{v_P} = \frac{1.8}{0.821} = 2.19 \text{ kg/s}$$

The latent load controls; therefore the room percentage saturation will be the maximum permitted, 60%, and the fresh air will be above the minimum.

(ii)   Primary coil $\dot{Q} = \dot{m}_P(h_O - h_W) = 2.34(64.2 - 30.7) = 78.5\,\text{kW}$

(iii)   From the chart, the room dew point is 13.2 °C so the air leaving the secondary coil must be at 16.2 °C. All the supply air passes through the secondary coil; the total mass flow rate can therefore be found from the zone sensible gains, i.e.

$$\dot{m}_T = \frac{\dot{S}}{c_{pma}(t_R - t_S)} = \frac{100}{1.02(21 - 16.2)} = 20.42\,\text{kg/s}$$

and the induction ratio is

$$\dot{m}_R = \frac{20.42 - 2.34}{2.34} = 7.73$$

(iv)   The state of the air entering the room coil can now be found by using the approximate mixing equation for the primary and secondary air:

$$t_M = \left(21 \times \frac{7.73}{8.73}\right) + \left(13 \times \frac{1}{8.73}\right) = 20.1\,°C$$

The summer cycle can now be completed and the secondary cooling coil load calculated as follows.

$$\text{Room coil } \dot{Q} = \dot{m}_T c_{pma}(t_M - t_S)$$
$$= 20.42 \times 1.02(20.1 - 16.2) = 81.2\,\text{kW}$$

(v)   In winter the primary air is still at 13 °C and the mixed temperature will therefore still be 20.1 °C. The primary air will be cooling the room, which will put an extra load on the secondary coils. The secondary coil heating load is therefore calculated as follows.

$$\text{Room cooling load} = \dot{m}_P c_{pma}(t_R - t_P)$$
$$= 2.34 \times 1.02(21 - 13) = 19.1\,\text{kW}$$
$$\text{Secondary coil load, } \dot{Q} = \dot{S} + \text{Room cooling} = 60 + 19.1$$
$$= 79.1\,\text{kW}$$

(vi)   The room state in winter is at 21 °C and 35% saturation, for which the specific humidity is 0.0055. With no change in latent load the summer moisture differential between the room air and the primary air of $(0.009\,45 - 0.007\,77) = 0.001\,68$ will be the same in winter. (Only the primary air can satisfy the latent heat load.) The specific humidity of the primary air is therefore

$$\omega_P = (0.0055 - 0.001\,68) = 0.003\,82$$

This enables point P to be established on the chart (Fig. 8.53(b)) together with points O, D, R and M. The new supply temperature can be calculated from Eqn [1.37], i.e.

$$t_S = t_R + \frac{\dot{S}}{\dot{m}_a c_{pma}} = 21 + \frac{60}{20.42 \times 1.02} = 23.9\,°C$$

The cycle can now be completed and the steam flow rate calculated as,

$$\dot{m}_P(\omega_P - \omega_O) = 2.34(0.003\,82 - 0.003) = 0.0019\,\text{kg/s}$$

If this had been an induction system instead of a fan–coil system the mass flow rate of primary air and steam and the cooling and heating loads would have been exactly the same. The induction ratio, however, would have been 6.1.

8.11 VARIABLE AIR VOLUME SYSTEMS (VAVs)

In all the multiple-zone systems dealt with so far, the mass flow rate of air has remained constant and changes in room sensible heat gains have been dealt with by changing the air supply temperature. In a variable air volume system the supply temperature is kept constant and the mass flow rate of air reduced as the sensible heat gains fall.

There are two basic ways of delivering a reduced flow rate of air to a room. Illustrated in Fig. 8.54(a) is a variable-velocity terminal. The diagram shows the air being throttled adjacent to the terminal by expanding a bellows, but throttling may take place upstream so that a number of fixed diffusers can be supplied by one device.

Figure 8.54 VAV terminals

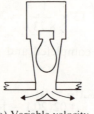

(a) Variable velocity

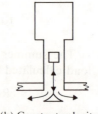

(b) Constant velocity

The second method is illustrated in Fig. 8.54(b). The air enters the room via movable diffusers which close as the room sensible heat gains fall. This is known as a constant-velocity terminal.

Central plant

A basic VAV system requires only mixing dampers, a cooling coil and a fan, as shown in Fig. 8.53. A heater and steam humidifier may be added for winter operation in temperate climates but the heater is only used to pre-heat the building prior to occupancy.

Figure 8.55 Basic VAV system

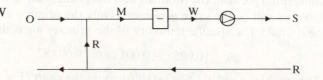

In normal operation a VAV system is therefore always cooling. The system is ideally suited to internal zones of a building which have sensible heat gains throughout the year. When used for perimeter zones it can be combined with a low-pressure hot water radiator system.

The distribution ductwork is normally high-velocity with reducers to give a lower velocity to terminal branches. This, together with the pressure drops across the terminals, results in temperature gains across the fan of 2–3 K.

Turndown ratios

A turndown ratio is defined as the ratio of the maximum to minimum flow rates. However, in a VAV system it is important to distinguish between the turndown ratio of the system and that of the individual zones or diffuser terminals. ASHRAE[8.6] quotes values of the latter of up to 4:1 but most systems will operate with lower values than this.

$$\text{Terminal turndown ratio} = \frac{\text{Design terminal flow rate}}{\text{Minimum mass flow rate}} \qquad [8.5]$$

The flow rates are normally expressed as volumes, but with a constant supply state the ratio of the mass flow rates is the same as the ratio of the volume flow rates. Since different zones will be operating at different percentages of their design loads (if they are not, then VAV is probably the wrong choice of system), the system turndown ratio will always be lower than the value for the terminals.

$$\text{System turndown ratio} = \frac{\text{Central plant design flow rate}}{\text{Central plant minimum flow rate}} \qquad [8.6]$$

ASHRAE[8.6] quotes a maximum system turndown ratio of 2:1. If any source uses the term 'turndown ratio' without qualification, it should be understood as referring to the terminals.

One advantage of a VAV system is that it can save power by using variable-volume fans. The most satisfactory method of varying the volume flow rate is to use a variable-speed motor; this gives a saving in fan power proportional to the average system turndown ratio. The fan speed can be controlled by a pressure sensor in the distribution ductwork which reduces the motor speed if the pressure rises because terminals require less air. The most satisfactory drive is of course also the most expensive in capital cost.

A very simple and inexpensive system may use a constant-volume fan and discharge some of the air into the exhaust duct with no saving in fan power. Between the extremes of a variable-speed drive and a discharge system, the volume flow rate may be varied using variable inlet vanes or variable-pitch fans. These are a compromise between energy saving and capital cost. Further details of variable-volume control systems can be found in Ref. 8.8.

The turndown ratio of terminals is limited by the mixing of the supply and room air. Terminals are invariably installed in ceilings with the airstream directed across the ceiling, mixing with room air as it progresses. The tendency of the air to cling to the ceiling is due to the *Coanda effect* and the distance it moves across the ceiling is known as the *air throw*. Ideally these should remain constant throughout the operating range of the terminal but in practice they will decrease as the air flow rate decreases. If the air flow rate falls too low the cold air will drop without mixing and create local cold areas. This is known as *air dumping*. Air dumping is more likely with variable-velocity terminals than with constant-velocity terminals. The latter can therefore operate with higher turndown ratios. They are of course more expensive. The *CIBSE Guide*[8.7] contains a section on room air diffusion.

With any system, mixing will be most satisfactory at the design air flow rate. It is therefore doubly important not to oversize VAV systems since any oversizing will result in the terminals always operating below the design flow rates.

Fresh air requirements

All zones must be supplied with adequate fresh air when the terminals are fully turned down, and the need from the most critical zone will dictate the re-circulation ratio which can be used in the central plant. This results in lower re-circulation ratios than could be used with alternative systems and the oversupplying of most zones with fresh air. The higher the turndown ratio, the lower will be the re-circulation ratio; this is a penalty which must be paid for using terminals with high turndown ratios.

Part-load humidities

Figure 8.56(a) shows the psychrometrics for a system with a constant mass flow rate and variable supply temperature. This compares with the system shown in Fig. 8.56(b) with a constant supply temperature and a variable mass flow rate; both design and part-load conditions are shown. The two systems are the same for the design case and the room ratio lines must have the same slope for any particular part-load case. For the constant-volume operation, as long as the latent heat gain remains constant then increasing the supply temperature to S_1 and S_2 as the sensible load falls results in the same room state R. However, if the supply temperature remains constant and the mass flow rate varies, the room humidity must progressively rise to the values corresponding to R_1 and R_2 as the sensible heat load falls. This can also be deduced from Eqns [1.37] and [1.38]:

$$\dot{S} = \dot{m}_a c_{pma}(t_R - t_S)$$
$$\dot{L} = \dot{m}_a(\omega_R - \omega_S)h_g$$

There is therefore no control of humidity in individual zones.

Figure 8.56 VAV systems: (a) constant volume; (b) variable volume

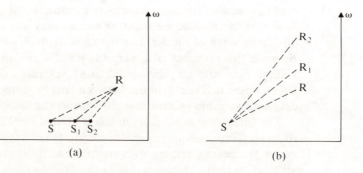

(a)   (b)

Example 8.20

A zone of a VAV air conditioning system has a turndown ratio of 2:1 and a minimum fresh air requirement of 1.86 m³/s. The zone design state is 21 °C and 50% saturation when the outside air is at 29 °C db and 23 °C wb. The sensible heat gain is 100 kW and the latent heat gain 25 kW. If the air leaves the cooling coil at 10 °C and then undergoes a 2 K temperature rise across the fan, calculate:

(i)   the re-circulation ratio;
(ii) the cooling coil capacity;
(iii) the minimum sensible heat gain required to maintain the room temperature;
(iv) the room percentage saturation for case (iii).

Solution

(i) The central plant is as illustrated in Fig. 8.55, (page 468) and the psychrometric cycle is shown in Fig. 8.57. The supply temperature is 12 °C, which is 2 K above the off-coil temperature of 10 °C. The temperature leaving the fan is 12 °C and the specific volume of the air will be about 0.815 m³/kg. (When the cycle is completed this can be checked—in this case the value is 0.816 m³/kg—an inspection of the chart shows that it is not difficult to establish a sufficiently accurate estimate of specific volume, knowing only the temperature, so that there is no need to modify the calculations.)

$$\text{Fresh air } \dot{m}_a = \frac{\dot{V}}{v_S} = \frac{1.86}{0.815} = 2.28 \text{ kg/s}$$

$$\text{Maximum } \dot{m}_a = \frac{\dot{S}}{c_{pma}(t_R - t_S)} = \frac{100}{1.02(21 - 12)} = 10.9 \text{ kg/s}$$

Figure 8.57 Example 8.20

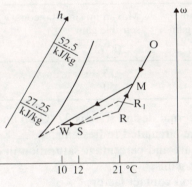

With a turndown ratio of 2:1,

$$\text{Minimum } \dot{m} = 10.9 \times 0.5 = 5.45 \text{ kg/s}$$

The fresh air requirement must be satisfied at the minimum flow rate. Therefore

$$\text{Re-circulation ratio} = \frac{5.45 - 2.28}{2.28} = 1.39$$

(ii) The supply state can now be fixed by either calculating the supply specific humidity from Eqn [1.38] or using the room ratio line as follows

$$\text{Room ratio line} = \frac{\dot{S}}{\dot{S} + \dot{L}} = \frac{100}{100 + 25} = 0.8$$

The room and outside air states can be established on the chart and point M located by proportioning line OM to line MR in the ratio of 1.39:1. The *RRL* can be drawn from R to cut the 12 °C db line at S which locates the supply state, S. The off-coil state W has the same specific humidity as S and a temperature of 10 °C. Joining M to W completes the cycle. Taking enthalpy values from the chart:

$$\text{Cooling coil load, } \dot{Q} = \dot{m}(h_M - h_W) = 10.9(52.5 - 27.25) = 275 \text{ kW}$$

(iii)　For minimum mass flow rate:

$$\text{Minimum gain } \dot{S} = 5.45 \times 1.02 \times (21 - 12) = 50\,\text{kW}$$

If the sensible heat load falls below 50 kW, the room temperature will fall below 21 °C.

(iv)　New $RRL = \dfrac{50}{50 + 25} = 0.67$

If the new RRL is drawn from S it cuts the 21 °C db line at R_1 and the percentage saturation is 56%.

Example 8.21

A variable air volume air conditioning system with a turndown ratio of 2.5:1 attempts to maintain two zones in a building at 22 °C with an 8 K temperature difference between the room and supply air. The air leaving the cooling coil has a specific humidity of 0.0076 and there is a temperature rise across the fan of 2 K. Details of the design heat gains are as follows.

Zone	Maximum sensible gain (kW)	Maximum simultaneous sensible gain (kW)	Constant latent gain (kW)	Minimum fresh air (kg/s)
A	50	15	8	0.64
B	50	50	8	0.82

If the summer external design state is 27 °C db and 22 °C wb, calculate:
(i)　the ratio of re-circulated to fresh air;
(ii)　the temperature and percentage saturation in each zone during maximum simultaneous gains;
(iii)　the cooling coil contact factor;
(iv)　the maximum cooling coil load.

(Wolverhampton Polytechnic)

Solution
(i)　The plant is as shown in Fig. 8.55, (page 468), and the psychrometric cycle is shown in Fig. 8.58. With the same maximum sensible heat gain and the same turndown ratio, each zone will have the same maximum and minimum mass flow rates.

$$\text{Maximum } \dot{m}_a = \frac{\dot{S}}{c_{pma}\,\Delta t} = \frac{50}{1.02 \times 8} = 6.13\,\text{kg/s}$$

Figure 8.58　Example 8.21

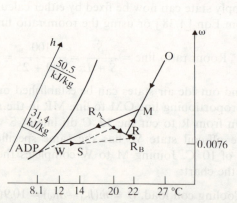

$$\text{Minimum } \dot{m} = \frac{6.13}{2.5} = 2.45 \text{ kg/s}$$

Since both zones have the same minimum mass flow rate the zone requiring the most fresh air will control the re-circulation ratio.

$$\text{Re-circulation ratio} = \frac{2.45 - 0.82}{0.82} = 2$$

(ii)   Supply temperature $= 22 - 8 = 14\,°C$

Zone A: $RRL = \dfrac{15}{15 + 8} = 0.65$

Zone B: $RRL = \dfrac{50}{50 + 8} = 0.86$

The minimum load to maintain zone A at $22\,°C$ is $50/2.5 = 20\,kW$; with a load of only $15\,kW$ the temperature must fall below this.

Zone A: $t_R = t_S + \dfrac{\dot{S}}{\dot{m}c_{pma}} = 14 + \dfrac{15}{2.45 \times 1.02} = 20\,°C$

Since zone B is at its maximum sensible gain, it will be at the design temperature of $22\,°C$.

The two room ratio lines can now be drawn from the supply state to establish the room states R_A and R_B. From the chart, zone A is 61% saturation and zone B at 48% saturation.

(iii)   The return air state R will lie along a line joining R_A and R_B. It can be fixed by proportioning the lines or calculating the mixed temperature, i.e.

$$t_R = \left(\frac{2.45}{2.45 + 6.13} \times 20\right) + \left(\frac{6.13}{2.45 + 6.13} \times 22\right) = 21.43\,°C$$

The psychrometric cycle can now be completed using the re-circulation ratio of 2:1 to establish point M:

$$t_M = \left(27 \times \frac{1}{3}\right) + \left(21.43 \times \frac{2}{3}\right) = 23.3\,°C$$

and locating point W 2 K to the left of S and extending line MW to locate the ADP.

The cooling coil contact factor can now be calculated:

$$\text{Contact factor} = \frac{t_M - t_W}{t_M - t_{ADP}} = \frac{23.3 - 12}{23.3 - 8.1} = 0.74$$

(iv)   The total mass flow rate through the central plant is $(6.13 + 2.45) = 8.58$ kg/s.

$$\therefore \quad \text{Cooling coil } \dot{Q} = \dot{m}_T(h_M - h_W) = 8.58(50.5 - 31.4) = 164\,kW$$

If a terminal reheat system had been used, the mass flow rate through the cooling coil would have been $(6.13 + 6.13) = 12.26$ kg/s and reheat would have been required for zone A. However, the comparison is not quite so simple because a constant-volume system could have used a higher re-circulation ratio.

Variable air volume with reheat

These are VAV systems with reheaters fitted to each zone. As the sensible load falls, the first step in control is to reduce the flow rate of air. Only when the terminals are fully turned down and the room temperature is still falling will the reheater be switched on.

These systems are designed for maximum simultaneous gains in the same way as basic VAV systems and therefore have the same advantages. The reheaters may be required only in areas with intermittent loads.

Example 8.22

A building is air-conditioned by a VAV system which uses terminals with a turndown ratio of 3:1. One zone of the building is a conference room which has a sensible heat gain of 24 kW when fully occupied which falls to 4 kW when unoccupied. The zone design temperature is 22 °C but this is allowed to fall to 20 °C when unoccupied. If the supply temperature is 14 °C, calculate the size of reheater required.

Solution

$$\text{Design } \dot{m}_a = \frac{\dot{S}}{c_{pma}(t_R - t_S)} = \frac{24}{1.02(22 - 14)} = 2.94 \text{ kg/s}$$

$$\text{Minimum } \dot{m}_a = \frac{2.94}{3} = 0.98 \text{ kg/s}$$

At minimum load and minimum mass flow rate, the room temperature without reheat would be:

$$t_R = t_S + \frac{\dot{S}}{\dot{m}_a c_{pma}} = 14 + \frac{4}{0.98 \times 1.02} = 18 \text{ °C}$$

The necessary reheat is therefore:

$$\dot{Q} = 0.98 \times 1.02 \times (20 - 18) = 2 \text{ kW}$$

8.12 ANNUAL ENERGY CONSUMPTION

The calculation of heating and air conditioning loads was covered in Chapter 4. It must be emphasized that the values calculated in Chapter 4 were the room or building loads. In the case of a simple LPHW heating system the steady-state output from the plant (less any distribution losses) is the same as the steady-state room sensible heat load for both design and part-loads. This considerably simplifies the prediction of annual energy consumptions for simple heating systems.

The situation is quite different for air conditioning systems where both the internal air temperature and the internal humidity are controlled. For an air conditioning system, the plant design and part-loads are quite different from the building sensible and latent heat loads. In Example 8.1, the design room loads were 20 kW sensible heat gain and 5 kW latent heat gain, but the design cooling coil load was 66 kW. A further complication for an air conditioning system is that two systems with the same design cooling load can have quite different plant part-loads for the same room loads and external air state. This was also demonstrated in Example 8.1 where, for a part-load case with room loads of 10 kW sensible heat gain and 5 kW latent heat gain, an off-coil temperature (dew point) control system required a plant cooling load of 60 kW but a sequence heating and cooling control system required a plant cooling load of only 32 kW. In addition to the higher cooling load, the off-coil temperature system of control also had a reheater load of 10 kW.

The plant and room loads are also different in winter. In Example 8.2 the room winter design loads were a 15 kW sensible heat loss and a 5 kW latent heat gain. Using a spray humidifier, the design plant loads were: frost coil, 23 kW: pre-heater, 62 kW; and after-heater, 36 kW.

Cooling Coil Annual Loads

One way of determining the energy consumption for a cooling coil is to analyse the frequency of occurrence of external air states and the corresponding room loads, and then calculate the cooling coil part-loads throughout the year. Weather frequency data will give the duration of each part-load which enables the annual energy consumption to be calculated. A cooling coil load is calculated from the equation $\dot{Q} = \dot{m}_a(h_O - h_W)$ where h_O is specific enthalpy of the outside air and h_W is the specific enthalpy of the air leaving the coil. If $\dot{m}_a$ is constant the cooling coil load depends only on the change in specific enthalpy Δh across the coil.

Since cooling coil loads depend on changes in specific enthalpy, it is enthalpy data, rather than temperature data, which must be used when analysing the frequency of a particular part-load. Collingbourn and Legg[8.9] have analysed weather frequency data for 23 meteorological stations throughout the UK. One of the sets of data they produced is the frequency of occurrence of specific enthalpy values for six different timespans during the day. Most of the data comprise the averages for a period of 18 years between 1960 and 1977. An extract from Ref. 8.9 for Birmingham airport (Elmdon) is reproduced here as Table 8.3 by permission of The Polytechnic of the South Bank.

The frequency data, expressed as a percentage of the total timespan, are given at intervals of 2 kJ/kg for each timespan. For example, the column headed 00.00–23.00 cover $24 \times 365 = 8760$ hours and the specific enthalpy between 52.0 and 53.9 kJ/kg (i.e. 52 to 54 inclusive) occurs for 0.14% of 8760 = 12.26 hours during an average year. Figure 8.59 shows data from Table 8.3 for a 24-hour period plotted on a psychrometric chart. The specific enthalpy interval used in the figure is 4 kJ/kg, which is reasonable for hand calculations. The frequency values shown are the sums of two adjacent table values, for example between 52 and 56 kJ/kg the frequency is 0.14 + 0.10 = 0.24%. It should be noted that, for the same enthalpy increment, between 0700 and 1800 the frequency is 0.39% and between 1900 and 0600 the frequency is 0.09%. This reflects the higher enthalpy values during the day compared with at night. Also, the mean 24-hour frequency is the average of the two, i.e. (0.39 + 0.9)/2 = 0.24%.

Summer outside design states normally lie within the upper bands of the weather envelope. It is clear that they are only reached or exceeded for a very small percentage of the year. Also, the frequency values are still quite small over a large area of the total envelope well below the enthalpy value

Table 8.3 Enthalpy frequency data for Birmingham airport

Elmdon (Birmingham airport) Specific enthalpy 2 kJ/kg intervals −20.0 to +79.9 kJ/kg
Percentages for period 1/1/1960 to 31/12/1977

Specific enthalpy (kJ/kg)	Hours (GMT) 0000–2300	0700–1800	1900–0600	1000–1700	1800–0100	0200–0900
Below limits	0.0	0.0	0.0	0.0	0.0	0.0
−20.0 to −18.1	0.0	0.0	0.0	0.0	0.0	0.0
−18.0 to −16.1	0.0	0.0	0.0	0.0	0.0	0.0
−16.0 to −14.1	0.0	0.0	0.0	0.0	0.0	0.0
−14.0 to −12.1	0.0	0.0	0.01	0.0	0.0	0.01
−12.0 to −10.1	0.01	0.01	0.01	0.0	0.01	0.01
−10.0 to −8.1	0.01	0.01	0.01	0.0	0.01	0.01
−8.0 to −6.1	0.03	0.01	0.04	0.0	0.02	0.05
−6.0 to −4.1	0.05	0.03	0.07	0.0	0.04	0.11
−4.0 to −2.1	0.11	0.07	0.16	0.01	0.11	0.22
−2.0 to −0.1	0.22	0.14	0.31	0.06	0.21	0.41
0.0 to 1.9	0.44	0.28	0.59	0.15	0.43	0.73
2.0 to 3.9	0.77	0.53	1.00	0.35	0.79	1.16
4.0 to 5.9	1.33	1.00	1.67	0.75	1.37	1.88
6.0 to 7.9	2.21	1.68	2.73	1.33	2.42	2.87
8.0 to 9.9	3.76	3.34	4.19	2.95	3.80	4.54
10.0 to 11.9	4.27	3.75	4.78	3.41	4.39	4.99
12.0 to 13.9	5.01	4.36	5.67	4.03	5.16	5.85
14.0 to 15.9	5.83	5.23	6.42	4.98	6.23	6.27
16.0 to 17.9	6.36	5.02	6.71	5.87	6.49	6.72
18.0 to 19.9	6.51	6.24	6.78	6.27	6.40	6.86
20.0 to 21.9	6.45	6.05	6.84	6.07	6.49	6.78
22.0 to 23.9	6.54	6.12	6.96	6.09	6.70	6.84
24.0 to 25.9	6.37	6.20	6.54	6.13	6.34	6.63
26.0 to 27.9	6.39	6.12	6.65	6.15	6.42	6.58
28.0 to 29.9	6.19	6.16	6.22	6.22	6.18	6.17
30.0 to 31.9	5.89	5.95	5.82	5.88	5.96	5.82
32.0 to 33.9	5.45	5.75	5.15	5.78	5.55	5.02
34.0 to 35.9	4.83	5.55	4.12	5.63	4.49	4.39
36.0 to 37.9	4.04	4.71	3.36	5.02	3.95	3.15
38.0 to 39.9	3.26	4.02	2.50	4.37	3.19	2.20
40.0 to 41.9	2.68	3.43	1.93	3.82	2.39	1.82
42.0 to 43.9	1.93	2.67	1.20	3.04	1.75	1.01
44.0 to 45.9	1.27	1.79	0.75	2.11	1.19	0.50
46.0 to 47.9	0.77	1.13	0.41	1.37	0.71	0.22
48.0 to 49.9	0.47	0.74	0.20	0.93	0.37	0.11
50.0 to 51.9	0.27	0.43	0.11	0.55	0.22	0.04
52.0 to 53.9	0.14	0.22	0.06	0.27	0.12	0.02
54.0 to 55.9	0.10	0.17	0.03	0.22	0.07	0.01
56.0 to 57.9	0.04	0.06	0.01	0.08	0.03	0.0
58.0 to 59.9	0.01	0.02	0.0	0.03	0.01	0.0
60.0 to 61.9	0.0	0.01	0.0	0.01	0.0	0.0
62.0 to 63.9	0.0	0.0	0.0	0.01	0.0	0.0
64.0 to 65.9	0.0	0.0	0.0	0.0	0.0	0.0
66.0 to 67.9	0.0	0.0	0.0	0.0	0.0	0.0
68.0 to 69.9	0.0	0.0	0.0	0.0	0.0	0.0
70.0 to 71.9	0.0	0.0	0.0	0.0	0.0	0.0
72.0 to 73.9	0.0	0.0	0.0	0.0	0.0	0.0
74.0 to 75.9	0.0	0.0	0.0	0.0	0.0	0.0
76.0 to 77.9	0.0	0.0	0.0	0.0	0.0	0.0
78.0 to 79.9	0.0	0.0	0.0	0.0	0.0	0.0
Above limits	0.0	0.0	0.0	0.0	0.0	0.0
Missing data	0.0	0.0	0.0	0.0	0.0	0.0

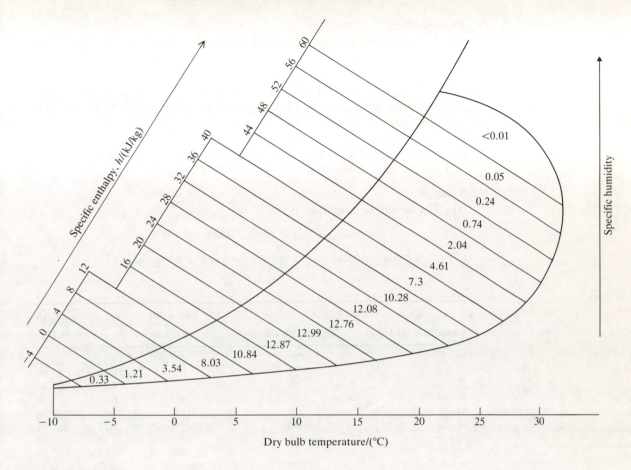

Figure 8.59 24-hour enthalpy frequency, 1960–1977, Birmingham airport

corresponding to a typical summer design state. This should be taken into account when selecting systems and their controls. There may be little point in providing a very sophisticated system to cater for very high enthalpy values when these occur so infrequently.

If the air passing through a cooling coil undergoes a change in specific enthalpy, Δh, for a percentage of the total hours (24, 12 or 8 multiplied by 365) in the year, f, then the product $f\Delta h$ is proportional to the energy consumed during that time. For a full year $\sum f$ is 100%; therefore the average annual value of Δh is:

$$\text{Average annual } \Delta h = \frac{\sum f\Delta h}{100} \qquad [8.7]$$

If a cooling coil operates continuously, Eqn [8.7] is satisfactory. However, most cooling coils only operate for part of the year. If only the weather bands when the plant is running are considered and the corresponding operating frequencies, f_o, are used, then the operating time is proportional to $\sum f_o$ and the equation for the average operating Δh is:

$$\text{Average operating } \Delta h = \frac{\sum f_o \Delta h}{\sum f_o} \qquad [8.8]$$

The use of enthalpy frequency data to calculate annual cooling loads is demonstrated in the following example.

Example 8.23

An air conditioning plant operates for 12 hours during the day providing a constant 2 kg/s of conditioned air. There is no re-circulation of room air. The plant has an off-coil temperature (dew point) system of control and the enthalpy of the air leaving the coil may be taken as a constant 28 kJ/kg. The outside design specific enthalpy is 60 kJ/kg which corresponds to 27 °C db and 20.8 °C wb. Calculate:

(i)   the design cooling coil load;
(ii) the average cooling coil load throughout the cooling season;
(iii) the annual cooling load.

Solution

(i)   The design cooling load is:

$$\dot{Q} = \dot{m}(h_{\mathrm{O}} - h_{\mathrm{W}}) = 2(60 - 28) = 64\,\mathrm{kW}$$

(ii) A tabular calculation is used to analyse the seasonal load.

Range of h_{o}/(kJ/kg)	Mean h_{o}/(kJ/kg)	$\Delta h = \dfrac{(h_{\mathrm{o}} - 28)}{(\mathrm{kJ/kg})}$	f_{o} %	$\dfrac{f_{\mathrm{o}}\Delta h}{(\mathrm{kJ/kg})}$
28–32	30	2	12.11	24.22
32–36	34	6	11.30	67.80
36–40	38	10	8.73	87.30
40–44	42	14	6.10	85.40
44–48	46	18	2.92	52.56
48–52	50	22	1.17	25.74
52–56	54	26	0.39	10.14
56–60	58	30	0.08	2.40
>60	60	32	0.01	0.32
	Totals		42.81	355.88

For any part-load the change in specific enthalpy across the coil is given by $\Delta h = (h_{\mathrm{o}} - 28)\,\mathrm{kJ/kg}$. When the specific enthalpy value of the outside air is 28 kJ/kg there is no load and when the outside air reaches or exceeds 60 kJ/kg the plant is operating at its maximum load and no further increase in Δh is possible. These values of h_{o} provide the limits to the table and between them convenient increments of h_{o} are selected. Increments of 4 kJ/kg will give reasonable accuracy. The increments do not have to be equal and it is often necessary to use smaller increments at the start and end of the table; for example, if the off-coil air had had a specific enthalpy of 27 kJ/kg the first increment would have been 27–28 kJ/kg and if the design value of h_{o} had been 59 kJ/kg the last three increments would have been 56–58, 58–59 and above 59 kJ/kg. The last increment is always all values above the design value of h_{o}.

The second column in the table is the mean value of h_{o} over the increment chosen. The third column is the mean change in specific enthalpy over the coil, Δh. The example states that the plant operates for 12 hours per day: therefore the appropriate column in Table 8.3 for the frequency data is the one headed 0700–1800 hours. The fourth column is the frequency of occurrence, f_{o}, of the specific enthalpy increment when the cooling coil is on; for example, between

36 and 40 kJ/kg the frequency is $4.71 + 4.02 = 8.73\%$. The last column is the product $f_o \Delta h$. The last two columns are then totalled to give $\sum f_o$ and $\sum f_o \Delta h$. The first total shows that the cooling coil was operating for 42.81% of the total possible hours.

The average load when the cooling coil is on can now be calculated using Eqn [8.8] as follows:

$$\text{Average operating } \Delta h = \frac{\sum f_o \Delta h}{\sum f_o} = \frac{355.88}{42.81} = 8.313 \text{ kJ/kg}$$

$$\text{Average operating } \dot{Q} = \dot{m}_a \Delta h = 2 \times 8.313 = 16.626 \text{ kW}$$

The average cooling coil load is therefore 16.6 kW, compared with the design cooling coil load of 64 kW. As a percentage this is 26% which is a very low value compared with, say, a typical boiler.

(iii)   The average operating load is now used to calculate the annual cooling load:

$$\text{Annual cooling load} = 16.626 \times \frac{42.81}{100} \times 365 \times 12 \times 3600 = 112.2 \text{ GJ}$$

Figure 8.60 shows graphs of Δh and f both plotted on a base of h_o. The area under the frequency graph is 100% with the area to the right of the 28 kJ/kg line being 42.81%. This graph illustrates the very low frequency of occurrence of very high and very low specific enthalpy values. The top graph, showing the change in specific enthalpy through the coil, reaches a maximum value for the design case and then remains constant. The above tabular calculation procedure is a numerical integration of the lower graph and the product of points from each graph.

Annual power consumption

Establishing an annual cooling load may be sufficient to compare two possible systems but it does not give the energy input required. For a cooling load the energy input is in the form of power to the refrigeration plant, and a knowledge of the plant coefficient of performance is required. Ideally the coefficient of performance for each part-load case should be used, which extends the tabular calculation. Example 7.13 in Chapter 7 demonstrated how part-load coefficients of performance can be calculated. If only a mean value of coefficient of performance is available the energy input to the plant can be calculated, as shown in the following example.

Example 8.24

If the average overall coefficient of performance of the refrigeration plant servicing the air conditioning plant in Example 8.23 is 3.2, calculate the average energy input and the annual energy consumption.

Solution

$$\text{Average energy input} = \frac{\dot{Q}}{\text{COP}} = \frac{16.626}{3.2} = 5.2 \text{ kW}$$

Figure 8.60 Example 8.23

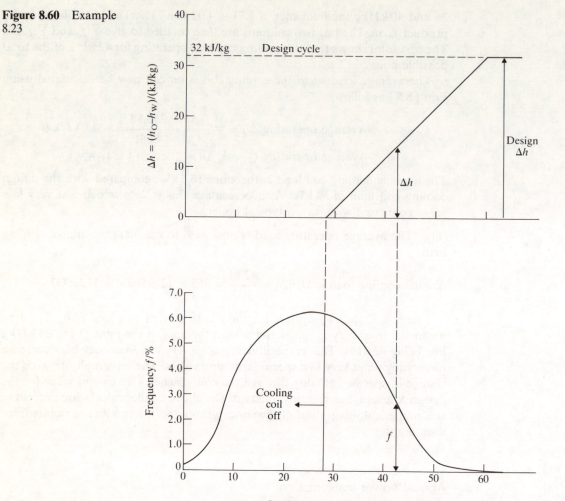

Outside air specific enthalpy, h_O/(kJ/kg)

$$\text{Annual consumption} = 5.2 \times \frac{42.81}{100} \times 365 \times 12 = 9750 \text{ kW h}$$

The annual energy consumption is therefore 9750 kW h.

Plant Annual Loads

The examples in this section are based on a plant with an off-coil temperature control system, which simplifies calculations for two reasons: firstly, because a relatively constant specific enthalpy value for the air leaving the coil is maintained throughout the year; secondly, because the room load does not affect the cooling coil load (the reheater controls the room temperature—see Section 8.5). For more complex plants the enthalpy value of the air leaving the coil may change and changes in the room load must be taken into account.

This section has only covered cooling coils and the seasonal performance of the other items of equipment is required for a complete plant analysis. For the summer operation of an off-coil (dew point) plant the reheater loads, which depend on the room loads, would also have to be calculated, as would the energy consumption of fans and pumps.

Dry bulb frequency data, which are found in the CIBSE *Guide*[8.10] as well as in Ref. 8.9, can be used to analyse the seasonal performance of heaters and frost coils. To calculate the annual energy consumption for a complete plant requires analysis of psychrometric and refrigeration cycles throughout the year and the use of appropriate banded weather frequency data.

A computer program is the ideal aid to analysing annual energy consumptions, provided the user is aware of the limitations of the program.

8.13 SYSTEM SELECTION

One of the most difficult tasks for Building Services Engineers is the selection of the most appropriate air conditioning system for a particular application. They must start with a thorough understanding of the operation, advantages and disadvantages of all the possible options. They should also try to start with an open mind, without prejudice from previous jobs or systems they may have particular expertise in designing. A selection process may become very involved and only a brief outline of the minimum work required is given in this section.

Specification

This lays down the criteria for the design. It should be written after discussions involving all the interested parties—the client, the Building Services Engineer and the Architect are the minimum. The specification should clearly identify each feature as being either essential or desirable. Care must be taken in identifying any feature as essential since this may restrict the choice of system. Examples where essential features restrict the choice of system include operating theatres in hospitals which must use 100% fresh air to prevent the possible spread of infection; libraries for which it may be mandatory not to use water services within the conditioned spaces because of the possibility of damage to expensive books if leaks occur; city centre sites in which the use of wet cooling towers may be banned because of the fear of Legionnaires' disease.

Some of the features which might be included in the specification are the following.

(1) Capital cost.
(2) Operating cost.
(3) Space temperature control limits (using a higher design temperature in summer than in winter can save a lot of energy).
(4) Humidity control limits (close control—say $50 \pm 5\%$—uses much more energy than comfort control—say 30% to 70%).
(5) Outside design conditions for summer and winter (weather data from the nearest meteorological station should be analysed and tolerable periods when the weather will be outside the design limits established).
(6) Occupancy (activities and fresh air requirements, occupancy times).
(7) Space noise limits.
(8) Location of plant rooms and condensers—any weight limitations?

(9) Floor/underfloor/ceiling void spaces available for equipment.
(10) Space flexibility—might the proposed usage change? (This may be very important for buildings which will be rented.)
(11) Aesthetics
(12) Water supply—can water from a river, a canal or the sea be used for heat pumps or cooling condensers?
(13) Electricity supply (are there any limitations? Ice banks can be used to reduce the refrigeration plant maximum demand.)

Once the specification has been agreed the designer knows what he is trying to achieve and what is available to him to design a satisfactory system.

Evaluation

A careful study of the specification will eliminate the possibility of using some systems and the designer will be left with a more manageable number of possible systems which could all satisfy the essential requirements. It is now necessary to evaluate the remaining options. An evaluation worksheet is a useful aid in this process. An example of an evaluation worksheet is reproduced here by permission of the CIBSE from Ref. 8.11 (Table 8.5).

The table lists system features in the left-hand column; for each system there are three columns headed SC (score); WTG (weighting or relative importance); and TOT (total weighted score). The features (items) listed are only typical of those which might be considered and any particular job would have its own individual list. Features need not be included if they are the same for all the systems.

The weighting is a mark out of 10 which reflects how important a feature is and will be the same for all systems. (A modification to the table shown is to have the weightings in a single column.) For example, if close control of humidity had been specified as desirable but not essential, a weighting of 7 might have been allocated.

The score is also a mark out of 10 which reflects how a particular system would perform. For the feature humidity, a plant with an off-coil temperature control system might merit a score of 10; whereas a plant with a sequence heating and cooling control system with a high level humidity override might warrant only 5.

The weighted score is then the product of the weighting and the score. The plant with the off-coil temperature control would have a weighted score of $10 \times 7 = 70$, and the plant with the sequence heating and cooling system of control and high level humidity override, a weighted score of $5 \times 7 = 35$. The weighted scores are then added together to give a total weighted score and the system with the highest total is selected.

ASHRAE[8.12] gives details of a similar approach but with the added complication that the designer makes an original evaluation and the client then makes a re-evaluation. The designer's evaluation marks each feature out of 100 and these figures are then multiplied by the client's 'relative importance' factor before being totalled. The 'relative importance' factors in ASHRAE are fractions which total 1 for all the features.

The use of evaluation worksheets allows the designer to concentrate on one feature at a time instead of trying to compare complete systems. The decision-making process is more objective and easier to justify than an 'inspired' selection.

Table 8.5 Evaluation worksheet

Item	System 1 SC*	WTG	TOT	System 2 SC	WTG	TOT	System 3 SC	WTG	TOT
Economic initial costs									
System	..	..	...	..	..	...	..	..	...
Related services	..	..	...	..	..	...	..	..	...
Related building	..	..	...	..	..	...	..	..	...
O & M costs									
Energy	..	..	...	..	..	...	..	..	...
Materials	..	..	...	..	..	...	..	..	...
Labour	..	..	...	..	..	...	..	..	...
Total (Economics)			...			...			...
Quality (Subjective)									
Appearance	..	..	...	..	..	...	..	..	...
Maintainability	..	..	...	..	..	...	..	..	...
Reliability	..	..	...	..	..	...	..	..	...
Flexibility	..	..	...	..	..	...	..	..	...
Comfort									
Temperature	..	..	...	..	..	...	..	..	...
Humidity	..	..	...	..	..	...	..	..	...
Air distribution	..	..	...	..	..	...	..	..	...
Air quality	..	..	...	..	..	...	..	..	...
Noise	..	..	...	..	..	...	..	..	...
Control	..	..	...	..	..	...	..	..	...
Fire/smoke control	..	..	...	..	..	...	..	..	...
Space suitability	..	..	...	..	..	...	..	..	...
Hazards									
Pollution	..	..	...	..	..	...	..	..	...
Liquid leakage	..	..	...	..	..	...	..	..	...
Public relations	..	..	...	..	..	...	..	..	...
Special (list)									
....	..	..	...	..	..	...	..	..	...
....	..	..	...	..	..	...	..	..	...
Totals (Quality)			...			...			...

* SC = score (out of 10)
WTG = weighting or relative importance (on a scale of 0 to 10)
TOT = total weighted score (SC multiplied by WTG)

PROBLEMS

8.1 (a) An air conditioning plant with no re-circulation of room air has sensible and latent heat loads of 30 kW and 8 kW respectively for the summer design state of 27 °C db and 21 °C wb. The cooling coil contact factor is 0.75 and the ADP 8 °C. If the room temperature is 22 °C and the temperature rise across the fan is 1.25 K, calculate the design cooling coil load and the room percentage saturation.

(b) For a part-load case, the room sensible heat gain falls to 16 kW with no change in the latent heat gain when the outside air is at 24 °C db and 21 °C wb. If a sequence heating and cooling system of control is used, calculate the new ADP, cooling coil load and room percentage saturation.

(c)   If a high limit humidity detector had been fitted to reset the controls to an off-coil (dew point) control system for the conditions in part (b), calculate the new values of the ADP, cooling coil load, room percentage saturation and after-heater load. Assume the coiling coil contact factor is constant and that the off-coil temperature is the same as for the design case.
(100 kW, 54%; 14 °C, 60.3 kW, 71%; 9 °C, 93.7 kW, 58%, 14 kW)

8.2   (a)   Define the following terms. Where appropriate, use sketches to illustrate your answers.
(i)   Room ratio line.
(ii)   Adiabatic humidification.
(iii)   Outdoor air design temperature.
(iv)   Wet bulb temperature, sling and screen.
(b)   In an air conditioning system, outdoor air is mixed with re-circulated air. The mixed air then passes through a water spray humidifier followed by a sensible heating coil. If the final air temperature is 30 °C, then using the data below and the psychrometric chart provided, calculate the heater load for the following conditions:
(i)   humidifier contact factor 1;
(ii)   humidifier contact factor 0.75.

Data
Outdoor air: dry bulb temperature, 2 °C; percentage saturation, 50%; air mass flow rate, 3.0 kg/s. Re-circulated air: dry bulb temperature, 22 °C; wet bulb temperature, 15 °C; air mass flow rate, 4.5 kg/s. Humid specific heat, 1.02 kJ/kg K.
   Assume no miscellaneous heat gains from fans and ducts.

(CIBSE)

(157 kW; 148 kW)

8.3   An art gallery is required to be maintained at 22 °C db, 50% saturation throughout the year by the air conditioning system shown in Fig. 8.61. If a constant dew point control system is used, determine the peak loads on the cooling coil and reheater and the percentage of re-circulation air in winter.

Figure 8.61   Problem 8.3

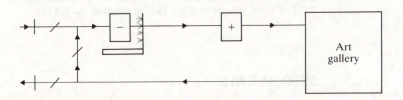

Data
Cooling coil contact factor 0.95; minimum fresh air, 0.2 kg/s. Summer: sensible heat gain, 10 kW; latent heat gain, 1 kW; maximum temperature difference between room and supply, 8 K; outdoor air condition, 27 °C db, 20 °C wb. Winter: sensible heat loss, 15 kW; latent heat gain, 1 kW; outdoor air condition, −3 °C db, 95% saturated. Humidifier efficiency, 100%.

(Newcastle upon Tyne Polytechnic)

(17 kW, 29 kW, 70%)

8.4 A room is maintained at internal conditions of 20 °C db and 50% saturation. Air is extracted from the room and part re-circulated to be mixed with fresh air. The mixed air is then cooled and de-humidified before being reheated and supplied to the room. Determine:
(i)   the condition of the air supplied to the room;
(ii) the condition of the mixed air;
(iii) the apparatus dew point of the cooling coil;
(iv) the load on the cooling coil; and
(v)   the load on the heating coil.

Design data
Humid specific heat of air = 1.026 kJ/kg K. Sensible heat gain to the room = 21.9 kW. Latent heat gain to the room = 9.4 kW. Minimum flow rate of fresh air = 0.5 kg/s. Outside condition = 27 °C db and 21.5 °C wb. Maximum temperature difference between the room and supply air = 6 K. Cooling coil contact factor = 0.85.

(University of Liverpool)
(14 °C and 64%; 21 °C and 52%; 6.7 °C; 60.5 kW; 18.2 kW)

8.5 A room is air-conditioned by a simple single-zone re-circulating plant in which the heater and cooler are controlled in sequence by a room thermostat, i.e. non-reheat is to be used. Determine the room percentage saturation which will arise under the following conditions:

Outside air, 26 °C db and 19 °C wb; inside air temperature, 20 °C; supply differential t_S (max.) 9 K; sensible gain, 60 kW; latent gain, 8 kW; minimum fresh air, 25%; coil contact factor, 0.9.

If the room sensible gain drops to 30 kW and all other conditions remain constant, calculate the level to which the percentage saturation will rise if the coil capacity is controlled by:
(a)   diverting chilled water;
(b)   face and by-pass dampers—assume apparatus dew point and contact factor remain constant.

(CIBSE)

(55%; 75%; 60%)

8.6 The plant shown in Fig. 8.62 maintains a room at 21 °C. Calculate the plant loads and room percentage saturation for the design and part-load condition given below.

Design

Outside air	28 °C db	19 °C wb
Room loads	100 kW sensible	25 kW latent

Part-load

Outside air	23 °C db	18 °C wb
Room loads	50 kW sensible	10 kW latent

Maximum supply—room temperature differential	9 K
Design cooling coil contact factor	0.8
Part-load cooling coil contact factor	$1 - e^{k/\dot{V}}$

where k is a constant and $\dot{V}$ is the volume flow rate from the coil.

(Polytechnic of the South Bank)

(256 kW and 52%; 109 kW and 62%)

Figure 8.62 Problem 8.6

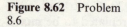

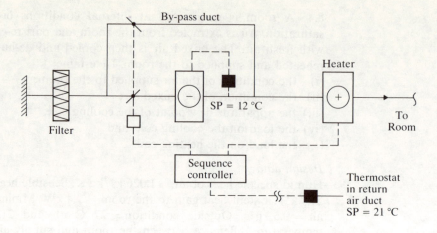

8.7 The internal design conditions for a building are 21 °C and 40% saturation in winter and 23 °C and 60% saturation in summer. The plant consists of a mixing chamber, cooling coil, pan humidifier and after-heater. The cooling coil has a contact factor of 0.8. The ratio of re-circulated to fresh air is 2:1 by mass for both the summer and winter design cases.

If fan and duct gains are negligible, use the data below to calculate:

(i)   the mass flow rate of supply air;
(ii) the cooling coil apparatus dew point;
(iii) the cooling coil load;
(iv) the supply temperature in winter;
(v)   the pan humidifier effectiveness;
(vi) the after-heater load.

Data
Summer ambient 29 °C db, 25 °C wb; winter ambient −4 °C saturated. Summer loads: 80 kW sensible gain; 20 kW latent gain. Winter loads: 60 kW sensible loss; 20 kW latent gain. For moist air $c_{pma} = 1.02$ kJ/kg K.

(Wolverhampton Polytechnic)
(9.8 kg/s; 12.5 °C; 182 kW; 27 °C; 30%; 156 kW)

8.8 (a) The proposed arrangement for an air conditioning system is shown in Fig. 8.63. The air flow rate which is extracted through the ceiling-mounted light fittings is 80% of the air mass flow rate supplied to the room and this extract air removes 60% of the installed lighting load. Using the data below and the psychrometric chart provided, calculate the following:
(i)   the supply air mass flow rate;
(ii) the cooling coil design load;
(iii) the heating coil design load (based on the design sensible heat loss).
(b) On Fig. 8.63, draw an outline control system to maintain the room design conditions. Briefly describe the function of each of the control elements shown.
(c) Show clearly on the psychrometric chart the air condition processes for the summer and winter design conditions appropriate to this control system.

Data
Sensible heat gains: solar and transmission, 27 kW; occupancy, 12 kW; installed lighting load, 20 kW. Sensible heat loss at the winter design condition, 20 kW;

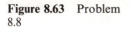

Figure 8.63 Problem 8.8

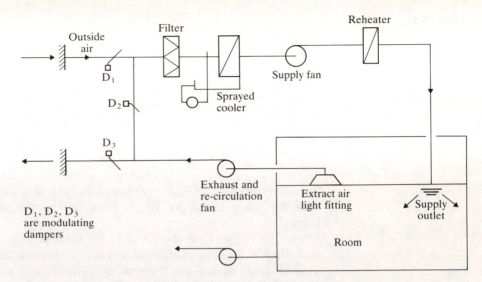

Outside air

Filter

Reheater

D_1

D_2

D_3

D_1, D_2, D_3 are modulating dampers

Sprayed cooler

Supply fan

Exhaust and re-circulation fan

Extract air light fitting

Supply outlet

Room

constant year-round latent heat gain, 10 kW. Maximum design temperature differentials: for cooling, 8 K; for heating, 12 K. Outside air design conditions: summer 30 °C db, 21 °C wb; winter, −2 °C db and saturated. Room air conditions: summer, 21 °C and 55% saturation; winter 19 °C and 45% saturation. *Maximum* re-circulation rate is 70% of the supply air mass flow rate. Sprayed cooling coil contact factor, 0.8. Specific heat of humid air, 1.02 kJ/kg K. Assume zero fan and duct gains.

(Polytechnic of the South Bank)

(5.76 kg/s; 98 kW; 97.5 kW)

8.9 An air conditioning plant consists of a mixing chamber, a cooling coil, an after-heater and an electric steam humidifier. The maximum ratio of re-circulated to fresh air is 3:1 by mass. The fan and duct gains are 2 K for the summer design case but for the winter design case the fan heat gains are equal to the duct heat losses.

The building is at 21 °C and 50% saturation in winter and summer. If in summer the supply air is at 12 °C, calculate:
(i)   the cooling coil design load and contact factor;
(ii) the winter supply temperature;
(iii) the mass flow rate of steam from the humidifier;
(iv) the maximum after-heater load;
(v)   the minimum outside temperature (assume saturated) which would allow the room design percentage saturation to be achieved without humidification.

Data
Summer design ambient air, 29 db and 22 wb; winter design ambient air, −5 °C saturated. Summer design loads: 100 kW sensible gain; 20 kW latent gain. Winter design loads: 40 kW sensible loss; 20 kW latent gain. Specific heat of moist air, 1.02 kJ/kg K.

(Wolverhampton Polytechnic)

(212 kW, 0.87; 24.6 °C; 27.4 kg/h; 112 kW; 3.8 °C)

8.10 Figure 8.64 shows an air conditioning plant which uses re-circulated air to avoid reheating. The following may be considered constant:

Figure 8.64 Problem 8.10

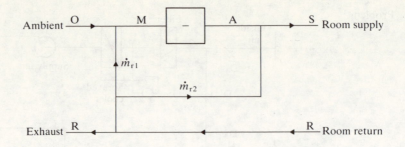

Room temperature, 21 °C; latent heat gain, 15 kW; minimum fresh air, 2 kg/s; mass flow rate of supply air; cooling coil contact factor; apparatus dew point.

For the design case the room has a sensible heat gain of 70 kW and a percentage saturation of 50% when the outside air is 28 °C db and 22 °C wb. This results in an off-coil temperature of 10 °C and a supply temperature of 13 °C. If the fan and duct gains are negligible, establish:

(i) the cooling coil contact factor and apparatus dew point;
(ii) four equations for which new values of m_{r1} and m_{r2} could be calculated when the outside air is at 23 °C db and 19 °C wb and the room sensible heat gain falls to 40 kW;
(iii) four equations which could be used to calculate the new room specific humidity for (ii).

It is not necessary to simplify or solve the equations, and c_{pma} and h_g may be taken as 1.02 kJ/kg K and 2540 kJ/kg respectively.

(Wolverhampton Polytechnic)

(0.8; 6.6 °C)

8.11 (a)   Figure 8.65 indicates diagrammatically an air conditioning system. On test, the system gave the following results from instruments located as noted. Using the psychrometric chart provided, calculate the following:

(i)   the cooling coil load;
(ii)   the cooling coil contact factor;
(iii) the after-heater loads.

You may assume no duct leakage and no fan/duct heat gains. Humid specific heat = 1.02 kJ/kg K.

Instrument position	Volume flow rate [m³/s]	Dry bulb temperature (°C)	Wet bulb temperature (°C)
1	—	30	22
2	—	12	11
3	1.4	18	—
4	1.8	12	—

(b)   Using Fig. 8.65, draw a suitable outline control schematic for the air conditioning system to maintain the conditions in the room at levels suitable for thermal comfort. Explain briefly the function of each element shown.

(CIBSE)

(125 kW; 0.82; 0 and 10.3 kW)

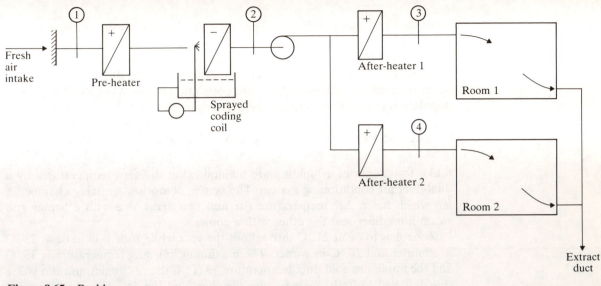

Figure 8.65 Problem 8.11

8.12 An air conditioning plant serves two rooms A and B, as shown in Fig. 8.66. The supply air temperature to each room is dictated by room thermostats controlling the adjacent reheater. The relative humidity in room A is dictated by a humidistat controlling the chilled water on to the main cooling coil. The humidity in room B is left to find its own level. Using the data given below, determine, using the psychrometric chart provided:

(i) The air mass flow rate to each room;
(ii) The relative humidity in room B when:
  (a) both rooms are occupied;
  (b) room A only is occupied;
  (c) room B only is occupied.
(iii) The maximum duties (in kilowatts) handled by each reheater coil and by the cooling coil.

Figure 8.66 Problem 8.12

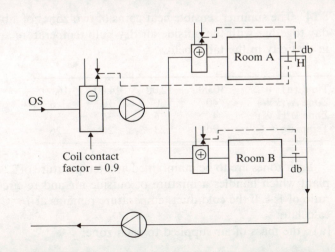

Data

Sensible and latent gains per person = 50 W and 40 W; outside design condition = 30 °C db, 22 °C wb; room A sensible load (excluding occupants), 20 kW; room B sensible load (excluding occupants), 30 kW; minimum air supply temperature, 15 °C; room A design conditions = 22 °C db, 50% saturation; room B design conditions = 22 °C db; population (room A) = 0 to 250 people; population (room B) = 0 to 200 people. Ignore fan heat.

(University of Liverpool)

(4.55 kg/s; 5.6 kg/s, 48%, 45%, 54%, A 25 kW, B 35 kW, 348 kW)

8.13 Two zones of a building are maintained at different temperatures by a dual-duct air conditioning system. The system comprises a mixing chamber, a fan which has a 2 K temperature rise and two ducts, one with a heater and steam humidifier and the other with a cooler.

Zone A is to be at 21 °C throughout the year while zone B is to be at 23 °C in summer and 20 °C in winter. The maximum hot duct temperature is 35 °C and the minimum cold duct temperature 13 °C. If the re-circulation ratio is 3:1 (re-circulated to fresh) by mass, use the chart and the data below to:

(i)   calculate the mass flow rate through each duct during maximum simultaneous gains;

(ii) calculate the maximum mass flow rate in the hot duct;

(iii) *sketch* the psychrometric cycle for the winter design case.

Data

Zone	Max. $\dot{S}$ (kW)	Max. simultaneous $\dot{S}$ (kW)	Constant $\dot{L}$ (kW)	Min. $\dot{S}$ (kW)
A	100 gain	100 gain	6	60 loss
B	80 gain	50 gain	8	20 loss

Summer external design state, 28 °C db and 23 °C wb; winter external design state, −5 °C, saturated; c_{pma} for moist air, 1.02 kJ/kg K.

(Wolverhampton Polytechnic)

(17.7 and 2.4 kg/s; 8.8 kg/s)

8.14 The summer sensible heat gains of two zones of a building on the design day together with the outside air dry bulb temperature are given at four times in that day in the table below.

Time/(h)	10.00	12.00	14.00	16.00
Zone A/(kW)	40	30	10	6
Zone B/(kW)	4	6	36	50
Temp. db/(°C)	20	23	26	26

Both zones are to be maintained at a temperature of 22 °C db by a dual-duct plant which handles a mixture of outside air and re-circulated room air at a ratio of 1:4. If the cold duct temperature remains at 16 °C throughout the day, calculate:

(a)   the mass of air supplied to each zone;

(b)   the mass of air flowing in the hot duct at 12.00 and 14.00 hours (assume the hot duct heater battery is turned off);

(c)   the dry bulb temperature of the air in the hot duct at 10.00 and 16.00 hours.

Given

Specific heat of humid air = 1.02 kJ/kg K
Neglect fan and duct gains.

(CIBSE)

(A 6.54 kg/s, B 8.17 kg/s; 8.55 kg/s, 6.3 kg/s; 20.6 °C, 22.8 °C)

8.15 A hot-deck/cold deck air conditioning system maintains two zones at 21 °C db, delivering 2.4 kg/s to each. The cold deck temperature is maintained by controlling a chilled water coil (contact factor 0.85) and the re-circulation dampers in sequence. An enthalpy controller closes the re-circulation dampers to minimum (40%) fresh air when the outside air enthalpy exceeds the room enthalpy.

At an intermediate summer condition, when the outside air is at 22 °C db and 19 °C wb, the cooling loads in the two zones are as follows:

	Sensible	Latent
	(kW)	(kW)
Zone A	20	6
Zone B	5	6

A compensator controller has reset the cold duct to 12 °C and the heater in the hot duct is off. If fan and duct gains are negligible calculate the percentage saturation in each zone.

(61% in A and 74% in B)

8.16 The primary plant of a non-change over induction system consists of a pre-heater, sprayed cooling coil, after-heater and fan. The sprayed coil has a constant contact factor of 0.8 for both cooling and humidifying. For the summer design case, the apparatus dew point is 8 °C and the air leaving the secondary coils must be 4 °C above the room dew point. If an off-coil (dew point) temperature detector maintains a constant off-coil temperature throughout the year, use the data below to calculate:

(i)   the maximum primary and secondary cooling coil loads;
(ii) the induction ratio;
(iii) the summer supply state;
(iv) the winter design supply state and room percentage saturation;
(v)   the mass flow rate of water supplied to the secondary cooling coil when the temperature rise of the water is 5 K and the sensible heat loss falls to 50% of the winter design value. Assume the state of the primary air is the same as for the design case.

Data

Summer design, 28 °C db, 23 °C wb; winter design, −3 °C saturated; summer sensible heat gain, 300 kW; winter sensible heat loss, 75 kW; constant latent gain, 25 kW; summer internal state, 22 °C, 55% saturation; winter internal temperature, 21 °C; minimum fresh air, 8 kg/s; fan and duct gain, 3 K. For moist air, $c_{pma} = 1.02$ kJ/kg K. For water, $h_g = 2540$ kJ/kg:

(Wolverhampton Polytechnic)

(404 kW, 217 kW; 3.5; 16.4 °C and 78%; 22.4 °C and 42%, 47%; 1.8 kg/s)

8.17 (a)   Compare briefly the advantages and disadvantages of two-pipe and four-pipe induction systems for air conditioning of a building.

(b)   The summer design criteria for an induction unit air conditioning system are as follows.

Outside air dry bulb temperature (°C), 26; outside air saturation (%), 70; inside air dry bulb temperature (°C), 21; inside air saturation (%) 60; primary air dry bulb temperature (°C), 10; room sensible heat gain (kW), 70; room latent gain (kW), 10.

The air conditioning system is to be a two-pipe induction system. In each induction unit secondary air passes through a coil before mixing with the primary air and returning to the room. In summer, the secondary air is to be at 90% saturation after passing through the unit coil.

The primary air is 100% outside air. The primary air plant consists of a filter cooling coil, heater, steam injection humidifier and fan. The cooling coil has a contact factor of 0.9. Draw the summer cycle on the psychrometric chart provided, and submit the chart with your answer.

Calculate the following quantities:

(i)   the secondary to primary air induction ratio;
(ii)   the primary air flow rate (kg/s);
(iii)   primary and secondary cooling coil loads (kW).

(University of Manchester)

(3.7:1; 2 kg/s; 71 kW, 48 kW)

8.18   A modular office building is to be air-conditioned using a non-changeover induction unit system to achieve a design condition of 22 °C and 50% saturation in summer, with the temperature being allowed to fall to 20 °C in winter. The system comprises a filter, pre-heater,, sprayed chilled water coil, reheater and supply fan.

The chilled water coil and pre-heater are controlled in sequence by a thermostat located downstream of the cooling coil with a set point selected to meet the summer design duty. Humidification in winter is achieved by spraying the cooling coil with the chilled water off.

The unit room coil must not have an air leaving temperature of less than 16 °C and must maintain a surface temperature above the room dew point.

Data

Summer outside design condition, 29 °C dry bulb, 21 °C wet bulb; winter outside design condition, −1 °C dry bulb, −1 °C wet bulb; summer maximum sensible heat gain, 536 kW; winter maximum sensible heat loss, 308 kW; latent heat gain, 50 kW; minimum ventilation air quantity, 9 m³/s; primary cooling coil apparatus dew point, 6.5 °C; cooling coil contact factor (cooling), 0.9; cooling coil contact factor (spraying) 0.8; maximum primary air temperature, 48 °C; fan and duct gain to primary air, 3 K.

Using the data provided above.

(a)   Sketch the proposed system showing the chilled water circuit and the essential controls, explaining briefly their operation.

(b)   Determine:
(i)   a suitable primary fan volume air flow rate;
(ii) a suitable induction ratio for the room units;
(iii) the maximum primary and secondary cooling duties;
(iv) the average room humidity in winter;
(v)   the maximum pre-heater and after-heater duties;
(vi) the temperature of the air entering the rooms from the units in both extreme conditions.

(CIBSE)

(9.1 m³/s; 6.1; 393 kW and 419 kW; 51%; 175 kW and 403 kW; 15.4 °C and 23.8 °C)

8.19   (a)   Compare and contrast the operating characteristics of a variable-volume (VAV) system with that of a constant flow rate, zonal reheat system.
(b)   In a VAV system, the supply air is maintained at a dry bulb temperature of 13 °C and a moisture content of 0.008 kg/kg dry air.

For the following design criteria, determine the rise in room relative humidity when the sensible heat gain in one of the air-conditioned rooms falls to 30 kW, the latent gain remaining constant.

Design heat gains: sensible, 90 kW; latent, 15 kW. Room to supply air temperature differential, 9 K; terminal unit turndown ratio, 1:3; humid specific heat, 1.025 kJ/kg K; enthalpy of water vapour, 2540 kJ/kg K.

(Engineering Council)

(From 51.5 to 58.5%)

8.20   A variable-volume system with a turndown of 50% attempts to maintain three zones of a building at 21 °C. The supply air is at a temperature of 15 °C and specific humidity of 0.009. The temperature rise across the fan is 1.5 K. The summer outside design state is 27 °C db and 22 °C wb. Details of the three zones are as follows.

	Max. $\dot{S}$	Max. simultaneous $\dot{S}$	Constant $\dot{L}$	Min. outside air
	(kW)	(kW)	(kW)	(kg/s)
A	90	90	10	1.0
B	60	60	20	2.0
C	50	20	12.5	0.8

Calculate:
(i)   the ratio of re-circulated air to fresh air;
(ii) the temperature and percentage saturation in each zone for maximum simultaneous gains.
(iii) the maximum cooling coil load;
(iv) the temperature and percentage saturation in zone A if the machines contributing to the sensible gains are switched off and the new sensible heat load is 25 kW.

Take the specific heat of moist air as 1.02 kJ/kg K.

(Wolverhampton Polytechnic)

(1.45; A 21 °C and 59%, B 21 °C and 62%, C 19.8 °C and 71%; 478 kW; 18.3 °C and 73%)

8.21 A variable-volume air conditioning system has terminals which can reduce the supply air to 40% of the maximum. The supply air is at 13 °C and a moisture content of 0.007. The summer design state is 28 °C db and 23 °C wb. If the system attempts to maintain a temperature of 22 °C in each zone, use the data below to calculate:

(i) the ratio of re-circulated to fresh;
(ii) the temperature and percentage saturation in each zone when maximum simultaneous gains occur;
(iii) the temperature rise across the fan and ductwork if the cooling coil contact factor is 0.8;
(iv) the maximum cooling coil load.

Data

	Max. $\dot{S}$	Max. simultaneous $\dot{S}$	Constant $\dot{L}$	Min. outside air
	(kW)	(kW)	(kW)	(m^3/s)
A	60	40	10	0.75
B	70	25	10	0.8
C	100	100	15	1.0

For moist air $c_{pma} = 1.02$ kJ/kg K.

(Wolverhampton Polytechnic)
(1.85; A 22 °C and 49%, B 21 °C and 53%, C 22 °C and 47%; 2.7 °C; 430 kW)

8.22 A variable-volume air conditioning system serves two zones of a building which are maintained at 23 °C and 50% saturation when the maximum heat gains are as follows:

Zone 1   100 kW sensible   15 kW latent
Zone 2   125 kW sensible   19 kW latent

The outside air conditions at the time are 27 °C db, 20 °C wb. The supply volume to each zone can be reduced to 50% of its maximum. In order to maintain sufficient fresh air to the zones at times of minimum volume, the ratio of outside air to re-circulated room air entering the plant is 3:1 (3 outside to 1 room). This ratio remains constant. The supply air from the plant is at a constant 13 °C db temperature at all times.

(a) Calculate the cooling coil duty under the maximum conditions.

(b) On another occasion when the outside air conditions are at 18 °C db and 16 °C wb, the heat gains to the zones are as follows:

Zone 1   40 kW sensible   0 kW latent
Zone 2   55 kW sensible   0 kW latent

If the cooling coil apparatus dew point remains constant, calculate the dry bulb temperatures in the two zones and the duty of the cooling coil. Take the specific heat of humid air as 1.02 kJ/kg.

(CIBSE)

(445 kW; 21 °C and 21.8 °C; 110 kW)

8.23 An air conditioning plant operates for 12 hours during the day providing a constant 1.5 kg/s of conditioned air. There is no re-circulation of room air. The plant has an off-coil (dew point) system of control and the enthalpy of the air leaving the coil may be taken as a constant 32 kJ/kg. The outside design specific enthalpy is 57 kJ/kg. If the refrigeration plant has a mean overall coefficient of performance of 3, calculate the annual energy consumption in kilowatt hours. Use the weather data for Birmingham airport from Table 8.3. (4572 kW h)

8.24 An air-conditioned building near Birmingham has a plant which operates for 24 hours per day providing a constant 3 kg/s of conditioned air. An off-coil temperature control is used which maintains the specific enthalpy of the air leaving the coil at 30 kJ/kg. The specific enthalpy of the outside air is 58 kJ/kg for the design case. Calculate the annual cooling loads assuming:

(i) there is no re-circulation of room air;
(ii) an enthalpy detector is fitted which controls dampers so that the plant uses 100% outside air only when the enthalpy of the outside air is less than the room specific enthalpy of 44 kJ/kg; above this value there is a mixing of re-circulated room air and outside air in the ratio of 2 to 1 by mass.

(203.4 GJ, 191.6 GJ)

REFERENCES

8.1 CIBSE 1986 *Guide to Current Practice* volume A7, Tables A7.1 and 2
8.2 ASHRAE 1976 *Handbook of Fundamentals*
8.3 CIBSE 1986 *Guide to Current Practice* volume A1, Table A1.5
8.4 CIBSE 1986 *Guide to Current Practice* volume B.11
8.5 Health and Safety Executive January 1987 *Guidance Note EH 48 Legionnaires' Disease* Environmental Hygiene Series
8.6 ASHRAE 1976 *Systems Handbook* chapter 3
8.7 CIBSE 1986 *Guide to Current Practice* volume B.3
8.8 ASHRAE 1976 *Handbook and Product Directory*
8.9 Collingbourn R H and Legg R C 1987 *The Annual Frequency of Occurrence of Hourly Values of Outside Air Conditions at 23 Meteorological Stations in the UK* Institute of Environmental Engineering, Polytechnic of the South Bank, London
8.10 CIBSE 1986 *Guide to Current Practice* volume A2, Table A2.5
8.11 CIBSE May 1989 *The Specifier's Guide to Air Conditioning*
8.12 ASHRAE 1976 *Systems Handbook* Chapter 1

GENERAL READING

Carrier Air Conditioning Company 1965 *Handbook of Air Conditioning System Design* McGraw-Hill
Stoeker W F and Jones J W 1982 *Refrigeration and Air Conditioning* McGraw-Hill
Jones W P 1973 *Air Conditioning Engineering* Edward Arnold
Jones W P 1980 *Air Conditioning Applications and Design* Edward Arnold

9 EFFICIENT USE OF ENERGY

Mankind's requirement for energy is increasing at an alarming rate as the world's population grows and the standard of living of the majority increases year by year. Most of the energy currently produced comes from the burning of fossil fuels such as coal, oil and gas, and the nuclear fission of uranium. Two major problems confront us: firstly, the supply of fossil fuel is finite and is being depleted rapidly; secondly, pollution of the environment is beginning to affect the stability of the Earth's atmosphere and oceans. Short-wave radiation from the sun passes through the Earth's atmosphere with negligible absorption but the long-wave radiation emitted from the Earth's surface to space is absorbed by the atmosphere, thus creating the so-called 'greenhouse effect'. The accelerating build-up in the atmosphere of gases (such as carbon dioxide) which have high absorptivities to long-wave radiation is therefore leading to global warming. This in turn will affect sea levels and lead to climatic changes affecting crops, etc.

Nuclear fission does not contribute directly to the greenhouse effect but the containment of nuclear radiation, the safe disposal of spent fuel, and the de-commissioning of obsolete reactors are major problems. Also, the threat of a disaster of the scale of Chernobyl continually hangs over us.

Energy from uranium or fossil fuels is *non-renewable* but there are also sources of *renewable energy* stemming directly or indirectly from the energy of the sun and moon. Direct solar radiation can be exploited to heat water or generate power; indirect solar radiation causes atmospheric movement which can be exploited by means of wind turbines, or wave energy devices in the sea; indirect solar radiation also leads to hydro-electric power schemes using the potential energy of water evaporated from the sea by the sun; the pull of the moon on the earth causes the tides, whose energy can be harnessed using tidal barrages. Crop, wood, and animal waste are also sources of renewable energy, but unfortunately the combustion of such waste creates carbon dioxide since the major constituent is carbon.

Development of schemes utilizing renewable energy sources is expensive; judged by normal commercial standards the rate of return on capital in many cases does not justify the scheme, but the time is rapidly approaching when long-term solutions to the world's energy problem must be faced, regardless of short-term commercial considerations.

In the meantime it is essential that energy is used more efficiently. Fortunately, since a saving of energy almost always leads to a saving in money, there is a strong incentive to introduce energy-saving measures.

The efficient use of energy is a major subject and in this short chapter it is only possible to give a brief treatment of some aspects; a fuller treatment of energy efficiency for engineering and technology students is given by Eastop and Croft.[9.1]

9.1 TOTAL ENERGY SCHEMES

In any power production system, heat is provided from an energy source at a high temperature and heat is rejected to a sink at a lower temperature. The Second Law of Thermodynamics states that no power-producing device can produce power while exchanging heat with an energy reservoir at one fixed temperature. For the most efficient power production systems currently in use, the ratio of the work done to the heat supplied is no higher than 30–40%. Better use can be made of the energy from the high-temperature source if the heat rejected at the lower temperature is used as a source of low-temperature energy, say for heating hot water. Schemes designed to use more of the energy input in this way are sometimes known as total energy systems. A definition of a total energy scheme is as follows:[9.1]

'A scheme in which the total energy requirements of a plant in the form of power and heat are provided from a supply of primary fuel and in which the energy wastage is reduced to a minimum.'

Examples of total energy schemes are many and varied: any system in which gas or liquid is rejected to waste at a higher temperature than the atmosphere can be made more effective by using a total energy approach. Probably the best known example is *Combined Heat and Power*, CHP (known also as *Cogeneration*), in which electrical power production is combined with the provision of domestic hot water and hot water for space heating. Major schemes covering districts or complete cities are in use in many countries and are being introduced more slowly into the UK. When heating on its own is provided from a central source for a district or town it is known as *District Heating*. For a more extensive treatment of this topic, Ref. 9.1 should be consulted.

In the examples given in this chapter it is assumed that the reader has a knowledge of basic Thermodynamics, to the level of Ref. 9.2, for example.

Example 9.1

(a) Derive expressions for the fractional fuel savings in a total energy scheme for the case when

$$\sigma \leqslant \eta_H/\eta_W \quad \text{and} \quad \text{when} \quad \sigma \geqslant \eta_H/\eta_W$$

where σ is the heat:power ratio, η_H the useful heat produced for each unit of fuel energy supplied, and η_W is the electrical power generated for each unit of fuel energy supplied.

(b) Given the values below, show graphically the relationship between the fractional fuel savings and the heat:power ratio, where η_B is the boiler efficiency, and η_G the efficiency of the public electricity generating system.

$$\eta_B = 0.80; \quad \eta_G = 0.30$$

Diesel: $\eta_W = 0.40; \quad \eta_H = 0.15$
Gas turbine: $\eta_W = 0.20; \quad \eta_H = 0.50$

Steam turbine: $\eta_W = 0.15$; $\eta_H = 0.65$

(Newcastle upon Tyne Polytechnic)

Solution

(a) A total energy scheme is shown diagrammatically in Fig. 9.1 where the useful power, $\dot{W}$, is $\eta_W \dot{E}$ and the useful rate of heat supply, $\dot{Q}$, is $\eta_H \dot{E}$, where $\dot{E}$ is the rate of energy supplied from the fuel.

Figure 9.1 Diagram of a total energy scheme

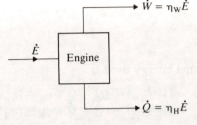

Note: The rate of heat rejected which is not put to useful purposes is given by $\dot{E} - \eta_W \dot{E} - \eta_H \dot{E} = \dot{E}(1 - \eta_W - \eta_H)$.

The heat:power ratio is

$$\sigma = \dot{Q}/\dot{W} = \eta_H/\eta_W$$

This system will provide the heat output for the given efficiencies for a ratio of $\sigma = \eta_H/\eta_W$. When smaller amounts of useful heat are required for the same work output, then more heat must be rejected as waste heat. When larger amounts of useful heat are required, then the additional quantity must be provided by a boiler.

(i) Useful heat supplied entirely from plant, $\sigma \leqslant \eta_H/\eta_W$:

The energy required to produce the same work and useful heat from the grid and from a boiler is

$$(\dot{W}/\eta_G) + (\dot{Q}/\eta_B)$$

Hence the fractional fuel saving, f, is given by

$$\frac{(\dot{W}/\eta_G) + (\dot{Q}/\eta_B) - (\dot{W}/\eta_W)}{(\dot{W}/\eta_G) + (\dot{Q}/\eta_B)}$$

$$= 1 - \frac{(1/\eta_W)}{\{(1/\eta_G) + (\sigma/\eta_B)\}} \qquad [9.1]$$

Note: The fractional fuel saving, f, will be zero when

$$1/\eta_W = (1/\eta_G) + (\sigma/\eta_B)$$

i.e. when $\sigma = \{(1/\eta_W) - (1/\eta_G)\}\eta_B \qquad [9.2]$

(ii) Useful heat supplied partly by plant and partly from a boiler, $\sigma \geqslant \eta_H/\eta_W$:

The system is shown diagrammatically in Fig. 9.2. In this case the useful heat output is

$$\dot{Q} + \dot{Q}' = \eta_H \dot{E} + \eta_B \dot{E}'$$

i.e. $\dot{E}' = (\dot{Q} + \dot{Q}')/\eta_B - (\eta_H \dot{E}/\eta_B)$

Figure 9.2 Modified
total energy scheme

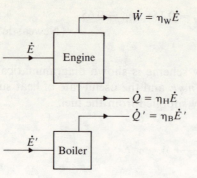

Therefore the total fuel energy used is

$$\dot{E} + \dot{E}' = (\dot{W}/\eta_W) + (\dot{Q} + \dot{Q}')/\eta_B - (\eta_H \dot{E}/\eta_B)$$
$$= (\dot{W}/\eta_W) + (\dot{Q} + \dot{Q}')/\eta_B - (\eta_H \dot{W}/\eta_W \eta_B)$$

Using the grid to provide $\dot{W}$, and a boiler to provide $(\dot{Q} + \dot{Q}')$, the energy used is,

$$(\dot{W}/\eta_G) + (\dot{Q} + \dot{Q}')/\eta_B$$

Then the fractional fuel saving, f, is given by

$$1 - \frac{\{(\dot{W}/\eta_W) + (\dot{Q} + \dot{Q}')/\eta_B - (\eta_H \dot{W}/\eta_W \eta_B)\}}{(\dot{W}/\eta_G) + (\dot{Q} + \dot{Q}')/\eta_B}$$
$$= 1 - \frac{\{(1/\eta_W) + (\sigma/\eta_B) - (\eta_H/\eta_W \eta_B)\}}{\{(1/\eta_G) + (\sigma/\eta_B)\}} \qquad [9.3]$$

(b)   Diesel engine:

Critical value of $\sigma = \eta_H/\eta_W = 0.15/0.4 = 0.375$

At this point all the heat required is supplied by the engine. From Eqn [9.1], at this value of σ we have

$$f = 1 - \frac{(1/0.4)}{\{(1/0.3) + (0.375/0.8)\}}$$
$$= 0.343$$

In this case if we calculate a value of σ for $f = 0$ from Eqn [9.2], we find that it is negative. This is because with η_W greater than η_G, there must always be some saving from a diesel engine scheme.

Hence we calculate a value for f when $\sigma = 0$, i.e. $f = 1 - (\eta_G/\eta_W) = 0.25$.

Between $\sigma = 0.375$ and $\sigma = 0$ the potential heat output from the diesel engine cannot be utilized fully; some heat must be rejected, together with the other losses.

For values of f when $\sigma \geqslant 0.375$ the boiler is required and Eqn [9.3] must be used. For example, for $\sigma = 1$, say, we have

$$f = 1 - \frac{\{(1/0.4) + (1/0.8) - (0.15/0.4 \times 0.8)\}}{\{(1/0.3) + (1/0.8)\}}$$
$$= 0.248$$

A graph of the fractional fuel saving, f, against the heat:power ratio, σ, can be drawn as shown in Fig. 9.3.

Figure 9.3 Diesel
engine for Example 9.1

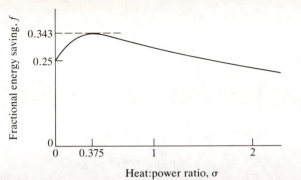

Gas turbine:

$$\text{Critical value of } \sigma = 0.5/0.2 = 2.5$$

From Eqn [9.1], at this value of σ we have

$$f = 1 - \frac{(1/\eta_W)}{\{(1/\eta_G) + (\sigma/\eta_B)\}}$$

i.e.

$$f = 1 - \frac{(1/0.2)}{\{(1/0.3) + (2.5/0.8)\}}$$

$$= 0.226$$

In this case with $\eta_G > \eta_W$, there must be some value of σ for which $f = 0$.
When $f = 0$, from Eqn [9.2], $\sigma = \{(1/\eta_W) - (1/\eta_G)\}\eta_B$, we have:

$$\sigma = \{(1/0.2) - (1/0.3)\}0.8 = 1.333$$

Values of f at $\sigma \geqslant 2.5$ are given by Eqn [9.3].
A graph of f against σ is shown as Fig. 9.4.
Steam turbine:

$$\text{Critical value of } \sigma = 0.65/0.15 = 4.333$$

The value of f at this value of σ is given by Eqn [9.1] as

$$f = 1 - \frac{(1/\eta_W)}{\{(1/\eta_G) + (\sigma/\eta_B)\}}$$

$$= 1 - \frac{(1/0.15)}{\{(1/0.3) + (4.333/0.8)\}}$$

$$= 0.238$$

Figure 9.4 Gas turbine
for Example 9.1

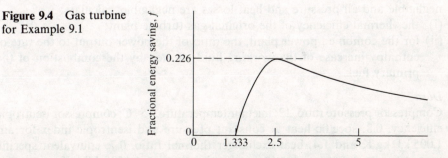

Figure 9.5 Steam
turbine for Example 9.1

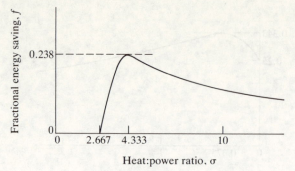

Heat:power ratio, σ

Also, when $f = 0$ from Eqn [9.2], $\sigma = \{(1/\eta_W) - (1/\eta_G)\}\eta_B$, we have:

$$\sigma = \{(1/0.15) - (1/0.3)\}0.8 = 2.667$$

Values of f at $\sigma \geqslant 4.333$ are given by Eqn [9.3].
A graph of f against σ is shown as Fig. 9.5.

It can be seen by reference to Figs 9.3–9.5 that the saving in fuel in any combined heat and power scheme depends on the required heat:power ratio. For the typical efficiencies given in Example 9:1, the diesel engine scheme is best for heat:power ratios from 0 to 4; the gas turbine system would be more costly than using power from the grid and a boiler, for heat:power ratios below 1.333, and is best in the heat:power range of 2–8, say; the steam turbine system would be more costly for heat:power ratios below 2.667, and is best in the heat:power range of 4–10, say. The efficiencies given in the question are representative of small-to-medium power units.

Example 9.2

A gas turbine plant consisting of a compressor, combustion chamber, turbine and heat exchanger develops 25 MW net power output. The data for the plant are given below.

It is proposed to adapt the plant for use as part of a combined heat and power plant by dispensing with the heat exchanger and passing the exhaust from the turbine through a heat recovery boiler. The steam raised is used to develop additional power using a steam turbine.

Assuming that the overall thermal efficiency of the steam cycle is 35%, the exhaust gases leave the heat recovery boiler at 80 °C, the mass flow rate of exhaust gases from the turbine is unchanged, the mass flow rate of fuel is negligible and all pressure and heat losses are negligible, calculate:

(i) the thermal efficiency of the original gas turbine plant;
(ii) for the combined power plant, the ratio of the power output to the rate of enthalpy increase of the working fluid effected by the combustion of the primary fuel.

Data

Compressor pressure ratio, 12; inlet air temperature, 15 °C; compressor isentropic efficiency, 0.8; specific heat at constant pressure and isentropic index for air, 1.005 kJ/kg K and 1.4; heat exchanger thermal ratio, 0.7; equivalent specific heat at constant pressure for the combustion process, 1.150 kJ/kg K; maximum cycle temperature in gas turbine plant in both cases, 950 °C; specific heat at constant pressure and isentropic index for combustion products, 1.150 kJ/kg K and 4/3; gas turbine isentropic efficiency, 0.82.

(CIBSE)

Solution

(i) The original gas turbine plant is shown diagrammatically in Fig. 9.6(a) with the corresponding state points shown on a $T–s$ diagram in Fig. 9.6(b). The $T–s$ diagram shows pressure drops in the heat exchanger and combustion chamber. In this example the pressure drops are to be neglected and hence points 2, 3 and 4 are at the compressor pressure, and points 5, 6 and 1 are at atmospheric pressure.

Figure 9.6 Gas turbine plant for Example 9.2

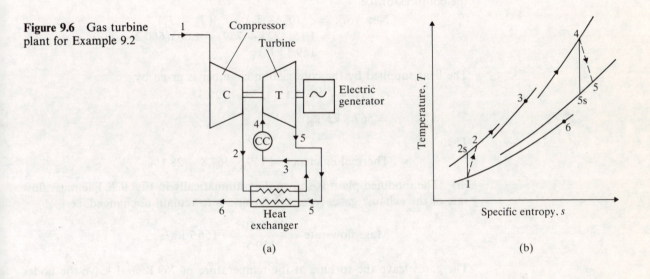

(a)   (b)

Using the isentropic temperature–pressure relationship for a perfect gas, we have

$$T_{2s} = T_1(r_p)^{(\gamma - 1)/\gamma}$$

where r_p is the pressure ratio across the compressor.

$$\therefore \qquad T_{2s} = 288(12)^{0.2857} = 585.8 \text{ K}$$

Then, using the isentropic efficiency given,

$$T_2 = T_1 + \frac{(T_{2s} - T_1)}{0.8} = 288 + \frac{(585.8 - 288)}{0.8}$$

$$= 660.2 \text{ K}$$

The maximum cycle temperature, T_4, is given as 1223 K, and therefore,

$$T_{5s} = T_4/(r_p)^{(\gamma - 1)/\gamma}$$

(Note that in this case the pressure losses between states 2 and 4, and between 5 and the ambient pressure, are neglected.)

$$T_{5s} = 1223/(12)^{0.25} = 657.1 \text{ K}$$

Then, using the isentropic efficiency of the turbine,

$$T_5 = T_4 - (T_4 - T_{5s}) \times 0.82$$
$$= 1223 - (1223 - 657.1) \times 0.82$$
$$= 759.0 \text{ K}$$

The thermal ratio of the heat exchanger is given by

$$0.7 = \frac{T_3 - T_2}{T_5 - T_2}$$

i.e.

$$T_3 = 660.2 + (759 - 660.2) \times 0.7$$
$$= 729.3 \text{ K}$$

The net work of the unit is given by the turbine work less the work input to the compressor, i.e.

$$\text{Net work} = c_{pg}(T_4 - T_5) - c_{pa}(T_2 - T_1)$$
$$= 1.15(1223 - 759) - 1.005(660.2 - 288)$$
$$= 159.5 \text{ kJ/kg}$$

The heat supplied by the combustion chamber is given by

$$c_{pg}(T_4 - T_3) = 1.15(1223 - 729.3)$$
$$= 567.8 \text{ kJ/kg}$$

Then,

$$\text{Thermal efficiency} = 159.5/567.8 = 28.1\%$$

(ii)　The modified plant is shown diagrammatically in Fig. 9.7. The mass flow rate of the exhaust gases from the turbine is to remain unchanged, i.e.

$$\text{Mass flow rate} = \frac{25 \times 10^3}{159.5} = 156.7 \text{ kg/s}$$

The gases leave the turbine at the temperature of 759 K and leave the boiler at $(80 + 273) = 353$ K. Therefore,

$$\text{Heat supplied in boiler} = 156.7 \times 1.15(759 - 353)$$
$$= 73\,163 \text{ kW}$$

Figure 9.7　Combined power plant of Example 9.2

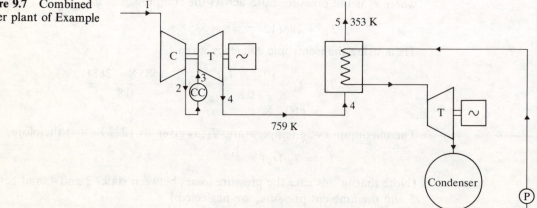

Since the thermal efficiency of the steam cycle is 35%, we have

$$\text{Power output from steam turbine} = 73\,163 \times 0.35$$
$$= 25\,607\,\text{kW}$$

In this case the heat supplied from the fuel to the turbine combustion chamber is higher because there is no heat exchanger, i.e.

$$\text{Heat supplied} = 156.7 \times 1.15(1223 - 660.2)$$
$$= 101\,419\,\text{kW}$$

Therefore the required ratio of power output to equivalent fuel energy input is

$$\frac{25\,000 + 25\,607}{101\,419} = 0.499$$

It can be seen that the combined plant is almost twice as efficient as the gas turbine plant on its own.

Micro-CHP

Small systems for combined heat and power, up to about 150 kW electrical power output, are generally known as micro-CHP schemes. The most common type consists of a diesel engine or gas engine with a complete energy recovery system, recovering energy from the lubricating oil, the cooling water and the exhaust gases. The engine is coupled to an electric generator which can be operated independently or in parallel with the national grid. Power can be exported to the grid or bought from the grid as necessary.

More and more such schemes are being introduced in a wide range of buildings such as hospitals, hotels, residential homes, leisure centres and colleges. The basic criteria for economic operation are sufficiently large and relatively constant heat and power requirements for a large part of the year. The Department of Energy publishes details of schemes which have been installed (see for example Ref. 9.3). Eastop and Croft[9.1] give further details of some typical micro-CHP systems.

Steam Systems

In many buildings steam is required for process work (e.g. laundries, breweries, hospitals, etc.) and if power is also required then a combined steam system may be more economical than using a boiler alone to generate the steam. Steam can also be used for the heating system in such cases, either directly or by producing hot water in a calorifier.

Example 9.3

(a)   A factory requires dry saturated steam at a pressure of 1.2 bar for process work, the process load being 15 000 kW. The steam is completely condensed during the process but the condensate is non-recoverable. Since the factory also has a power requirement of 3000 kW it is decided to satisfy both requirements using a back-pressure turbine. Assuming that the feed water to the boiler is at

15 °C, that the turbine isentropic efficiency is 0.8, and neglecting all other losses, calculate:
(i)   the rate of steam flow required;
(ii) the steam supply conditions at entry to the turbine;
(iii) the ratio of the useful energy output to the total energy input, neglecting feed pump work, and taking a boiler efficiency of 80%.

Use the h–s chart for steam.
(b)   Comment briefly on the above method of achieving the process and power requirements, stating under what circumstances an alternative approach would be preferable.

(CIBSE)

Solution
(a)   The system is shown diagrammatically in Fig. 9.8(a) and the expansion process in the turbine is shown on a sketch of the h–s chart in Fig. 9.8(b).
(i)   The process steam is completely condensed and hence the energy used is the enthalpy of vaporization of steam at 1.2 bar. From tables[9.4] we have $h_{fg} = 2244$ kJ/kg.

Figure 9.8 Back-pressure steam system of Example 9.3

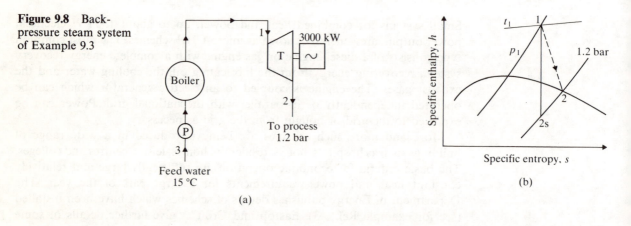

Therefore,

$$\text{Mass flow rate of steam required} = \frac{15\ 000}{2244} = 6.68 \text{ kg/s}$$

(ii)   The enthalpy drop through the turbine times the mass flow rate of steam gives the power output of the turbine, i.e.

$$\text{Enthalpy drop, } h_1 - h_2 = \frac{3000}{6.68} = 448.8 \text{ kJ/kg}$$

The enthalpy of the steam at point 2 can be read from tables or from the h–s chart, i.e. $h_2 = h_{g2} = 2683$ kJ/kg.
Therefore,

$$h_1 = h_2 + 448.8 = 2683 + 448.8 = 3131.8 \text{ kJ/kg}$$

Then, using the isentropic efficiency of the turbine given, we have

$$h_1 - h_{2s} = 448.8/0.8 = 561 \text{ kJ/kg}$$

Therefore,

$$h_{2s} = 3131.8 - 561 = 2570.8 \text{ kJ/kg}$$

Point 2s is then located on the h–s chart at 1.2 bar and $h_{2s} = 2570.8 \text{ kJ/kg}$. A vertical line can then be drawn from point 2s until it cuts the horizontal line representing the enthalpy value, $h_1 = 3131.8 \text{ kJ/kg}$. Point 1 is now located on the chart and the pressure, p_1, and the temperature, t_1, can be read off. Thus, from the chart, $p_1 = 17$ bar and $t_1 = 342\,°C$

(iii)   The total energy input is the fuel energy input to the boiler to heat feed water at 15 °C to steam at 17 bar and 342 °C. That is,

$$\text{Energy input} = \frac{\dot{m}(h_1 - h_3)}{0.8} = \frac{6.68(3131.8 - 62.9)}{0.8}$$

$$= 25\,625 \text{ kW}$$

where, from tables, h_f at 15 °C is 62.9 kJ/kg.

$$\therefore \quad \frac{\text{Useful energy output}}{\text{Useful energy input}} = \frac{3000 + 15\,000}{25\,625} = 0.702$$

(b)   It is interesting to note that the system in this example is very similar to the steam turbine total energy scheme of Example 9.1. Using the same notation as Example 9.1, we have

$$\sigma = 15\,000/3000 = 5$$

and $\eta_w = 3000/26\,625 = 0.117$, $\eta_H = 15\,000/26\,625 = 0.585$

The fractional fuel energy saving is given by Eqn [9.1], i.e.

$$f = 1 - \frac{(1/0.117)}{[(1/0.3) + (5/0.8)]} = 0.108$$

where the efficiency of power generation from the grid is taken as 30%, as before.

One of the disadvantages of this system is that, if the demand for process steam falls but the power requirement is unchanged, then the excess steam flow at 1.2 bar must be exhausted to waste.

When the power requirement increases with the process steam requirement unchanged, power can be purchased from the grid. When the power requirement is less than the design value with the process requirement unchanged, the steam supply to the turbine can be throttled, but this will cause the steam to become more highly superheated and hence the system will not provide dry saturated steam at 1.2 bar without external cooling. By suitable advance arrangement with the electricity company a better solution is to sell excess power to the grid.

An alternative way of obtaining power and process steam is by using a pass-out turbine as shown diagrammatically in Fig. 9.9(a) with the expansion process on the h–s chart as in Fig. 9.9(b). The pressure p_3 can be fixed at a suitable value depending on the temperature of the available condenser cooling water (e.g. about 0.05 bar).

The power output is then given by

$$\dot{W} = \dot{m}(h_1 - h_2) + (\dot{m} - \dot{m}_P)(h_2 - h_3)$$

Figure 9.9 Pass-out steam system

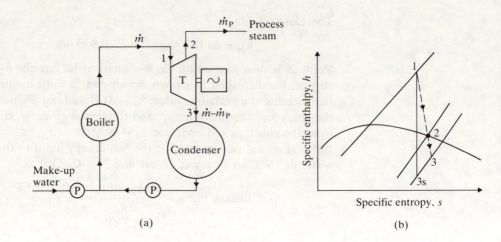

(a)   (b)

The process heat load is given by

$$\dot{Q} = \dot{m}_P h_{fg}$$

[assuming that the steam is required to be dry saturated and that it is completely condensed, as in part (a)].

The system in part (a) using a back-pressure turbine is more suited to high values of heat:power ratio. Higher values of the ratio can be obtained by designing for a lower turbine stop valve pressure and hence a lower power output. From the $h-s$ chart it can be seen for the case in part (a) that for a heat:power ratio of 10 the turbine entry conditions would be 5.5 bar and 230 °C with all other conditions unchanged.

The system using a pass-out turbine is more suited to low heat:power ratios. For this system the heat:power ratio can be varied for fixed supply and condenser conditions by varying the ratio of pass-out steam to boiler steam, $\dot{m}_P/\dot{m}$. At part-load conditions similar losses must be accepted as for the back-pressure case; when the process requirement falls, the excess pass-out steam can be blown to waste, or the boiler operated at part-load; when the power requirement falls with the process demand unchanged, excess steam can be blown off at the pass-out point so as to reduce the power from the second stage of the turbine, or if possible the excess power can be sold to the grid.

Energy from Waste

Waste disposal is one of the problems of any developed society; the more prosperous the country, the more waste is created. Many methods are available depending on the character of the waste, including re-cycling, incineration, and the use of land-fill. Re-cycling of waste, e.g. metals or paper, is becoming more necessary to avoid a reduction of valuable resources.

When waste is incinerated, the energy released can be used to generate steam or provide hot water. Alternatively, if the waste is taken to a land-fill site, then the methane formed as the waste biodegrades can be pumped off to be burned as a fuel.

It has been calculated that if the total amount of municipal, industrial and medical waste produced in the UK were incinerated, then the energy reclaimed would be about 20% of the total national energy consumption.

Taking the calorific value of oil as 44 MJ/kg and of coal as 26 MJ/kg, comparative values for some wastes are:

Paper, 15 MJ/kg; Wood, 18 MJ/kg; Rubber, 35 MJ/kg.

An average value for municipal waste is about 11 MJ/kg; industrial waste is usually higher, depending on the type of industry, say about 16 MJ/kg; hospital waste which generally has a lower moisture content can have as high a value as 19 MJ/kg.

Great care must be taken over the disposal of toxic and other hazardous wastes, and in the case of incineration any dangerous gases must be chemically neutralized before discharge of the gases to the atmosphere.

Incineration with energy recovery is the most effective method of disposal. The waste can be processed into pellets or briquettes (known as refuse-derived fuel, RDF), and used to supplement coal in a coal-fired boiler, or pre-sorted and fired directly in a special furnace. More information on the various methods of combustion is given by Eastop and Croft.[9.1]

Agricultural residues from crops, wood and animal waste can also be exploited by incineration (examples of calorific values: straw, 18 MJ/kg; animal waste, 18 MJ/kg). On a larger scale, municipal authorities in other countries have used the energy available from sewage plants in the form of methane (see Problem 9.11, Fig. 9.27).

Example 9.4

A hospital produces 450 tonnes of combustible waste per year. It is proposed to install an incineration plant with a waste heat boiler to replace an existing coal-fired boiler and separate incinerator. Using the data given, calculate the simple pay-back time for the new plant.

Data

Efficiency of waste heat boiler, 55%; efficiency of coal-fired boiler, 75%; gross calorific value of waste, 17 MJ/kg; gross calorific value of coal used, 28 MJ/kg; cost of coal, £70 per tonne; capital cost of new incinerator with waste heat boiler, £80 000; scrap value of existing plant, £15 000.

Solution

Energy available from waste $= 450 \times 10^3 \times 17 \times 0.55 = 4207.5$ GJ/year

i.e.
$$\text{Coal saved} = \frac{4207.5 \times 10^3}{28 \times 0.75} = 200.36 \times 10^3 \text{ kg/year}$$

Annual saving $= (200.36 \times 10^3 \times 70)/1000 = £14\,025$

Therefore,
$$\text{Pay-back time} = \frac{(80\,000 - 15\,000)}{14\,025} = 4.64 \text{ years}$$

9.2 ENERGY RECOVERY

A well-designed building will have a good standard of insulation and the ventilation rate will be controlled at the correct level for the necessary comfort conditions. Previous chapters deal with methods of calculating heat losses due

Table 9.1 *U*-values from Building Regulations, April 1990

Building fabric	*U*-value (W/m² K)
Walls in dwellings, and in industrial and commercial buildings	0.45
Roofs in dwellings	0.25
Roofs in industrial and commercial buildings	0.45
Floors	0.45

to heat transfer through the fabric and through ventilation both for steady and transient conditions.

The Building Regulations lay down recommended values for *U*-values for new houses and offices: at April 1990 these were as shown in Table 9.1.

Buildings of much higher insulation are now being designed in the UK, and have been designed for some time in other countries. *U*-values of 0.3 W/m² K and below for walls, roofs and floors, are easily achievable with modern materials.[9.5] Windows are one of the main causes of heat loss; typical single glazed windows have a *U*-value of 5.6 W/m² K and double-glazed windows a value of about 3.0 W/m² K; double-glazed windows filled with argon and with a low-emissivity coating can have a *U*-value as low as 1.9 W/m² K.

Glass can be incorporated into a building design to exploit solar radiation input and hence reduce effective energy loss; this is known as *passive solar design*.[9.6]

Ventilation heat loss can be reduced to a minimum by controlling the fresh air input to the recommended values for comfort. Values of outdoor air supply rates for different types of buildings and occupancy are given by CIBSE;[9.7] some of the values are given in Table 9.2.

Table 9.2 Some values of recommended outdoor air supply rates

Type of space	Air supply rate (litre/s per person)
Open-plan office (some smoking)	8
Private office (heavy smoking)	12
Conference rooms (some smoking)	18
Boardrooms (heavy smoking)	25

Heating and air conditioning systems should be designed to avoid any unnecessary energy consumption: re-circulation of the room air must be used to the maximum possible extent; unnecessary heating or cooling of air must be avoided.

When all possible measures have been taken to provide maximum insulation and minimum ventilation loss, the possibility of energy recovery should be considered. In winter, any air, water or gases which leave the building at a higher temperature than the ambient can be used to pre-heat air, or water, entering the building. In summer, air leaving the building at a temperature below the ambient can be used to pre-cool the air entering. In complex buildings in which many processes occur (e.g. a brewery), a complete analysis of the whole system should be made using *process integration*, also known as *pinch technology*. This technique was first suggested by Linnhof;[9.8] a simple treatment of the basic method is given by Eastop and Croft.[9.1]

For simpler systems a straightforward choice of energy recovery devices can be made.

If the temperature of the exhaust fluid is high enough to heat air or water directly, and is reasonably adjacent to the air or water inlet, then a heat exchanger can be used. For exhaust gas to air a *recuperative plate-fin type* can be used, or a *regenerative thermal wheel*, or a *static regenerator*: for exhaust water, either a *corrugated plate type*, or *shell-and-tube type* can be used. These heat exchangers are shown diagrammatically in Figs 9.10–9.14.

If the temperature of the exhaust fluid is not high enough to warrant the use of a heat exchanger, energy recovery may still be possible using the exhaust fluid as the source of a heat pump. If the exhaust fluid is not adjacent to the inlet air or water to be heated, then a *run-around coil* can be used (see Fig. 9.15), or the exhaust fluid used as the heat source for a heat pump.

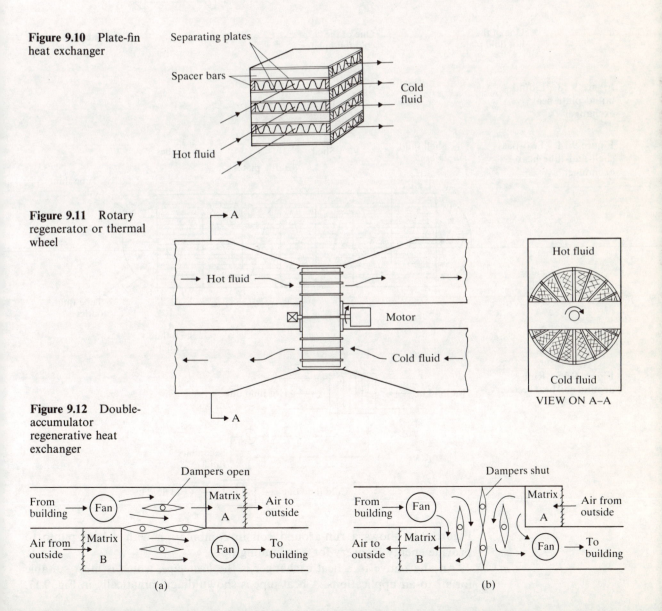

Figure 9.10 Plate-fin heat exchanger

Figure 9.11 Rotary regenerator or thermal wheel

Figure 9.12 Double-accumulator regenerative heat exchanger

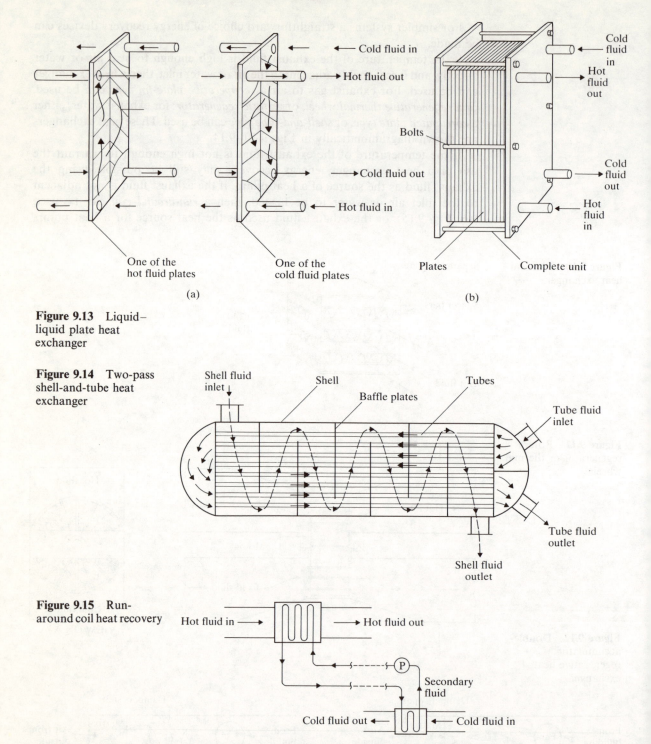

Figure 9.13 Liquid–liquid plate heat exchanger

Figure 9.14 Two-pass shell-and-tube heat exchanger

Figure 9.15 Run-around coil heat recovery

Figure 9.16 shows a run-around coil in combination with a heat pump to provide energy recovery for a swimming pool.

An alternative to a heat exchanger is the *heat pipe*, which is most suitable for air-to-air applications. A heat pipe is shown diagrammatically in Fig. 9.17.

Figure 9.16 Heat recovery for a swimming pool using a heat pump and run-around coil

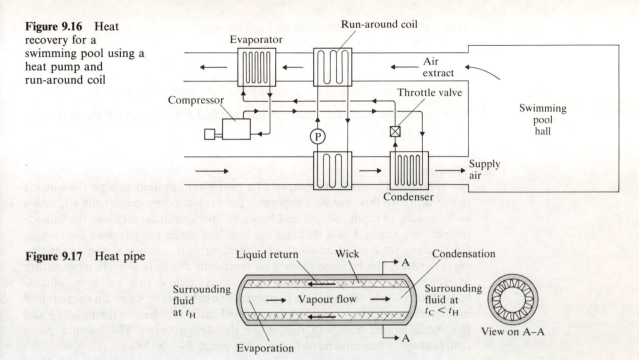

Figure 9.17 Heat pipe

A heat pipe tube is completely sealed as shown, and the tube usually has a finned outside surface in heat recovery applications. A bank of heat pipe tubes are then placed such that one end of each tube is in the hot fluid duct and the other end of each tube is in the cold fluid duct. A heat pipe gives good energy recovery in air-to-air applications when flow and exhaust ducts are immediately adjacent, but the capital cost is high.

Descriptions and analyses of all the above devices are given by Eastop and Croft.[9.1]

Example 9.5

(a) Discuss briefly the disadvantage of a heat pump used for heating a building when the source of energy is the atmosphere.

(b) A process room requires a supply of air at 30 °C at a rate of 0.4 kg/s; the air is exhausted from the room at 28 °C at the same mass flow rate. The air is supplied from an atmosphere at 10 °C, and is heated electrically. To reduce the electricity requirements it is proposed to use an electric heat pump with R12 as refrigerant, with the evaporator coils in the exhaust air stream and the condenser coils in the supply air stream; the evaporator saturation temperature is 5 °C and the condenser saturation temperature is 45 °C.

The following may be assumed:

— the refrigerant leaves the evaporator as a dry saturated vapour;
— the compression is isentropic;
— there is no undercooling in the condenser;
— the combined mechanical and electrical efficiency of the compressor drive is 85%;
— the heat pump provides the total energy input required.

Sketch the lay-out of the plant and calculate:
(i)   the mass flow rate of refrigerant;
(ii) the annual cash saving achieved by using the heat pump when the plant is in use for 6000 h per year and the electricity costs 6 p/kW h;
(iii) the temperature of the air leaving the heat pump evaporator.

Take the specific heat of air throughout as 1.005 kJ/kg K, and use the properties for R12 from tables.[9.4]

(CIBSE)

Solution
(a)   Figure 9.18 shows the output of a heat pump plotted against the outside temperature. In this case the evaporator fluid is heated by the outside air, which is the source of input energy, and hence the performance varies as the outside temperature varies. For a building the heat loss varies linearly with the outside temperature for a constant mean inside temperature, as shown. Where the two curves cross is the balance point; this represents the only outside temperature at which the building heat loss is satisfied. Below this value of outside temperature the heat pump fails to provide enough heat to maintain the required inside temperature; above this value excess heat is supplied to the building and the inside temperature will rise above the design value. The balance point temperature is recommended to be in the range 2–3 °C.[9.9]

Figure 9.18 Heat pump balance point

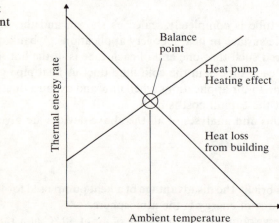

When a heat pump is used for heating a building with the outside temperature as the source of energy, it is normal to use a boiler to supply supplementary heating below the balance point. This is known as a *bivalent system*: the supplementary heat can be provided either as an addition to the heat pump output below the balance point, or as an alternative to the heat pump below the balance point.

It can be seen that the heat pump is not ideally suited to the role of heating a building when the source is the outside temperature. If a source at approximately constant temperature is available, then the heat pump is much more effective; for example groundwater has been suggested as a suitable source where the temperature has been shown to vary in the range 13 °C (October), to 9 °C (April), for a typical location in the UK.

A hierarchy of use for heat pumps drawn up by BSRIA[9.9] is as follows:
(a) dehumidification; (b) combined heating and cooling (e.g. hypermarkets);
(c) heat recovery systems; (d) heating only. The use for heat recovery is growing
as more designers realise the need for energy-saving. A suitable source of waste
heat can be exploited as demonstrated by the example in part (b).

(b)   The lay-out of the plant is shown diagrammatically in Fig. 9.19.

Figure 9.19 Heat
pump heat recovery
system for Example 9.4

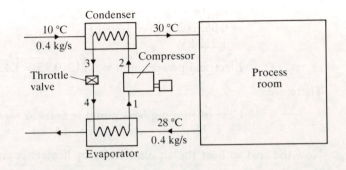

(i)   The refrigeration cycle is shown on a sketch of the T–s diagram in
Fig. 9.20 with the state point numbers corresponding to those on Fig. 9.19.

Figure 9.20 T–s
diagram for Example 9.4

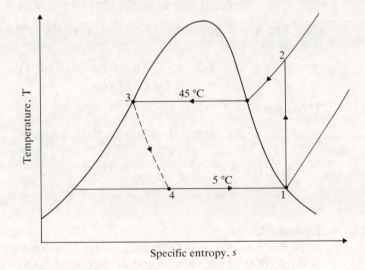

From tables,

$$h_1 = h_g \text{ at } 5\ ^\circ\text{C} = 189.66 \text{ kJ/kg}$$
$$h_3 = h_f \text{ at } 45\ ^\circ\text{C} = 79.71 \text{ kJ/kg}$$
$$s_1 = s_2 = 0.6943 \text{ kJ/kg K}$$

By interpolating,

$$h_2 = 204.87 + \frac{(0.6943 - 0.6811)}{(0.7175 - 0.6811)}(216.74 - 204.87)$$
$$= 209.18 \text{ kJ/kg}$$

The heat input required to the air from the condenser is given by

$$0.4 \times 1.005 \times (30 - 10) = 8.04 \text{ kW}$$

i.e.   Mass flow rate of refrigerant $= \dfrac{8.04}{(209.18 - 79.71)}$

$$= 0.0621 \text{ kg/s}$$

(ii)   Power input to refrigerant in the compressor

$$= \dot{m}_{ref}(h_2 - h_1)$$
$$= 0.0621(209.18 - 189.66)$$
$$= 1.212 \text{ kW}$$

i.e.   Electrical power input $= 1.212/0.85 = 1.426 \text{ kW}$

Therefore,

Cost of running heat pump $= £(1.426 \times 6000 \times 6/100)$
$$= £513.4$$

Now the cost to heat the air electrically by heaters is given by

$$0.4 \times 1.005(30 - 10) \times 6000 \times 6/100$$
$$= £2894.4$$

Hence,

Annual cost saving $= £2894.4 - £513.4 = £2381$

(iii)   The heat supplied by the extract air to the refrigerant in the evaporator is given by

$$\dot{m}_{ref}(h_1 - h_4) = 0.0621(189.66 - 79.71)$$
$$= 6.83 \text{ kW}$$

Therefore,

Air temperature leaving the evaporator

$$= 28 - \dfrac{6.83}{0.4 \times 1.005} = 11 \text{ °C}$$

Example 9.6

A thermal wheel is used to transfer energy from an air stream at 25 °C flowing at a rate of 0.5 kg/s, to a second air stream at 10 °C also flowing at 0.5 kg/s. The mass of the wheel is 6 kg and it rotates at 10 rev/min.

Taking the specific heat of the matrix material as 0.8 kJ/kg K, and the specific heat of air as 1.005 kJ/kg K, and using the graph given as Fig. 9.21, calculate:

(i)   the rate of energy recovery and the temperatures of the air leaving each side for an equivalent counter-flow heat exchanger of effectiveness 0.7;
(ii) the effectiveness of the thermal wheel;
(iii) the rate of energy recovery and the temperature of the air leaving each side of the wheel;
(iv) the energy recovery if the rotational speed of the wheel is reduced to 5 rev/min.

Figure 9.21 Influence of matrix rotational speed on efficiency

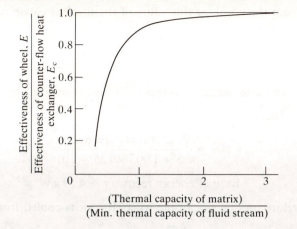

Solution

(i) For an equivalent simple counter-flow case, we have

$$0.5 \times 1.005(25 - t_{H2}) = 0.5 \times 1.005(t_{C2} - 10)$$

where t_{H2} and t_{C2} are the temperatures of the hot and cold fluids at their respective outlets. That is,

$$t_{H2} + t_{C2} = 35 \tag{1}$$

The effectiveness of an equivalent counter-flow heat exchanger, E_c, is given by

$$\frac{\text{Actual temperature drop}}{\text{Maximum possible temperature drop}}$$

i.e. $E_c = (25 - t_{H2})/(25 - 10) = 0.7 \text{ (given)}$

∴ $t_{H2} = 25 - \{0.7(25 - 10)\} = 14.5\,°C$

and from Eqn [1],

$$t_{C2} = 35 - 14.5 = 20.5\,°C$$

Also,

Energy recovery $= \dot{m}c_p(t_{H2} - t_{H1})$
$= 0.5 \times 1.005(25 - 14.5) = 5.28\,\text{kW}$

(ii) The thermal capacity of the matrix is given by

Mass of matrix × Specific heat × Revolutions per unit time
$= (6 \times 0.8 \times 10)/60 = 0.8\,\text{kW/K}$

The thermal capacity of the air flow is

Mass flow rate × Specific heat
$= 0.5 \times 1.005 = 0.5025\,\text{kW/K}$

The ratio of the thermal capacity of the matrix to that of the air is therefore $0.8/0.5025 = 1.592$.

From the graph at this value of the thermal capacity ratio we can read off $E/E_c = 0.87$, i.e.

$$\text{Effectiveness of thermal wheel} = 0.87 \times 0.7$$
$$= 0.609$$

Note: The term 'efficiency' is frequently used instead of 'effectiveness'.

(iii)   Then,

$$0.609 = \frac{\text{Rate of energy recovery}}{0.5 \times 1.005 \times (25 - 10)}$$

i.e.   Rate of energy recovery $= 4.59 \text{ kW}$

Therefore on one side of the wheel the air is cooled from 25 °C to

$$25 - \frac{4.59}{0.5 \times 1.005} = 15.87 \text{ °C}$$

and on the other side the air is heated from 10 °C to

$$10 + \frac{4.59}{0.5 \times 1.005} = 19.13 \text{ °C}$$

(iv)   When the rotational speed of the wheel is reduced from 10 rev/min to 5 rev/min, the ratio of the thermal capacities in Fig. 9.21 is reduced to $1.592/2 = 0.796$. Therefore, from Fig. 9.21, we have $E/E_c = 0.83$.

i.e.   Effectiveness $= 0.83 \times 0.7 = 0.581$

Then,

$$\text{Rate of energy recovery} = 0.581 \times 0.5 \times 1.005 \times (25 - 10)$$
$$= 4.38 \text{ kW}$$

It can be seen from the above example, and from Fig. 9.21, that altering the revolutionary speed of the wheel does not have an appreciable effect on the performance of the wheel above about 5 rev/min. It is better to control air temperature if necessary by allowing some of the air to by-pass the wheel.

The thermal wheel for the temperature range in the above example would probably have a mesh of oxidized aluminium mounted in a galvanized steel frame. For thermal wheels used for heat recovery from high-temperature gases (in boiler flues, for example), the matrix material is usually a steel alloy or a ceramic material. In the above example it is assumed that there is no transfer of moisture between the two streams and therefore only 'sensible heat' is transferred. One of the advantages of the thermal wheel for building services use is the possibility of transferring additional energy through the enthalpy of vaporization of the water vapour in the warmer air stream. If the matrix surface on passing through the warmer air stream is at a temperature sufficiently below the dew point of the warm air stream to allow mass transfer to occur, then additional energy will be transferred from the warm air to the matrix due to the condensation process. Since the matrix revolves slowly through both streams its temperature will vary cyclically but always at a higher temperature than the cold stream, and at a lower temperature than the hot stream. In order to determine whether energy of vaporization can be transferred by a particular wheel for air streams at given temperatures and humidities, it is necessary to have information about the thermal capacity and heat transfer characteristics of the matrix at its design range of speeds. It is possible to ensure the transfer of moisture between the air streams by using a specially designed matrix coated with an absorbent such as lithium chloride. The process of moisture transfer is then by physical sorption, which occurs due to the difference in specific humidity between the air stream and the equilibrium value of specific humidity at the matrix surface; the sorbent material reduces the pressure of the water vapour in contact with it below the saturation value. When the wheel revolves into the stream with the lower specific humidity, the equilibrium specific humidity of the matrix surface is now higher than that of the air stream and moisture is released from the matrix surface into the stream. Such a wheel is sometimes known as a *hygroscopic wheel*.

The efficiency of the overall heat and moisture transfer process can be defined in terms of the specific enthalpies. Thus,

$$\eta_o = \frac{h_2 - h_2'}{h_2 - h_1} \qquad [9.4]$$

where the subscript 2 refers to the fresh air intake; the subscript 2 refers to the fresh air leaving the wheel; the subscript 1 refers to the room exhaust air entering the wheel.

The efficiency of the moisture transfer process can be defined in terms of the specific humidities:

$$\eta_\omega = \frac{\omega_2 - \omega_2'}{\omega_2 - \omega_1} \qquad [9.5]$$

Example 9.7
A building is sited in a location where the design conditions are:

> Summer   26 °C, 60% saturation
> Winter   −2 °C, 100% saturation

The room conditions are controlled at 20 °C, 50% saturation by an air conditioning plant using 100% fresh air at a mass flow rate of 2 kg/s. Energy recovery is achieved by using a hygroscopic thermal wheel operating between the fresh air intake duct and the exhaust duct from the building. The overall efficiency with moisture transfer is 70%, and the moisture transfer efficiency is 65%.

Calculate:

(i) the rate of energy recovery for both summer and winter;
(ii) the temperature of the fresh air leaving the wheel in summer and in winter.

Solution
(i) The thermal wheel is shown in Fig. 9.22.

Figure 9.22 Thermal wheel for Example 9.7

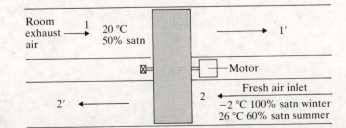

Room exhaust air — 1 — 20 °C 50% satn — 1′

Motor

Fresh air inlet
2 — −2 °C 100% satn winter / 26 °C 60% satn summer

2′

Summer:
From the psychrometric chart,

$$h_1 = 38.8 \text{ kJ/kg}; \quad h_2 = 58.9 \text{ kJ/kg}$$

Then using Eqn [9.4],

$$\text{Overall efficiency} = \frac{h_2 - h_2'}{h_2 - h_1} = 0.70$$

$$\therefore \quad h_2' = 58.9 - \{0.70(58.9 - 38.8)\} = 44.8 \text{ kJ/kg}$$

i.e.   Energy recovery $= \dot{m}(h_2 - h_2')$
$$= 2(58.9 - 44.8) = 28.2 \text{ kW}$$

Winter:
From the psychrometric chart,

$$h_1 = 38.8 \text{ kJ/kg}; \quad h_2 = 6.0 \text{ kJ/kg}$$

Then using Eqn [9.4],

$$\text{Overall efficiency} = \frac{h_2 - h_2'}{h_2 - h_1} = 0.70$$

$$\therefore \quad h_2' = \{6.0 + 0.70(38.8 - 6.0)\} = 29.0 \text{ kJ/kg}$$

i.e.   Energy recovery $= 2(29.0 - 6.0) = 46.0 \text{ kW}$

(ii)   The state of the fresh air leaving the wheel can be found using the moisture transfer efficiency.
Summer:
From Eqn [9.5],

$$0.65 = \frac{\omega_2 - \omega_2'}{\omega_2 - \omega_1} = \frac{0.0129 - \omega_2'}{0.0129 - 0.0074}$$

$$\therefore \quad \omega_2' = 0.0093$$

The enthalpy at point 2' from part (i) is 44.8 kJ/kg; therefore point 2' is fixed on the psychrometric chart and $t_2' = 21.0\,°\text{C}$. The processes are shown on a sketch of the psychrometric chart in Fig. 9.23(a).

Figure 9.23
Psychrometric chart for Example 9.6: (a) summer; (b) winter

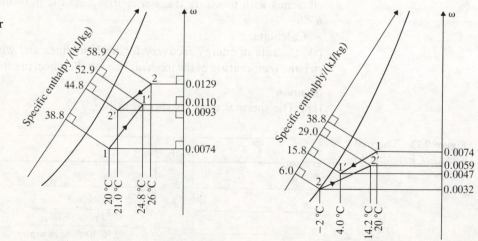

Winter:
Using Eqn [9.5],

$$0.65 = \frac{\omega_2 - \omega_2'}{\omega_2 - \omega_1} = \frac{\omega_2' - 0.0032}{0.0074 - 0.0032}$$
$$\therefore \qquad \omega_2' = 0.0059$$

From part (i), $h_2' = 29.0 \text{ kJ/kg}$; therefore point $2'$ can be fixed on the chart at $t_2' = 14.2 \,^{\circ}\text{C}$. The processes are shown on a sketch of the psychrometric chart in Fig. 9.23(b).

The thermal wheel makes considerable energy savings, but note that a similar order of magnitude of saving could be made by re-circulating some of the room exhaust air to mix with the fresh air. The capital cost of a thermal wheel would be greater than that of a mixing box; also the exhaust and fresh air ducts must be adjacent for a thermal wheel to be practicable.

If the air from the room is contaminated (as in certain hospital applications, for example), then recirculating the air may not be allowable. In such cases a thermal wheel is a possibility although there is still a risk of cross-contamination of the streams. A run-around coil would therefore be a better solution.

In the above example, the wheel will not be effective in winter when the specific humidity of the fresh air entering the wheel approaches 0.0074. Therefore the wheel will not transfer moisture, and hence work to its full efficiency, if the temperature of the saturated fresh air approaches the dew point of the air in the room, i.e. 9.5 °C. The wheel will continue to transfer heat until the fresh air temperature approaches 20 °C.

Note also that, for the winter design conditions given, there is a strong possibility that the lowest matrix temperature will be below the dew point (9.5 °C) of the room air entering the wheel, and therefore condensation may occur. Condensation is not necessarily desirable since it may lead to corrosion of the wheel material, particularly in cases where the warm air stream is very humid and also contains other gases such as chlorine—conditions which occur in most swimming pool halls. For such cases a heat pump, or a combination of heat pump and either run-around coil or plate–fin heat exchanger, gives a better solution.

Example 9.8

A factory measuring 60 m × 60 m × 10 m high requires a ventilation rate of 10 air changes per hour. The ventilation system includes a ducted central fresh air supply system and a de-centralized roof-top extract leading to a common extract duct with filtration plant.

Determine energy savings possible by the installation of a run-around coil ventilation heat recovery system in this factory. Coil row combinations of 2/2, 4/4, 6/6 and 8/8 should be compared using the tables below, and a recommendation made on the basis of a *net* energy cost ratio of 1:3 for boiler fuel:electricity. The following assumptions may be made in order to make a recommendation:

Factory occupation, 8640 h per year; extract air condition, 24 °C, 50% saturation; control temperature above which heat recovery is not required, 16 °C, with the pump controlled accordingly; air velocity at coil, 2.5 m/s; water flow velocity in coil, 1.0 m/s; air density, 1.2 kg/s; supply and extract air flow rates equal.

(Newcastle upon Tyne Polytechnic)

Energy recovered by run-around coil system:
8640 h run; exhaust air at 20 °C, 50% saturation; control temperature 16 °C

	Energy recovered per kg/s air flow					
	(GJ)					
	Air velocity 2.5 m/s			Air velocity 5 m/s		
	Water velocity			Water velocity		
	(m/s)			(m/s)		
Coil row combination	0.5	1.0	2.0	0.5	1.0	2.0
2/2	76	85	92	50	59	66
2/4	92	101	106	63	73	80
2/6	98	106	111	69	79	86
2/8	101	109	113	72	82	89
4/4	116	123	127	85	96	103
4/6	126	131	133	96	106	112
4/8	131	134	136	102	112	117
6/6	138	140	141	110	119	124
6/8	144	144	143	119	127	130
8/8	150	149	146	128	139	137

Correction factors* for other exhaust temperatures:

| Exhaust temperature | Control temperature | | |
| (°C) | (°C) | | |
	12	16	20
16	0.79	0.69	—
24	1.14	1.24	1.37

* *Note:* The energy recovered is the energy recovered from the previous table multiplied by the above correction factor

Correction factors* for other running hours:

| | Control temperature | | |
| | (°C) | | |
Hours run	12	16	20
3600	0.35	0.35	0.36
2580	0.25	0.26	0.26

* *Note:* The energy recovered from the first table above is to be multiplied by the correction factor in this table

Additional fan energy required by coil system:
8640 h run

Total number of coil rows*	Energy required per kg/s air flow (GJ)	
	Air velocity at coil (m/s)	
	2.5	5.0
4	3.0	11.6
6	4.3	17.5
8	5.8	23.4
10	7.3	29.2
12	8.8	35.0
14	10.3	40.8
16	11.7	46.7

Assumptions: fan total efficiency, 70%; motor drive efficiency, 80%; exhaust air at 20 °C, 50% saturation.

* *Note:* The total number of coil rows is the number in the supply coil plus the number in the exhaust coil

Basic pump energy consumption:
8640 hours run; no control.

Total number of coil rows	Energy required per kg/s air flow (GJ)			
	Air velocity (m/s)			
	2.5		5.0	
	Water velocity (m/s)		Water velocity (m/s)	
	1.0	2.0	1.0	2.0
4	4.9	30.8	2.5	15.5
6	6.7	44.1	3.5	21.9
8	8.8	57.1	4.4	28.6
10	10.7	70.2	5.3	35.1
12	12.7	83.2	6.4	41.6
14	14.5	96.3	7.4	48.1
16	16.5	109.4	8.3	54.7

Assumptions: pump electrical efficiency, 75%; pump mechanical efficiency, 50%; water velocity in interconnecting pipe work, 1 m/s; control valve authority, 0.5

Correction factors for controlled pump:
8640 hours run

Control temperature (°C)	Correction factor
12	0.61
16	0.82
20	0.94

Solution

The mass flow rate of air is given by the volume multiplied by the air change rate and the mean density, i.e.

$$\dot{m} = (60 \times 60 \times 10) \times 10 \times 1.2/3600 = 120 \text{ kg/s}$$

The tables give the energy recovery for an exhaust temperature of 20 °C and these must therefore be corrected for the actual exhaust temperature of 24 °C; the correction for other running hours is not required in this case. The correction factor for a control temperature of 16 °C is 1.24 for all coil combinations.

We therefore have, for water and air velocities of 1 m/s and 2.5 m/s (given),

For 2/2:   Energy recovered = $85 \times 1.24 \times 120$ = 12 648 GJ
For 4/4:   Energy recovered = $123 \times 1.24 \times 120$ = 18 302 GJ
For 6/6:   Energy recovered = $140 \times 1.24 \times 120$ = 20 832 GJ
For 8/8:   Energy recovered = $149 \times 1.24 \times 120$ = 22 171 GJ

The additional fan and pump energy required for the run-around coil system must be offset against the energy recovered. The fan energy is given for exhaust air at 20 °C but, since a mean density of 1.2 kg/s has been taken, the difference in the energy required due to the temperature being 24 °C instead of 20 °C will be ignored.

Therefore,

For 2/2:   Fan energy required = $3 \times 120/0.7 \times 0.8$ = 643 GJ
For 4/4:   Fan energy required = $5.8 \times 120/0.7 \times 0.8$ = 1243 GJ
For 6/6:   Fan energy required = $8.8 \times 120/0.7 \times 0.8$ = 1886 GJ
For 8/8:   Fan energy required = $11.7 \times 120/0.7 \times 0.8$ = 2507 GJ

For the pump there is a correction factor for the control temperature of 16 °C of 0.82 for all coil combinations.

Therefore,

For 2/2:   Pump energy required = $4.9 \times 120 \times 0.82/0.75 \times 0.5 = 1286$ GJ
For 4/4:   Pump energy required = $8.8 \times 120 \times 0.82/0.75 \times 0.5 = 2309$ GJ
For 6/6:   Pump energy required = $12.7 \times 120 \times 0.82/0.75 \times 0.5$
                                          = 3333 GJ
For 8/8:   Pump energy required = $16.5 \times 120 \times 0.82/0.75 \times 0.5$
                                          = 4330 GJ

The fan and pump energy is supplied by electricity, whereas the energy saving from the run-around system is of boiler fuel; the weighting of 1:3 must therefore be taken into consideration in any comparison. For example, for the 2/2 combination, we have:

$$\text{Equivalent energy saving} = 12\ 648 - (643 + 1286) \times 3$$
$$= 6861 \text{ GJ}$$

The table below gives the equivalent energy saved for all the coil combinations.

Summary of equivalent energy cost saving:

Coil combination	Equivalent energy saving (GJ)
2/2	6861
4/4	7646
6/6	5175
8/8	1660

It can be seen from the table that the 4/4 coil combination gives the greatest net energy cost saving.

9.3 THERMAL PERFORMANCE MONITORING

The heating requirements for a building, and hence the cost of heating, vary with the weather conditions. To monitor the performance of a given building it is therefore necessary to allow for weather variations through the heating season and from year to year. A given type of building of a particular size can be compared with a set standard and the performance then measured against this standard; this method of comparison will be discussed later in this section, but first the effect of weather changes is considered.

Degree Days

The weather conditions are defined in terms of the daily difference between a base temperature and the 24-hour mean outside temperature when the base temperature is higher than the maximum daily temperature. An accumulated total for each month gives a degree day total for that month.

The 24-hour mean outside temperature is usually taken as the average of the minimum and maximum outside temperatures. On days when the base temperature is less than the maximum outside temperature, then a slight variation in the method of calculation is necessary.[9.1]

The normal base temperature in the UK is 15.5 °C; this is chosen on the assumption that internal gains in a building allow the required internal temperature of 18 °C, say, to be maintained when the outside temperature is 15.5 °C. For hospitals a higher figure is chosen to allow for a higher internal design temperature.

The Department of Energy publishes monthly figures of degree days for 17 regional areas of the UK together with a 20-year monthly average. These figures are reproduced by CIBSE in the journal *Building Services*. Typical examples are given in Table 9.3.

Table 9.3 Degree days for three regions of the UK

Month	Degree days					
	1989			20-year average		
	Midlands	SE	West Scotland	Midlands	SE	West Scotland
January	301	311	257	374	365	383
February	277	271	291	354	340	347
March	257	238	308	322	311	329
April	277	252	280	242	232	245
May	115	90	152	159	147	168
June	88	74	105	84	73	92
July	27	21	45	44	39	57
August	40	36	59	49	44	63
September	64	45	115	91	82	111
October	132	107	152	177	161	191
November	295	280	310	277	268	299
December	336	295	421	325	327	345
Totals	2209	2020	2495	2498	2389	2630

It can be seen from the table that 1989 was warmer on average than the average year over a 20-year period up to 1989. The south-east of the country is usually warmer than the Midlands, which is in turn warmer than the West of Scotland. Over the country as a whole the warmest area is the South-West and the coldest is the North-East of Scotland; the ratio of degree days for these two regions over the normal heating season is about 1.5 and therefore the fuel costs are 50% higher in a building in Aberdeen, say, compared with Plymouth, for instance.

A building with a controlled inside temperature has a space-heating load which varies directly with the outside temperature during the heating season, and hence with the number of degree days. The demand for hot water, on the other hand, tends to remain constant throughout the year. When the monthly fuel consumption is plotted against degree days for a particular building, a graph of the form shown in Fig. 9.24 is obtained. If the best straight line is fitted to the points using the least-squares method, then the line obtained cuts the vertical axis at the value of the base hot water plus process load.

Figure 9.24 Monthly fuel consumption for the heating season plotted against degree days

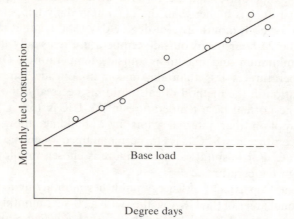

A graph such as that of Fig. 9.24 shows the performance of the building over a given heating season. If changes are made to the building, then these should cause a change in the line in the following heating season. For example, an improvement to the insulation of the building should reduce the slope of the line; installation of more efficient boilers should move the whole line downwards, leaving it parallel with the original line.

Example 9.9

(a)   A building has the monthly gas consumption figures shown in the table opposite with the corresponding 20-year average degree days for the location and the degree days for the year in which the gas consumption was measured.

Plot the gas consumption against degree days and hence find the average annual hot water supply base load. Calculate also the total gas consumption that should be expected in a typical year, assuming that the heating season is from September to May and that the hot water supply load is constant throughout the year. The boiler which supplies both hot water and heating has an efficiency of 70%.

Gas consumption figures for the building of Example 9.9:

Month	Gas consumption (GJ)	Current year degree days	20-year av. degree days
September	630	64	91
October	810	132	177
November	1145	295	277
December	1290	336	325
January	1200	301	374
February	1100	277	354
March	1100	257	322
April	1140	277	242
May	710	115	159
Totals		2054	2321

(b)   In a summer period the building insulation is upgraded and it is estimated that this measure will reduce the heat loss by 10%. At the same time, new boilers are also installed, improving the boiler efficiency from 70% to 80%.

Draw a graph of the expected gas consumption against degree days.

Solution

(a)   The gas consumption is plotted against the current year's degree days in Fig. 9.25. Using the method of least mean squares,[9.1] the base load consumption is found to be 467.5 GJ/month, and the slope of the line is 2.394 GJ/degree day. Therefore, for a typical year,

$$\text{Hot water base load} = 467.5 \times 12 \times 0.7 = 3927 \text{ GJ}$$

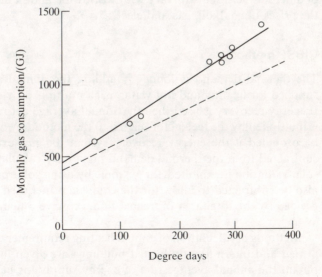

Figure 9.25 Gas consumption figures for Example 9.7

Also for a typical year,

$$\text{Gas consumption for heating} = 2.394 \times 2321 = 5556.5 \text{ GJ}$$

where the total degree days for the heating season of a typical year over a 20-year period are 2321, from the table.

(b)   The initial average annual heating load is $5556.5 \times 0.7 = 3889.5$ GJ. Due to the improvements to the insulation this is estimated to be reduced to $0.9 \times 3889.5 = 3500.6$ GJ.

i.e.   New average annual gas consumption $= 3500.6/0.8 = 4375.7$ GJ

This gives a graph of slope $4375.7/2321 = 1.885$ GJ/degree day. The new base load gas consumption is $3927/(0.8 \times 12) = 409.1$ GJ/month.

The new line to be expected is plotted as a broken line on Fig. 9.25. Gas consumption figures for each month can be plotted against this line as the heating season progresses and should not depart markedly from it.

Energy Demands

Thermal and electrical demands for a particular building can be calculated using the method adopted by CIBSE,[9.10] and then compared with a target value for a building of the same type. In the CIBSE method all the energy demands are related back to a primary fuel energy base by multiplying by a factor which allows for the energy loss in going from the original fuel to the manufactured fuel at point of use; for electricity the factor includes the efficiency of the power station. The figures are based on calculations by the Building Research Establishment. For example, factors for some fuels are as follows:

Coal, 1.03; Natural gas, 1.07; Oil, 1.09; Electricity, 3.82

This method gives a fixed basis for comparison of buildings of the same type. Of course the running costs of a building depend also on the prevailing cost of different fuels and will vary with worldwide and national factors controlling the prices of coal, oil, gas and electricity.

CIBSE method

The thermal demand is found by adding the annual average rate of energy supplied for heating and hot water, each multiplied by the relevant fuel factor. If energy recovery is used then the annual average net rate of energy recovered is multiplied by the fuel factor related to the method by which this energy would be provided if the energy recovery system did not exist. This figure is then subtracted from the thermal demand previously calculated. In buildings where solar radiation is exploited, for example by using solar panels, then a term must also be subtracted to allow for the equivalent fuel used. The final total is then divided by the total area of treated floors to give a figure, usually expressed as W/m^2.

Methods of calculation of annual heating requirements for both continuously heated and intermittently heated buildings are given in Chapters 3 and 4. To obtain the annual average rate of energy supply for heating it is necessary to use the 24-hour mean outside temperature averaged over the heating season for the locality in which the building is placed (see Example 4.2, page 179). For intermittently heated buildings CIBSE[9.10] give an approximate expression for the rate of energy supply as follows:

$$\{\Sigma(UA_o) + \rho n c_p V\}(\bar{t}_m - \bar{t}_{ao}) \qquad\qquad [9.6]$$

where $\bar{t}_m$ is the 24-hour mean dry resultant temperature during the heating season; $\bar{t}_{ao}$ is the 24-hour mean outside temperature averaged over the heating season; the other symbols are as defined in Chapter 3.

The temperature $\bar{t}_m$ is defined by the following equation:

$$\bar{t}_m - \bar{t}_{ao} = \frac{Nf_r}{\{(Nf_r) + (24 - N)\}}(t_c - \bar{t}_{ao}) \qquad [9.7]$$

where N is the number of hours per day of heating; f_r is the response factor defined by Eqn [4.17], page 177; t_c is the design dry resultant temperature.

This approximate method tends to overestimate the heating demand; a comparison of the method of Chapter 4 and the approximate method is given by Eastop and Croft.[9.1]

The thermal demand for hot water supply is based on a standardized daily consumption per person which is then multiplied by the number of occupants and a diversity factor. The diversity factor averaged over the year is taken as 0.88, and typical values of hot water consumption per person are given by CIBSE.[9.10] For example:

Offices, shops and factories   4 litre per person per day
General hospitals   110 litre per person per day
Educational buildings   7 litre per person per day
The annual hot water demand in kJ is then given by

$$\rho_w c_w (t_{HWS} - 10) N_P f_{HWS} \dot{V} N_w / \eta_{HWS} \qquad [9.8]$$

where t_{HWS} is the temperature of the hot water supply in °C; N_P is the number of occupants; f_{HWS} is the diversity factor; $\dot{V}$ is the volume flow rate in litres per person per day; N_w is the number of working days per year; c_w is the mean specific heat of water in kJ/kg K; η_{HWS} is the overall efficiency of the hot water supply system given by the boiler efficiency multiplied by the efficiency of distribution; ρ_w is the density of water in kg/litre; note that the cold supply temperature is taken as 10 °C.

The electricity demand is due to the lighting load plus the load from the fans and pumps associated with the heating, ventilating and hot water supply systems. Lifts and escalators are not included.

The lighting load depends on the amount of daylight which can be used. For the known proportional glazed area either in the walls or as rooflights, a diversity factor can be found from graphs[9.10] or from a CIBSE table.[9.11] The diversity factor, f_L, is the number of hours in the year for which the illuminance is below the minimum required level, divided by the number of working hours in the year. When the lights are controlled to come on automatically when the illuminance falls below the minimum level, then the diversity factor can be applied with accuracy to calculate the lighting load.

Utilization factors can also be used for fans and pumps; see Table 2.5.11.[9.10]

Example 9.10

A naturally ventilated office building is occupied for 10 hours per day (from 0800 h to 1800 h), for 5 days per week for 250 working days in the year. The building is heated intermittently, 12 hours on and off, during the heating season of 145 days. The heating and hot water are provided by gas-fired boilers. The lighting is controlled to go off and on to give the desired illuminance as daylight alters.

Using the data below and neglecting casual gains, heat recovery, and solar radiation, calculate the annual thermal and electrical demands in primary fuel units.

Data

$\Sigma(UA_o) = 1200$ W/m^2; $\Sigma(YA) = 3800$ W/m^2; total floor area, 800 m^2; ventilation rate, 0.6 m^3/s; design inside dry resultant temperature, 18 °C; 24-hour mean outside temperature averaged over the heating season, 6 °C; temperature of hot water supply, 40 °C; number of occupants, 50; diversity factor, 0.88; hot water supply rate, 4 litre per person per day; boiler efficiency, 0.8; efficiency of heating and hot water distribution system, 0.97; required illuminance 750 lx; daylight factor, 3%; absorbed lighting power, 12 W/m^2; pump full load power, 1.2 kW; pump utilization factor, 0.2; primary fuel factors: natural gas, 1.07; electricity, 3.82.

An extract from the CIBSE daylight availability table[9.11] is given as Table 9.4.

Table 9.4 Annual availability of daylight for working day of different duration[9.11]

Sky illuminance (klx)	Hours illuminance is below the stated value*		
	Starting of working day (clock time)		
	0700 h	0800 h	0900 h
	Finish of working day (clock time)		
	1900 h	1800 h	1900 h
0.5	476	215	374
1.0	559	264	420
5.0	1123	706	801
10.0	1779	1243	1319
20.0	2948	2241	2290
25.0	3385	2657	2683
70.0	4380	3650	3650

*Note: The hours when the illuminance is below the stated value for 70 klx and above is the total annual hours of 365 days at 10 or 12 hours per day

Solution

The response factor, f_r, for the building is given by Eqn [4.17], and nV is given as 0.6 m^3/s, i.e.

$$f_r = \frac{\Sigma(YA) + \rho n c_p V}{\Sigma(UA_o) + \rho n c_p V} = \frac{3800 + (0.6 \times 1.2 \times 1005)}{1200 + (0.6 \times 1.2 \times 1005)}$$
$$= 2.35$$

Then, substituting in Eqn [9.7],

$$\bar{t}_m - \bar{t}_{ao} = \frac{12 \times 2.35}{\{(12 \times 2.35) + (24 - 12)\}}(18 - 6)$$

$$= 8.42 \text{ K}$$

Substituting in Eqn [9.6],

Rate of energy supplied for space heating
$$= \{\Sigma(UA_o) + \rho n c_p V\}(\bar{t}_m - \bar{t}_{ao})$$
$$= \{1200 + (1.2 \times 0.6 \times 1005)\} \times 8.42$$
$$= 16\,197 \text{ W}$$

Therefore,

Primary energy rate for heating

$$= \frac{\text{Rate of energy supplied for space heating} \times \text{Primary fuel factor}}{\text{Boiler efficiency} \times \text{Distribution efficiency}}$$

$$= 16\,197 \times 1.07/0.8 \times 0.97 \times 1000$$

$$= 22.33\,\text{kW}$$

Therefore,

Annual primary energy for heating

$$= 22.33 \times 145 \times 24 \times 3600/1000 \times 800$$

$$= 349.7\,\text{MJ/m}^2$$

and

Average annual primary energy rate for heating

$$= 22.33 \times 1000 \times 145/365 \times 800$$

$$= 11.09\,\text{W/m}^2$$

The annual hot water thermal demand is given by Eqn [9.8], i.e.

Annual hot water energy rate

$$= \rho_{\text{w}} c_{\text{w}} (t_{\text{HWS}} - 10) N_{\text{P}} f_{\text{HWS}} \dot{V} N_{\text{w}} / \eta_{\text{HWS}}$$

$$= \frac{1 \times 4.186 \times (40 - 10) \times 50 \times 0.88 \times 4 \times 250}{0.8 \times 0.97 \times 1000}\,\text{MJ}$$

$$= 7120.5\,\text{MJ}$$

Then,

Annual hot water primary energy rate

$$= 1.07 \times 7120.5 = 7619\,\text{MJ}$$

$$= 7619 \times 10^6/365 \times 24 \times 3600 \times 800$$

$$= 0.30\,\text{W/m}^2$$

and

Annual hot water demand in primary units

$$= 7619/800 = 9.52\,\text{MJ/m}^2$$

i.e.

Total rate of use of primary energy for heating and hot water

$$= 11.09 + 0.3$$

$$= 11.39\,\text{W/m}^2$$

The total thermal demand in primary energy terms is $349.7 + 9.52 = 359.2\,\text{MJ/m}^2$.

To calculate the lighting load it is necessary to find the limiting daylight illuminance in the building:

$$\text{Limiting illuminance} = \frac{\text{Required illuminance}}{\text{Fractional daylight factor}}$$

$$= 750/(1000 \times 0.03) = 25\,\text{klx}$$

From the table, for an illuminance of 25 klx and a 10-hour day from 0800 h to 1800 h, we have

Electric lighting is required for 2657 h

i.e.

Diversity factor for lighting $= 2657/3650 = 0.73$

Then,

> Total annual electrical demand due to lighting
> = Absorbed lighting power × Diversity factor
>   × Working days per year × Hours of occupancy per day
> = $(12 \times 3600 \times 10^{-6}) \times 0.73 \times 250 \times 10$
> = 78.84 MJ/m^2

The electrical load for the pumps is

$$\frac{\text{Pump power} \times \text{Utilization factor}}{\text{Total floor area}}$$
$$= 1.2 \times 1000 \times 0.2/800 = 0.3 \text{ W/m}^2$$

Hence the annual load is

$$0.3 \times 3600 \times 24 \times 365 \times 10^{-6} = 9.46 \text{ MJ/m}^2$$

Therefore,

> Total annual electrical demand in primary fuel energy
> = $(78.84 + 9.46) \times 3.82 = 337.3 \text{ MJ/m}^2$

Then,   Total annual demand of primary fuel energy
> = Thermal demand + Electrical demand
> = $359.2 + 337.3 = 696.5 \text{ MJ/m}^2$

In the above example the energy gains inside the building due to people, lighting and solar radiation have been neglected. CIBSE[9.10] recommend that casual gains should be neglected when there is no control of the heating by room thermostat. When a room thermostat does control the heating, then casual gains should be calculated and subtracted from the thermal demand; CIBSE recommends that the allowance for casual gains should be limited to 10 W/m^2.

Energy Targets

When the thermal, electrical and total demands in primary energy have been calculated for a particular building, the values can then be compared with target values for that type of building. The CIBSE method of obtaining targets[9.10] is to assume, for a range of types of building, values for the full range of factors used in the calculation for thermal and electrical demand.

For example:

— For heating demand, the factors assumed for different building types are
(i) number of days in the heating season;
(ii) heating system efficiency;
(iii) boiler efficiency;
(iv) response factor;
(v) hours of heating;
(vi) mean U-values for walls, glazing, roof and floor;
(vii) floor to ceiling height;
(viii) the temperature difference between the inside dry resultant temperature and the 24-hour mean outside temperature is taken as 14 K.

— For the hot water supply demand, values for various building types are assumed directly.

— For the electrical demand, the following are assumed for different building types:

(i)   given levels of illuminance in lux;

(ii)   a lighting efficacy of 55 lm/W;

(iii)   number of hours of occupancy per day;

(iv)   number of working days in the year;

(v)   diversity factor;

(vi)   target load for lighting in W/m^2.

— For electrical demand from pumps and fans, this is assumed to be directly proportional to the heating and hot water demand and is taken as 10% of the thermal demand in primary energy terms.

This method of arriving at targets involves a considerable number of assumptions about building types and their construction. The Energy Efficiency Office has produced target figures for different buildings in cash terms, by surveying large numbers of buildings of different types. These figures can be converted into energy terms using the prices prevailing at the time of the survey (see for example Ref. 9.1).

For instance, for offices the Energy Efficiency Office survey gives approximately:

— For thermal demand,
    Good $<640 \, MJ/m^2$;   satisfactory, $640–700 \, MJ/m^2$;
    very poor $>1400 \, MJ/m^2$

— For electrical demand,
    Good $<85 \, MJ/m^2$;   satisfactory, $85–105 \, MJ/m^2$;
    very poor $>245 \, MJ/m^2$

The comparable figures in Example 9.10 are:

$$\text{Thermal demand} = 359.2/1.07 = 336 \, MJ/m^2$$

which is therefore very satisfactory since it is well below the target of $640 \, MJ/m^2$ for a good building.

$$\text{Electrical demand} = 337.3/3.82 = 88.3 \, MJ/m^2$$

which is satisfactory since it is just above the value of $85 \, MJ/m^2$ for a good building.

9.4 ENERGY MONITORING AND AUDITING

The recording of all data relevant to the energy use of a building is known as energy monitoring. The detailed consumption of electricity, gas, oil, coal and water should be recorded either continuously or at regular intervals, and the temperatures of all relevant parts of the building also measured and recorded. The use of electricity, or a fuel, for special purposes, e.g. for process steam or machinery, should be monitored separately; it is also desirable to monitor different parts of the same building separately when the use varies. The suppliers of gas and electricity will install a meter for the whole site but additional metering by the consumer may be desirable.

The objective of effective monitoring is to allow an energy audit to be carried out. An energy audit sets out the exact inputs of energy and the exact amounts consumed over fixed periods of time. The results of a careful audit will show up possible areas for energy saving, and will enable comparison from year to year, or month to month, thus indicating any changes in energy consumption due to, for example, lack of proper maintenance. Weather influences can be compensated by using degree days or a similar method (see Section 9.3).

When an energy target is known for a building, either from the CIBSE method[9.10] or from energy surveys of similar building types, then the energy audit should show whether the target is being met.

In many cases careful housekeeping and systematic maintenance will lead to annual energy savings of as much as 5% of the energy bill. The installation of a control system which also records energy use, known as an energy management system, will very often lead to direct savings. It is pointed out in Section 9.3[9.10] that when a room thermostat is used to control heating, for example, then casual gains from people, lighting and solar radiation can be used to offset the heat input required for that room. The capital cost of any energy-saving measure must of course be recouped within a reasonably short time period (under three years, say) through the annual saving in fuel or electricity.

Tariffs

For the supply of utilities such as electricity, gas and water, the supplier charges the consumer not only for the amount consumed but also for the cost of distribution and metering. (In the UK at present domestic water is not metered and hence a fixed charge is made based on the value of the property and not on the amount consumed; this may change in the near future.) For electricity and gas the tariff varies according to the amount consumed by the customer. Electricity tariffs, and to a lesser extent gas tariffs, are complex and in many cases the bill can be reduced by simply changing from one tariff to another. Energy managers should look closely at the tariff on which their electricity and gas is being charged to make sure that the most cost-effective one is being used.

For electricity there are basically three tariffs: domestic, block and maximum demand. A detailed discussion of these three tariffs is given by Eastop and Croft;[9.1] the following is a brief summary.

(a)   Domestic tariff

This is available for customers with a consumption below 25 kW maximum. There is a fixed quarterly charge and a charge per kilowatt hour (kWh) of energy consumed. Customers may opt for a separate night meter (operating during a seven-hour period between 2200 h and 0900 h), which is charged at a lower rate (about 35% of the normal day-time rate), but with a corresponding increase in the day-time rate of about 5%.

(b)   Block tariff

This applies to non-domestic customers with loads up to 50 kVA (apparent power). As with the domestic tariff this comprises a fixed quarterly charge and a charge for units consumed, but in this case the first 1000 kW h each quarter are charged at a fixed rate with the remainder at a rate of about 75% of this. A cheap night-time metering system can be opted for in a similar way to the domestic tariff.

(c) Maximum demand tariff

In this method a *chargeable capacity* for the customer is fixed first of all. This is the higher of (i) a negotiated capacity in kVA when the supply is first connected to the premises, or (ii) the highest maximum demand recorded in a previous period of 11 months.

There are then four basic charges made to the customer as follows.

(i) An *availability charge* which is expressed in terms of the chargeable capacity defined above.

(ii) A *demand charge* which is based on the maximum demand recorded in a single month averaged over a half-hour period; the maximum demand is measured by a special meter which meters consumption every half-hour period and then re-sets itself.

(iii) A *unit charge* which varies according to day-time or night-time (seven-hour period, as before) use.

(iv) A *fuel price adjustment* is used to alter the unit charge when the price of the fuel burned at the power station varies above or below a fixed datum price; this adjustment is usually very small.

Two further items complicate the tariff structure: (a) the availability, demand, and unit charges depend on whether the supply voltage is above or below 1000 V; (b) one of three scales must be selected as follows:

Scale 1   The unit charge varies for day- and night-time use but the availability and demand charges are based on any time.

Scale 2   The unit charge is as in scale 1 but the availability and demand charges apply only to day-time periods.

Scale 3   The unit charge is fixed at one value and the availability and demand charges are based on any time; this scale applies only to low-voltage supplies.

It can be seen that a customer must choose carefully which tariff applies and which scale applies if on the maximum demand tariff. It is not surprising that discussion with the electricity supply body is usually necessary before the decision is made.

For large users it is generally true that a high load factor will lead to lower costs, where the load factor is defined as the average annual rate of consumption divided by the maximum demand rate. Since the maximum demand charge is based on the average rate of use over a half-hour period, a microprocessor-based system can be installed which compares the actual metered maximum demand against a reference value and gives a warning when the reference value is exceeded. Any possible measures should be taken to reduce maximum demand and spread the energy use more evenly. Since night tariffs are cheaper, systems which use energy storage for space and water heating are generally more cost-effective; any process which can be performed automatically at night will also save fuel costs (e.g. dish-washing in a hotel or hospital, washing of clothes in a laundry, etc.).

Private Generation

It is possible to sell power to the supplier of electricity and therefore there are tariffs laid down for electricity exported to the national grid; the prices paid to the private generator vary according to the time of day and the time of year since the level of demand nationally affects the value of the electricity being exported to the grid. The supplier charges private generators a different tariff for the supply of electricity and also reserves the right to increase the charge considerably if the electricity exported to the grid ceases for any reason.

Power Factor

It will be seen from the above that for the maximum demand tariff the chargeable capacity is expressed as apparent power in kVA and not the actual power in kW. Any inductive or capacitive load causes a phase difference between the current and voltage of the supply and hence the real power is a fraction of the apparent power in kVA; an inductive load causes the current to lag the voltage whereas a capacitive load causes the current to lead the voltage. For a phase angle of Φ between the current, I, and the voltage, V, the real power is given by $IV\cos\Phi$, and the apparent power by IV.

A power factor is then defined as

$$\frac{\text{Real power}}{\text{Apparent power}} = \cos\Phi \qquad [9.9]$$

Circuits which contain only electrical resistance, with minimum inductance or capacitance, such as heaters, have a power factor of unity. Equipment with a high inductance, such as induction motors, have a low power factor indicating an apparent power in kVA much higher than the actual power in kW. A power supply at a given voltage to a customer with a low power factor requires a larger current than if the power factor were unity; this in turn implies more expensive switchgear, transformers and cables. This explains why the supplier uses a maximum demand in kVA as the basis for the fixed charge.

Clearly, a customer can make great savings in electricity costs by increasing the power factor by avoiding high inductive loads or by reducing the phase angle, Φ, by adding a capacitive load into the circuit to offset the inductive load. The latter can be achieved easily by installing a bank of capacitors either at the inductive item, or at the input of power to the building. Induction motors at part-load have a much lower power factor; hence it is very important to install motors of the correct capacity for the load required, and not to run them for long periods at part-load.

Example 9.11

A customer has an electricity supply at below 1000 V and has very little demand for electricity at nights or at week-ends. The annual consumption is measured at 80 000 kW h, the average power factor is 0.85, and the maximum demand is 42.5 kW; the consumption is the same in every month.

When the building was opened, the negotiated chargeable capacity was 45 kVA.

Neglecting the fuel price adjustment, and using the data given, find which tariff is the best for this particular customer.

Data

Block tariff: quarterly charge, £8.5; unit charge for first 1000 kW h, 8.9 p/kW h; unit charge thereafter, 6.7 p/kW h.

Maximum demand tariff: availability charge for the first 50 kVA of chargeable capacity, £1.1/kVA; availability charge thereafter, £0.87/kVA: demand charge, March to October nil, November and February £2.3/kVA, December and January £7.4/kVA: unit charge, 5.1 p/kW h.

Solution

Block tariff:

$$\text{Quarterly fixed charge} = £8.5$$

The unit charge is made up of 1000 kW h in each quarter at the higher rate plus the remaining 76 000 kW h at the lower rate, i.e.

$$\text{Annual unit charge} = \frac{1000 \times 4 \times 8.9}{100} + \frac{76\,000 \times 6.7}{100}$$

$$= £5448$$

$$\text{Total annual charge} = £(4 \times 8.5) + £5448 = £5482$$

Maximum demand tariff:

The chargeable capacity is the greater of 45 kVA or the actual monthly maximum demand, which is $42.5/0.85 = 50$ kVA. Therefore the chargeable capacity is 50 kVA.

Then,

$$\text{Annual availability charge} = 50 \times 1.1 \times 12 = £660$$

$$\text{Demand charge} = (2 \times 2.3 \times 50) + (2 \times 7.4 \times 50)$$

$$= £970$$

$$\text{Unit charge} = \frac{5.1 \times 80\,000}{100} = £4080$$

i.e.

$$\text{Total annual charge} = £660 + £970 + £4080$$

$$= £5710$$

It can be seen that the block tariff is better.

Natural Gas Tariffs

Gas tariffs are much simpler than those for electricity. There are basically three types of consumer: the domestic customer who pays a quarterly fixed charge and a charge per therm used; the commercial customer who uses over 25 000 therm per year; the commercial customer who uses over 200 000 therm per year.

Note: 1 therm = 29.307 kW h = 105.506 MJ.

For a customer who uses more than 25 000 therm per year but less than 200 000 therm per year the charge per therm used decreases as the quantity used increases.

When the annual consumption exceeds 200 000 therm, the customer has the choice of accepting an interruptible supply instead of the normal fixed supply, the cost per therm being less for an interruptible supply. An interruptible contract means that the gas supplier has the right to stop supplying gas for a certain number of days in the year, not necessarily concurrent; the notice required to be given to the customer of the interruption in the supply is only six hours. The customer must be able to use an alternative fuel (e.g. oil) in the periods when no gas is supplied. The interruption to the gas supply will tend to occur in cold periods in winter when the national demand rises.

There are three possible interruptible gas contracts with varying numbers of total days when the supply may be stopped. These are as follows:

Short period   Interruption between 7 days and 35 days
Medium period   Interruption between 7 days and 63 days
Long period   Interruption between 7 days and 90 days

The cost per therm is about 7% less for a medium-period contract compared with a short-period contract, and about 25% less for a long-period contract compaed with a short-period contract. The cost per therm is also less for greater annual consumptions, as for the firm supply case.

The cost per therm for a large consumer of more than 10^6 therm per year, say, with an interruptible supply for the longest period, is less than half that for a customer who uses only between 25 000 and 50 000 therm per year with a firm supply. In many cases dual-fuel burners can be used so that changing from gas to oil is relatively easy given six hours notice, but if there are units in the building which must have gas (e.g. kitchens using gas for cooking), then a firm supply is essential.

Example 9.12

The thermal demand of a large building used entirely for office work is suppled by boilers using natural gas. The existing supply of gas is firm and the total annual consumption in an average year is 300 000 therm. The heating season is for 250 days, and the gas consumption in the heating season represents 85% of the total annual thermal demand.

It is proposed to convert the boilers to dual-fuel burning and apply to have an interruptible gas supply over the medium period of from 7 to 63 days interruption.

Using the data below, assuming that the boiler efficiency is the same when burning oil or gas, calculate the approximate annual cost saving to be set against the capital cost of the alterations required.

Data
Tariff for firm supply, 31.25 p/therm.
Tariff for interruptible supply (medium period), 26.60 p/therm. Price of oil, 15 p/litre; gross calorific value, 40 MJ/litre.

Solution
Firm supply:

$$\text{Annual cost} = 31.25 \times 300\,000/100 = £93\,750$$

Interruptible supply:

Gas cost for minimum period of interruption

$$= \frac{26.6 \times (250 - 7) \times 0.85 \times 300\,000}{250 \times 100} + \frac{26.6 \times 0.15 \times 300\,000}{100}$$

$$= £77\,901$$

Oil cost for a minimum period of interruption

$$= \frac{7 \times 15 \times 0.85 \times 300\,000 \times 105.506}{250 \times 40 \times 100} = £2825$$

Total cost $= £77\,901 + £2825 = £80\,726$

Gas cost for maximum period of interruption

$$= \frac{26.6 \times (250 - 63) \times 0.85 \times 300\,000}{250 \times 100} + \frac{26.6 \times 0.15 \times 300\,000}{100}$$

$$= £62\,707$$

Oil cost for maximum period of interruption

$$= 63 \times 2825/7 = £25\,425$$

Total cost $= £62\,707 + £25\,425 = £88\,132$

Therefore the approximate annual saving in fuel cost varies between £$(93\,750 - 80\,726) = £13\,024$ and £$(93\,750 - 88\,132) = £5\,618$.

It can be seen that a capital cost of £13 000, say, could be recouped in approximately one to two years. The capital cost would include the cost of modifying the boiler for dual-fuel burning and adding the necessary pipework, pump and storage tank.

PROBLEMS

9.1 A total energy scheme uses a diesel engine with an overall efficiency of power output to fuel energy input rate of 38%. Heat recovered from the oil and water cooling systems, and from the exhaust gases, is used for a hot water heating system. It is estimated that 25% of the energy rejected by the engine can be recovered for the heating system. Assuming that additional heat can be obtained using a boiler of efficiency 80%, calculate:
(i) the maximum heat: power ratio available from the engine without using the boiler;
(ii) the ratio of boiler fuel energy input to diesel engine fuel energy input for a heat:power ratio of 1.5.
(0.408; 0.123)

9.2 An open-circuit gas turbine is to be combined with a steam plant operating on the Rankine cycle using the exhaust gas to generate steam in a waste heat boiler. The steam condition is 20 bar and 450 °C. The gas exhaust temperature from the boiler is 35 K above the evaporating temperature for the steam. The condenser pressure is 0.1 bar and the isentropic efficiency of the steam turbine is 85%.

With reference to a sketch of the plant, neglecting pressure losses, feed pump work, and using the data below, determine:
(i) the net power output for the gas turbine;
(ii) the mass flow rate of steam;
(iii) the steam turbine power output;
(iv) the overall thermal efficiency of the plant.

Data

Mass flow rate of air, 56 kg/s; mass flow rate of fuel, 1.26 kg/s; calorific value of fuel, 30 MJ/kg; combustion efficiency 100%; air inlet pressure and temperature, 1.0 bar and 15 °C; compressor delivery pressure, 8 bar; compressor isentropic efficiency, 81%; gas turbine isentropic efficiency, 85%. Specific heats at constant pressure: for the combustion process, 1.13 kJ/kg K; for the turbine expansion and combustion products, 1.16 kJ/kg K; for the compression process, 1.005 kJ/kg K. Ratio of specific heats, γ: for air, 1.4; for combustion products, 1.333.

(Newcastle upon Tyne Polytechnic)

(9677 kW; 4.67 kg/s; 4161 kW; 36.6%)

9.3 (a) Briefly discuss the relative merits of using back-pressure turbines and extract turbines for combined heat and power plant.

(b) A back-pressure turbine plant is to be used to provide 5400 kW of electricity and 40 000 kW of heat. The electrical generator conversion efficiency is 90% and steam is supplied to the turbine at 15 bar.

The exhaust from the turbine is supplied to a variety of heaters as saturated vapour at 2 bar. There is no under-cooling of the condensate in any of the heaters. Only 70% of the condensate is returned to the hot well and this falls in temperature by 30 K in the pipework. If make-up is provided at 20 °C, calculate:

(i) the degree of superheat of the steam supplied to the turbine;
(ii) the turbine isentropic efficiency;
(iii) the ratio of useful output to the heat supplied to the plant.
Use the h–s chart for steam.

(Wolverhampton Polytechnic)

(100 K; 80.1%; 0.91)

9.4 A combined power and process plant supplies 5 MW of electrical power, 2 kg/s of steam at 14 bar, and 5 kg/s of steam at 2 bar. A double extraction pass-out turbine is to be used which is supplied with steam at 70 bar and 400 °C, and exhausts to a condenser at 0.08 bar. The isentropic efficiency of each stage of the expansion is 78%. The extraction steam is used for process heating and the condensate from the heat exchangers is returned to the boiler feedline as saturated liquid at its extraction pressure.

The boiler is coal-fired and has a combustion efficiency of 85% when using coal of calorific value 25 MJ/kg. Neglecting the feed pump work and other losses, calculate the daily coal consumption. Use the h–s chart and property tables as appropriate and assume the efficiency of the electrical generator as 100%.

(Wolverhampton Polytechnic)

(88 640 kg/day)

9.5 In a total energy scheme a diesel engine is used to drive the electricity-generating plant. The engine exhaust gases are passed through a waste heat boiler to generate steam. The steam is used to drive a steam turbine before being exhausted to a condenser at a pressure of 0.1 bar, the condensate being pumped back into the boiler. Assuming that the turbine isentropic efficiency is 80% and ignoring the feed pump work, estimate the engine exhaust gas temperature, the mass flow rate of steam, the turbine power output, and the overall thermal efficiency of the plant.

Data

Diesel engine:

Shaft power output, 1 MW; thermal efficiency, 37.5%; percentage of energy supplied which is rejected in the exhaust gas, 30.5%; datum temperature for energy balance, 298 K; air:fuel ratio by mass, 28; air and fuel inlet temperature, 298 K; calorific value of fuel, 42 MJ/kg; specific heat at constant pressure of exhaust gases, 1.04 kJ/kg K.

Boiler:

Exhaust gas outlet temperature, 160 °C; steam pressure, 10 bar; steam outlet temperature, 400 °C.

(Newcastle upon Tyne Polytechnic)

(451 °C; 0.18 kg/s; 129.4 kW; 42.4%)

9.6 (a) 'Development of District Heating in the UK has lagged well behind that in other countries and the potential for large scale CHP has not been exploited.'

Discuss the above statement in the context of the UK electricity generation system and the current state of development of large scale CHP schemes in the UK.

Give your views on a future national energy policy, taking into account the impact of privatization of the electricity supply industry.

(b) A small hotel currently uses gas-fired boilers for its thermal requirements and buys all its electricity from the grid. It is proposed to install a micro-CHP unit which will be sized to satisfy completely the summer thermal and electrical loads; in winter the existing gas boilers will supply the additional thermal loads; the extra electricity requirements will be imported from the grid.

On this basis a micro-CHP unit of 32 kW electrical output and 60 kW thermal output is required which will be run at full-load continuously for 16 hours per day for 350 days per year. The unit is able to use low-grade thermal energy for the hotel swimming pool and hence 88% of the gas energy input is able to be converted into heat and power. The existing gas boilers have an efficiency of 75%.

The capital cost of the unit is £15 500 and there will be an additional annual maintenance cost of £1000. Taking the cost of gas as 36 p/therm and the cost of electricity as 3.7 p/kW h, estimate the simple pay-back period.

Note: 1 therm = 29.307 kW h.

(Wolverhampton Polytechnic)

(3.93 years)

9.7 A chemical plant shown in Fig. 9.26 has an electrical power requirement and a process steam requirement which remain constant during the working day. These requirements are provided by a residual-fuel-fired turbo-charged, slow-speed diesel engine, and a waste-heat boiler supplemented with residual-fuel oil.

Air drawn from the atmosphere is compressed in a turbine-driven centrifugal compressor, and passed through the air cooler to the engine cylinders. The exhaust from the engine cylinders passes through a turbine which develops just enough power to drive the compressor; the gases then pass to the boiler where additional residual-fuel oil is burned.

The dry saturated steam generated in the boiler goes to the process; the condensate is non-returnable. The make-up feed water passes through a heat

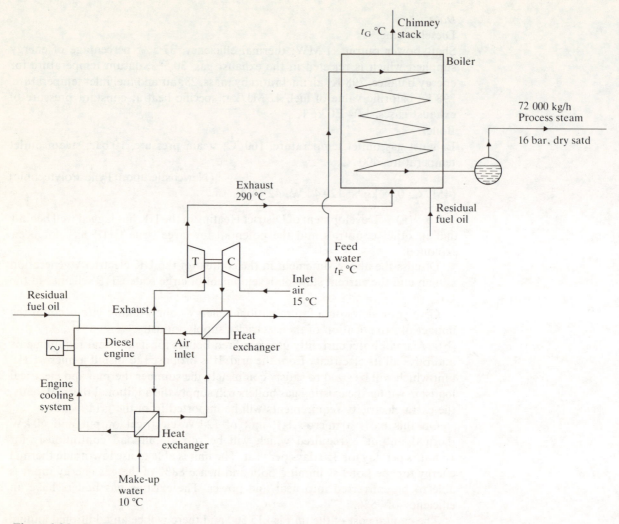

Figure 9.26 Power and process system for Problem 9.7

exchanger which cools the cooling water for the engine cylinders, piston and lubricating oil, and then through a heat exchanger which cools the water from the air cooler. The feed water is then delivered to the boiler. Using the data below and making suitable assumptions, calculate:
(i) the temperature of the feed water entering the boiler;
(ii) the temperature of the gases entering the chimney stack.

Data
Air flow to engine, 213 000 kg/h; air:fuel ratio of engine, 42; atmospheric temperature, 15 °C; pressure ratio of compressor, 2.7; isentropic efficiency of compressor, 0.7; temperature of the air leaving the air cooler, 40 °C; total rate of energy dissipated from engine coolant system, 1.7 MW; process steam mass flow rate, 72 000 kg/h; process steam condition, dry saturated at 16 bar; temperature of make-up water, 10 °C; specific heat and isentropic index for air, 1.005 kJ/kg K and 1.4; specific heat for exhaust gases in boiler, 1.1 kJ/kg K; temperature of exhaust gases entering the boiler from the engine, 290 °C; flow

of residual-fuel oil to boiler, 4400 kg/h; calorific value of residual-fuel oil, 41 500 kJ/kg; combustion efficiency, 75%.

(CIBSE)

(108.7 °C; 155.6 °C)

9.8 A plant combining electrical power generation with the supply of process steam operates with boiler steam conditions at 30 bar, 400 °C, a condenser pressure of 0.04 bar, and a dryness fraction at turbine exit of 0.89.

The power requirements are steady at 10 MW but the process steam demand fluctuates between 18 000 kg/h and 14 400 kg/h. The process steam is taken off between turbine stages at a pressure of 1.2 bar, and the process condensate is not returned to the system; make-up water at 10 °C is supplied to the hot well. During the periods of running with low process steam demand the excess steam at 1.2 bar is used to heat the feed water leaving the hot well. A closed feed heater is used with the drain pumped into the line downsteam of the heater.

Sketch the plant and calculate:
(i) the mass flow rate of boiler steam;
(ii) the heat supplied in the boiler for the cases of maximum and minimum process steam demand.

Use the h–s chart and assume that the condition line for the expansion process is straight. Neglect pump work and other losses.

(CIBSE)

(12.71 kg/s; 39 920 kW, 37 280 kW)

9.9 (a) A factory has a requirement for power and process steam and it is proposed to satisfy the demand by using a steam plant with a back-pressure turbine. The steam leaving the turbine is supplied to the process and the condensate from the process is pumped back to the boiler. The ratio of process energy to power output is 10.

Neglecting feed pump work and pressure and thermal losses, calculate:
(i) the required pressure and temperature of the steam supplied by the boiler;
(ii) the power output of the turbine;
(iii) the overall efficiency of the plant defined as the total useful energy divided by the energy input from the fuel.
Use the data below and the h–s chart for steam.

Data
Pressure of steam supplied to process, 2.0 bar; dryness fraction of steam supplied to process, 0.98; pressure of saturated water returned from process, 1.9 bar; mass flow rate of steam supplied by the boiler, 25 000 kg/h; boiler efficiency, 80%; isentropic efficiency of steam turbine, 85%.

(b) Describe an alternative way of supplying power and process requirements using a steam plant and steam turbine, and discuss briefly any technical and/or financial reasons for choosing one method instead of the other.

(CIBSE)

(7.6 bar, 215 °C; 1503.5 kW; 80%)

9.10 A gas turbine plant consists of a compressor, combustion chamber, and turbine mounted on the same shaft as the compressor. The exhaust gases from the turbine pass through a heat exchanger to heat water for space heating before passing to the chimney. Using the data below, neglecting heat losses, and assuming that the mass flow rate of fuel is negligible in comparison with the air flow rate, calculate:

(i)   the temperature of the gases leaving the turbine;
(ii)   the mass flow rate of the exhaust gases;
(iii)   the temperature of the gases entering the chimney;
(iv)   the overall efficiency of the system defined as the useful energy output divided by the energy input from the fuel.

Data
Net electrical power output of gas turbine unit, 2 MW; combined mechanical and electrical efficiency of unit, 90%; combustion efficiency, 99%; compressor pressure ratio, 10; isentropic efficiency of compressor, 80%; isentropic efficiency of turbine, 83%; inlet air conditions, 1.013 bar and 15 °C; pressure drop in combustion chamber, 0.20 bar; pressure drop of gases in heat exchanger, 0.15 bar; pressure drop in chimney, 0.05 bar; maximum cycle temperature for gas turbine unit, 1100 K; specific heat at constant pressure for air, 1.005 kJ/kg K; isentropic index for air, 1.4; specific heat at constant pressure for gases throughout, 1.150 kJ/kg K; isentropic index for gases, 4/3; water return temperature from space heating system, 60 °C; water flow temperature to space heating system, 80 °C; mass flow rate of water to space heating system, 200 000 kg/h; mean specific heat of water at 70 °C, 4.191 kJ/kg K.

(CIBSE)

(453.8 °C; 24.05 kg/s; 285.4 °C; 50%)

9.11   In a sewage plant the raw sludge is processed using the cooling water from the condenser of a steam plant as shown diagrammatically in Fig. 9.27.

The digester gas produced from the sewage processing is used in a gas turbine combustion chamber (not shown in the figure), and the exhaust from the gas

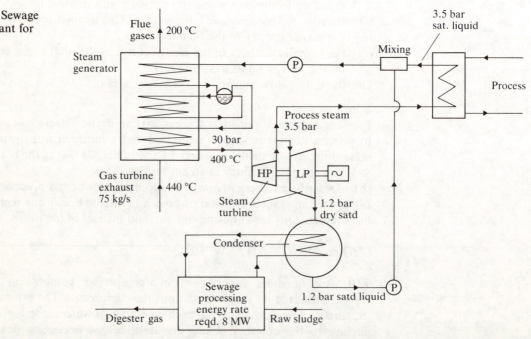

Figure 9.27 Sewage processing plant for Problem 9.11

turbine is passed to a steam generator. At the exhaust from the high-pressure steam turbine some steam is extracted for process use, leaving the process heat exchanger as a saturated liquid. The remaining steam expands through the low-pressure turbine to the condenser and is then pumped to the intermediate steam pressure where it mixes with the condensate from the process. The mixed feed water is then pumped through the steam generator.

Using the h–s chart, steam tables, and the data below, calculate, neglecting heat losses and pump work:

(i)   the mass flow rate of steam through the low-pressure turbine;
(ii)   the mass flow rate of steam through the high-pressure turbine;
(iii) the total power output from the steam turbines;
(iv)   the heat supplied to the process;
(v)   the minimum temperature difference between the hot gases and the steam in the steam generator.

Data
Gases:
Temperatures entering and leaving the steam generator, 440 °C and 200 °C; average specific heat at constant pressure, 1.15 kJ/kg K; mass flow rate, 75 kg/s.
Steam:
Condition leaving the generator, 30 bar and 400 °C; pressure of extract steam and steam entering the low-pressure turbine, 3.5 bar; steam condition entering the condenser, 1.2 bar dry saturated; condensate leaving the process heat exchanger, 3.5 bar saturated liquid; for the overall expansion process from 30 bar to 1.2 bar take the condition line as straight on the h–s chart.
Sewage:
Rate of energy required for processing, 8 MW.

(CIBSE)

(3.565 kg/s; 7.625 kg/s; 3.521 MW; 9.180 MW; 9.7 K)

9.12 A heat pump using freon 12 is used to extract heat from a sewage system for the purpose of central heating. The compressor is to be driven by a diesel engine having a thermal efficiency of 38%, and a further 40% of the calorific value of the fuel is to be recovered by utilizing heat from the engine-cooling water and from the exhaust gases. The mean sewage temperature is 10 °C and a temperature difference of 10 K between the sewage and the refrigerant would keep the evaporator down to a reasonable size. Compression to 900 kN/m² starts with dry saturated freon 12 and is followed by de-superheating, condensation and sub-cooling to 25 °C. The heat pump cycle is then completed by adiabatic throttling. Assume that the actual thermal advantage will be 70% of that for the theoretical cycle.

Calculate, using tables and/or the p–h chart, the percentage saving in fuel compared with a simple oil-fired heating plant of 80% thermal efficiency using a fuel similar to that used by the oil engine.

(University of Glasgow)

(67.4%)

9.13 A hypermarket has an annual space-heating requirement of 29 000 GJ. It produces 600 tonnes of combustible waste per year and it is proposed to install an incinerator with waste heat boiler to reduce the load in the existing oil-fired boilers. Using the data below, calculate:
(i) the percentage of the heating load that can be satisfied by waste incineration;

(ii) the capital that the firm could spend on the scheme if the simple pay-back period is to be no greater than $3\frac{1}{2}$ years.

Assume that maintenance costs are unchanged.

Data

Efficiency of incinerator and waste heat boiler, 59%; efficiency of existing oil-fired boiler, 79%; gross calorific value of oil, 43.4 MJ/kg; gross calorific value of waste, 12 MJ/kg; cost of oil, 12 p/litre; density of oil, 0.94 kg/litre.
(14.65%; £55 360)

9.14 A hygroscopic thermal-wheel heat exchanger has an overall efficiency of 68% and a moisture transfer efficiency of 63%. The extract air from the building enters the wheel at 20 °C dry bulb, 50% saturation. There is no re-circulation of room air.

Determine the condition of the supply air leaving the thermal wheel for summer and winter design conditions of:
(i) 30 °C dry bulb, 20 °C wet bulb;
(ii) -1 °C, 90% saturation.
(22.8 °C dry bulb, 16 °C wet bulb; 14.0 °C dry bulb, 9.8 °C wet bulb)

9.15 (a) Explain the characteristics and operation of a rotary air pre-heater for recovering the heat in flue gases by pre-heating the combustion air.
(b) List and briefly explain the advantages and disadvantages of a rotary air pre-heater compared to a fixed-surface exchanger.
(c) Combustion air initially at 20 °C is pre-heated in the rotary air pre-heater shown in Fig. 9.28 so that the temperature of the outlet combustion air is 280 °C. The heat is provided by a flue gas stream of 4200 kg/h which enters the air pre-heater at 400 °C.

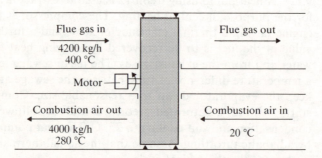

Figure 9.28 Thermal wheel for Problem 9.15

Calculate the flue gas outlet temperature when the leakage from the inlet combustion air stream via the air pre-heater to the outlet flue gas stream is:
(i) zero;
(ii) 7% of the inlet combustion air flow;
(iii) 12% of the inlet combustion air flow.

The specific heats of air and flue gas are 1.04 kJ/kg K and 1.10 kJ/kg K respectively.

(Wolverhampton Polytechnic)

(165.9 °C; 156.6 °C; 149.9 °C)

9.16 (a) Write a short answer to the question: why choose a heat pump? Your answer should include a survey of types of heat pump, types of application, and should discuss relative costs.

(b) An older type of swimming pool has 100% fresh air intake with gas-fired boilers providing hot water to heating batteries in the air inlet duct.

It is proposed to modify the design as shown in Fig. 9.29 by introducing a run-around coil, an electric water-to-water heat pump, and a re-circulatory air system. The chilled water coils from the heat pump evaporator de-humidify the pool extract air allowing 60% of this air to be re-circulated.

Figure 9.29 Swimming pool heat recovery system for Problem 9.16

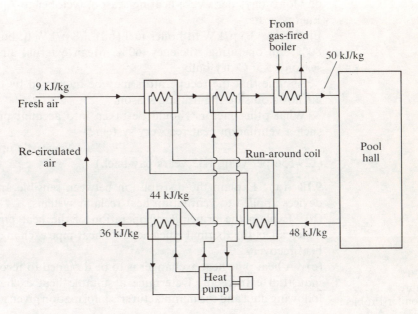

The enthalpies of the air per kilogram of dry air are shown on key points in the system in the figure. The coefficient of performance of the heat pump including thermal losses is 1.9, and the overall efficiency of the gas boiler is 75%; neglect thermal losses in the run-around coil.

Calculate the percentage saving in energy supplied for the pool hall air in using the new system, assuming that:

(i)   pump work, fan work, and all other losses are negligible,
(ii) the rate of mass flow of dry air is the same at inlet to and outlet from the pool,
(iii) the flow rate of air to the pool is unchanged in the new system.

(c) Describe briefly the net present value method of economic appraisal using the scheme in part (b) as an example of how it might be applied.

(Wolverhampton Polytechnic)

Note: For part (c) refer to Eastop and Croft.[9.1]
(74.2%)

9.17 It is estimated that an engineering factory will require a ventilation rate of around six air changes per hour in order to maintain an acceptable working environment. The factory dimensions are 60 m by 80 m by 5 m high.

Evaluate the possible energy savings and overall economics of a ventilation heat recovery system, comprising aluminium plate heat recuperators, from the following data:

Plate heat recuperators:

Supply air flow rate, 4 m³/s; extract air flow rate, 4 m³/s; installation cost, £3900 per unit; pressure drop through heat exchanger, 400 N/m²; sensible efficiency, 65%.

Factory operation:

Internal design temperature, 18 °C dry bulb; working schedule, 24 hours per day for 5 days per week; heating season, October to April (30 weeks).

Energy costs:

Electricity, 3.5 p/kW h; boiler fuel (oil), 1.8 p/kW h; boiler thermal efficiency, 80%; fan operating efficiency, 60%. Mean external air temperature (heating season), 5.5 °C dry bulb.

Assume that the recuperators may be by-passed by the supply and extract air flows outside the heating season.

What other factors should be taken into account prior to recommending such a ventilation heat recovery system?

(Newcastle upon Tyne Polytechnic)

(£29 556 per year; 1.32 years pay-back)

9.18 (a) Explain the distribution between sensible and total heat recovery devices applied to ventilation heat reclaim systems.

(b) Explain the principle of operation of the heat pipe and discuss factors which affect the thermal efficiency of a heat pipe exchanger used for ventilation heat recovery.

(c) A heat pipe heat exchanger is to be designed to recover sensible heat from industrial extract air. Determine a suitable heat exchanger design from the following data and the manufacturer's information given as Figs 9.30(a) and (b):

Figure 9.30 Manufacturer's information for Problem 9.18

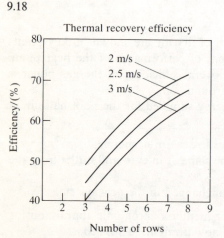

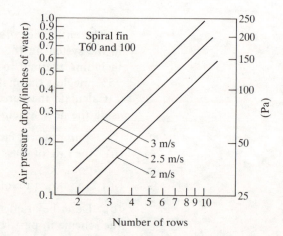

Correction factors

Fins per inch	14	12	10	8	6	4
Multiply efficiency by	1.05	1.00	0.94	0.87	0.75	0.67

(a)

Correction factors (spiral fin)

Fins per inch	14	12	10	8	6	4
Multiply pressure drop by	1.18	1.00	0.85	0.71	0.58	0.47

(b)

Air flow rate, supply and extract, 3 m³/s; face velocity, supply and extract, 2 m/s; heat exchanger type: spiral; 10 fins per inch; 7 rows.

Calculate the net energy savings available from this heat exchanger at the mean winter design condition and the corresponding extract air temperature available from the building.

Mean winter external air temperature, 5 °C; extract air temperature, 18 °C; fan overall efficiency, 60%.

Take air density and specific heat at constant pressure as 1.2 kg/m³ and 1.005 kJ/kg K respectively.

(Newcastle upon Tyne Polytechnic)

Note: For part (b) refer to Eastop and Croft.[9.1]
(30.1 kW; 9.7 °C)

9.19 (a) Describe how degree days may be used to predict the seasonal energy consumption for heating a building.
(b) In an actual building the measured monthly fuel consumptions for the first four months of a season are tabulated below, together with the measured monthly degree day totals for that site. Find the trend line for predicting the monthly fuel consumption, and estimate the mean monthly consumption for domestic hot water.

Month	Fuel consumption (GJ)	Degree days
October	124	174
November	172	280
December	192	319
January	169	274

(University of Manchester)

(45 GJ)

9.20 (a) Describe the methods by which an existing building's energy consumption may be compared with an energy performance indicator using Part 4 of the CIBSE Energy Code.[9.10]
(b) An office block is provided with a gas boiler system (efficiency 72%) providing heat and hot water service. The heating season is September to May inclusive; in the remainder of the period the boilers provide domestic hot water supplies only. It is proposed to replace the existing system with a separate boiler for heating (efficiency 79%) and a gas circulator (efficiency 80%) for domestic hot water supplies. Utilizing the following data, determine:
(i) the average energy consumption of the space-heating system for a typical year;
(ii) the simple pay-back period for the proposed scheme for a typical year.

Monitored gas consumption (30 day period):

	J	F	M	A	M	J	J	A	S	O	N	D
(GJ)	860	770	780	590	470	100	93	110	330	480	712	841
(DD)	380	340	332	259	193	78	43	52	108	183	300	364

Data

Degree days for the area (20-year average), September to May inclusive, 2330.
Replacement plant costs, £12 700.

(Newcastle upon Tyne Polytechnic)

(4660 GJ; 7.22 years)

9.21 A factory's energy consumption has been monitored at regular four-weekly periods over a year. The factory uses oil for space heating, domestic hot water supplies and process; electricity is used for lighting and process. The process load may be considered relatively stable and a function of factory occupancy. Utilizing the following information, and the data and table below:

(a) make appropriate estimates and thus produce a percentage break-down of site energy use in prime energy units for each of the above services;

(b) briefly state how you may be able to improve the accuracy of your estimation without recourse to further monitoring.

Period	Oil consumption (GJ)	Electricity consumption (GJ)
1	1890	270
2	1770	252
3	1746	240
4	1531	238
5	1330	210
6	718	200
7	378	100
8	794	237
9	1077	241
10	1304	250
11	1652	269
12	1841	272
13	900	136

Data

Floor area, 15 000 m^2; population, 300 (average diversity 0.88); lighting, 500 lx, 10 W/m^2, auto-control; roof lights giving average daylight factor of 10%; factory holiday closure, 2 weeks in period 7 and 2 weeks in period 13; domestic hot water consumption, 4 litre/day per person; boiler efficiency, 72% (seasonal average); heating season, periods 1 to 5 and 9 to 13; domestic hot water system efficiency, 80%; domestic hot water supply temperature, 60 °C; prime energy fuel factors: oil, 1.09; electricity, 3.82; factory occupancy, 12 h per day (0700–1900), for 5 days per week.

(Use also Table 9.4, page 530)

(Newcastle upon Tyne Polytechnic)

(Lighting, 5.1%; heating, 29.0%; hot water service, 0.3%; process, 65.6%)

9.22 A building has a total electrical power consumption for the year of 60 000 kW h. The maximum demand is highest in the months of November, December, January and February, and in these months is 78 kW, 82 kW, 85 kW, and 83 kW. The negotiated demand is 110 kVA.

The average power factor is 0.7.

The electricity tariff chosen is the maximum demand tariff for a voltage supply under 1000 V, and there is no significant night-time use of power.

Using the data below calculate the annual cost saving and the percentage cost saving, by increasing the power factor to 0.95 by installing capacitor banks.

Data
Availability charge for the first 50 kVA of chargeable capacity, £1.1 per kVA; availability charge thereafter, £0.87 per kVA; demand charge March to October, nil; demand charge in November and February, £2.3 per kVA; demand charge in December and January, £7.4 per kVA; unit charge, 5.1 p/kVA.
(£603.8; 9.09%)

REFERENCES

9.1   Eastop T D and Croft D R 1990 *Energy Efficiency* Longman

9.2   Eastop T D and McConkey A 1986 *Applied Thermodynamics for Engineers and Technologists* 4th edn Longman

9.3   FEC Consultants 1986 Totem total energy system in a hotel. *Extended Report*: *Energy Demonstration Scheme* Energy Efficiency Office

9.4   Rogers G F C and Mayhew Y R 1988 *Thermodynamic and Transport Properties of Fluids* 4th edn Basil Blackwell

9.5   Sherratt A F C (ed) 1987 *High Insulation: Impact on Building Services* CICC Publications

9.6   Watt Committee 1989 *Passive Solar Energy in Buildings* Watt Committee Series No. 17 Elsevier

9.7   CIBSE 1986 *Guide to Current Practice* volume A1 table A1.5

9.8   Linnhof B, Townsend D W, Boland D, Hewitt G F, Thomas B E A, Guy A R and Marsland R H 1982 *A User Guide to Process Integration for the Efficient Use of Energy* IChemE

9.9   Kew J 1985 *Heat Pumps for Building Services* BSRIA TN 8

9.10 CIBSE Building Energy Code:
Part 1 *Guidance towards Energy Conserving Design of Buildings and Services* 1977
Part 2 *Calculation of Energy Demands and Targets for the Design of New Buildings and Services* 1981
Part 3 *Guidance Towards Energy Conserving Operation of Buildings and Services* 1979
Part 4 *Measurement of Energy Consumption and Comparison with Targets for Existing Buildings and Services* 1982

9.11 CIBSE 1986 *Guide to Current Practive* volume A2 table A2.18

INDEX

6192

CINCINNATI TECHNICAL COLLEGE LRC